Man at High Altitude

Professor Peter Harris with an alpaca he had subjected to cardiac catheterization at La Raya (4200 m) in August 1979 during an expedition to the Peruvian Andes with the authors. This indigenous mountain animal has a low pulmonary arterial pressure, although exposed to the chronic hypoxia of high altitude. This is a feature of adaptation as contrasted to acclimatization (Chapter 27). As can be seen, there were no ill feelings.

Man at High Altitude
THE PATHOPHYSIOLOGY OF ACCLIMATIZATION AND ADAPTATION

Donald Heath
DSc (Sheffield), MD (Sheffield), PhD (Birmingham), FRCP, FRCPath

George Holt Professor of Pathology, University of
Liverpool; Honorary Professor of Pathology, Cayetano Heredia
Medical School, Lima, Peru; Member of the National Academy
of Medicine of Peru; Consultant to the Royal Observatory,
Edinburgh, on Health at High Altitude

David Reid Williams
Department of Pathology, University of Liverpool. Consultant to
the Royal Observatory, Edinburgh, on Health at High Altitude

Foreword by
Sir Cyril Astley Clarke
KBE, MD, DSc, FRCP, FRS

With a chapter on myocardial metabolism by
Peter Harris
MD, PhD (London) FRCP

Simon Marks Professor of Cardiology, Cardiothoracic Institute,
University of London; Physician, National Heart Hospital,
London

SECOND EDITION

CHURCHILL LIVINGSTONE
EDINBURGH LONDON MELBOURNE AND NEW YORK 1981

CHURCHILL LIVINGSTONE
Medical Division of Longman Group Limited

Distributed in the United States of America by Churchill
Livingstone Inc., 19 West 44th Street, New York, N.Y. 10036,
and by associated companies, branches and representatives
throughout the world.

First edition 1977
Second edition 1981

ISBN 0 443 02081 7

British Library Cataloguing in Publication Data

Heath, Donald
    Man at high altitude. - 2nd ed.
    1. Altitude, Influence of
    2. Mountain sickness
    3. Physiology, Pathological
    I. Title   II. Williams, David Reid
    III. Harris, Peter *b. 1923 (May)*
    616.9'893'07   RC103.A4

Printed and bound in Great Britain by
William Clowes (Beccles) Limited, Beccles and London

THIS BOOK IS DEDICATED TO
CLAUDIA WILLIAMS

AND TO THE MEMORY OF
FLORENCE HEATH

Foreword to the Second Edition

The new Table of Contents shows how an already superb work has increased in stature. My higher education has correspondingly benefited and any moment now I am due for the training certificate. However, I remain unconvinced by the Yeti and still have Dragon Stones to contend with, but the 'Third Man' of extreme altitude is constantly in my mind. He is a real phenomenon and I look forward to understanding more about him in the next edition.

Liverpool, 1981   Sir Cyril Astley Clarke

Foreword to the First Edition

In 1968 Professor Heath left Birmingham and came to Liverpool, and it was then that he and I became next door neighbours in our respective Departments. Donald soon took me in hand. First he brushed up my morbid anatomy. In emphysema it was pulmonary hypoxia not the destruction of lung parenchyma that was the cause of cor pulmonale. Clinical pharmacology followed. I learnt that *Crotalaria spectabilis*, a leguminous plant allied to ragwort, contained pyrrolizidine alkaloids which could cause pulmonary hypertension in animals. Might a dietetic cause also sometimes account for this in man? The hunt was on, an epidemic occurred in Germany, and an appetite reducer was found to be a possible cause.

But morbid anatomists must have an emblem and so on to the carotid bodies, and, when I had appreciated their dynamism, my basic training was completed. I could see clearly the next step. Donald took to the mountains, realizing that Nature had a ready-made experiment for him, 'man at high altitude'.

He and David Williams have put together a fascinating story of acclimatization and adaptation to hypoxia, enlivened by many personal experiences in the Andes. The book gives encyclopaedic information on the effects of altitude under every conceivable circumstance, and the contrast between what can happen in man and animals is particularly fascinating.

It was a very great honour to be invited to write the foreword to this splendid book, for I personally am devoted to sea level. Maybe I was asked because for some mysterious reason I sired an Everest climber—perhaps an example of opposites, Donald's 'overshoot phenomenon'. It is good that Everest can still contribute to science, for who would have thought before 1975 that man could endure a night at 8600 metres without additional oxygen? 'Great things are done when men and mountains meet.'

Royal College of Physicians,
London, 1976   Sir Cyril Astley Clarke

Preface

On November 17, 1921, Barcroft embarked from Liverpool for the Andes to carry out his classic studies on high altitude physiology. Travelling by Chosica, Matucana and La Oroya, he arrived at Cerro de Pasco on Boxing Day. Half a century later we were privileged to follow in his footsteps by the same route to Cerro where many of the data included in this volume were collected. Our link with Liverpool is stronger for the specimens collected on our expeditions to Central Peru were subsequently studied in the laboratories of the Department of Pathology of the University of Liverpool. In his book, *The Respiratory Function of the Blood: Lessons from High Altitudes*, published in 1925, Barcroft acknowledges the cooperation he received from Peruvian colleagues and we enjoyed the same kindness and generosity fifty years on. We are especially grateful to Dr Hever Krüger, Professor J. Arias-Stella, Professor L. Eguren, Dr R. Guerra-Garcia, Professor A. Hurtado, Professor C. Monge, Dr D. Peñaloza and Dr F. Sime.

We have now carried out field work in the Andes on five occasions and our appreciation of the beauty of Peru has become enriched by affection for her people. Several Peruvians have made outstanding contributions to the study of man at high altitude which have helped to reveal the important medical and biological implications of the adjustments of the human body to conditions of oxygen deprivation. If we see the patient with chronic lung disease as acclimatizing to hypoxia we come to realise that his physiological problems and his methods of dealing with them have much in common with those of the Quechua of the Andes. Our understanding of the patho-physiology of the pulmonary and systemic circulations, the blood, the carotid bodies and virtually every system in the body becomes that much clearer when illuminated by the Andean experience.

We owe a profound debt of gratitude to Professor Peter Harris who made our visits to South America possible and who has been intimately associated with our investigations there in the mountains. We are delighted that Sir Cyril Clarke has written a foreword for this second edition and we are grateful to Miss Linda Byron for all her skill and patience in the typing of the manuscript.

Liverpool 1981

D.H.
D.R.W.

Acknowledgements

The following colleagues have kindly provided us with clinical and pathological material from which the figures indicated have been prepared, or were co-authors in papers from which the illustrations have been taken.

Professor J. Arias-Stella
Figs. 8.8, 8.9, 12.6, 16.9, 16.10, 27.5

Dr Y. Castillo
Figs. 8.8, 8.9, 12.6

Chester Zoo (North of England Zoological Society)
Fig. 34.5

Dr C. Clarke (British Everest Expedition 1975)
Fig. 31.5

Dr J. Dickinson
Figs. 3.9, 14.10

Dr C. Edwards
Figs. 8.5–8.7, 8.8, 8.9, 8.12, 8.13

Mrs E. Ellison and Mr W. Butt
Fig. 17.6

Dr U. Garcia
Fig. 11.5

Dr E. Gradwell
Figs. 25.2, 25.3

Professor P. Harris
Figs. 5.7, 8.5–8.7, 8.8, 8.9, 8.12, 8.13, 12.6, 27.5

Professor H. H. Hecht
Fig. 13.1

Professor A. Hurtado
Fig. 12.6

Dr F. Jackson
Figs. 18.2, 18.3

Dr A. H. Kombe
Figs. 12.8–12.10

Dr H. Krüger
Figs. 8.8, 8.9, 12.6, 15.3, 27.5

Dr K. Marsh
Figs. 25.4, 34.1–34.4

Dr C. C. Monge
Fig. 16.1

Dr W. Mooi
Figs. 30.7, 30.8

Dr H. Moosavi
Fig. 15.4

Dr F. Murphy
Figs. 24.7–24.9

Dr D. Peñaloza
Figs. 11.7, 15.2, 16.4, 16.11

Professor J. Rüttner
Fig. 34.7

Dr F. Sime
Fig. 15.2

Dr P. Smith
Figs. 5.7, 11.10, 12.1, 12.8–12.10, 15.4, 27.5, 30.7, 30.8

Miss L. Taggart
Fig. 24.3

Dr D. Weinman
Figs. 24.4–24.6

Dr W. Whitaker
Fig. 18.4

We are indebted to the Editors of the following journals for permission to reproduce the illustrations listed below which were previously published by them.

The Alpine Club
Fig. 18.2

American Journal of Cardiology
Fig. 15.2

British Heart Journal
Fig. 12.6

British Journal of Diseases of the Chest
Figs. 12.8–12.10

Experimental Cell Biology
Fig. 5.7

Investigative and Cell Pathology
  Figs. 15.5, 30.6

Journal of Pathology
  Figs. 8.8, 8.9, 8.12, 8.13, 11.10, 30.7, 30.8

Journal of Virology
  Figs. 24.7–24.9

Popperfoto
  Fig. 29.1

Thorax
  Figs. 8.5–8.7, 12.1, 27.5

Contents

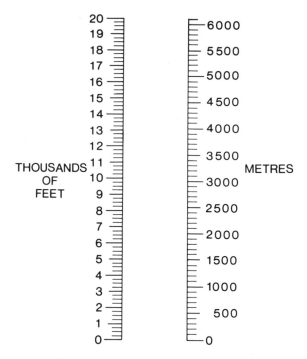

Above: Nomogram relating feet and metres. Throughout the text altitudes are expressed in metric units but may be readily converted above to the more familiar 'thousands of feet'.

Below: To facilitate rapid interpretation of the histograms the conventions shown in this Figure are employed consistently throughout the book with only one or two exceptions. Conventions other than those shown are used infrequently and do not have a uniform meaning.

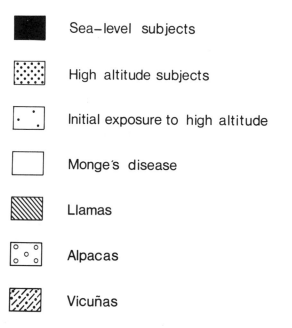

The importance of high altitude studies

For centuries man has been aware that breathing in high mountainous areas is difficult. According to Whymper (1892) Father Joseph de Acosta (1590) was in no doubt that the nausea and vomiting and other symptoms of acute mountain sickness which he suffered while on a journey through the Andes were related in some way to the thin air. 'I therefore persuade myself that the elements of the air is there so subtle and delicate, as it is not proportionable with the breathing of man, which requires a more gross and temperate air, and I believe it is the cause that doth so much alter the stomach and trouble all the disposition'. Whymper (1892), the conqueror of the Matterhorn who opened up the way for the conquest of peaks throughout the world, was under no illusion that all that was required was expertise in climbing and determination. After an expedition into the Andes he concluded that 'from the effects on respiration none can escape. In every country, and at all times, they will impose limitations upon the range of man; and those persons in the future, who, either in pursuit of knowledge or in quest of fame, may strive to reach the loftiest summits in the earth, will find themselves confronted by augmenting difficulties which they will have to meet with constantly diminishing powers.' In his account of his ascent into the mountains of South America, Whymper (1892) describes intense headache, a marked acceleration of the pulse and 'an indescribable feeling of illness pervading the whole body' while they were 'preoccupied by the paramount necessity of obtaining air'. He also makes what amounts to one of the earliest references to acclimatization in reporting that they 'became *somewhat* habituated to low pressures'.

In this book, however, we are not primarily concerned with the severe physiological stresses endured by high altitude climbers striving 'to reach the loftiest summits in the earth'. We consider the problems of exposure to extreme altitudes (Chapter 30) but only in so far as they illustrate the ultimate physiological responses to hypoxia. Excellent accounts of the biological stresses in high altitude climbers are available (Bonington, 1971; Ward, 1975). In this book we are concerned rather with the changes in form and function in the human body that occur on acute and chronic exposure to altitudes where people live permanently.

Large populations in different parts of the world live at these altitudes and they exhibit physiological and microanatomical changes not only in their thoracic organs, where they might be anticipated under such conditions, but in most systems of the body including the endocrine and reproductive organs. We do not accept these changes in form and function as normal even though they occur in millions of people throughout the world. We think they are to be regarded as pathophysiological responses to deprivation of oxygen. Indeed, if these responses become exaggerated, as they do in a minority of subjects, they induce disease states which may prove fatal.

High altitude diseases

Hence in addition to the features of acclimatization to life at high altitude we shall consider diseases caused by this environment, namely acute mountain sickness, high altitude pulmonary oedema, Monge's disease and, in cattle, brisket disease. Sea-level subjects, or native highlanders after a period at low altitude, who ascend high mountains and undertake physical exertion may develop high altitude pulmonary oedema. This serious condition is a hazard for the high altitude climber and may prove fatal (Chapter 15). Experience from military operations in the Himalayas shows that acute mountain sickness may be a matter of considerable military importance, if large numbers of troops are moved from sea level to mountainous areas (Chapter 14). Diminished barometric pressure may have deleterious effects on the performance of sea-level athletes who compete in international sports meetings held at high altitude venues like Mexico City (Chapter 29).

Acute mountain sickness and the tourist industry

Acute mountain sickness is likely to prove an increasing problem for the travel industry for the introduction in recent years of trekking and 'adventure holidays' in remote mountainous areas raises the likelihood of tourists being suddenly exposed to the hypoxia of high altitude after rapid transit from sea level. Every day in July and August about 3600 people visit the summit of Pike's Peak, Colorado, at 4300 m (*Lancet* Annotation, 1976). Hackett and his colleagues (1976) studied the incidence of acute mountain sickness in tourist-hikers, as opposed to expedition-mountaineers, who were ascending through the village of Pheriche (4240 m) in the Himalayas of Nepal to visit the Everest Base Camp at 5500 m. Up this one tiny, remote mountain path some 648 tourists hiked in a mere four weeks from October 10th to November 10th, 1975. There are, moreover, plenty of such peaks from which the tourist may choose. Some 568 Himalayan peaks are higher than 6100 m according to Mordecai (1966), and, excluding Antarctica, about 2.5 per cent of the world's land surface lies above 3050 m (*Lancet* Annotation, 1976). The number of tourists visiting high mountain areas is increasing explosively and it seems undisputed that thousands more people will suffer from acute mountain sickness every year with increasing mortality. In the opinion of Hackett and his colleagues (1976) this should be a matter of concern both to the tourist industry and to the medical profession. Many physicians who accompany trekking groups into the mountains are often totally and tragically ignorant of acute mountain sickness. Ignorance of the nature, prevention, and treatment of diseases facing those who acutely expose

themselves to high altitude has been stressed in an annotation in the *Lancet* (1976) grimly entitled 'See Nuptse and die'.

Volcanoes are another great tourist attraction throughout the world and some of them are high enough to threaten with acute mountain sickness the holiday-makers struggling up their slopes. Mount Fuji (3775 m), the sacred mountain of Japan (Fig. 1.1), is visited each year by more than a hundred thousand people. However, the vast majority of

Fig. 1.1 Mount Fuji (3775 m) the sacred volcano of Japan.

visitors ascend two thirds of the way up by motor coach and only a few climb the summit as a novelty or as a pilgrimage. Under these circumstances the risk of acute mountain sickness is slight. This is not, however, the case with Mount Teide (3720 m) on Tenerife. This is ascended by numerous holiday-makers who are first driven up rapidly in buses from hotels on the coast to the foot of the volcano. They are then transported almost to the summit extremely quickly by cable car leaving only a final, brisk steep walk to the top. The authors have witnessed the summit of this volcano swarming with tourists many of whom exhibit symptoms which the initiated will realise lie on the very threshold of the serious acute high altitude diseases to be described later in this book.

Telescopes in high places

Occasionally man has to work on the summits of volcanoes for periods of months or years rather than visit them fleetingly. Thus astronomers operate the new infrared telescope built by the British Science Research Council on the summit of the volcano, Mauna Kea, on the island of Hawaii (Fig. 1.2). This telescope is situated at an altitude of 4210 m and can be reached by the Saddle Road from the port of Hilo with a lodge for sleeping and acclimatization at Hale Pohaku (2750 m) (Fig. 1.3). Such working

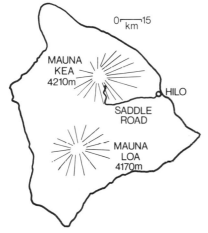

Fig. 1.2 Line diagram of the island of Hawaii showing the situations of its two volcanoes, and demonstrating the remarkable accessibility of the summit of Mauna Kea by the Saddle Road from the port of Hilo. The newcomer from sea level can reach an altitude of 4210 m in one and a half hours.

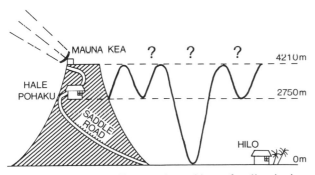

Fig. 1.3 Diagram to illustrate the problems of acclimatization and avoidance of high altitude pulmonary oedema in astronomers working on the telescope on the summit of Mauna Kea (4210 m), sleeping at Hale Pohaku (2750 m), and having their families in residence at Hilo (sea level). The problem is to determine the periods which may be spent at the three elevations without inducing pulmonary oedema. This diagram illustrates one important and unusual practical application of high altitude studies.

conditions present a most unusual set of circumstances in that the astronomers oscillate between two elevations at high altitude and one at the coast

(Fig. 1.3). The effect of such oscillation is of considerable interest for it may interfere with acclimatization or even predispose to high altitude pulmonary oedema. At the time of writing studies are being carried out to evaluate the influence of such oscillation on acclimatization.

A minority of lowlanders has to take up prolonged or even permanent residence in mountainous areas as, for example, in the employment of mining companies. To such people a knowledge of the presentation and treatment of, and prophylaxis against, acute mountain sickness (Chapter 14), high altitude pulmonary oedema (Chapter 15), and cerebral oedema (Chapter 14) is of importance.

Clinical applications of high altitude studies

The study of the acclimatization of man to high altitude has the wider implication of revealing the responses of living tissues to chronic hypoxia without the complicating factor of superadded disease. Chronic hypoxia and hypoxaemia are not confined to life at high altitude but are to be found in common and important heart and lung diseases such as chronic bronchitis and emphysema. Thus the results of investigations into high altitude physiology and pathology can be applied to much human cardiopulmonary disease.

To give one example, our studies demonstrating that the carotid bodies are enlarged in cases of pulmonary emphysema complicated by chronic hypoxia (Edwards et al, 1971) were stimulated by the demonstration of Arias-Stella (1969) that the carotid bodies of the Quechua Indians of the High Andes, exposed to the chronic hypoxia inherent in diminished barometric pressure, were larger than those of Mestizos living on the coast. Up to this period the interest of pathologists in the carotid body was largely confined to its tumour, the chemodectoma, but the implications of the behaviour of chemoreceptors at high altitude have led to an awakening of interest in the pathology of glomic tissue in states of chronic hypoxaemia in various diseases (Chapter 8). It is clear that this particular study of high altitude pathology still has considerable areas of clinical application to be developed. Thus cases of cyanotic congenital heart disease are characterized by hypoxaemia and right ventricular hypertrophy, the very condition known to be associated with the enlargement and development of characteristic ultrastructural changes in the carotid bodies at high altitude. Yet to our knowledge a study of the pathology of the carotid bodies in cyanotic congenital heart disease is still to be undertaken. The interrelationships between the heart and carotid bodies have yet to be determined. This is one example of how our knowledge of high altitude pathology has enhanced concepts in general pathology and opened the door to new fruitful areas for research.

We may select a second, clinical example. In Chapters 4, 16 and 23 we describe how haemorrhages in the finger nails occur respectively in native highlanders, in sufferers from Monge's disease and in those exposed acutely to extreme altitudes. Splinter haemorrhages have been classically associated with sub-acute bacterial endocarditis. The frequency of nail-haemorrhages in

various groups of people at high altitude suggested to us that their causation needed further investigation. In Chapter 23 we refer to subsequent observations that we made to elucidate the clinical significance of such haemorrhages.

Since the end of the last century a large volume of data has been accumulated on high altitude medicine and pathology. Much of this material is widely scattered in specialist journals and in this book we have attempted to bring some of it together. In doing this it has become clear that some widely accepted views of life and disease at high altitudes are anecdotal and not based on fact and we have preferred not to include these subjective opinions. Nevertheless, by integrating many of these clinical, physiological, pathological and experimental data we have attempted to present a readable account of the characteristics of man at high altitude.

REFERENCES

de Acosta, Father Joseph (1590) *Historia Natural y Moral de las Indias*.

Arias-Stella, J. (1969) Human carotid body at high altitudes. In: *69th Programme and Abstracts of the American Association of Pathologists and Bacteriologists*. San Francisco, Item 150.

Bonington, C. (1971) *Annapurna South Face*. London: Cassell.

Edwards, C., Heath, D. & Harris, P. (1971) The carotid body in emphysema and left ventricular hypertrophy. *Journal of Pathology*, **104**, 1.

Hackett, P. H., Rennie, D. & Levine, H. D. (1976) The incidence, importance and prophylaxis of acute mountain sickness. *Lancet*, **ii**, 1149.

Lancet Annotation (1976) See Nuptse and die. *Lancet*, **ii**, 1177.

Mordecai, D. (1966) *The Himalayas*. Calcutta.

Ward, M. (1975) *Mountain Medicine. A Clinical Study of Cold and High Altitude*. London: Crosby Lockwood Staples.

Whymper, E. (1892) *Travels Amongst the Great Andes of the Equator*. London: John Murray.

2

Physical factors at high altitude

The concept of 'high altitude'

The term 'high altitude' has no precise definition. To many in Western Europe the Swiss Alps epitomize great heights but it is salutary to realize that in the High Andes of Peru and Bolivia active economic life is carried on at altitudes exceeding the summit of the Matterhorn (4480 m). The wide range of elevations covered by the term are illustrated by the situation of various 'high altitude' research stations throughout the world (Fig. 2.1). They range from Jujuy in Argentina at 1260 m to Morococha in Peru at 4540 m. Throughout this book we shall take 'high altitude' to mean an elevation of 3000 m or more. We select this height because at this point, in the majority of subjects ascending high mountains, unequivocal signs and symptoms associated with the ascent appear. Above this altitude biochemical, physiological and anatomical features of acclimatization become progressively more pronounced. However, it will be appreciated that any definition of 'high altitude', as applied to its effects on man living in this environment, will be unable to take into account the considerable individual variation in response to such conditions. Furthermore, it should be realized that the biological effects readily discernible at 3000 m will start to occur at much lower elevations. These effects only become readily apparent at this elevation and it is in this sense that the reader should interpret the term 'high altitude' throughout this book.

High altitude areas of the world

On the basis of this definition there are many areas of the world where man lives at high altitude. Areas of high elevation are shown in Figure 2.2 and include:

Mauna Kea and Mauna Loa (volcanoes on Hawaii)
The Rocky Mountains of Canada and the United States
The Sierra Madre of Mexico
The Andes of South America
The Pyrenees lying between France and Spain

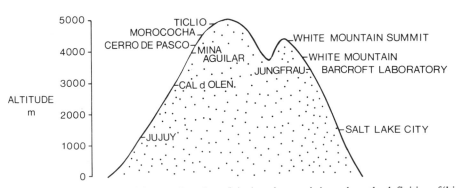

Fig. 2.1 The elevation of various 'high altitude' research stations. It is clear that much depends on the definition of 'high altitude'. In this book it is defined as 3000 m and above, while 'extreme altitude' exceeds 5800 m (Chapter 30).

Fig. 2.2 The world distribution of areas of high altitude exceeding 3000 m. (1) Mauna Kea and Mauna Loa (volcanoes on Hawaii); (2) Rocky Mountains; (3) Sierra Madre; (4) Andes; (5) Pyrenees; (6) Mountain ranges of Eastern Turkey, Persia, Afghanistan and Pakistan; (7) Himalayas; (8) Tibetan Plateau and Southern China; (9) Mount Teide (volcano on Tenerife); (10) Atlas Mountains; (11) High Plains of Ethiopia; (12) Kilimanjaro; (13) Basutoland; (14) Antarctica; (15) Mount Fuji (volcano in Japan).

The mountain ranges of Eastern Turkey, Persia, Afghanistan and Pakistan
The Himalayas
The Tibetan Plateau and Southern China
Mount Teide (volcano on Tenerife)
The Atlas Mountains of Morocco
The High Plains of Ethiopia
Kilimanjaro
Basutoland
The mountains of Antarctica
Mount Fuji (volcano in Japan)

Hypoxia

Inherent in life at high altitude is exposure to physical factors not operative at sea level. The most important of these is hypoxia. The amount of oxygen in the atmosphere remains constant at 20.93 per cent up to an altitude of 110 000 m (Frisancho, 1975). The percentage of oxygen in the ambient air is the same at high altitude as it is at sea level. However, gas is compressible and this means that the number of molecules a unit volume contains is greater at sea level than at high altitude. In other words the barometric pressure, which depends upon the molecular concentration of the air, decreases with increase in altitude. This in turn means that the partial pressure of oxygen in the ambient air at high altitude, (PB_{O_2}) is reduced.

At sea level the barometric pressure is 760 mmHg and hence the partial pressure of oxygen in the air is 20.93 per cent of that value, namely 159 mmHg. When air is breathed into the bronchial tree, it becomes saturated with water vapour which exerts a pressure of 47 mmHg so the partial pressure of oxygen in inspired air as contrasted to ambient air is 20.93 per cent of (760 − 47) mmHg, i.e. 149 mmHg. The relation between altitude, barometric pressure and air (and oxygen) pressure as a percentage of that at sea level is shown in Figure 2.3. As an example, at 3500 m the barometric pressure is 493 mmHg and P_{O_2} is 103 mmHg which is 65 per cent of the sea-level value (Fig. 2.3). Throughout this book we shall be largely concerned with the biochemical, physiological and microanatomical changes which occur during acclimatization and adaptation to the chronic hypoxia of high altitude.

Cold

Another environmental hazard that faces man at high altitude is cold. Temperature falls with increasing altitude to the extent of approximately 1°C for every 150 m and this is independent of latitude. However, latitude does influence the temperature of mountainous areas for in tropical climes there is very little seasonal change but much diurnal variation in temperature. The reverse is true

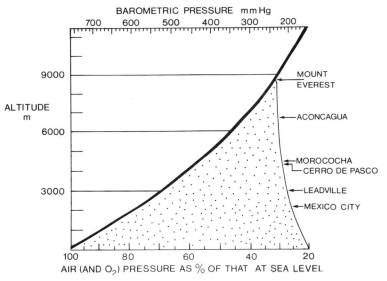

Fig. 2.3 The relation between altitude, barometric pressure, and air (and oxygen) pressure as a percentage of that at sea level (after Frisancho, 1975).

of areas of higher latitude. At night a black-bulb thermometer exposed to the sky records a lower temperature than a sheltered instrument. This is a measure of the intense re-radiation of heat to the night sky at high altitude. Hence the environment for nocturnal animals is very cold, forcing foxes, snow leopards and similar animals to shelter in the ground where the daily temperature variation is much narrower. In the Andes we have been impressed by the striking difference in temperature which occurs in shadow as contrasted with direct sunlight. This difference has been documented by Swan (1961) (Fig. 2.4). He recorded a temperature of 33.3°C on the sunlit rock surface at 5490 m; the air temperature in the shade was at the same time 12.8°C. With increasing altitude there is less atmospheric absorption of transmitted radiant energy and less effective cloudiness. Swan (1961) refers to Jung Marmet and Ernst Schmidt, who climbed Everest in 1956 and removed their down-filled clothing because of the heat at 8530 m. Such conditions occur only in sunshine and with a low wind velocity (see below).

In view of all these variations it is impossible to

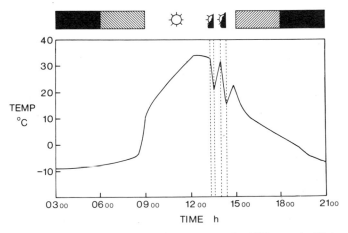

Fig. 2.4 The effect of direct sunlight on the 'exposed-black-bulb temperature' at 4880 m on the Ripimu Glacier in eastern Nepal as measured by Swan (1961). The hours indicated by filled blocks are those of darkness, while those shown by hatching are dawn and dusk. Note that a temporary obscuring of the sun by cloud induces an immediate and significant fall in temperature (after Swan, 1961).

make definitive statements concerning altitude and temperature, and average values have to be assumed. The nature of man's immediate response to cold and long term acclimatization to cold are described in Chapter 22. The pathological changes in the tissues induced by exposure to the severe cold of extreme altitudes are considered in Chapter 30.

Exposure to wind is a factor which has to be considered closely in association with cold at high altitude for with increasing wind velocity the effective skin temperature drops. In essence the insulating layer of warm air around the skin is blown away. This is known as the wind-chill factor (Ward 1975) and we shall refer to it further in Chapter 30. Unsworth (1977) relates ascending levels of the wind-chill factor (K) to various body sensations ranging from warmth to freezing of exposed flesh in 60 seconds (Fig. 2.5). At a value of 1400 for K

Karakorums one can sweat profusely. The combination of low temperature and low relative humidity can be subjectively very unpleasant. Sensitive areas such as the lips can dry and crack in a matter of hours due to the dehydration caused by sweating. At extreme altitudes the low humidity and consequent excessive sweating leads to dehydration which may in turn play a rôle in initiating thrombosis (Chapter 30). The general dehydration consequent upon low humidity has been held by some as a contra-indication to the use of frusemide in the prophylactic treatment of acute mountain sickness and high altitude pulmonary oedema (Chapters 14 and 15).

Solar radiation and physical geography

Paradoxically enough, although temperature falls

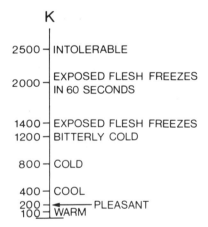

Fig. 2.5   The relation between ascending values of the wind-chill factor (K) to various bodily sensations. Also shown are various combinations of temperature and wind velocity which produce a value for K of 1400 at which exposed flesh freezes (after Unsworth, 1977).

exposed flesh freezes and Unsworth gives examples of how this value may be reached by various combinations of temperature and wind velocity (Fig. 2.5).

Humidity

As we shall see in Chapter 22 humidity is of importance as a factor influencing the loss of heat from the body by evaporation. In areas of high humidity sweating and heat regulation are more difficult. Conversely in arid areas of low humidity such as the western slopes of the Andes or in the

with increasing altitude, there is increased exposure to solar radiation. According to Ward (1975) at sea level the average amount of solar radiation absorbed is 963 kJ/m²/h. On the other hand at 5790 m in clear weather Pugh (1963) found the solar heat absorbed by the surface of the clothed human body was 1465 kJ/m²/h. The solar heat absorbed by the body depends on several factors not attributable to the environment such as the clothing (dark clothing absorbing more than white) and the position of the individual. However, there are environmental factors which will affect considerably the amount of solar radiation to which man at high altitude is exposed. The clear air of many mountainous areas will more easily permit the passage of solar radiation to the earth's surface. Another important factor at high altitude which increases the level of solar radiation to which man is exposed is snow. In Western Europe the altitude of the permanent snowline varies from one mountain range to another. On the northern slopes of the Pyrenees it is at 2750 m while on Mont Blanc it is at 2990 m. Permanent snow reflects solar radiation so that the effect is enhanced. Such reflectivity of solar radiation, known as the *albedo*, is less than 25 per cent if there is no snow but rises to a range of 75 to 90 per cent in its presence. The factor is greatest at high altitude in polar regions.

Ultraviolet radiation

Ultraviolet radiation is a segment of solar radiation. This part of the electromagnetic spectrum extends from 200 to 400 nm. It is often considered in three ranges, u.v.–A (400 to 315 nm), u.v.–B (315 to 280 nm), u.v.–C (280 to 200 nm). Ultraviolet radiation is somewhat increased at high altitude and we are indebted to Dr Thomas of the Appleton Laboratory of the Science Research Council for providing the data showing this in Table 2.1. We

Table 2.1 Ultraviolet spectral intensity distribution* (values refer to overhead sun)

Wavelength λ (nm)	Intensity at ground level (ergs cm² s Å)	Factor by which intensity increased at 4000 m
400	83	1.47
380	58	1.58
360	49	1.68
340	39	1.78
320	18	1.97
300	0.43	2.50

* Provided by Dr L. Thomas from data in Tousey (1966), and Elterman (1964).

note there that at an altitude of 4000 m ultraviolet radiation of wavelength 300 nm is increased by a factor of 2.5. It should be kept in mind, however, that the primary factor which determines the level of ultraviolet radiation in any area is the amount of sunlight it receives.

Thus an investigation by Robertson (1969) compared the exposure to ultraviolet radiation in Cloncurry and Brisbane in Australia at sea level, on the one hand, and Goroka (1585 m) in New Guinea. Values at Cloncurry were 1.3 times the average values for Brisbane. During the summer months at no time did the maximum monthly exposure of ultraviolet radiation at Goroka reach the levels found at Cloncurry despite its higher altitude. During the winter months, however, the monthly exposure at Goroka is higher. Hence to put the matter into perspective there is some increase in the levels of ultraviolet radiation at high altitude but this factor is easily counterbalanced by the amount of sunlight received by the area in question.

As we have seen, reflection of solar, including ultraviolet, radiation by snow enhances its effects at high altitude. Snow can reflect up to 90 per cent of incident ultraviolet radiation compared to 9 to 17 per cent reflection from ground covered by grass or heather (Buettner, 1969). Hence on snow-covered ground at high altitude the combination of incident and reflected ultraviolet radiation may be more formidable.

It is widely accepted that prolonged exposure to sunlight, and consequently to ultraviolet radiation, may lead to the development of skin cancer in man in countries like Australia (Gordon and Silverstone, 1969; Urbach, 1969). Since, as we have seen above, the level of ultraviolet radiation is somewhat increased at high altitude (Table 2.1) one might anticipate that the incidence of skin cancer might be increased at high altitude. In fact, as we shall see in Chapter 23, Krüger and Arias-Stella (1964) found no such increase in a community in the High Andes of Peru. Jones (1975) is also sceptical of the increased level of ultraviolet radiation in the mountains as a hazard increasing the risk of skin cancer. He states that in Denver, Colorado (1600 m) the ultraviolet intensity at 295 nm is increased by 125 per cent, while in the State as a whole the percentage of clear days (70 per cent) is above the national average for the United States. However, in spite of these factors in the physical environment, Colorado has one of the lowest skin cancer fatality rates of all the States. The entire desert region of the South West of the U.S.A. is exposed to an ultraviolet radiation some 125 per cent greater than in New England yet the death rate from skin cancer in New England is 24 per cent higher than it is in the South West (Jones, 1975). It should be noted that

Jones bases his argument on death rates from skin cancer which relate to squamous cell carcinoma and malignant melanoma. Any increased incidence of basal cell carcinoma would not be reflected in such data. In Chapter 23 we consider the rôle of ultraviolet radiation in producing 'high altitude dermatopathy'.

Ionizing radiation

Man is subjected to natural ionizing radiation which may be cosmic or arise from his terrestrial environment. At high altitude there is an increase in the levels of cosmic radiation which consists of extremely penetrating rays from outer space. High energy protons from outside the earth's atmosphere interact with the nuclei of the atmosphere to produce π mesons and a small number of other particles. Some of these π mesons decay to form μ mesons which in the main survive down to ground level to form the hard, penetrating component of the radiation. Some of the μ mesons decay to give rise to electrons which form the soft component of the radiation. In addition to π and μ mesons cosmic radiation comprises heavier K mesons and hyperons. At an altitude of 3000 m there is a threefold increase in the sea-level value of approximately 24 millirads a year (Baikie, 1970). A rad is the dose of radiation equalling an absorption of 100 ergs per gram of substance irradiated. The intensity of radiation also varies with geographical location (Baikie, 1970). Background radiation depends on radioactive elements in soil, rocks and some building materials notably granite. Because of this latter factor the houses built of this material in Aberdeen produce a mean dose rate of 85 millirads a year and the levels of radiation occurring at high altitude can be put into perspective by comparing them with these high levels of background radiation at sea level.

The total dose rate for background radiation from all natural sources varies from 50 to 200 millirads per year (Baikie, 1970). Such natural radioactivity would seem to provide an ideal opportunity to study the biological effects of low levels of radiation. Epidemiological studies of the incidence of diseases such as leukaemia and bone cancer could be studied in relation to the level of ionizing radiation to which the population was exposed. However, changes induced by doses of radiation within the range of natural exposure are so infrequent that it is not feasible to obtain data on the effects of such low levels by epidemiological means. For these reasons it is not possible to identify any deleterious effects due to increased cosmic radiation at high altitude.

The influence of tissue hypoxia on the biological effects of ionizing radiation

The mechanism by which ionizing radiation affects cells is not clearly understood. In the past it was believed that most of its effects were due to direct action on the intracellular structures causing either ionization or excitation of the molecule. It is now thought that such direct action is less important than the indirect effects mediated by the cell water (Baikie, 1970). Direct action kills viruses more easily than indirect action, although the latter is more effective in the inactivation of enzymes (Walters and Israel, 1974).

The main radiochemical effect on tissues appears to be the formation of 'hot radicals' such as HO_2 and H_2O_2. Tissue PO_2 appears to have an important influence on the formation of such radicals. Normally an increased PO_2 increases radiosensitivity and a decrease in PO_2 such as occurs at high altitude, reduces the effects of radiation. Support for this theory comes from the fact that some reducing substances give a degree of protection against the biological effects of radiation. These include cyanides, nitrites, 5-hydroxytryptamine, adrenaline and, most effective of all, derivatives of cysteine. Although these compounds are thought to be protective by producing tissue hypoxia, the actual mode of action is not clear. For instance, ascorbic acid is a powerful reducing agent but has little protective effect. Several alternatives have been suggested. One is that the protective chemicals compete for the free radicals. As HO_2, a powerful oxidising agent, is found only in the presence of oxygen this would explain why hypoxia and chemical protection are related. It seems certain that hypoxia protects against the effects of ionizing radiation. Consequently the hypoxic environment of high altitude is likely to offer some protection against the level of radiation which, as we have seen above, is raised not excessively above sea level values.

Radiation and genetic effects

High doses of radiation cause inhibition of mitosis and DNA synthesis and the production of chromosomal aberrations. These abnormalities have been studied in man after doses of radiation such as are used in radiotherapy or as occur in radiation accidents. Consequently they are far in excess of what might be found at high altitude. To what extent natural radiation contributes to any harmful mutations is not known but it is conceivable that even small doses of radiation may cause damage to the chromosomes and give rise to harmful mutations. Several investigations into the genetic effects of radiation at levels found in the environment have been made in the state of Kerala on the southwest coast of India. The high natural background radiation in Kerala arises from the presence of varying amounts of monazite which contains 8 to 10.5 per cent of thorium with a long half life (Ahuja et al, 1973). There is an emission of α, β and γ rays, the latter penetrating radiation being most likely to be concerned with genetic effects. Two areas particularly rich in monazite are Manavalakurichi and Kayamkulum-Neendkara, both on the Malabar coast. The γ radiation range three feet above ground is 0.03 to 1.3 millirads/hr with a mean of 0.33 mr/hr or 2.9 rads/year. Over 30 years this represents a dose of 87 rads compared to an average dose on a worldwide basis of 3 rads.

Nevertheless, this high level of background radiation appears to bring about no demonstrable harmful effects. Thus no abnormalities were found in wild rats or their litters in the area (Baikie, 1970). Grüneberg et al (1966) had previously surveyed a rat population in Kerala with respect to several qualitative and quantitative traits but were unable to detect any effect attributable to background radiation. Ahuja et al (1973) studied dermatoglyphic parameters from the Araya people who have lived in the area for over a thousand years. These parameters are known to be extremely sensitive to numerical and structural changes in the karyotype (Penrose, 1968). Dermatoglyphic factors studied included total finger ridge count and other specialised indices and, as in the case of the previous rat studies, they showed no abnormality in the presence of excessive background radiation (Ahuja et al, 1973). On such evidence it seems unlikely that increased environmental radiation at high altitude exerts a deleterious effect on the highlander. An increase in the incidence of chemodectomas at high altitude has certainly been reported (Saldaña et al, 1973) but, as we discuss in Chapter 8, such tumours are more likely to be the results of hyperplasia of the chief cells due to the effects of hypoxia rather than radioactivity.

REFERENCES

Ahuja, Y. R., Sharma, A., Nampoothiri, K. U. K., Ahuja, M. R. & Dempster, E. R. (1973) Evaluation of effects of high natural background radiation on some genetic traits in the inhabitants of Monazite Belt in Kerala, India. *Human Biology*, **45**, 167.

Baikie, A. G. (1970) *A Companion to Medical Studies*, 2 Chap. 33. Oxford: Blackwell Scientific Publications.

Buettner, K. J. K. (1969) The effects of natural sunlight on human skin. In: *The Biologic Effects of Ultraviolet Radiation with Special Emphasis on the Skin*, p. 237. Edited by F. Urbach. Oxford: Pergamon Press.

Elterman, L. (1964) *Atmospheric Attenuation Model, 1964, in the Ultraviolet, Visible and Infra-red Regions for Altitudes to 50 km.* U.S. Air Force Cambridge Research Laboratories. Environmental Research Paper No. 46. AFCRL–64–740.

Frisancho, A. R. (1975) Functional adaptation to high altitude hypoxia. *Science*, **187**, 313.

Gordon, D. & Silverstone, H. (1969) Deaths from skin cancer in Queensland, Australia. In: *The Biologic Effects of Ultraviolet Radiation with Special Emphasis on the Skin*, p. 625. Edited by F. Urbach. Oxford: Pergamon Press.

Grüneberg, H., Bains, G. S., Berry, R. J., Riles, L., Smith, C. A. B. & Weiss, R. A. (1966) A search for genetic effects of high natural radioactivity in South India. *Medical Research Council Special Report Series* No. 307, H.M.S.O. London.

Jones, A. (1975) Ozone depletion and cancer. *New Scientist*, **68**, 14.

Krüger, H. & Arias-Stella, J. (1964) Malignant tumours in high altitude people. *Cancer*, **17**, 1340.

Penrose, L. S. (1968) Medical significance of fingerprints and related phenomena. *British Medical Journal*, ii, 321.

Pugh, L. G. C. E. (1963) Tolerance to extreme cold at altitudes in a Nepalese pilgrim. *Journal of Applied Physiology*, **18**, 1234.

Robertson, D. F. (1969) Correlation of observed ultraviolet exposure and skin cancer incidence in the population of Queensland and New Guinea. In: *The Biologic Effects of Ultraviolet Radiation with Special Emphasis on the Skin*, p. 619. Edited by F. Urbach. Oxford: Pergamon Press.

Saldaña, M. J., Salem, L. E. & Travezan, R. (1973) High altitude hypoxia and chemodectomas. *Human Pathology*, **4**, 251.

Swan, L. W. (1961) The ecology of the high Himalayas. *Scientific American*, **205**, 68.

Tousey, R. (1966) The radiation from the sun. In: *The Middle Ultraviolet: Its Science and Technology*. Edited by A. E. S. Green. New York: Wiley.

Unsworth, W. (1977) *Encyclopaedia of Mountaineering*, p. 369. Harmondsworth: Penguin Books.

Urbach, F. (1969) Geographic pathology of skin cancer. In: *The Biologic Effects of Ultraviolet Radiation with Special Emphasis on the Skin*. p. 635. Edited by F. Urbach. Oxford: Pergamon Press.

Walters, J. B. & Israel, M. S. (1974) *General Pathology* 4th edn., p. 408. Edinburgh: Churchill Livingstone.

Ward, M. (1975) *Mountain Medicine. A Clinical Study of Cold and High Altitude*. London: Crosby Lockwood Staples.

3

Physical and human geography:
High altitude fauna and flora

Life at high altitude inevitably implies exposure to hypoxia and the other physical factors listed in the preceding chapter. However, climatic factors also influence the nature of the mountain environment. In particular the abundance or absence of precipitation so modifies the climate of the two great mountain ranges in the world, the Himalayas and the Andes, that they present very different habitats for the native highlanders of the two areas.

Rainfall in the Himalayas and Andes

The Himalayan range extends across the base of the Indian sub-continent for about 1700 miles from Assam in the south-east to Afghanistan in the west, and is up to 150 miles wide. This range comes under the influence of the Monsoon, which originates in the Indian Ocean (Fig. 3.1). Water-laden air flows from east to west across India, ascending the mountain slopes and cooling as it does so (Fig. 3.1).

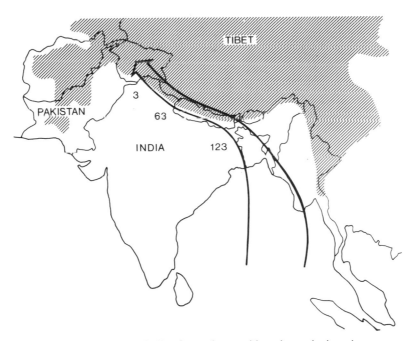

Fig. 3.1   As the Monsoon sweeps inland from the Indian Ocean, heavy with moisture, it gives rise to an annual downpour of 123 inches a year at Darjeeling with heavy snowfalls in the Nepal Himalayas. As it crosses the subcontinent it becomes drier so that the annual rainfall at Simla in the central Himalayas is 63 inches. By the time the wind reaches the western extremities of the Himalayan range in Kashmir it is dry and the rainfall is but 3 inches a year. Hence the Karakorums are arid in contrast to the Nepal Himalayas. Areas of high altitude are indicated by cross-hatching.

Eventually the water vapour condenses and rain pours in torrents between April and October over the eastern part of the Himalayan range. However, as the Monsoon passes to the west, it becomes progressively depleted of its content of water vapour. As a result the eastern Himalayas are drenched and the western Himalayas are arid. At Darjeeling, on the border of Sikkim, the rainfall is 123 inches a year. At Simla, north of Delhi, in the central Himalayan range it is 63 inches. In contrast the mountains of the western extremity of the Himalayas in Leh, Kashmir, has an annual rainfall of only three inches (Nicolson, 1975) (Fig. 3.1). Hence, while the mountains of the eastern Himalayas are covered with snow and ice the Karakorums (derived from Turki words giving the meaning 'black and crumbling rock') which divide Pakistan from China in the west, are arid. Indeed

one of the highest mountains in this range is called 'the naked mountain' (Nanga Parbat) because it is not permanently covered by snow. The other dry western extremity of the Himalayas is represented by the Hindu Kush ('Hindu Killer') which is 600 miles long and rises to 7620 m. It is situated mainly in Afghanistan. The Himalayas are the highest mountains in the world. Three of the peaks exceed 8530 m (Table 3.1) and there are thirteen others above 7920 m. This vast mountain range extends through India, Bhutan, Sikkim, Nepal, Pakistan, China, Tibet and Afghanistan and from them the Pamirs extend northwards.

The Andes extend the whole length of the western side of the South American continent from Ecuador to the Argentine (Fig. 3.2). They consist of three principal ranges or cordillera, running approximately from north-west to south-east in parallel with

Table 3.1　Heights of some mountains (to the nearest 5 m).

Peak	Height (m)	Situation
*Everest (Chomolungma 'Goddess Mother of the Earth'. Tibetan)	8850	Everest range, Tibet/Nepal
K2 (Godwin–Austen)	8610	Karakorams, (Disputed area between Sinkian and Kashmir)
Kangchenjunga ('Five treasures of the Eternal Snows')	8600	Everest range, Sikkim
Lhotse	8500	Everest range, Tibet/Nepal
Makalu	8470	Everest range, Nepal
Dhaulagiri	8170	Everest range, Nepal
Cho Oyu	8155	Tibet/Nepal border
Nanga Parbat ('Naked Mountain')	8125	Karakorams, Kashmir
Manaslu	8125	Everest range, Nepal
Annapurna I ⎫ ⎬ ('Giver of Life') Annapurna II ⎭	8080	Everest range, Nepal
Gasherbrum II ⎫ Gasherbrum IV ⎭	8070	Karakorams, Disputed area between Sinkian and Kashmir
Hidden peak	8065	Karakorams, Kashmir
Broad peak	8045	Karakorams, Disputed area between Sinkian and Kashmir
Shisha Pangma	8015	Tibet
Aconcagua	6960	Andes
Huascaran	6765	Andes
Mount McKinley	6195	Alaska
Kilimanjaro	5895	Africa
Mont Blanc	4810	France
Monte Rosa	4635	Switzerland
Weisshorn	4505	Switzerland
Matterhorn	4480	Swiss/Italian border
Mauna Kea	4210	Hawaii
Mount Fuji	3775	Japan
Mount Cook	3765	Australia

*In 1865 Sir Andrew Waugh, Director of the Indian Survey, suggested the highest mountain should be named after Sir George Everest, Surveyor-General of India from 1830 to 1843.

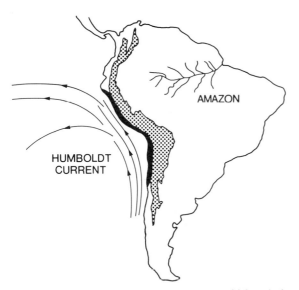

Fig. 3.2 The path of the cold Humboldt current which cools the overlying air, reducing its capacity to hold moisture, and thus ensuring the aridity of the coastal strip of Peru and Chile (solid area) and the western slopes of the Andes, which are indicated by stippling.

Fig. 3.3 Houses built on the wet, precipitous eastern slopes of the Andes. They form part of the remains of Majchu Picchu, (2440 m), the 'lost City of the Incas'.

the coast, and separated one from the other by high plateaux forming the Altiplano (Marett, 1969). The western range runs in a continuous chain along the whole continent but the central and eastern cordillera, while well defined in the north, and to a less extent in the south, become merged and lost in great knots of transverse mountain ranges (Marett, 1969). The coastal strip of Peru is desert and indeed it is commonly said 'It never rains in Lima'. Along the whole length of the Peruvian Pacific coast the effect of the cold Humboldt current is to cool the air above the sea and reduce the capacity to retain the moisture which in normal circumstances would have fallen on the land as rain (Fig. 3.2). Moreover, once the air passes over the land it is warmed up again, increasing its capacity to retain moisture and making precipitation impossible. The western slopes of the Andes are dry as dust, and cacti and eucalyptus trees abound. Only a few high mountain ranges such as that including Huascarán (6765 m) are snow-covered. The contrast with the icy conditions of the Himalayas is striking. On the other hand the eastern slopes of the Andes descend into the Amazon and are progressively more humid and tree-covered. On such wet slopes was built the fortress of Majchu Picchu which has come to be called the 'Lost City of the Incas' (Fig. 3.3).

Mountain zones in the Himalayas

As one ascends from sea level to high altitude in the eastern Himalayas one passes through five well-defined zones. They are the result of the interplay of Monsoon rain and increasing altitude. *Rain forest* persists to an altitude of about 2000 m (Fig. 3.4). Drenched by Monsoon rain the flora is composed of giant ferns, palms and vines with a luxuriance of flowering plants such as orchids. Just above 2000 m the jungle gradually opens out into a *temperate zone* in which conifers and oaks grow amidst a proliferation of magnolia, irises and rhododendrons (Nicolson, 1975) (Fig. 3.4). Above the tree line at about 3000 m the cool, green *alpine zone* commences and this rises up to 4500 m. Alpine plants such as gentians and saxifrages predominate. Above this elevation the hills of the *Tibetan plateau* become snow-dusted. Finally, above 5000 m one enters what Nicolson (1975) has aptly described as 'The Ice Kingdom'. Ice and snow predominate and the terrain becomes one of glaciers, ice walls and crevasses. Winds up to hurricane force and temperatures as low as −40°C have to be contended with. In fact the name 'Himalayas' means 'The Abode of Snow', although as we have already seen this description really applies only to the most easterly of its three ranges, the Everest range.

Mountain zones in the Alps

The Alps do not compare in elevation with the Andes or Himalayas (Table 3.1). Nevertheless they are the most accessible and familiar mountains in

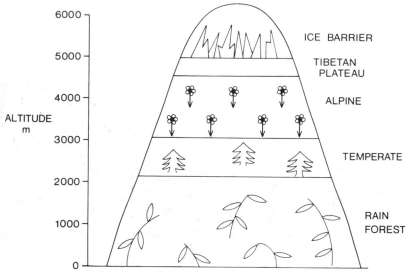

Fig. 3.4   The five zones of the mountain ranges of the Eastern Himalayas which result from the interplay of Monsoon rain and increasing altitude.

Europe and also present characteristic zones. They are areas of high precipitation and hence the pastures surrounding high villages such as Zermatt (1620 m) are well supplied by water from the not infrequent rainshowers and the full streams fed by the snows and glaciers of the surrounding mountains (Fig. 3.5). Pine forests extend up the steep hills of the narrow valleys to an altitude of about 2200 m. The meadows of these lower slopes

Fig. 3.5   Glacier (Gornergletscher) at the foot of the Monte Rosa seen from the Stockhorn (3530 m).

between the trees (Fig. 3.6) abound in a host of wild Alpine flowers while some cultivated plants are also characteristic of the area (Fig. 3.7). Above the tree line the terrain becomes rocky and vegetation is sparse. The permanent snow line is at about 3000 m and the snow-covered mountains include the Monte Rosa (4635 m), Matterhorn (4480 m) (Fig. 33.1), Weisshorn (4505 m) and Dom (4545 m). These peaks are surrounded by extensive glaciers such as the Findelngletscher, the Gornergletscher (Fig. 3.5), the Theodulgletscher and the Zmuttgletscher. Thus at 3000 m in the Valais massif one is in a world of ice and snow whereas at the same altitude in Tarma in the Peruvian Andes the terrain is dry as dust and abounds in cacti. This illustrates that altitude by itself does not determine the nature of the terrain and the associated flora and fauna. The level of precipitation is also of considerable importance.

Human geography

In general the mountain environment does not attract man as a permanent home. Even in Europe where the mountains are not very high, population becomes sparser with increasing elevation. In Switzerland areas above 1000 m contain only 5 per cent of the country's population and those over 1520 m only 0.5 per cent of the population (Perpillou, 1966). At really high altitudes the mountain environment becomes even less attractive because of the unpleasant physiological and physical conditions associated with the elevation. Demo-

Fig. 3.6 Alpine pastures and pine forest at the foot of the Matterhorn seen faintly in the background.

Fig. 3.7 *Eryngium alpinum*, a species of cultivated plant characteristic of the mountain valleys of the Valais massif.

graphic trends in this century confirm this flight from the mountains (Chapter 26).

Nevertheless, in spite of this general trend of depopulation, the density of population of a mountain range depends also on *accessibility* as well as an absolute level of altitude. In this respect the Andes differ greatly from the Himalayas. Thus much of the economic life of Peru goes on at high altitude because the mining areas and the routes to the Amazon are so accessible. In South America considerable mining activity goes on at Cerro de Pasco (4330 m), Potasi (4070 m) and Oruro (3760 m). This relative ease of access to the Andean heights has influenced the history of the region. The empire of the Incas extended throughout the Andes and had its capital at Cuzco (3400 m) which is in fact the Quechua word for 'navel'. This highland race built for itself at high altitude a fortress at Majchu Picchu at 2440 m on the edge of the Amazon to prevent the fierce jungle tribes penetrating the deep gorges to reach the sacred city of Cuzco. At this altitude the Inca people built a town with incredible terracing and lived on these precipitous mountain sides (Fig. 3.3). At Sacsayhuaman (3580 m) remains of fortresses and temples still exist as testimony that in the past large communities have existed at high altitude just as they do today (Fig. 3.8). In contrast Tibet and the Himalayas are largely empty above 4570 m. There are gold mines at Tok Dschalung at 4880 m at the headwaters of the Indus but such settlements are exceptional. In general, access to the Himalayas is very difficult.

Fig. 3.8 Part of a fortification at Sacsayhuaman (3580 m) showing the characteristic features of Inca architecture. Huge blocks of stone have been accurately shaped together to form a wall without the use of any form of cement. Their immense size may be judged from the proximity of one of the authors (DRW) engaged in cinefilm work at the site. It is impossible to insert a knife blade between the blocks of stone. Another remarkable aspect of such building is that the blocks of stone were transported to the building site at great altitudes without the use of the wheel which was unknown to the Inca people.

Accessibility to mountains and high altitude disease

Accessibility also has considerable import so far as acclimatization and the susceptibility to high altitude diseases are concerned. In Peru, tourists, lowlanders, and highlanders returning to their mountain home may reach altitudes exceeding 4000 m in very few hours by car or bus. Hence the risks of developing acute mountain sickness or even high altitude pulmonary oedema are considerable. In contrast to this, high altitude climbers journeying to ascend a peak in the Himalayas have to trek by foot through high mountain passes for two or three weeks to reach the peak selected for assault. Such inaccessibility means that climbers arriving at high altitude in Nepal or Tibet are acclimatized (Chapter 27). In recent years, however, the increasing numbers of trekkers in the Himalayas not infrequently fly to a high altitude to start their trek and thereby constitute a growing problem of cases of acute mountain sickness in holidaymakers to these remote and potentially dangerous areas.

In the Alps accessibility is available only to moderately high altitudes barely exceeding 3000 m. However, the profusion of cable-cars and mountain railways in the Swiss Alps means that one can ascend to the borderline of significant high altitude very rapidly indeed. Thus in little more than an hour one can ascend by mountain railway from Zermatt (1620 m) to Gornergrat station (3130 m) and thence to the Stockhorn station (3405 m) by cable-car. Such rapidity of ascent precludes any chance of acclimatization and the Swiss are either fortunate or far-seeing in that their cable-cars end just at the altitude above which physiological troubles could be expected. Another interesting example of even more rapid ascent, to which we refer in Chapter 1 and which is of practical import, is the manning of the new infrared telescope at Mauna Kea situated on a volcano on the island of Hawaii at an altitude of 4210 m. The ascent to this telescope can be achieved by car in little more than one hour.

In tropical areas the *moderation of temperature* inherent in living at high altitude attracts communities. In Saudi Arabia, a country of high temperatures, Sana, the capital city of the Yemen, is situated at an altitude of 2130 m. In Ethiopia, where the majority of the country is about 2000 m, there are many large towns. The capital city, Addis Ababa, is situated at an altitude of 2400 m and Ankaber at 2600 m.

Characteristics of high altitude plants

The outstanding characteristic of high altitude vegetation is the absence of trees and shrubs. Indeed at great heights the aerial parts of the plant species present are rarely perennial. The atmosphere is too cold and most plants survive by persisting as underground strong runners, or rhizomes. Characteristically the aerial parts develop rapidly in the spring only to wither in the autumn; even then they are dwarfs in comparison to low altitude plants and keep close to the ground.

Plants, unlike animals, are not affected by the diminished PO_2 of the ambient air. They are, however, very susceptible to the aridity of mountains. Hence many develop the characteristics of xerophytes designed to preserve water from being lost by transpiration to the mountain air. In contrast to the xerophytes of deserts which need long tap roots, the high altitude plant tends to have superficial fibrous roots to occupy the thin layer of soil covering the rock. Above 5000 m the major factor limiting plant life is the lack of water. Even lichens, so often considered a characteristic feature of high altitudes, were found by Swan (1961) to be scarce, small and often dead above 5490 m in the Himalayas. At these great altitudes plants have to congregate in areas where there is water. Thus they grow where there is some sub-surface drainage of water from a higher snow field or in niches at rock bases. Characteristically high altitude plants are to be found at the edge of melting snow; as this edge moves progressively up the mountainside in spring the zone of flowering moves with it.

In the Everest range of the Himalayas each mountain zone (Fig. 3.4) has its own typical flora (Table 3.2). Mani (1974) gives lists of species to be found in the Himalayas, Pamirs, Andes and African mountains. The highest altitude at which a plant has been found appears to be 6140 m at which elevation Swan (1961) found the cushion plant, *Stellaria decumbens*. In contrast to such plants striving to survive on the precious water, some species such as *Saussurea* have a form designed for protection against the excessive rainfall of the Monsoon.

Characteristics of high altitude insects

A detailed study of insect life at high altitude has been made by Mani (1968, 1974). Many species of insect found at such heights show structural and physiological features which appear to enable them to adjust to life on the mountains. Thus in contrast to the insects of the tundra, where the vegetation is similar, the mountain species show heavy body pigmentation in dark colours such as black, reddish-brown, and deeper tones of orange and yellow than are customary in sea-level species. In

Table 3.2 Typical flora of the mountain zones of the Himalayas (after Nicolson, 1975).

Mountain zone	Typical altitude range	Typical flora
Rain forest	0 to 2000 m	Rain forest vegetation. Giant ferns. Palms. Vines. Jungle flowering plants. Orchids
Temperate zone	2000 to 3000 m	Spruce. Juniper. Fir. Pine. Oaks. Magnolia. Irises. Rhododendrons
Alpine zone	3000 to 4500 m (Tree line at 4100 m)	(Above tree line.) Potentillas. Anemones. Aconite. Sandwort. Edelweiss. Saxifrages. *Saussurea*. Asters. Sagebush. Himalayan trumpet gentians. Contoneasters. Rock jasmine. Sedges. Primulas
Tibetan plateau	4500 to 5000 m	Himalayan blue poppy
Ice barrier	5000 to 6150 m	Cushion plant (*Stellaria decumbens*)

mountain insects the wings too are darker. Pale-coloured insects are decidedly uncommon at high altitude with the minor exceptions of a few soil-inhabiting species. Mani (1974) believes this melanism protects the body-tissues of insects from the high levels of ultraviolet radiation. The pigmentation may also aid the absorption of solar radiation to maintain warmth.

A second peculiarity of high altitude insects is the reduction and loss of wings and of the power of flight (Mani, 1974). This flightlessness is probably a protection against the high wind velocities of mountainsides. There is also a pronounced tendency for a reduction in body size of insects with increase in altitude, although medium-sized and even large species do occur in mountainous areas. The period for growth and the rate of growth in insects is certainly diminished by the coldness of the high altitude environment. Insects at these heights are able to survive at far lower temperature than sea-level species. However, this is not to say that cold does not have a striking inhibitory effect on the activity of even mountain species. Thus, when the sun is shining, they are remarkably active and swarms may be found on warm rock surfaces exposing the maximum surface to the sunlight. However, in cloud the temperature drops and they behave erratically. Swan (1961) refers to bumble-bees lying torpid by rocks and butterflies lying on their sides. In the cold of the mountain night flying insects disappear into crevices and holes. Insects at high altitude for this reason tend to comprise species which live in the protecting soil rather than on the exposed plants. Nocturnal species are unknown. The low ambient temperatures have the advantage of diminishing the risk of desiccation. Mountain snows form the major source of moisture for insects and also form a useful cover for hibernation. In the Himalayas up to an altitude of 4880 m a wide variety of insect species is to be found, including various species of ants, bees, wasps, flies, butterflies, moths, beetles, aphids, grasshoppers and so on. For a detailed account of the orders and species of insect found at high altitude readers are referred to the writings of Mani (1968, 1974).

Summit-seeking insects

High and prominent mountain tops seem in some unexplained manner to attract strongly some species of insect especially in the autumn (Mani, 1974). In contrast to 'Aeolian derelicts' (see below) these species are exclusively winged and active, and include winged ants, beetles and moths. The insects congregate in enormous groups under some sheltered rock. Mani (1974) describes having seen at 4260 m one group of *Coccinella septempunctata* and *Hippodamia* numbering about two million. At present the reason for these mass gatherings is obscure. The majority die on the mountain and form a source of nitrogen both to higher animal species and to plants.

Aeolian derelicts

Particulate matter is constantly transported passively by air-currents from the plains to be deposited on the sides of mountains. Since Aeolus is the god of winds Mani (1974) has coined the delightful term 'aeolian derelicts' to describe this wind-wafted material. It comprises organic debris of various sorts including seeds, pollen, fungal spores, leaves and bark. Living material may also get caught up in these currents. Thus hapless insects from the plains may be carried up by warm air currents, chilled in the upper layers of air, and eventually dropped frozen and dead on the mountainside (Mani, 1974). Such insects include spiders, mites, aphids, dragonflies and butterflies. Even locusts have been so transported to an icy grave. Mani (1974) reports that in 20 minutes he observed over 400 dead insects deposited on an area of 10 m² at an elevation of 3500 to 4000 m. The deposited particles become trapped in snow and glacier ice. Much is deposited on the lee sides of rocks as high as 6100 m. These aeolian derelicts are of considerable biological significance because they are broken down by fungi to form the first stage of a food chain providing nutrition for a community of extreme altitude dwellers, comprising the so-called 'Aeolian Community'.

The Aeolian Community

The first suspicion that some lower forms of life may permanently colonise the extraordinarily inhospitable environment of extreme altitude came in 1924. In that year members of the British Everest Expedition reported that they had seen jumping spiders of the family *Salticidae* at elevations as high as 6710 m. Swan (1961) refers to such extreme altitude dwellers as 'the Aeolian Community'. They are members of the orders *Thysanura* and *Collembola* and thus represent the oldest and most primitive insects. The most characteristic member of the community is the springtail, a crawling and jumping insect of the order *Collembola* which has a trigger-like mechanism under its body capable of throwing it a few inches into the air. *Anthymyiid* flies are also found, and they are eaten by the *Salticid* jumping spiders observed by the British climbers in 1924. According to Swan (1961), herbivorous and predaceous mites and small centipedes may also range above 5940 m. This supra-alpine Aeolian community survives at an altitude incapable of supporting flowering plants. In addition to the wind-blown debris it requires periods of time when its supporting rock-base niche attains a temperature above freezing so that the moisture required for survival is provided.

Species of animal at high altitude

A great variety of groups of animals belonging to many different phyla like Protozoa, Platyhelminths, Nemanthelminths, Arthropods, Molluscs, and Reptiles have successfully evolved into high altitude forms. Since consideration of these is not a purpose of this volume readers are referred to a more detailed treatment of these species by Mani (1974). Sightings of high altitude reptiles are given in Table 3.3.

Mammals indigenous to high altitude

Like man, most mammals cannot exist permanently at extreme altitudes exceeding 5800 m (Chapter 30). There are indigenous high altitude species such as the Llama (*Lama glama*), Alpaca (*Lama pacos*), Vicuña (*Lama vicugna*) and Guanaco (*Lama guanicoe*) in the Andes, and the Yak (*Bos grunniens*) in the Himalayas (Fig. 3.9). Such species are *adapted*

Fig. 3.9 Yak (*Bos grunniens*) in the Himalayas. In the background is Taweche (altitude 6710 m to 7010 m). This photograph was taken by Dr John Dickinson of the Shanta Bhawan Hospital, Kathmandu, Nepal.

biologically and biochemically to high altitude as we describe in Chapter 27. However, many species of animal have been seen at great heights and a list of sightings of mammals above 3810 m on the slopes of the Himalayas as provided by Napier (1976) are

Table 3.3 Sightings of reptiles at high altitude (after Mani, 1974).

Altitude	Area	Species	Type of reptile concerned
5500 m	Himalayas	*Leiolopisma ladakense*	Lizard
	Himalayas	*Leiolopisma himalayanum*	Lizard
4900 m	Peru	*Liolaemus multiformes*	Lizard
4800 m	Himalayas	*Ancistrodon himalayanum*	Snake
4570 m	Mexico	*Pseudoeurycea godovii*	Salamander
4530 m	Citlapetl	*Sceloporus microlepidotus*	Lizard
	Citlapetl	*Crotalus triseriatus*	Snake
4400 m	Andes	*Bufo*	Amphibia
	Junin	*Batrachophrynus*	
	Lake Titicaca	*Telmatobius culeus*	

Table 3.4 Sightings of mammals above 3810 m on the north and south slopes of the Himalayas (after Napier, 1976).

Altitude	Common name	Scientific name
6100 m	Yak	*Bos grunniens*
	Bharal (blue sheep)	*Pseudois nayaur*
	Pika (mouse hare)	*Ochotona roylei*
6035 m	Woolly hare	*Lepus olostolus*
5945 m	Ibex	*Capra sibirica sacin*
5790 m	Woolly wolf	*Canis lupus chanco*
	Argali (Tibetan wild sheep)	*Ovis ammon hodgsoni*
5640 m	Hill fox	*Vulpes vulpes montana*
5490 m	Lynx	*Lynx lynx isabellinus*
	Kiang (wild ass)	*Equus hemionus kiang*
	Chiru (Tibetan antelope)	*Panthalopus hodgsoni*
	Goa (Tibetan gazelle)	*Procapra picticaudata*
	Marmot	*Marmota bobak*
	Marco Polo's sheep	*Ovis ammon poli*
	Snow leopard	*Panthera uncia*
5180 m	Weasel	*Mustela altaica*
	Field vole	*Pitymys leucurus everesti*
5060 m	Brown bear	*Ursus arctos isabellinus*
	Tibetan sand fox	*Vulpes ferrillata*
4880 m	Himalayan weasel	*Mustela sibirica*
4570 m	Pallas' cat	*Felis manul*
3810 m	Musk deer	*Moschus moschiferus*

shown in Table 3.4. Such mammals, like the Sherpas accompanying mountaineering expeditions, and the Tibetan pilgrims, make only temporary visits to these great altitudes.

Mountain birds

A few species of bird are native to high mountainous regions. Usually they are birds of prey and carrion, for the perils of life on the mountainsides ensure a constant supply of carcasses (Nicolson, 1975). In the Himalayas, Griffon vultures (*Gyptäetus himalayensis*) spiral on wind currents up to 7620 m. Lammergeyers (*Gyptäetus barbatus*) are also found. Many snowline birds are seasonal commuters leaving their summertime sojourn in the snows to fly down to forests as low as 1520 m in the winter. Birds such as the Chough (*Pyrrhocorax graculus*) have visited the high camps of early Everest expeditions at 8230 m. In the spring, the Bar-headed goose (*Anser indicus*) migrates from the lakes of India at sea level to the lakes of Tibet across the summit of Mount Everest (8850 m). This is a bird of great interest when one considers the various aspects of adaptation and acclimatization (Chapter 27). The Andes have their own striking birds of prey such as the terrifying Condor (*Vultur gryphus*).

Man at high altitude

Having considered the physical and climatic factors prevailing at high altitude and the animals and plants that survive in these conditions we may now study the features of men who make the high mountains their permanent home.

REFERENCES

Mani, M. S. (1968) *Ecology and Biogeography of High Altitude Insects*. The Hague, Netherlands: Dr. W. Junk N.V.

Mani, M. S. (1974) High altitude insects. In: *Fundamentals of High Altitude Biology*, p. 67. New Delhi: Oxford and IBH Publishing Co.

Marett, R. (1969) In: *Peru* p. 25 London: Ernest Benn.

Napier, J. (1976) *Bigfoot. The Yeti and Sasquatch in Myth and Reality*. Abacus edition. London: Cox and Wyman.

Nicolson, N. (1975) *The Himalayas. The World's Wild Places*. Amsterdam: Time-Life Books.

Perpillou, A. V. (1966) In: *Human Geography,* p. 348. London: Longman.

Swan, L. W. (1961) The ecology of the High Himalayas. *Scientific American*, **205**, 68.

4

The native highlander

The two major regions of the world where large communities exist at high altitudes are the Andes and the Tibetan plateau. In both areas the native population is of Mongoloid stock and the usual explanation for this is that the remote ancestors of the American Indians emigrated from Asia by way of the Bering Strait in late glacial times. It is likely that the physique of these high altitude natives is largely determined genetically by their ethnic origin and one must be cautious before ascribing too readily physiological advantages to some of their physical attributes. At the same time it is reasonable to accept that these peoples may have undergone some anatomical or biochemical adjustments to the extreme environment in which they have lived for many thousands of years. Proof of such prolonged undisturbed residence in their mountain habitat is provided by the finding of traces of man at Lauricocha in Peru with a radio-carbon date of about 7500 BC (Marett, 1969). It should be noted, however, that long though this history is of settled communities at high altitude, it is probably much shorter than the time over which the Sherpa people have been resident in the Himalayas. We shall consider the significance of this later in Chapter 27. One example of an anatomical feature thought to be an adaptation to life at high altitude is commonly taken to be the development of a barrel-shaped chest which is thought to bring beneficial effects on respiration and which we consider later in this chapter. We shall examine the physical features of the present day Quechua Indians of the Peruvian Andes to see if they support the concept of a distinct 'high altitude man' whose physique fits him specifically for life at high altitude. Before proceeding with this, however, we shall first briefly consider the blood groups of these highlanders.

Blood groups in the Andean Indian

There is a remarkable predominance of blood group O among the Indian population of the Andes. While studies of the ABO blood groups in Liverpool show that half of the population is of blood group O (Mourant et al, 1958), in the Andes of Northern Peru 100 per cent of the population are of this group (Arce Larreta, 1930) (Fig. 4.1). Intermediate incidence of the group is found in the mestizos of Lima where intermarriage of Spanish stock and Andean Indians is commonplace (Escajadillo, 1948) (Fig. 4.1). A high proportion of the population with blood group O is also found in the Andean town of Junin (San Martin, 1951) (Fig. 4.1). Blood group O has also been reported in 95 per cent of Bolivian Indians (Quilici, 1968), in 85 per cent of Quechuas in Milpo and Colquijirca in Peru (4000 m) (Ruiz and Peñaloza, 1970), and in 93 per cent of the Aymaras of the Andean altiplano (Durand, 1971). Arias-Stella (1971) studied a native group called the Lamistas in a province of Peru near the jungle and found that 100 per cent of them were of group O.

Such a predominance of one blood group emphasises the importance of ethnic considerations in the native population of the Andes and supports the view that the physical features of the Quechua and Aymara peoples are largely racial rather than exemplifying the body build of a 'high altitude man'. Furthermore we should note that group O is not common to all highlanders. Thus in Nepal only 33 per cent of subjects are of this blood group (Macfarlane, 1937; Mourant et al, 1958) and in Tibet the incidence of blood group O is only 42 per cent of the population (Büchi, 1952). Hence blood group appears to be related to the origin of the race rather than to any adaptation to life at high altitude.

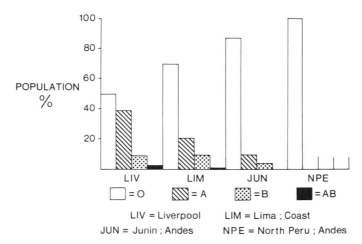

Fig. 4.1 The distribution of the ABO blood groups at Liverpool, in the mestizo population of Lima and in the largely Quechua communities of the Andes. Note the predominance of blood group O in the Indian population of high altitude. (From data of Mourant et al, 1958).

However, this view is not universally accepted. In particular Peñaloza (1971) believes that the present predominance of group O in the indigenous race of the High Andes is the result of a genetic mutation occurring as an adaptive phenomenon over the centuries. Such a belief arises from the study of the blood groups in Peruvian mummies.

Blood groups in Peruvian mummies

Peñaloza (1971) referred to the fact that studies of blood stains found in five South American mummies from the pre-Columbian period revealed blood group A in 3 of 5 cases (Gilbey and Lubran, 1952). In the study of Boyd and Boyd (1937) six mummies were found to be of group B and two of group AB. These findings, in face of the predominance of blood group O in most present-day Andean highlanders, has led some to believe that the finding of A, B or AB groups is artefactual and perhaps due to contamination of mummy tissues by bacteria (Hübener-Thomsen-Friedenreich phenomenon).

This view is not shared by Allison et al (1976) who believe that ABO group antigens have palaeo-serological value since they are not restricted to red cells but are distributed throughout the body to include white cells. Thus bone and hair can be used to determine ABO groups and Allison et al (1976) have studied material from dry muscle of forearm or calf, apex of heart, and abdominal skin. They

determined the blood groups of Peruvian mummies of known origin by agglutination-inhibition, induction of antibody production, and mixed cell agglutination. Their results indicate unequivocally the presence of A, B and AB blood groups in America prior to known European contact.

It does seem likely that group B has been eliminated from the Andes by natural selection. Thus in the province of Pisco in the most northerly area of Peru the incidence of blood group O in mummies is only 53.1 per cent (Table 4.1). As one moves south to Arequipa (Table 4.1) groups B and AB disappear

Table 4.1 ABO blood group distribution based on geographic location of cemetery where mummies were found. The provinces listed are progressively more southerly situated (after Allison et al, 1976).

Province	Blood Group (%)			
	A	B	AB	O
Pisco and Ica	28.6	6.1	12.2	53.1
Nazca	16.6	0	11.2	72.2
Arequipa and Tarapaca	7.2	0	0	92.8

and Boyd (1959) suggests that the B group may have been eliminated by natural selection. In the Ica culture (Table 4.2) group O was present in 57.1 per cent but in the Incas it rose to 92.4 per cent (Allison et al, 1976) (Table 4.2). The Murga colonial people had three times as many A individuals as the Incas but the group B of the Ica culture had gone (Table 4.2). One must be careful before ascribing too

Table 4.2 Cultural groups and carbon 14 dating of mummies – agglutination-inhibition technique (after Allison et al, 1976).

Culture	Carbon 14 dating	ABO blood group distribution (%)			
		A	B	AB	O
Paracas	265-85 BC	28.5	7.1	21.4	43.0
Nazca	95 BC - 700 AD	0	0	40.0	60.0
Huari	890-1235 AD	26.7	0	20.0	53.3
Ica	1350-1550 AD	32.2	7.1	3.6	57.1
Inca	1245-1650 AD	7.6	0	0	92.4
Tarapaca colonial	1550-1600 AD	10.0	0	0	90.0
Murga colonial	1580-1700 AD	23.6	0	3.7	72.7
Total		22.9	2.2	7.8	67.1

readily blood group O to mummies to realise that A and B antigens may simply have been destroyed by time so that α and β agglutinins are not absorbed.

The physical features of the Quechua Indian

Facially the Quechua resembles the Asiatic mongol (Fig. 4.2a, b, c). The eyes are almond-shaped with thick epicanthic folds (Fig. 4.2b). The skin is pigmented and of a dark olive hue and forms a

Fig. 4.2 (a) The facial features of a Quechua, 66 years of age, from La Raya (4200 m) in the Peruvian Andes. The skin is pigmented (see Chapter 23). The lips were of a deep russet-red colour due to the high haematocrit (see Chapter 6). (b) The eyes of a Quechua, 23 years of age, from Cerro de Pasco (4330 m): note the thick epicanthic folds. (c) The facial features of the young Quechua Indian shown in Fig. 4.2b compared with those of one of the authors, D.R.W., (left).

striking contrast to the russet red colouration of the mucous membranes, the lips, and the lobes of the ears which is an expression of the high level of haemoglobin described in Chapter 6. Superimposed on this dark red colour is associated central cyanosis consequent upon the systemic arterial oxygen unsaturation due to the hypoxia of diminished barometric pressure. Cyanosis is usually detected when the systemic arterial oxygen saturation falls below 80 per cent and the systemic arterial tension has fallen to 50 mmHg or less (Cumming and Semple, 1973). Such conditions are readily met at high altitude and cyanosis becomes apparent because of the elevated haematocrit in the Quechua Indian. Some 4 to 5 g of deoxyhaemoglobin need to be present per dl of blood for cyanosis to be detected and since the normal level of haemoglobin in the healthy highlander is 20 g/dl (Chapter 6) the necessary conditions are readily met. The combination of erythraemia and cyanosis also produces

(a)
(b)
(c)

suffusion of the conjunctival vessels which are both tortuous and prominent. The conjunctiva has a creamy-yellow appearance rather than the clear white opaque appearance seen in the lowlander in good health. The lips tend to be bulbous and the nose straight. The hair is straight and black.

Clubbed and beaked fingers

Reference has been made above to the cyanosis which is consequent upon the reduction of oxygen saturation of systemic arterial blood at high altitude. At sea level diseases causing such unsaturation commonly give rise to clubbing of the fingers. Hurtado came to the conclusion that similar clubbing is to be found in the Quechua, from his classical early study of the physical anthropology of this mountain people (Hurtado, 1932). He made a clinical study of 950 males between the ages of 4 and 75 years and he stated that clubbing is present in 59 per cent of Quechua children between 4 and 9 years of age and in 94 per cent of adults. Its degree varied widely from individual to individual and in many instances it was limited to the thumb, index and little fingers. Our personal observations in the Andes lead us to believe that clubbing of the fingers in the healthy Indian population is far less common than indicated by the figures of Hurtado. It seems to us that the fingers are commonly what British clinicians would term 'beaked' rather than 'clubbed' (Fig. 4.3). In our opinion true clubbing of the

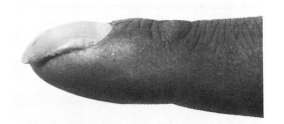

Fig. 4.3 One of the fingers of the young Quechua shown above. It shows 'beaking' in contrast to the clubbing of the fingers in cases of Monge's disease and illustrated in Fig. 16.2d.

fingers is more likely to be encountered in patients with 'Monge's disease' (Chapter 16) (Fig. 16.2d) in whom the degree of systemic arterial oxygen unsaturation and cyanosis are greater.

In our experience small haemorrhages within the fingernails are common in the Indian population (Fig. 23.1) and they are probably related to repeated minor trauma to the nails just as they are at sea level as we discuss at length in Chapter 23. Haemorrhages within the finger nails are more striking in patients with Monge's disease (Chapter 16) (Fig. 23.2) and furthermore their distribution is different from those found in either residents at sea level or in healthy highlanders. As we discuss in Chapter 23 we believe that the nail-haemorrhages of patients with Monge's disease are related to a greatly elevated haematocrit level. They are also found in patients with cyanotic congenital heart disease with secondary polycythaemia (Chapter 23).

Height

The Quechua is shorter than the mestizo from the coast or the Caucasian (Figs. 4.4 c and d). This difference is already apparent in early childhood (Fig. 4.5). In a study of 21 high altitude natives, 10 years of age, born and living at Morococha (4540 m), the average height in 1932 was 122.2 cm and was some 8.5 per cent less than a comparable low altitude group in North America (Hurtado, 1932). Since that time improved social conditions in North America and Europe have led to an increase in body height and weight in children and young adults. In the region of the Andes similar improvement in the conditions has not taken place. Hence the contrast between these native children and a Caucasian population of the present day is now even more pronounced. The average height of the present day Caucasian boy of 10 years is 136.8 cm, a difference of 10.7 per cent from the figures given by Hurtado (1932). The difference in height between Quechua children of four to 19 years of age (Hurtado, 1932) and British children of comparable age in 1965 (Tanner et al, 1966) is shown in Figure 4.5. This difference in height continues on into adulthood. As early as 1932 Hurtado found that the average height of young adult Quechua Indians was 159 cm compared with an average height of 168 cm in Caucasian Peruvian students living at sea level. With the improvement of modern social conditions the difference in height from young Caucasians is even greater.

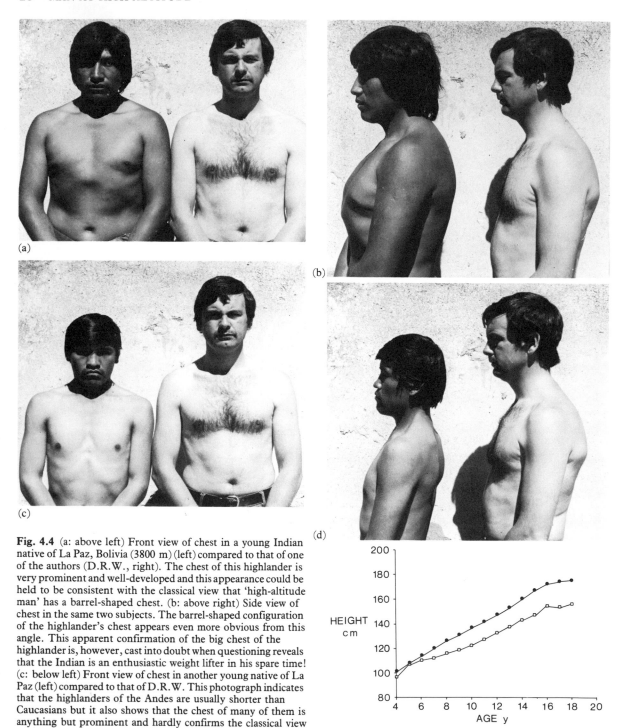

Fig. 4.4 (a: above left) Front view of chest in a young Indian native of La Paz, Bolivia (3800 m) (left) compared to that of one of the authors (D.R.W., right). The chest of this highlander is very prominent and well-developed and this appearance could be held to be consistent with the classical view that 'high-altitude man' has a barrel-shaped chest. (b: above right) Side view of chest in the same two subjects. The barrel-shaped configuration of the highlander's chest appears even more obvious from this angle. This apparent confirmation of the big chest of the highlander is, however, cast into doubt when questioning reveals that the Indian is an enthusiastic weight lifter in his spare time! (c: below left) Front view of chest in another young native of La Paz (left) compared to that of D.R.W. This photograph indicates that the highlanders of the Andes are usually shorter than Caucasians but it also shows that the chest of many of them is anything but prominent and hardly confirms the classical view referred to in the text. (d: below right) Side view of chest in the same two subjects. It shows how flat is the thorax in some highlanders. It will be readily appreciated from Fig. 4.4 that with such variation in chest size and shape it is hazardous to make any generalization as to the internal surface area of the lung in native highlanders.

Fig. 4.5 The height (in cm) of Quechua children between the age of 4 and 18 years in 1932 (Hurtado) and British children of comparable age in 1965 (Tanner, Whitehouse and Takaishi, 1966).

Weight

Total body weight is also less in the Quechua than in his sea-level Caucasian counterpart. In Figure 4.6 are plotted the average weights in kg of Quechua children between the age of four and 18 years (Hurtado, 1932) and British children of comparable age in 1965 (Tanner et al, 1966). The average body

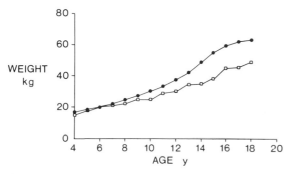

Fig. 4.6 The weight (in kg) of Quechua children between the age of 4 and 18 years in 1932 (Hurtado) and British children of comparable age in 1965 (Tanner et al, 1966). The symbols are the same as in Fig. 4.5.

weight in Quechua Indians is 55.4 kg in the age range 20 to 34 years, 56.8 kg between 35 and 49 years, and 55.2 kg between 50 and 75 years. Obesity is a rarity in native highlanders of the Andes.

Body growth and development in highlanders other than Quechuas

As we have just seen, highlanders in the Andes show a slow growth in body size and delayed skeletal maturation. They also show late onset of secondary sexual maturation. These changes can be inferred to be due to the growth-retarding effects of high altitude hypoxia since dietary intake in selected highland samples does not appear to be related to growth (Frisancho, 1978). Furthermore, intra-population differences in subcutaneous fat do not parallel differences in growth in height. However, retarded body growth and maturation do not appear to be features of all highlanders. Thus studies of Ethiopian highland and lowland children show that the *former* grow and mature faster (Frisancho, 1978). The difference is probably related to the higher incidence of malaria and intestinal parasitism in the lowlanders. In this area highland and lowland-adults do not differ in stature suggesting

that the retarded growth of lowlanders is probably of recent origin and not a permanent characteristic. Measurements of chest size suggests that highlanders tend to have a greater thoracic volume than lowlanders.

Growth of body size among Himalayan Sherpas is comparable to that of lowland Tibetans. On the other hand Sherpas are more retarded with regard to skeletal age and tooth eruption (Frisancho, 1978). Adult Sherpas are shorter and leaner than lowland Tibetans. The thinner skinfolds of Sherpas suggests that they have poorer socio-economic conditions and this makes it difficult to decide if the growth-retarding effect is due to high altitude hypoxia or genetic factors. Similarly in the Tien Shan populations in the U.S.S.R. possible hypoxic effects are obscured by socio-economic factors. In an annotation the *Lancet* (1975) refers to child-growth surveys in Peru, Ethiopia and Nepal which illustrate the pitfalls of high altitude studies. Such comparative studies show that patterns of growth are population-specific and that it is not possible to extrapolate the effects of environmental stress from one population to another.

The chest

The shape and size of the chest of the native highlander have attracted much attention in the past and this is not surprising since it is in this area that one might anticipate finding the greatest structural adaptation to provide an increased ventilatory reserve compatible with a healthy and active life at high altitude. The point of view that has come to be widely accepted as the authoritative opinion was that put forward as early as 1932 by Hurtado following his studies in physical anthropology of the Quechuas of the Peruvian Andes. He stated that 'Since early childhood there is a definite appearance of a prominent and large chest as proportionally compared with the rest of the body dimensions, and at this age it is often continuous in front with a prominent abdomen. Later in adolescence this outstanding appearance of the chest is more evident on account of the thin lower entremities. This is particularly true in adult life. The native at this time, in the vast majority of cases, gives the impression that practically all the body mass lies on the chest'. We feel that this overstates the case and certainly takes no account of the considerable variation in the size and shape of the chest that is to be found in native highlanders (Fig. 4.4). On a visit to La Paz (3800 m) in 1979 we came across a young Indian native of the city whose massive chest (Figs. 4.4 a and b) appeared to confirm the classical view put forward by Hurtado. Unfortunately close questioning revealed that his spare-time was devoted to the hobby of weight-lifting! In contrast to this other natives of the Bolivian capital have a flat chest (Figs. 4.4 a and d). It is clear that due attention must be paid to the considerable variation in the size and shape of the chest in native highlanders and it will be readily appreciated that is difficult to make any generalization as to the internal surface area of the lung in these subjects.

The difference in the configuration of the chest of the native highlander from those of sea-level man are more subtle and they have required for their demonstration different methods of measurement which have varied considerably in their complexity and validity. In 1932 Hurtado attempted to calculate the so-called 'chest volume' in his cases from the somewhat crude physical measurements of 'chest width, depth and height'. He defined 'width' as the distance between the midaxillary lines at the level of sixth ribs. 'Depth' was the distance from the middle of the sternum to the vertebral column at the level of which width was determined. 'Height' was the distance between the upper level of the sternum and its lower tip. He found that in men between 20 and 34 years of age the chest volume was greater in Andean highlanders than in non-native residents at high altitudes, or in white Peruvian college students, in spite of the much taller structure of the latter two groups. A physical anthropological index based on such data and giving some idea of the shape of the chest is the so-called 'Chest index' defined as $\frac{depth}{width} \times 100$. In the child at sea level the index falls so that the chest changes from being rounded to ellipsoidal. At high altitude the chest in childhood maintains a more rounded shape. There is a widening of the costal angle.

Monge and Monge (1966) compared physical measurements of the chest in 120 subjects from Lima (150 m) and 53 subjects from altitudes of between 3500 m and 4500 m in Bolivia. The average anteroposterior diameter in the highlanders was increased to 213 mm from 203.6 mm in the lowlanders while the average transverse diameter in both was 283 mm. As a result of this the chest was rounded rather than ellipsoidal as described above. The height of the sternum in the high altitude natives was also increased to 199 mm from 183.3 mm in sea-level subjects so that the chest as a whole was more voluminous.

However, results of studies in the Himalayas and Tien Shan mountains demonstrate that the enlargement of the thorax is not a general characteristic of all highland populations. Hence it would appear that the factors associated with an increase in chest size in the Andean regions are not the same as those in Asiatic populations (Frisancho, 1978).

The configuration of the chest and the internal surface of the lung

Whether the rounded chest of the Peruvian highlander is associated with alterations in internal surface area of the lung beneficial to life at high altitude is controversial. This is because unfortunately no quantitative morbid anatomical studies have been carried out on the lungs of highlanders. Hurtado (1932) attempted to assess the size of the lung in the large chest by percussion but this is a very crude, clinical method. The area where study is greatly needed is in the reliable macroscopic and histological methods of tissue morphometry which are now available for the determination of the internal surface area of the lung and of the number of alveolar spaces it contains.

We have employed these methods in the study of the emphysematous lung (Hicken et al, 1966) and their applications to the lung of high altitude man is long overdue. It would be of considerable interest to ascertain how the internal surface area of the lung is related to the rounding of the chest. The most accurate quantitative histological guide to the size of the air spaces in the lung is the 'mean linear intercept'. This is a measurement in tissue morphometry which may be determined from sections of lung which was inflated and fixed by formalin steam by the method of Weibel and Vidone (1961). In essence a special eyepiece micrometer is used to measure on such sections the average distance between the walls of alveolar spaces and ducts. This mean linear intercept is used to determine the internal surface area of the lung. Hasleton (1972) has shown that there is a significant negative correlation between body height and the 'mean linear intercept' (Fig. 4.7). This would

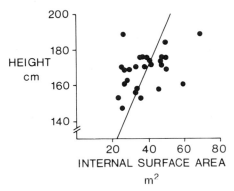

Fig. 4.7   The relation between height and internal surface area of the lung (after Hasleton, 1972).

suggest that the shortness of the Quechua would tend to be associated with a smaller internal surface area but the rounded configuration of the chest would tend to have the opposite effect. What is required is actual measurement of the internal surface area of the lungs in native highlanders by the methods referred to above.

Vital capacity

Early observations by Hurtado (1932) revealed that the vital capacity is increased in native highlanders of the Andes. The mean body surface area in 612 Quechua Indians between the ages of 20 and 75 years was 1.56 m² (Hurtado, 1932). Mean vital capacity in 478 Indians in the age group of 20 to 34 years was 2720 ml m² BSA. This compares with 2610 ml m² BSA in non-native residents at high altitude and 2460 ml m² BSA in sea-level Caucasian Peruvian students. Vital capacity depends on age as well as body surface area and in high altitude natives it fell from a mean value of 2720 ml m² BSA in 478 Quechuas between 20 and 34 years of age to a mean value of 2370 ml m² BSA in 29 Indians between 50 and 75 years of age (Fig. 4.8).

The vital capacity in subjects living at high altitude depends on the time of life when acclimatization takes place (Frisancho, 1975). The forced vital capacity (the maximum amount of air expired after

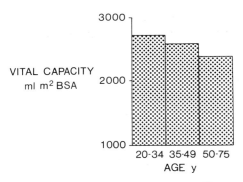

Fig. 4.8   Vital capacity (ml m² BSA) in three different age groups of high altitude natives at Morococha (4540 m). The mean values for the three age groups shown are based respectively on data from 478, 105 and 29 Quechua Indians, the smallest group being composed of the eldest subjects. (From data of Hurtado, 1932).

maximum inspiration) was measured by Frisancho et al (1973 a, b) in high altitude natives at 3400 m, in sea-level subjects acclimatized during growth, and in American subjects acclimatized as adults. Frisancho and his colleagues found that the lowland natives who were acclimatized to high altitude during growth, when adjusted for variations in body size, attained the same values of forced vital capacity as the highland natives (Fig. 4.9). In contrast,

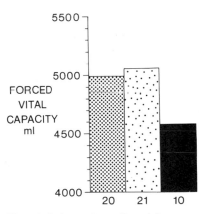

Fig. 4.9   Forced vital capacity, adjusted for age, weight and height in subjects tested at 3400 m. The heavily-stippled column is the mean value for 20 high altitude natives, the lightly-stippled column for 21 sea-level subjects acclimatized during growth, and the filled column for 10 Caucasian subjects from the United States acclimatized as adults (after Frisancho, 1975).

lowland subjects, both of Peruvian and American extraction who had acclimatized as adults had a significantly lower vital capacity than highland natives (Fig. 4.9). Frisancho (1975) postulates that the enlarged lung volume of the highland native is the result of changes occurring during growth and development. As we note below there is experimental evidence to show that young rats exposed to chronic hypoxia from birth show an increase in the number of alveolar spaces and in alveolar surface area and lung volume.

Hypoxia and the developing lung

There is some experimental evidence to suggest that the structure of the growing lung is influenced by the partial pressure of oxygen in the atmosphere. Burri and Weibel (1971) subjected three groups of rats to hypoxic, normal and hyperoxic atmospheres from the 23rd to the 44th day of postnatal development. The hypoxic group was exposed to a simulated altitude of 3450 m and a PO_2 of 100 mmHg. The 'normal' group was kept at normal barometric pressure with a PO_2 of 150 mmHg. The hyperoxic group was kept in an atmosphere of 40 per cent oxygen and 60 per cent nitrogen at a mean pressure of 730 mmHg and a PO_2 of 290 mmHg. The fourth group of rats of the same age and stock was killed on the 23rd day of postnatal development in order to provide control data.

The absolute lung volume measured by water displacement, on the 23rd day was 2.5 ml. On the 44th day this rose to 6.34 ml in the control group, 6.84 ml in the hypoxic group and 5.51 ml in the hyperoxic group. The animals in the hypoxic group weighed less than the other groups and so the lung volume per 100 g body weight is more relevant. The control group had a value of 4.5 ml/100 g compared with 5.5 ml/100 g in the hypoxic animals and 3.9 ml/100 g in the hyperoxic animals. This means that the lung volume per 100 g body weight in the hypoxic animals was some 20 per cent more than the controls (Fig. 4.10). Light and electron microscopy showed no significant quantitative changes in the relative composition of the lung. Hence the increase in lung volume could be attributed to increases in alveolar capillary and tissue volumes.

The importance of relating lung volume to body weight can be seen from the above results. Cook et al (1970) exposed young rats to a barometric pressure of 390 mmHg for 21 days to 91 days. They found that the growth rate of the lung was normal,

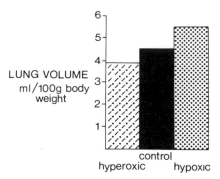

Fig. 4.10 The lung volumes (expressed as cm³/100 g body weight) in three groups of rats exposed to hyperoxic, normal and hypoxic atmospheres from the 23rd to the 44th day of postnatal development. The PO_2 to which the groups were exposed was 290, 150, and 100 mmHg respectively (from data of Burri and Weibel, 1971).

although the increase in body weight was reduced when compared with the controls. Hence the lung volume per 100 g body weight in their hypoxic animals was considerably increased. After a return to normal barometric pressure the animals gained weight and after one month the lung weights and body weights were normal. However, Bartlett (1970) conducted a similar experiment to Burri and Weibel (1971) when he exposed one-month-old rats to differing oxygen concentrations of 10.4 per cent, 18.3 per cent and 45.8 per cent at sea-level barometric pressure. His findings as to body weight were similar to Burri and Weibel (1971) but he found that lung development was not affected relative to body weight in the group exposed to a low oxygen concentration while in the group exposed to a high oxygen concentration body weight increased more rapidly than in controls and the lung growth was inhibited. Although at first his results with the group exposed to a low oxygen concentration appear to be in conflict with those of Burri and Weibel (1971), there are two important differences in his method. The rats used were one-month-old as opposed to 23 days and at such an early stage of postnatal life such a difference may be significant. Also it cannot be said with certainty that a low oxygen concentration at normal barometric pressure is the same stimulus as normal oxygen concentration at a low barometric pressure.

The nature of enlargement of the lung at high altitude

Brody and his colleagues (1977) studied the nature

of enlargement of the lung at high altitude. They were able to confirm from determination of forced vital capacity in Peruvian mountain dwellers between 17 and 20 years of age that the lung volume of highlanders is 30 to 35 per cent greater than that of lowlanders similar race, age, and body size. Vital capacities of Peruvian lowlanders were only 84 per cent of values found in Caucasians confirming that there are genetic differences in lung size between Aymara and Caucasian populations. Aymaras living in the mountains , however, had a vital capacity some 116 per cent of that of Caucasian values. Lung elastic recoil in highlanders and lowlanders is similar to that of Caucasian subjects of similar age, suggesting that neither increased muscle strength nor alterations in connective tissue properties of the lung are primarily responsible for the larger forced vital capacities in highlanders. However, in spite of the larger lungs of high altitude natives, absolute flow rates are not greater (Brody et al, 1977). In fact, expressed as a function of lung volume, flow rates are less in highlanders than in lowlanders. Brody et al (1977) believe the flow rate difference to be an expression of the manner in which hypoxia stimulates the growth of the lung.

The growth of lung associated with genetic or early fetal adaptation would be likely to be brought about by increasing the size and/or number of both airways and alveolar spaces (Fig. 4.11). Hence their volume matches the increase in alveolar mass, as occurs in the lungs of people of different sizes. On the other hand, the postnatal effects of environmental hypoxia on lung growth would not involve the airways (Fig. 4.11). Hence one would have the combination of the bronchial tree of a lowlander and increased alveolar numbers of a highlander (Fig. 4.11). In this situation cross-sectional area or volume of the airways would be the same in highlanders and lowlanders but would have to co-exist with increased lung volume in highlanders (Fig. 4.11). The data of Brody and his colleagues (1977) suggest that the hypoxia of high altitude brings about enlargement of the alveolar spaces but not of the airways.

The physique of 'high altitude man'

In summary we believe that the physique of the native highlanders in the Andes is basically determined by their ethnic origin. Superimposed on these racial characteristics are physical features resulting from acclimatization to high altitude. Thus, the cyanosis and high colour are the result of polycythaemia resulting from acclimatization (Chapter 6). The configuration of the chest appears to be an expression of Mongoloid origin but the lungs, when scientifically examined, will probably show an increased internal surface area which will represent a structural response to development in a hypoxic environment. We shall consider the important subjects of adaptation and acclimatization to high altitude at greater length in Chapter 27. In Chapter 8 we shall see that postnatal hypoxia, as well as affecting lung growth, has an important influence on the carotid bodies, blunting their response to hypoxia.

Body composition of highlanders

Bharadwaj et al (1973) studied the body composition of 30 young Tamilian soldiers from sea level using anthropometric techniques and compared these results from those on 45 Ladakhis at Leh (3660 m).

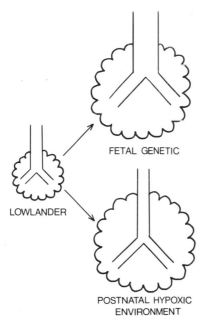

FETAL GENETIC

LOWLANDER

POSTNATAL HYPOXIC
ENVIRONMENT

Fig. 4.11 Diagram to illustrate the nature of the enlargement of the lung at high altitude. In an enlargement of fetal genetic basis both airways and alveolar lung tissue would be enlarged together. In fact in the highlander the bronchial tree is not enlarged but only the alveolar spaces, suggesting that the enlargement is due to postnatal effects of environmental hypoxia on lung growth.

No significant difference in body fat, water and cell solids was observed between the highlanders and sea-level residents (Fig. 4.12). Bone mineral was

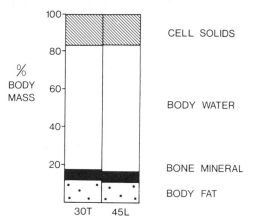

Fig. 4.12   Body composition using anthropometric techniques from 30 Tamilians at sea level (30T) and 45 Ladakhis at 3660 m (45L). There are no differences in body composition between lowlanders and highlanders (from data of Bharadwaj et al. 1973).

significantly greater in the Ladakhis and this was considered to be due to their wider bi-iliac, wrist, knee and ankle widths. Bharadwaj et al (1973) speculate that this increased width may be associated with a large volume of bone marrow. Earlier studies by Siri et al (1954) and Picon-Reategui et al (1961) had also indicated that there is no difference in body fat and cell solids between highlanders and sea-level residents. However, in contrast to the findings of Bharadwaj et al (1973) Picon-Reategui and his colleagues (1961) observed no difference in bone-mineral content between the two groups.

Longevity in the highlander

The question is frequently posed whether or not the high altitude native has a life expectancy similar to that of sea-level subjects. It will be readily appreciated that in remote, mountainous areas in underdeveloped countries mortality and morbidity statistics are either unavailable or unreliable. Furthermore in such areas there are many causes of sickness and death other than can be ascribed to the effects of high altitude. Thus in the Andes tuberculosis, infectious hepatitis, malnutrition and typhus are all common. In general terms, excluding

such infectious and nutritional diseases, it is clear that communities can acclimatize successfully to environments impoverished in oxygen. For example, in Peru important aspects of that nation's economic life are carried out at altitudes exceeding the summits of the Swiss Alps. There is, however, much individual variation in response to the chronic hypoxia of high altitude. Some subjects live into healthy retirement having spent all their lives at high altitude in an occupation demanding hard physical work such as mining and they are syptom-free. A much smaller minority may develop symptoms of chronic mountain sickness and are forced to leave their mountain home for the coast or else risk an early death.

There are remote mountainous areas in the world where people are alleged to live much longer and remain more vigorous in old age than in most modern societies. To investigate these claims Leaf (1973) visited the Andean village of Vilcabamba (the 'Sacred Valley' according to Davies, 1975), in Ecuador at an altitude of 2250 m (Fig. 4.13). He also

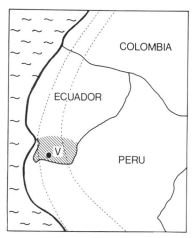

Fig. 4.13   The situation of the Andean village, Vilcabamba, in Ecuador which is famed for the longevity of its inhabitants.

visited the area inhabited by the Hunza in the Karakorum range of Kashmir (Fig. 4.14) and the Abkhazia region of the Caucasian mountains of Georgia in the Soviet Union rising to 2000 m (Fig. 4.15). Claims of longevity in such regions are hard to confirm because of the difficulty of establishing unequivocally the age of the subjects. In the absence of birth certificates church baptismal records are sometimes useful in supporting the age claimed. A

Fig. 4.14   The area inhabited by the Hunza in the Karakorum range of Kashmir.

Fig. 4.15   The areas of Abkhazia and Azerbaydzhan, famed for the longevity of their inhabitants.

census in 1971 suggested that in Vilcabamba no fewer than nine of the population of 819 were centenarians (Leaf, 1973). In the Hunza people the dating problem is difficult since in the high valleys between China and Afghanistan the language, Burushaski, bears no relationship to other tongues and there is no written form (Leaf, 1973). American investigators visiting Vilcabamba in 1978 were reported by *The Daily Telegraph* as having come to the opinion that the old people of the region 'lie outrageously about their age and have been doing so for years' (Ball, 1978). They found that a claim to great age enchanced their prestige in the community. Many claimed their father's baptismal records as their own.

Not only is it extremely difficult to substantiate the claims of longevity of these highlanders in remote areas because of the lack of authenticated

birth certificates, it is also impossible to unearth those factors in the way of life in these mountain folk which bestow long life. Indeed one can imagine the interest that would be aroused throughout the world, in both medical and non-medical quarters, should such a factor ever emerge.

Diet has, of course, been investigated as a factor. Davies (1975) dwells momentarily on the exotic possibility of trace elements in the soil, water and diet since considerable quantities of gold, iron, magnesium and cadmium are said to be found in Vilcabamba. Not surprisingly Leaf (1973) found the daily calorie intake in these old people from poor mountain societies in both Ecuador and the Karakorums to be low. In 55 adult males in the Hunza population there was an average intake of less than 2000 calories with 50 g protein, 36 g of fat and 354 g of carbohydrate and a low consumption of meat and dairy products. In Vilcabamba the average daily diet provided but 1200 calories, with 35 g protein, 12 to 19 g fat, and 200 to 260 g of carbohydrate. Under these circumstances it is not surprising that Leaf found that there were no obese subjects in these communities and it has long been appreciated that the life expectation of the obese is less than that of the thin.

However, any views Leaf may have formed on the relation between low calorie intake and longevity received a rude shock when he visited the Caucasus. There he found that everyone, including the aged, consumed a diet high in calories and rich in dairy and meat products. As a result many of the population were fat. One woman, said to be 107 years old, was obese: She said 'I became fat when I stopped having children. For 60 years I have been fat as a barrel and all my children are like me.' Furthermore many enjoyed intemperate habits. Leaf (1973) was impressed by the heavy alcohol intake of some of the centenarians at a meal he attended in Georgia. Davies (1975) also comments on the heavy alcohol intake of natives of Vilcabamba. One woman, said to be 130 years old, had smoked a packet of cigarettes a day for 62 years.

It is worthy of note that all highlanders undertake much hard physical labour from childhood onwards. Simply traversing the mountainous terrain each day sustains a high level of fitness. As we shall see in Chapter 18 myocardial ischaemia and infarction are rare in mountain communities and this must be a factor in a greater life expectation in these areas.

It is difficult to be certain of the importance of genetic factors. Probably they have an influence and Leaf (1973) notes that almost all the supposed centenarians had at least one parent or sibling who had enjoyed the same longevity. However, one should not think of these areas claimed to favour longevity as isolated, genetically pure communities. In the Caucasus there are many different ethnic groups such as Georgians, Azerbaydzhanis, Russians, Jews and Armenians. Davies (1975) also points out that La Thoma in Ecuador, famed for its centenarians, is a hotchpotch of humanity on the way to Peru.

One factor that must not be overlooked is the sociological structure of these communities in which the family unit is still of great importance. Not only that, but the aged still have an important rôle in the family group, a condition no longer obtaining in the so-called developed countries of the West. A feeling of being wanted and needed must be an incentive to survive but present evidence suggests that at the moment the need is to prove that the centenarians of the Andes actually exist before finding reasons for such supposed longevity.

REFERENCES

Allison, M. J. , Hossaini, A A., Castro, N., Munizaga, J., & Pezzia, A. (1976) ABO blood groups in Peruvian mummies. 1. An evaluation of techniques. *American Journal of Physical Anthropology*, **44**, 55.

Arce Larreta, J. (1930) Anales del Hospital de Lima, **3**, 74, quoted by Mourant et al. (see below).

Arias-Stella, J. (1971) In: *High Altitude Physiology: Cardiac and Respiratory Aspects*. p. 12. Ciba Symposium. Edited by R. Porter & J. Knight. Edinburgh: Churchill Livingstone.

Ball, I. (1978) Lies explode Andean claims for long life. p. 20 in *The Daily Telegraph* (London) March 18.

Bartlett, D. (1970) Postnatal growth of the mammalian lung: influence of low and high oxygen tensions. *Respiration Physiology*, **9**, 58.

Bharadwaj, H., Singh, A. P., & Malhotra, M. S. (1973) Body composition of the high-altitude natives of Ladakh. A comparison with sea-level residents. *Human Biology*, **45**, 423.

Boyd, W. C. (1959) A possible example of the action of selection in human blood groups. *Journal of Medical Education*, **34**, 398.

Boyd, W. C., & Boyd, L. G. (1937) Blood grouping tests on 300 mummies. *Journal of Immunology*, **32**, 307.

Brody, J. S., Lahiri, S., Simpser, M., Motoyama, E. K. & Velasquez, T. (1977) Lung elasticity and airway dynamics in Peruvian natives to high altitude. *American Journal of Physiology*: Respiratory, Environmental and Exercise Physiology, **42**, 245.

Büchi, E. C. (1952) Bulletin of the Department of Anthropology, India, **1**, 71, quoted by Mourant et al. (see below).

Burri, P.H., & Weibel, E. R. (1971) Environmental oxygen tension and lung growth. *Respiration Physiology*, **11**, 247.

Cook, C., Barer, G. R. , Shaw, J. W., & Clegg, E. S. (1970) Growth of the heart and lungs in normal and hypoxic rodents. *Journal of Anatomy*, **107**, 384.

Cumming, G., & Semple, S.J. (1973) *Disorders of the Respiratory System*. pp. 171 and 199. Oxford: Blackwell Scientific Publications.

Davies, D. (1975) 'The Centenarians of the Andes'. p. 113. London: Barrie and Jenkins.

Durand, J. (1971) In: *High Altitude Physiology: Cardiac and Respiratory Aspects*. p. 12. Ciba Symposium Edited by R. Porter & J. Knight. Edinburgh: Churchill Livingstone.

Escajadillo, T. (1948) *2° Curso sudamer. de transfus. y hermatol.* Santiago-Chile (1946) p. 58, quoted by Mourant et al. (see below).

Frisancho, A. R. (1975) Functional adaptation to high altitude hypoxia. *Science*, **187**, 313.

Frisancho, A. R. (1978) Human growth and development among high-altitude populations. In: *The Biology of High Altitude Peoples* p. 117, Edited by P.T. Baker. Cambridge: Cambridge University Press.

Frisancho, A. R. , Martinez, C., Velásquez, T., Sanchez, J., & Montoye, H. (1973a) Influence of developmental adaptation on aerobic capacity at high altitude. *Journal of Applied Physiology*, **34**, 176.

Frisancho, A.R., Velásquez, T., & Sanchez, J. (1973b) Influence of developmental adaptation on lung function at high altitude. *Human Biology*, **45**, 583.

Gilbey, B. E. & Lubran, M. (1952) Blood groups in South American Indian mummies. *Man*, **52**, 115.

Hasleton, P. S. (1972) The internal surface area of the adult human lung. *Journal of Anatomy*, **112**, 391.

Hicken, P., Brewer, D., & Heath, D. (1966) The relationship between the weight of the right ventricle of the heart and the internal surface area and number of alveoli in the human lung in emphysema. *Journal of Pathology and Bacteriology*, **92**, 529.

Hurtado, A. (1932) Respiratory adaptations in the Indian natives of the Peruvian Andes. *American Journal of Physical Anthropology*, **17**, 137.

Lancet annotation (1975) High living in Ethiopia. *Lancet*, ii, 17.

Leaf, A. (1973) Every day is a gift when you are over 100. *National Geographic Magazine*, **143**, 93.

MacFarlane, E. W. E. (1937) East Himalayan blood groups. *Man*, **37**, 127.

Marett, R. (1969) *Peru*, p. 33. London: Ernest Benn.

Monge, M.C., & Monge, C. C. (1966) In: *High-Altitude Diseases. Mechanism and Management*, p.14. Springfield, Illinois: Charles C. Thomas.

Mourant, A. E., Kopéc, A. C., & Domaniewska, -Sobczak, K. (1958) *The ABO Blood Groups*, pp.. 88 and 207. Comprehensive tables and maps of world distribution. Oxford: Blackwell Scientific Publications.

Peñaloza, D. (1971) In: *High Altitude Physiology: Cardiac and Respiratory Aspects*, p. 13. Ciba Symposium. Edited by R. Porter & J. Knight. Edinburgh: Churchill Livingstone.

Picon-Reategui, E., Lozano, R., & Valdivieso, J. (1961) Body composition at sea level and high altitudes. *Journal of Applied Physiology*, **16**, 589.

Quilici, J.K. (1968) Les altiplanides du corridor Andin. *Etude hémotypologique*. Toulouse: Center d'Hémotypologie.

Ruiz, L., & Peñaloza, D. (1970) *Altitude and Cardiovascular Diseases*, pp. 1–48. Progress Report to World Health Organisation.

San Martin, M. (1951) Anales de la Facultad de medicina Casilla 529, Lima, **34**, 276. quoted by Mourant et al (see above).

Siri, W. E. , Reynafarje, C., Berlin, N.I., & Lawrence, J. H. (1954). Body water at sea level and at altitude. *Journal of Applied Physiology*, 7, 333.

Tanner, J.M., Whitehouse, R. H., & Takaishi, M. (1966) Standards from birth to maturity for height, weight, height velocity and weight velocity: British Children, 1965. *Archives of Diseases of Childhood*. **41**, 454 and 613.

Weibel, E.R. & Vidone, R. B. (1961) Fixation of the lung by formalin steam in a controlled state of air inflation. *American Review of Respiratory Disease*, **84**, 856.

5

Ventilation and pulmonary diffusion

The stages of respiration

At rest the quantity of oxygen consumed by the tissues of the body each minute is 220 to 260 ml (STPD) (Harris and Heath, 1977). There is an inexhaustible reservoir of oxygen in the ambient air and the problem of providing an adequate supply of the gas to the mitochondria where it will be used lies rather in the complex nature of the route in the body along which the oxygen must travel to reach the cells. Respiration entails four distinct stages of conduction of oxygen from the atmosphere to the cells. In *ventilation* air flows through the trachea and bronchial tree to the alveolar spaces of the lung. *Pulmonary diffusion* is the stage by which oxygen in the alveoli comes into contact with the alveolar-capillary walls and passes through them to reach the blood. The gas is then carried by *blood transport* from the capillaries of the lung to those of the tissues. Finally in the fourth stage of *tissue diffusion* oxygen passes from the systemic capillaries to the respiratory enzymes of the intracellular mitochondria where it will be utilised.

Oxygen gradients

Respiration thus entails a complex combination of transport mechanisms and even at sea level various diseases may interfere with one or more of these stages, impairing the conduction of oxygen to the tissues. However, in healthy man living at sea level the complex operation of transporting the oxygen to the mitochondria functions efficiently because the difference between the partial pressure of oxygen in the atmosphere and that required in the mitochondria is so great. Thus, as we have already seen in Chapter 2, the P_{O_2} of ambient air at sea level is

159 mmHg whereas the critical P_{O_2} at the mitochondrial site for oxidative enzyme reactions is less than 3 mmHg (Mithoefer, 1966). At each of the four stages of respiration there is a fall in oxygen tension and the magnitude of these changes at sea level is shown in Figure 5.1. The succeeding falls in

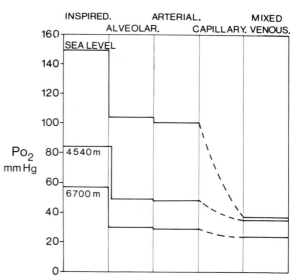

Fig. 5.1 Mean oxygen pressure gradients from inspired air to mixed venous blood in subjects native to sea level and 4540 m and in climbers at 6700 m. The 'oxygen cascade' in subjects at high altitude is seen to be much less steep than in sea-level residents. Thus, although the partial pressure of oxygen in the ambient air at high altitude is much less than at sea level, the final P_{O_2} achieved in the mixed venous blood of subjects at high altitude is not greatly diminished. Based on data from Hurtado (1964a) and Luft (1972).

P_{O_2} at each stage of respiration have sometimes been referred to, somewhat picturesquely, as 'the oxygen cascade'. At high altitude the barometric pressure is diminished and with it the pressure difference of oxygen available to transport the gas to the mitochondria. This is the central problem of respiration at high altitude which we shall consider in this and following chapters.

The oxygen cascade and acclimatization

There are two ways in which the shortfall in the PO_2 of ambient air could be compensated. The first is the modification of tissue metabolism so that metabolic demands are satisfied in spite of reduced availability of oxygen (Luft, 1972). The second is the adjustment by various means of the oxygen transport system so that the impact of the loss of oxygen pressure in the atmosphere is minimised for the tissues and particularly for the mitochondria (Luft, 1972). Most features of acclimatization fall within the second category and Figure 5.1 reveals how adjustments in the transport mechanisms for oxygen render 'the oxygen cascade' much less steep than would be anticipated if such adjustments did not take place. Thus in mountaineers climbing at 6700 m there is a diminution of 92 mmHg in the PO_2 of inspired air but the PO_2 of mixed venous blood is only 13 mmHg less than at sea level.

In this and succeeding chapters we shall consider some of the processes of acclimatization that bring about this remarkable alteration in the steepness of 'the oxygen cascade'. In this chapter we shall deal with changes in ventilation and pulmonary diffusion. In the next chapter we describe the important changes in the blood which occur at high altitude to facilitate oxygen transport to the systemic capillaries. In Chapter 7 we shall consider alterations in tissue diffusion. Changes of acclimatization in the systemic circulation are considered in Chapter 17. Biochemical changes at tissue level are discussed in Chapter 19. All of these changes together constitute respiratory modifications to the hypoxia of life at high altitude.

PO_2 of ambient, inspired, and alveolar air

It is clear that the partial pressure of oxygen in the alveolar spaces, PA_{O_2}, must be lower than atmospheric PO_2. As air passes through the bronchial tree to the alveoli it is humidified. As the partial pressure of water vapour in fully saturated air

at body temperature is 47 mmHg, the PO_2 of inspired air is 149 mmHg (20.93 per cent of $760 - 47$ mmHg) in contrast to that of ambient air which is 159 mmHg (20.93 per cent of 760 mmHg). Once in the alveolar spaces, oxygen diffuses through the alveolar walls to the pulmonary capillaries while carbon dioxide diffuses out from them into the alveoli. Hence PA_{O_2} at sea level approximates to 100 mmHg and already a third of the oxygen gradient transporting the gas to the mitochondria has been lost (Fig. 5.1). At each breath only some 15 per cent of the alveolar air is replaced by fresh ambient air. This slow replacement of alveolar air is important in preventing sudden changes in gas concentrations in the blood and stabilises the control of respiration and helps to prevent sudden and excessive changes in tissue oxygenation (Ward, 1975).

Ventilation at high altitude

As will be seen from Figure 5.1 the steepness of the gradient in PO_2 from ambient to alveolar air falls in subjects at high altitude. The reason is that they hyperventilate. Such hyperventilation replaces more of the alveolar air by freshly inspired air thus elevating PA_{O_2}. Hurtado (1964 a) states that the mean total ventilation for 103 sea-level subjects was 7.77 1 min^{-1} (s.d. 1.24) and for 80 highlanders (4540 m) was 9.49 (s.d. 1.77). Expressed per square metre of body surface area the figures were respectively 4.56 and 6.19. In general the native highlander hyperventilates some 25 to 35 per cent above the value for sea-level man. Pulmonary hyperventilation in the newcomer to high altitude occurs within a few hours of arrival and increases rapidly during the first week. It exceeds that of natives by some 20 per cent (Lenfant and Sullivan, 1971). With increasing duration of residence at high altitude this difference between the sojourner and the native diminishes. However, even this smaller difference in ventilatory response keeps the levels of PA_{O_2} and Pa_{O_2} elevated in the sojourner compared to the native.

Hyperventilation is an important ventilatory response to hypoxia and maintains an adequate oxygen tension in the alveolar spaces in the face of a low partial pressure of oxygen in the ambient air. It increases alveolar ventilation by 25 to 30 per cent, increasing PA_{O_2} and decreasing PA_{CO_2}. For example, at Morococha (4540 m) in the Peruvian Andes where ambient PO_2 is 84 mmHg the normal oxygen pressure differential from atmosphere to alveolar space of 50 mmHg found at sea level would result in a PA_{O_2} of 34 mmHg which would be critically low. In fact due to hyperventilation, PA_{O_2} at this altitude is 50 mmHg, the gradient from ambient to alveolar air having been improved by some 16 mmHg (Fig. 5.1). At the same time PA_{CO_2} falls from 40 to 30 mmHg.

The hyperventilation which occurs on acute exposure to high altitude is due to an increase in tidal volume rather than in respiratory rate. In fact, on visits to the Andes we have been impressed by the singular lack of increase in respiratory rate at rest. Thus the mean respiratory rate of one of us (DRW) at sea level was 12.5 per minute and at Cerro de Pasco (4330 m) was 13 per minute. However, on ascent to this altitude there was a steady increase in minute volume from 4.8 1 min^{-1} at sea level, to 5.6 1 min^{-1} at 3000 m, to 6.6 1 min^{-1} at 4330 m. It will be apparent that these values for resting minute volume are low but this is due to the slight stature of the subject studied. Ward (1975) agrees that on exposure to high altitude the initial change in ventilation appears to be an increase in tidal volume rather than in rate. He states that in experimental subjects at rest the minute volume increases with altitude above 3660 m but the respiratory rate does not increase significantly until an altitude of over 6000 m is reached. However, on exercise both tidal volume and respiratory rate increase much more at high altitude than at sea level. This increase is much greater in newcomers to the environment than in natives.

Initial hyperventilation and peripheral chemoreceptors

The initial hyperventilation in visitors ascending to high altitude is due to stimulation of the peripheral chemoreceptors by hypoxaemia. Moderate altitude will not bring about such stimulation immediately (Rahn and Otis, 1949), and Cumming and Semple (1973) also note that in acutely induced hypoxia it is necessary to lower Pa_{O_2} to about 55 mmHg before hyperventilation occurs. However, once acclimatization has taken place hyperventilation will be provoked by a Pa_{O_2} as high as 90 mmHg. Lowlanders who live for years at high altitude do not lose their hypoxic drive due to the sensitivity of their peripheral chemoreceptors to a diminished Pa_{O_2}. In striking contrast native highlanders, and indeed anyone exposed to severe hypoxia during the early part of his life, do not show the same degree of sensitivity to hypoxia. This fascinating problem is considered more fully in Chapter 8.

Respiratory alkalosis

The hyperventilation referred to above results in increased exhalation of carbon dioxide and thus to alkalaemia. This is gradually corrected by the renal excretion of excess bicarbonate restoring blood H^+ concentration towards normal. After this compensatory process has occurred the subject is left with a Pa_{CO_2} that is lower than normal and a reduced blood bicarbonate concentration (Monge and Monge, 1966; Mithoefer, 1966). Typical values for their reduced levels at an altitude of 4540 m are shown in Figure 5.2.

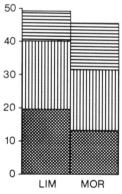

Fig. 5.2 Partial pressure of carbon dioxide and blood buffer base in systemic arterial blood in native residents at Lima (LIM) at 150 m and Morococha (MOR) at 4540 m. Based on data from Hurtado (1964a). Cross-hatched column = bicarbonate carbon dioxide (mmol/l); Vertical lines-column = partial pressure of carbon dioxide (mmHg); Horizontal lines-column = blood buffer base (mmol/l).

However, it must be appreciated that this is a slow compensatory process. Our personal experiences in the Andes show that on acute exposure to high altitude the pH of the urine shows little or no change during the first two or three weeks. In fact on further ascent from 3000 m to 4330 m we found the urine to become even more acid. Elsewhere we refer to the findings of Waterlow and Bunjé (1966) that during an expedition at high altitude in Colombia they found the urine to be alkaline on only three occasions during the testing of 500 specimens. Ward (1975) refers to this apparent paradox and believes it may be accounted for by the fact that many visitors to high altitude suffer from acute mountain sickness, anorexia and loss of weight. Acetone bodies may occur in the urine, masking the sluggish renal compensation.

The cerebrospinal fluid and respiration

Controversial evidence was brought forward by Severinghaus and his colleagues in 1963 to suggest that the pH of cerebrospinal fluid exerts an important influence on ventilatory acclimatization to high altitude. Within the medulla there is a receptor to levels of CO_2. This medullary CO_2 receptor is better termed a c.s.f. (H^+) receptor since it responds rapidly to an increased H^+ concentration in the cerebrospinal fluid by stimulating respiration which reduces PCO_2 and thereby returning the pH of the c.s.f. to normal. At high altitude, as we have seen above, the carotid bodies and peripheral chemoreceptors initially respond to the hypoxia and subsequent hypoxaemia by increasing ventilation. This hyperventilation reduces PA_{CO_2}, Pa_{CO_2}, and PCO_2 in the cerebrospinal fluid. This relative alkalinity of the fluid, as part of the general respiratory alkalosis described above, would inhibit respiration by its action on the medullary H^+ receptors. Such a depression of ventilation would reverse the advantageous hyperventilation following stimulation of the peripheral chemoreceptors. Severinghaus et al (1963) and Severinghaus and Mitchell (1964) presented evidence to suggest that this anti-acclimatizing alkalosis of the cerebrospinal fluid does not in fact occur. According to them active transport of bicarbonate ions out of the cerebrospinal fluid brings about a relative increase in its acidity stimulating ventilation and thus aiding acclimatization.

Their suggestion arose from a study of subjects taken to 3800 m. After a day at high altitude there was the expected fall in Pa_{O_2} and hyperventilation occurred and the arterial pH rose. PCO_2 in the cerebrospinal fluid also fell but its pH remained at the sea-level value of 7.32. This suggested to them that a mechanism exists for active transport of bicarbonate ions from the cerebrospinal fluid presumably via 'the blood–c.s.f. barrier', the choroid plexus. Once the pH of the cerebrospinal fluid had returned to normal the inhibition of the peripheral chemoreceptor is removed and ventilation increases. Davies (1978) has brought forward evidence for cerebral extracellular fluid H^+ as a stimulus during acclimatization to hypoxia.

However, this concept of the rôle of the pH of the cerebrospinal fluid as a homeostatic mechanism was soon challenged. Thus the conclusions of Severinghaus et al (1963) were not supported by Dempsey et al (1974) and Forster et al (1975) who found that the pH of the lumbar cerebrospinal fluid was more alkaline that at sea level in men stopping at 3100 m for three to four weeks and at 4300 m for only five to ten days. Weiskopf et al (1976) found that in six healthy men subjected to a simulated altitude of 4300 m for five days both lumbar and intracranial cerebrospinal fluid were significantly more alkaline than at sea level. These observations were extended and confirmed by Orr et al (1975) in unanaesthetized ponies. Bouvcrot and Burcau (1975) found no clear evidence from their studies that there is in acute or chronic hypoxia any better regulation of the pH in cerebrospinal fluid than in arterial blood. Finally, as a result of experiments carried out some years after the original study Crawford and Severinghaus (1978) accepted that, in steady states of respiratory acidosis and alkalosis, the pH of the cerebrospinal fluid is not perfectly regulated. Instead its steady-state value reflects the strength of peripheral chemoreceptor and other ventilatory drives. At altitude this strong hypoxic peripheral chemoreceptor drive is in fact partly offset by alkaline inhibition of the ·medullary chemoreceptors.

Sustained hyperventilation

As we have noted above the initial hyperventilation on exposure to high altitude is the result of stimulation of the carotid bodies and other peripheral chemoreceptors. Furthermore, this hypoxic drive is maintained in the newly-acclimatized for many weeks on exposure to high altitude (Michel and Milledge, 1963; Tenney et al, 1964). Even when lowlanders take up long term residence in the mountains they do not lose their sensitivity to a low PO_2. This is in striking contrast, however, to the loss of hypoxic drive in highlanders who were born and spent their early years at high altitude. It would appear that during the early period of acclimatization the increased hyperventilation is due to hypoxia.

At sea level ventilation is largely controlled by the level of carbon dioxide. The response of ventilation to the gas is influenced by the prevailing partial pressure of oxygen so that with increasing hypoxia the line of the response curve of ventilation to carbon dioxide steepens (Milledge, 1975). Hence using various levels of PO_2 a 'fan of curves' may be produced (Fig. 5.3). On ascent to high altitude the

blunted. Indeed some authorities believe that a heightened sensitivity to carbon dioxide is an important factor in sustaining hyperventilation. Thus Hurtado (1964b) reported an increased ventilatory response to carbon dioxide inhaled at high altitude as compared to sea level although Chiodi (1957) found no such response in two high altitude natives at 4520 m. Once this increased sensitivity has been induced in the acclimatized lowlander he will show only an insignificant fall in ventilation even when breathing oxygen which raises his Pa_{O_2} to equal or exceed that which was found at sea level (Cumming and Semple, 1973).

The respiratory sensitivity to carbon dioxide starts to increase after some 15 hours at high altitude and it continues to increase to the eighth day when no further changes occur. The magnitude of the increase in sensitivity is roughly proportional to the degree of altitude. After this early stage there is a slow decrease in ventilation which starts after a few

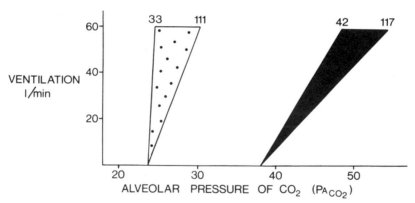

Fig. 5.3 Ventilatory response at sea level and at 5800 m. The numbers at the upper ends of the lines indicate the PO_2 value of that response line. At high altitude (stippled triangle) the respiratory centre responds to a level of carbon dioxide only half of sea-level values (filled triangle). Furthermore, the response is brisker to any increment in carbon dioxide. (From data of Milledge, 1975).

'fan of curves' moves to the left so that at an extreme altitude of 5800 m its origin moves from 38 mmHg to 23 mmHg. Hence the respiratory centre now responds to a level of carbon dioxide reduced by almost half. Furthermore, having started to respond, the respiratory centre responds more briskly to any increment in carbon dioxide. The influence of hypoxia on the response to carbon dioxide remains in principle the same as at sea level, so that a fall in PO_2 still increases the CO_2 response (Fig. 5.3). Hence in highlanders the sensitivity to carbon dioxide remains high but that to hypoxia is

weeks and continues over years. As we have seen newcomers, on ascent, sustain a higher ventilation than native highlanders whose resting ventilation, however, still exceeds that of sea-level man. As a result of this at any given altitude the resident highlander has a lower Pa_{O_2} and a higher Pa_{CO_2} than the acclimatized lowlander. Even if animals are deprived of their peripheral chemoreceptors, they still show some increase in ventilation when exposed to high altitude even though this is less and delayed (Cumming and Semple, 1973).

Carbon dioxide transport at high altitude

Carbon dioxide transport is modified in those living at high altitude. This is because the partial pressure of the gas in the alveolar spaces is abnormally low and because the haemoglobin content of the blood is increased. Carbon dioxide is carried in the blood in a variety of ways. In both plasma and cells small amounts are in solution but larger quantities are combined with base. At sea level, systemic arterial blood reaches the tissues with a PCO_2 of about 40 mmHg and since the PCO_2 of resting tissues is some 46 mmHg there is a diffusion of carbon dioxide into the blood. At high altitude the diminished barometric pressure of the ambient air and the hyperventilation responding to it both lead to a diminished partial pressure of carbon dioxide in the alveolar spaces. This leads to a fall in Pa_{CO_2}. Hurtado (1964a) found its mean value to fall from 40.1 mmHg in 80 subjects living at sea level to 33.0 mmHg in 40 subjects living at 4540 m. The fall in Pa_{CO_2} is compensated for by a proportional decrease in plasma bicarbonate (Fig. 5.2.) so that blood pH is maintained within normal limits. In the subjects referred to above the plasma pH was found to be 7.41 at sea level and 7.39 at 4540 m.

The increased fraction of red cells in the blood occurring during acclimatization is a second modifying influence on carbon dioxide transport for it leads to an increased participation of erythocytes in the process. Carbon dioxide combines with haemoglobin to form carbaminohaemoglobin, which may be expressed as Hb NH COOH. Reduced haemoglobin has a much greater power of forming carbaminohaemoglobin at any given level of PCO_2 than has oxyhaemoglobin. Carbon dioxide is taken up in the tissues and released in the lungs very rapidly on account of carbonic anhydrase in the erythrocytes. This catalyses, only in the red cell, the formation of carbonic acid from carbon dioxide and water, and the reverse reaction. As carbon dioxide from the tissues diffuses into the red cells to form carbaminohaemoglobin, oxygen leaves them. This reduces the affinity of haemoglobin for base because the isoelectric point of reduced haemoglobin is on the alkaline side of that of oxyhaemoglobin. At the same time carbon dioxide diffusing into the red cells under the action of carbonic anhydrase becomes carbonic acid. This acid acquires the base no longer

so firmly held by the haemoglobin. Some of the carbonic acid diffuses from the erythrocytes into the plasma where it combines with base and is transported in the blood. At the same time there is a reciprocal chloride shift into the red cells. In this way plasma base plays an important rôle in carbon dioxide transport.

In the pulmonary capillaries the reverse processes occur. Carbon dioxide dissociates from the carbamino compound and leaves the blood to enter the alveolar air. Oxygen enters the blood to form oxyhaemoglobin which combines more readily with base. The bicarbonate liberated splits into carbon dioxide and water which are lost to the alveoli. Bicarbonate passes from plasma to erythrocytes and chloride in the reverse direction.

The reduced levels of Pa_{CO_2}, blood bicarbonate concentration and blood buffer base at high altitude are shown in Figure 5.2. The various components of the transport of carbon dioxide in the blood in subjects at low and high altitude are shown in Figure 5.4. They include estimations of the total carbon dioxide and those portions free in the plasma and transported as bicarbonate and carbamino compound. It will be seen that in high altitude subjects while there is a reduction in total and free carbon dioxide and that bound as bicarbonate, the level of carbamino compound rises (Fig. 5.4). A normal balance between concentrations of anions and cations is maintained. The mean values of plasma

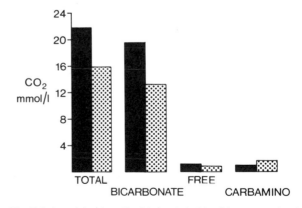

Fig. 5.4 Arterial carbon dioxide levels in 80 subjects at sea level (filled columns) and in 40 highlanders at 4540 m (stippled columns). In the high altitude residents total arterial carbon dioxide levels are decreased; this is true of free carbon dioxide and of that in the form of bicarbonate. In highlanders, however, carbamino-bound carbon dioxide is elevated compared to sea-level subjects (after Hurtado, 1964).

sodium, potassium and calcium are the same at high and low altitudes so that the decrease in bicarbonate is quantitatively compensated for by an increase in chloride. The buffering power for carbon dioxide is greater at high altitude than at sea level (Fig. 5.2) and this is because of the increase in haemoglobin concentration which we shall consider in the next chapter.

Suprapontine influences on hypoxic ventilatory control

The hypoxic ventilatory response is attenuated by sleep and by prolonged residence at high altitude. The basis for this may be a failure of the peripheral chemoreceptors as we discuss in Chapter 8, but Tenney et al (1971) think it possible that chronic hypoxia modifies a central mechanism influencing the respiratory centres in the pons and medulla. On the basis of experimental studies of cats exposed to a simulated high altitude they postulate that there is a cortical influence which is inhibitory and which can be removed to release an underlying facilitatory influence originating in the diencephalon. They think it possible that early in acclimatization 'arousal' by the reticular activating system accounts for the hyper-responsiveness and is the origin of symptoms such as insomnia and irritability which are common on exposure to high altitude (Chapter 14). The cortical inhibitory influence is presumed to develop more slowly but eventually it overrides the diencephalic facilitatory influence and then the blunted hypoxic drive becomes apparent. Tenney et al (1971) regard the orbital region of the frontal lobes, Walker's Area 13, as the likely region of the cortex from which the inhibitory stimuli arise.

Pulmonary diffusion

In the lung the oxygen tension in the alveolar spaces (PA_{O_2}) is higher than in the arterial blood (Pa_{O_2}). This A–a difference normally lies between 6 and 17 mmHg in sea-level subjects according to age (Harris and Heath, 1977). It is brought about by an anatomical diffusion barrier and a physiological shunt.

The anatomical diffusion barrier is the alveolar-capillary membrane. Due to the fact that the pulmonary capillaries lie on supporting connective tissue elements, one side of the membrane is thin-walled while the other is thick-walled (Fig. 15.6). The thin wall is concerned with the exchange of respiratory gases. It is composed in part of the flattened epithelium of type 1 membranous pneumocytes (Fig. 5.5) which is covered by surfactant material which is probably derived from granular type 2 pneumocytes found in the corners of alveolar spaces. The membranous pneumocytes lie on a 'fused basement membrane' which is common to it and the underlying endothelial cells of the pulmonary capillaries (Fig. 5.5). Thus oxygen must diffuse through alveolar epithelial cells and their covering of surfactant, fused basement membrane and pulmonary endothelial cells to reach the blood in the pulmonary capillaries. The thick wall of the alveolar-capillary membrane shows a layer of interstitial connective tissue between the separated membranes of the alveolar epithelial cells and the pulmonary endothelial cells (Fig. 15.6). The interstitial tissue is concerned with the transport of tissue fluid and is the site of J receptors. We consider this aspect of the alveolar-capillary wall and these receptors in relation to high altitude pulmonary oedema (Chapter 15).

There are also two physiological components which tend to reduce the oxygenation of the arterial blood. The first is an anatomical shunt of venous blood from bronchial and Thebesian veins or through pulmonary vessels which have no contact with alveoli. The second is the effective shunt caused by uneven ventilation:perfusion. The addition of the two represents the 'physiological shunt' or 'venous admixture' (Harris and Heath, 1977).

Clearly diffusion across the alveolar-capillary membrane is enhanced by a higher PA_{O_2} since this increases the difference in oxygen pressure from the alveolar air to that in the pulmonary capillary. A decreased internal surface area of the lung will diminish the capacity for pulmonary diffusion. The diffusion coefficient for oxygen is of importance since it is much less soluble a gas than carbon dioxide and will be affected much more readily by pathological thickening of the alveolar-capillary wall by oedema fluid or fibrosis. We may consider below each of these points in relation to the subject at high altitude. At low altitude the A–a difference is relatively unimportant since the quantity of oxygen

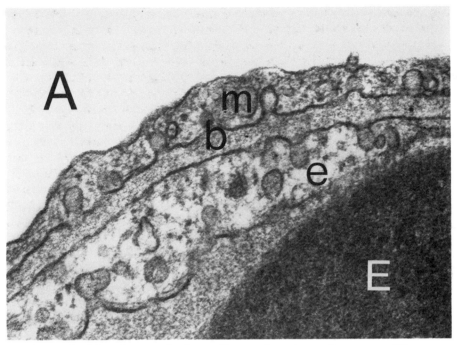

Fig. 5.5 Electron micrograph of the pulmonary alveolar-capillary membrane in a rat. Oxygen must diffuse across this barrier from alveolar air, A, to the erythrocyte, E, in the pulmonary capillary. The barrier comprises the attenuated cytoplasm of the membranous pneumocyte, m, the fused basement membrane, b, and the endothelial cell of the pulmonary capillary, e. This anatomical barrier is one factor in the production of an A–a gradient. ((Electron micrograph × 75 000).

carried by the blood does not fall substantially. However, at high altitude the A–a difference, were it maintained at sea-level magnitude, would assume significance as the fall in Pa_{O_2} on the steep slope of the oxygen-haemoglobin dissociation curve would cause a pronounced fall in the quantity of oxygen carried by the haemoglobin.

Pulmonary diffusing capacity in the newcomer to high altitude

There are no pronounced changes in pulmonary diffusing capacity shortly after ascent (Kreuzer and Compagne, 1965). During exercise there is some increase in diffusing capacity but it does not exceed that occurring under similar conditions of work at sea level. Such a lack of extra increase in diffusing capacity leads to a pronounced fall in arterial oxygen saturation during exercise despite the elevation in PA_{O_2} brought about by hyperventilation. Hence the A–a difference in newcomers to high altitude may be a limiting factor on exercise.

Pulmonary diffusing capacity in the native highlander

In contrast there seems to be general agreement that the A–a difference is reduced from its sea-level value of 6 to 17 mmHg to about 2 mmHg at 4270 m (Houston and Riley, 1947). According to Hurtado (1964b) the A–a difference in natives at Morococha (4540 m) is only 1 mmHg. Other evidence suggests that at rest the pulmonary diffusing capacity is some 20 to 30 per cent higher than the value for sea-level residents (De Graff et al, 1970; Remmers and Mithoefer, 1969). There appears to be only one dissenting voice in this general consensus. A study by Kreuzer et al (1964) did not establish a small A–a difference in high altitude natives and the value of 10 mmHg found by them led Mithoefer (1966) to express subsequently the opinion that a reduction in the A–a difference is not part of the process of acclimatization. As stated above, however, this is not the generally accepted view.

Factors which may operate in increasing the pulmonary diffusing capacity of native highlanders

An increased internal surface area of the lung might be one factor in increasing pulmonary diffusing capacity and reducing A–a difference. Hurtado (1964a) regards a barrel-shaped chest as characteristic of the physique of the Quechua Indians as we have discussed in Chapter 4. He finds that the high altitude native of the Andes shows an increase in the volume of some compartments of the lung. This applies to the volume of gas remaining in the lung at the end of a normal expiration (functional residual capacity) and that remaining after a forced expiration (residual volume) (Fig. 5.6). The lungs

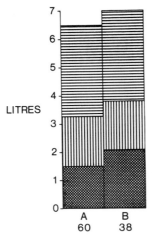

Fig. 5.6 Compartments of the total lung capacity (in litres) in adult males at sea level and at high altitude. Cross-hatched column= Residual volume; Vertical lines-column=Functional residual capacity; Horizontal lines-column=Total lung capacity. A=60 adult males, average age 23 yr at sea level. B=38 adult males, average age 23.3 yr at 4540 m. (Based on data from Hurtado, 1964a).

are thus held in a position of increased inflation in native highlanders and the same change can be detected in newcomers to high altitude during initial acclimatization (Tenney et al, 1953). It is possible that this hyperinflation of the lungs increases their internal surface area aiding the diffusion of oxygen. It is not possible to be dogmatic about the effect of such hyperinflation on the pulmonary blood flow or the distribution of ventilation and perfusion. Ward (1975) refers to an early observation by Barcroft (1925) that the ribs are more horizontal in native highlanders than in visitors. The average slope was 13° as against 21° in visitors. This is thought to be consistent with an increased antero-posterior diameter of the chest. We prefer to reserve judgement on these comparatively crude clinical measurements.

As far as we are aware methods of tissue morphometry to determine the internal surface area of the lungs, and the number and size of alveoli, such as we have employed in the study of pulmonary emphysema (Hicken et al, 1966), have not been applied to high altitude lungs. However, as we note in Chapter 4, rats reared from birth in a hypoxic environment develop lungs with an abnormally large internal surface area. Morphometric study of the lungs of high altitude natives would seem to be a fruitful area for research. It is possible that high altitude natives have an increased internal surface area of their lungs expanding the surface for gas exchange but this needs to be proven by the scientific methods of tissue morphometry now available.

Another factor that may operate in improving the pulmonary diffusing capacity at high altitude is the moderate degree of pulmonary hypertension that occurs in subjects at such elevations (Chapter 11). The elevation of pulmonary arterial pressure is possibly associated with a more uniform perfusion of the lung, especially of the upper zone which is underperfused at sea level. At high altitude there is an increase in total blood volume and this hypervolaemia associated with the mild pulmonary hypertension referred to above may distend patent capillaries and perhaps perfuse those unopened at sea level. This would enhance pulmonary diffusing capacity. A most important factor in achieving the best effective pulmonary diffusing capacity is the ratio of ventilation to perfusion. As we have seen in this chapter respiratory acclimatization to life at high altitude includes hyperventilation and increased diffusing capacity across the alveolar-capillary membrane but in the final analysis its efficiency depends on the matching of perfusion and ventilation throughout the lungs. As we have seen here and elsewhere in this book both pulmonary haemodynamics and ventilation are altered so one might anticipate some disturbance of the ratio of ventilation to perfusion. The increased alveolar ventilation is not associatied with an increase of total blood flow. There does not appear to be any change in distribution of the extra ventilation.

Alveolar surface tension

Castillo and Johnson (1969) claim to have demonstrated an increase in surface tension of lung extracts of mice after their acute exposure to a simulated altitude of 4270 m for 45 minutes. The technique of investigation used by these workers is troublesome for the sonication they employ would be expected to release much cellular lipid. Since surfactant lipids account for only 5 to 10 per cent of the total, they would tend to be swamped by the tissue lipids (Clements, 1976). Thus the surface tension-area loops shown in the paper of Castillo and Johnson (1969) are predictable from knowledge of the extraction procedure (Clements, 1976); they are not specifc to lung surfactant. Furthermore, blood plasma contains substances that can inhibit surface activity of lung surfactant at least *in vitro*. If lung fluid had increased acutely with hyperventilation at altitude, its admixture during mincing and sonication of the tissue might have caused the observed changes (Clements, 1976). These objections to the technique employed by Castillo and Johnson (1969) make it clear that before the suggestion that high altitude increases alveolar surface tension can be accepted, confirmation of their work must be forthcoming from other workers using techniques not open to the objections alluded to above.

Clara cells

Clara cells are non-ciliated cells found in the terminal portions of the bronchial tree. They are apocrine secretory cells, the apices of which are extruded into the bronchiolar lumen. Our experiments on rats suggested that there is an increase in the secretory activity in adults but not neonates, on acute exposure to hypoxia (Smith et al, 1974). Further studies in the Andes have demonstrated increased activity of the Clara cells in the llama (*Lama glama*) at 4720 m (Heath et al, 1976). The Clara cells of the llama proved to be very prominent in the terminal bronchioles and accounted for the great majority of epithelial cells there. They were easily identifiable with mushroom-shaped or club-shaped apical caps which projected into the lumen (Fig. 5.7). They had eosinophilic cytoplasm which was finely vesicular in the apical caps. In many Clara cells the club-shaped apical caps were attached to the basal, nucleated portion of the cell by only a long narrow isthmus. Large numbers of apical caps had been extruded into the bronchiolar lumen forming aggregates of cytoplasmic globules.

On electron microscopy the mushroom-shaped apical caps were obvious (Fig. 5.8). The cytoplasm was paler and looser than in the basal portion of the cell and the endoplasmic reticulum was prominent and showed vesicular dilatation in the cell apices.

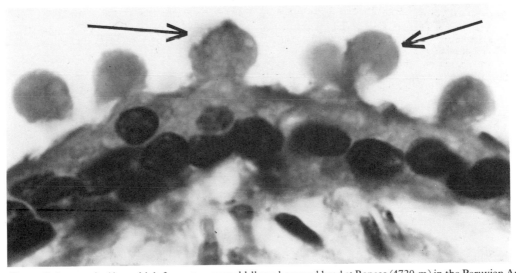

Fig. 5.7 Clara cells in a terminal bronchiole from a two-year old llama born and bred at Rancas (4720 m) in the Peruvian Andes. The apical caps of these cells (arrows) project into the lumen of the bronchiole. Subsequently they are extruded with liberation of their intra-cytoplasmic chemical substance, the function of which in the bronchial tree is unknown. The nuclei appear darkly-stained in the basal portions of the cells. (Haematoxylin and eosin, × 1500).

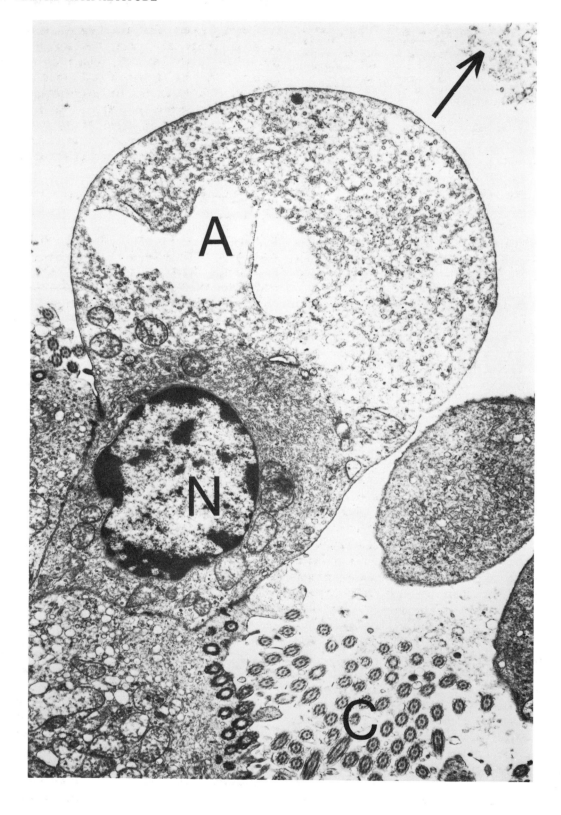

Fig. 5.8 (opposite) Electron micrograph of Clara cell from the llama referred to above. The apical portion of the cell, A, forms a mushroom-shaped cap which is extruded into the bronchiolar lumen (arrow). The basal portion of the cell with its nucleus, N, remains *in situ*. The cilia of the surrounding respiratory epithelium, C, are seen. (Electron micrograph, × 12 500).

Globules of cytoplasm were found to have been discharged into the bronchiolar lumen. These appearances are consistent with increased synthesis, storage and then apocrine secretion of some secretory product into the bronchiolar lumen and alveolar spaces. Since the llama is indigenous to high altitude, it is tempting to relate such hyperactivity of Clara cells with adaptation to chronic hypoxia. However, such a relation is not certain for the hyperactivity of this cell may be a feature of the species rather than of the environment.

As we have noted briefly above some of our previous studies have suggested that Clara cells respond to the stimulus of chronic hypoxia by an acceleration of the process of extrusion of the apical caps. In one experiment we subjected adult Wistar albino rats in a decompression chamber to a subatmospheric pressure of 265 mmHg for 12 hours; this produced an ambient PO_2 of 48 mmHg (Smith et al, 1974). This pressure simulated an altitude of 8840 m which is roughly equivalent to that of the summit of Mount Everest. In these rats there was an excess of smooth endoplasmic reticulum in the apical caps and an increased dehiscence of the caps into the bronchioles.

One possible association of acute hypoxia and hyperactivity of Clara cells arises from the studies of Etherton et al, (1973). Their autoradiographical studies have suggested that the extensive smooth endoplasmic reticulum of Clara cells contains accumulations of a phospholipid, dipalmitoyl lecithin. When tritiated palmitic acid is injected into mice, subsequent electron autoradiography of the lungs shows maximal labelling of the endoplasmic reticulum of Clara cells after three minutes. These are interesting findings for dipalmitoyl lecithin is a known pulmonary surfactant, lowering surface tension in the lung. If one wished to speculate on the functional significance of hyperactivity of Clara cells in the llama, one might postulate that, if the observations of Castillo and Johnson (1969) referred to above are true, permanent residence at high altitude might entail a tendency to increased surface

tension in the lung. This would entail a need for a constant perhaps extra supply of surfactant, hence the activity of the Clara cells. This is speculative. However, in closing our consideration of this subject we might recall that we have demonstrated a pronounced hyperplasia of Clara cells in chlorphentermine intoxication in Sprague-Dawley rats (Smith et al, 1974), a condition once again associated with the production of excess pulmonary surfactant.

Environmental and genetic factors in respiratory adjustment to high altitude

Lahiri and his colleagues (1976) have assessed the relative rôle of environmental and genetic factors in respiratory adjustment to high altitude. They studied the control of ventilation and the magnitude of lung volumes in different age groups from the neonatal period to adulthood, both at high altitude and sea level. Their results suggest strongly that adjustments to respiratory physiology at high altitude are acquired and determined by environmental rather than genetic factors. The relative changes in the ventilatory response to acute hypoxia and in the increased vital capacity from predicted sea-level volume are shown in Figure 5.9. The normal hypoxic ventilatory drive was developed in the native highlander up to the age of eight years before it was substantially lost during adult life.

Vital capacity was the same in neonates and young children at sea level and at high altitude but after the age of about nine years vital capacity in the highlander began to outspace that of sea-level children (121±13.7 per cent). Accelerated lung growth then persisted to native young adults 18 to 21 years of age (145±16.7 per cent). Thus the ventilatory response to hypoxia and vital capacity are similar in neonates and infants at sea level and at high altitudes. Vital capacity tends to increase in the post-natal period and continues to do so through the teenage period. The blunted hypoxic response, however, only begins to manifest itself in teenage life, becoming universal in adults. A genetic trait tends to be excluded on the grounds that offspring of lowlanders born and bred at high altitude showed the same phenomena as native highlanders.

Newborns at low and high altitude have similar ventilatory control mechanisms and lung volumes, suggesting that no genetic or intrauterine ac-

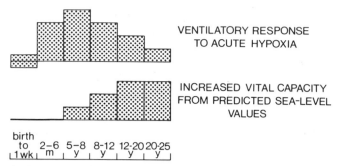

VENTILATORY RESPONSE
TO ACUTE HYPOXIA

INCREASED VITAL CAPACITY
FROM PREDICTED SEA-LEVEL
VALUES

birth
to   2–6   5–8   8-12 12-20 20-25
1 wk   m   y   y   y   y

Fig. 5.9 Respiratory acclimatization to high altitude from birth to the age of 25 years. Normal hypoxic ventilatory drive develops in the native highlander up to the age of eight years before becoming substantially lost during adult life. Vital capacity tends to increase throughout the teenage period (After data of Lahiri et al, 1976).

climatizing changes occur in these parameters in preparation for life at high altitude. It is presumably only after birth that high altitude neonates are exposed to greater hypoxic stimuli than at sea level. The adult pattern of responses and acclimatization to hypoxia begins to develop shortly after birth. In teenage life a variety of effective gas transport adjustments develop, such as the increased area for gas exchange provided by large lungs, and ventilatory hypoxic drive assumes a less important rôle.

Bronchial asthma at high altitude

Observations made in the Netherlands (Tromp and Bouma, 1974) have shown that asthmatics treated in low pressure climatic chambers at simulated altitudes above 1500 m, preferably 2000–2500 m, improve rapidly and usually with 60 to 100 treatments are 'cured'. Singh et al (1977) found that the morbidity rate for bronchial asthma in soldiers at high altitude fell (Fig. 5.10). There are sanatoria for the treatment of asthmatic subjects on the slopes of the volcanic Mount Teide on Tenerife in the Canary

Islands. A feature of the atmosphere at high altitude is the smaller content of pollen grains and other allergens and chemical pollutants.

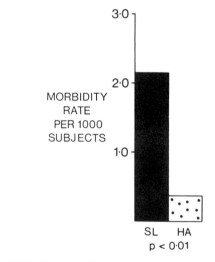

Fig. 5.10 The morbidity rate per thousand subjects from bronchial asthma among 20 000 soldiers stationed between altitudes of 3690 and 5540 m (stippled columns marked HA) and 130 700 men stationed on the plains (filled columns marked SL). (Based on data from Singh et al 1977).

REFERENCES

Barcroft, J. (1925) The respiratory function of the blood. In: *Lessons from High Altitude*. London: Cambridge University Press.

Bouverot, P., & Bureau, M. (1975) Ventilatory acclimatization and CSF acid-base balance in carotid chemodenervated dogs at 3550 m. *Pflügers Archiv (European Journal of Physiology)*, **361**, 17.

Castillo, Y. & Johnson, F. B. (1969) Pulmonary surfactant in acutely hypoxic mice. *Laboratory Investigation*, **21**, 61.

Chiodi, H. (1957) Respiratory adaptations to chronic high altitude hypoxia. *Journal of Applied Physiology*, **10**, 81.

Clements, J. A. (1976) Personal Communication.

Crawford, R. D. & Severinghaus, J. W. (1978) CSF pH and ventilatory acclimatization to altitude. *Journal of Applied Physiology*, **45**, 275.

Cumming, G. & Semple, S. J. (1973) In: *Disorders of the Respiratory System*, p. 236 Oxford: Blackwell Scientific Publications.

Davies, D. G. (1978) Evidence for cerebral extracellular fluid (H+) as a stimulus during acclimatization to hypoxia. *Respiration Physiology*, **32**, 167.

DeGraff, A. C. Jr., Grover, R. F. & Johnson, R. I. Jr. (1970)

Diffusing capacity of the lung in Caucasians native to 3100 m. *Journal of Applied Physiology*, **29**, 71.

Dempsey, J. A., Forster, H. V. & DoPico, G. A. (1974) Ventilatory acclimatization to moderate hypoxemia in man. *Journal of Clinical Investigation*, **53**, 1091.

Etherton, J. E., Conning, D. M. & Corrin, B. (1973) Autoradiographical and morphological evidence for apocrine secretion of dipalmitoyl lecithin in the terminal bronchiole of mouse lung. *American Journal of Anatomy*, **138**, 11.

Forster, H. V., Dempsey, J. A. & Chosey, L. W. (1975) Incomplete compensation of CSF (H⁺) in man during acclimatization to high altitude (4300 m). *Journal of Applied Physiology*, **38**, 1067.

Harris, P. & Heath, D. (1977) In: *The Human Pulmonary Circulation. Its Form and Function in Health and Disease*. Second Edition. Edinburgh, Churchill Livingstone.

Heath, D., Smith, P. & Harris, P. (1976) Clara cells in the llama. *Experimental Cell Biology*, **44**, 73.

Hicken, P., Brewer, D. & Heath, D. (1966) The relation between the weight of the right ventricle of the heart and the internal surface area and number of alveoli in the human lung in emphysema. *Journal of Pathology and Bacteriology*, **92**, 529.

Houston, C. S. & Riley, R. L. (1947) Respiratory and circulation changes during acclimatization to high altitude. *American Journal of Physiology*, **149**, 565.

Hurtado, A. (1964a) Some physiologic and clinical aspects of life at high altitudes. In: *Aging of the Lung*, p. 257, Edited by L. Cander & J. H. Moyer, New York: Grune and Stratton.

Hurtado, A. (1964b) Animals in high altitudes: Resident man. In: *Handbook of Physiology*, p.843. Section 4. Adaptation to environment. Washington: American Physiological Society.

Kreuzer, F., Tenney, M., Mithoefer, J. C. & Remmers, J. (1964) Alveolar arterial oxygen gradient in Andean natives at high altitude. *Journal of Applied Physiology*, **19**, 13.

Kreuzer, F. & Van Lookeren Compagne, P. (1965) Resting pulmonary diffusing capacity for CO and O_2 at high altitude. *Journal of Applied Physiology*, **20**, 519.

Lahiri, S., Delaney, R. G., Brody, J. S., Simpeer, M., Velasquez, T., Motoyama, E. K. & Polgar, C. (1976) Relative role of environmental and genetic factors in respiratory adaptation to high altitude. *Nature*, **261**, 133.

Lenfant, C. & Sullivan, K. (1971) Adaptation to high altitude. *New England Journal of Medicine*, **284**, 1298.

Luft, U. C. (1972) Principles of adaptations to altitude. In: *Physiological Adaptations. Desert and Mountain*. Edited by M. K. Yousef, S. M. Horvath & R. W. Bullard. New York and London: Academic Press.

Michel, C. C. & Milledge, J. S. (1963) Respiratory regulation in man during acclimatization to high altitude. *Journal of Physiology*, **168**, 631.

Milledge, J. S. (1975) Physiological effects of hypoxia. In: *Mountain Medicine and Physiology*, p.73. Edited by C. Clarke, M. Ward & E. Williams. Alpine Club.

Mithoefer, J. C. (1966) Physiological Patterns: The respiration of Andean natives. In: *Life at High Altitude*, p.21. Pan American Health Organization, Washington: Scientific Publication No. 140.

Monge, M. C. & Monge, C. C. (1966) *High-Altitude Diseases. Mechanisms and management*. Springfield, Illinois: Charles C. Thomas.

Orr, J. K., Bisgard, G. E., Forster, H. V., Buss, D. D., Dempsey, J. A. & Will, J. A. (1975) Cerebrospinal fluid alkalosis during high altitude sojourn in unanesthetized ponies. *Respiration Physiology*, **25**, 23.

Rahn, H., & Otis, A. B. (1949) Man's respiratory response during and after acclimatization to high altitude. *American Journal of Physiology*, **157**, 445.

Remmers, J. C. & Mithoefer, J. C. (1969) The carbon monoxide diffusing capacity in permanent residents at high altitudes. *Respiration Physiology*, **6**, 233.

Severinghaus, J. W., Mitchell, R. A., Richardson, B. W. & Singer, M. M. (1963) Respiratory control at high altitude suggesting active transport regulation of cerebrospinal fluid pH. *Journal of Applied Physiology*, **18**, 1155.

Severinghaus, J. A. & Mitchell, R. A. (1964) The role of cerebrospinal fluid in the respiratory acclimatization to high altitude in man. In: *The Physiological Effects of High Altitude*, p. 273. Edited by W. H. Weihe. Oxford: Pergamon Press.

Singh, I., Chohan, I. S., Lal, M., Khanna, P.K., Srivastava, M. C., Nanda, R. B., Lamba, J. S. & Malhotra, M. S. (1977). Effects on high altitude stay on the incidence of common diseases in man. *International Journal of Biometeorology*, **21**, 93.

Smith, P., Heath, D. & Moosavi, H. (1974) The Clara cell. *Thorax*, **29**, 147.

Tenney, S. M., Rahn, H., Stroud, R. C. & Mithoefer, J. C. (1953) Adaptation to high altitude: Changes in lung volumes during the first seven days at Mount Evans, Colorado. *Journal of Applied Physiology*, **5**. 607.

Tenney, S. M., Remmers, J. E. & Mithoefer, J. C. (1964) Hypoxic-hypercapneic interaction at high altitude. In: *The Physiological Effects of High Altitude*, p. 263. Edited by W. H. Weihe. Oxford: Pergamon Press.

Tenney, S. M., Scotto, P., Ou, L.C., Bartlett, D. & Remmers, J. E. (1971) Suprapontine influences on hypoxic ventilatory control. In: *High Altitude Physiology, Cardiac and Respiratory Aspects*, p. 89. Edited by R. Porter & J. Knight. Edinburgh: Churchill Livingstone.

Tromp, S. W. & Bouma, J. J. (1974) The biological effects of natural and simulated high altitude climate on physiological functions of healthy and diseased subjects (in particular Asthmatics). Monograph series. Biometeorological Research Centre XIII, 5.

Ward, M. (1975) *Mountain Medicine. A Clinical Study of Cold and High Altitude*. London: Crosby Lockwood Staples.

Waterlow, J. C. & Bunjé, H. W. (1966) Observations on mountain sickness in the Colombian Andes. *Lancet*, **ii**, 655.

Weiskopf, R. B., Gabel, R.A. & Fencl, V. (1976) Alkaline shift in lumbar and intracranial CSF in man after five days at high altitude. *Journal of Applied Physiology*, **41**, 93.

Transport and release of oxygen to the tissues

As we have noted in the previous chapter, at rest the tissues of the body need some 220 to 260 ml of oxygen per minute but on exercise this requirement may increase tenfold or even more (Thomas et al, 1974). The amount of oxygen transported per minute from the lungs to the rest of the body is a function of the cardiac output, the quantity of oxygen in systemic arterial blood and the affinity of haemoglobin for oxygen allowing the gas to be passed to the tissues. Increased requirements of oxygen can be met to some extent by an increase in cardiac output up to four times its normal level of 5 l/min and we refer to changes in the cardiac output in the early stages of acclimatization in Chapter 17. In this chapter we shall be concerned with the quantity of oxygen carried in the blood and with the ability of haemoglobin to release it to the tissues. The quantity of oxygen in the blood depends on the ability of the lungs to oxygenate it as we have already considered in Chapter 5, and also on the concentration of haemoglobin in the blood according to the following formula:

Oxygen supply = Cardiac output
  (ml/min)   (l/min)

     × Haemoglobin concentration × 1.34
              (g/l)

       × % oxygen saturation of arterial blood
            (ml O_2/g Hb).

At sea level haemoglobin leaving the lungs is 97 per cent saturated and in venous blood in the tissues this saturation has fallen to 70 per cent. The oxygen content of 100 ml blood leaving the lungs is 19.4 ml and that of blood leaving the capillaries of the tissues is 14.4 ml. Thus each 100 ml of blood delivers 5 ml of oxygen to the tissues. It will be apparent from these preliminary considerations that under the conditions of high altitude where the supply of oxygen is restricted by the diminished barometric pressure of the ambient air the haemoglobin concentration assumes considerable importance and we consider this problem now.

Haemoglobin concentration

The increased haemoglobin concentration which augments the oxygen supply to the tissues at high altitude was first demonstrated in animals living in the vicinity of the Andean town, Morococha (4540 m), by Viault (1891). However, as we shall see later in this chapter, there are interesting differences between the haematological findings in man and in animals indigenous to the mountains.

It is of interest to note that an early study by Hurtado (1932) led him to doubt the importance of an increase in haemoglobin as a factor in acclimatization. He stated at that time that 26 per cent of the highlanders he examined had a red cell count below 6×10^{12}/l, and in 11 per cent the erythrocyte count was lower than the normal average at sea level. He describes pronounced individual variation from 4.8 to 10.4×10^{12}/l and concluded that a normal sea-level count was not incompatible with health in such high regions. Similarly he found that the average haemoglobin level in one hundred highlanders was 15.93 g/dl, a value only slightly higher than the normal at sea level. He concluded that 'if we know that the native lives at high altitudes with a lower saturation of his arterial blood and with less circulating oxygen, as compared with the recent arrival, and yet is much more physically efficient than the latter, we have great difficulty in accepting the assumption that an increase of haemoglobin is of great importance'. Later workers have since drawn attention to the fact that there are considerable differences between individuals in the increase in red cell mass at altitude even if iron deficiency is excluded as a limiting factor. Different individuals reach an equilibrium with a haematocrit that may be anywhere in the range of 50 to 70 per cent. One factor that may operate here is whether the subject obtains relief from the chronic exposure to hypoxia by descending to lower altitude from time to time. On a recent visit to the Andes we were impressed by the not greatly elevated haematocrit of a man in late middle age who had worked as a miner at 4300 m for thirty years but who had taken frequent visits to the coast. Working with him was a young man of 23 years who had never left his mountain home; he had a haematocrit that was so high as to give concern that he was likely to develop chronic mountain sickness (Chapter 16).

Following later studies Hurtado (1964) came to accept that usually in man living at high altitude there is a definite increase in the number of circulating red blood cells, in the haemoglobin level and in the haematocrit. He compared haematological data from 250 sea-level residents at Lima and from 83 highlanders from Morococha (4540 m). This was the same town where Viault (1891) first demonstrated the rise in haemoglobin level in animals at high altitude. The mean value for the haematocrit rose from 46.6 per cent at sea level to 59.5 per cent at high altitude. The haemoglobin level rose from 15.6 to 20.1 g/dl. The red cell count rose from 5.1 to 6.4 × 10¹²/l. The reticulocyte count increased from 17.9 to 45.5 × 10⁹/l. Studies at the same Andean town were carried out by Merino (1950) who found the average haematocrit in the highlanders to be 66.7 per cent. The average haemoglobin level was stated by him to be 22.6 g/dl but this figure is rather high and approaches the level of 23 g/dl which has been held by some to represent the level at which Monge's disease may be diagnosed (Chapter 16). Merino (1950) also gives a high average value for the red cell count, namely 7.88 × 10¹²/l. It is important to give comparable data collected at the same altitude since haematological values are related to the elevation at which the subject lived. Average values as given by Reynafarje (1966) for various parameters in the peripheral blood of 50 human subjects at Morococha (4540 m)

are shown diagramatically and compared to those found in animals indigenous to high altitude (4200 to 4300 m) in Figures 6.1 to 6.5. The parameters selected are haemoglobin level (Fig. 6.1), haematocrit (Fig. 6.2), erythrocyte number (Fig. 6.3), mean corpuscular volume (Fig. 6.4) and mean corpuscular haemoglobin (Fig. 6.5). Average values as given by Peñaloza et al (1971) for the haemoglobin levels and haematocrits of 25 healthy highlanders living permanently at 4330 m are shown in Figures 16.6 a, b where they are compared with corresponding parameters in 10 subjects with Monge's disease. Monge and Monge (1966) gave somewhat lower values for natives of Huancayo (3260 m) who went to Morococha (4540 m) for 15 days. They found the average haemoglobin level of these visitors to the

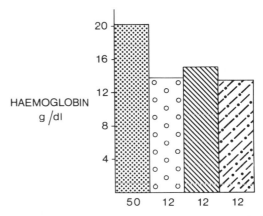

Fig. 6.1 Haemoglobin levels in the Quechua Indian and in three species of animals indigenous to the Andes: the alpaca, the llama and the vicuña. The numbers of individuals studied are shown at the foot of each column. In this figure and in Figures 6.2 to 6.5 stippled columns represent values for man, columns with open circles represent alpacas, hatched columns llamas, and columns with dash-dot, vicuñas. (Figures 6.1 to 6.5 are based on data from Reynafarje, 1966).

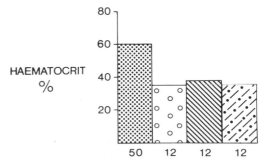

Fig. 6.2 Haematocrit percentages in the Quechua Indian and in the same three species of high altitude animal.

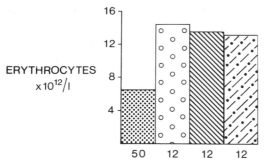

Fig. 6.3 Level of erythrocytes in the Quechua Indian and in the same three species of high altitude animal.

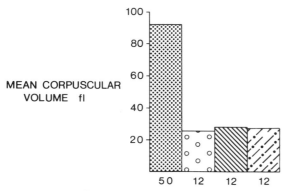

Fig. 6.4 Mean corpuscular volume in femtolitres (fl) in the Quechua Indian and in the same three species of high altitude animal.

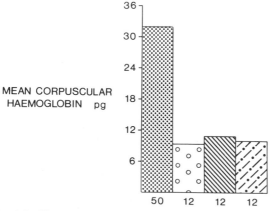

Fig. 6.5 Mean corpuscular haemoglobin concentration in picograms (pg) in the Quechua Indian and in the same three species of high altitude animal.

town to be 17.98 g/dl. The number of erythrocytes was 6.05×10^{12}/l and the haematocrit level was 54.43 per cent.

While the bone marrow shows a hyperplasia of the erythroid elements the myeloid cells and megakaryocytes remain normal in number and maturation (Merino and Reynafarje, 1949; Huff et al, 1951). The erythroid hyperplasia with increase in red cells with normal white cells differentiates this condition of high altitude polycythaemia from polycythaemia vera. We consider the peripheral and differential white cell count in Chapter 9 and platelets at high altitude in Chapter 10.

Blood volume

The pronounced polycythaemia at high altitude rapidly leads to an increase in the red cell volume (Fig. 17.2). This leads to an increase in the total blood volume (Fig. 17.2) but the plasma volume is reduced (Fig. 17.2). This aspect is dealt with more fully in Chapter 17, where we consider the systemic circulation at high altitude.

Erythrokinetics at high altitude

During the first one or two weeks at high altitude there is an initial decrease of 15 to 20 per cent in plasma volume which results in an increase in circulating haemoglobin of 1 to 2 g (Surks et al, 1966). The high haemoglobin level described above is maintained by a daily red cell production some 30 per cent higher than that of lowlanders at sea level (Reynafarje, 1964). A proportionally higher blood destruction maintains the equilibrium. The life span of erythrocytes in both native highlanders and in lowlanders acclimatized to high altitude is within normal limits (Berlin et al, 1954).

The remarkable sensitivity of the regulatory mechanisms which control erythropoietic activity is demonstrated by the fact that within two hours of exposure to high altitude red cell production increases. After this short period of time the iron turnover rate (in mg/day/kg body weight) increased on average from 0.37 to 0.54 in eight lowlanders taken to 4540 m (Reynafarje, 1964). The iron turnover rate rises to a maximum of 0.91 in one to two weeks so that erythropoietic activity is about three times higher than at sea level. Thereafter the iron turnover rate falls but even after six months it is still elevated. Even after a year's residence at high altitude the erythropoietic balance is still not achieved, for while the red cell mass increases the total blood volume is higher (Reynafarje, 1964).

Hence we see that the change in red cell mass is slow. The rise in haemoglobin level is one of the slower components of acclimatization.

The degree of polycythaemic response is related to the level of altitude. Up to an elevation of 3660 m the haemoglobin level increases in a linear relation to altitude. Above this height the haemoglobin concentration rises more rapidly with altitude than previously. There is, however, a limit. Under conditions of extreme hypoxia such as occurs at around 6000 m there begins a decrease in the formation of erythrocytes and haemoglobin (Hurtado et al, 1945). When the systemic arterial oxygen saturation is 60 per cent, erythropoietic hyperactivity decreases.

High altitude, haematocrit and age

Whittembury and Monge (1972) showed that there is a relation between haematocrit and the age of permanent residents at different altitudes. They collected data from healthy volunteers at Morococha (4540 m), Cerro de Pasco (4330 m) and Puno (3800 m). They find that a haematocrit of 75 per cent is accompanied by symptoms of chronic mountain sickness. This level is predictable at an age of 30 years at Morococha but at 63 years at Cerro de Pasco. It is of considerable interest to recall that the weight of the carotid bodies also increases with age at high altitude. As we shall see in Chapter 8 the chief cells of the carotid bodies have the ultrastructural features of APUD cells, which are known secretors of endocrines. There would appear to be grounds here for further investigation of the claims by Tramezzani et al (1971) that the carotid bodies may be an additional source of erythropoietin.

Erythropoietin

High altitude polycythaemia is the result of increased erythropoietin activity which occurs within two hours of exposure to hypoxia according to Reynafarje (1966). The degree of erythropoietic stimulation is proportional to the severity of the hypoxia within the limits just referred to. The initial rise in plasma and urinary erythropoietin which occurs within hours of the first ascent to high altitude falls in a few days to a value intermediate between the initial pronounced response and that at sea level (Siri et al, 1966; Faura et al, 1969). This secondary fall in the level of erythropoietin is related to the gradual rise in Pa_{O_2} which follows hyperventilation and the other mechanisms of acclimatization described elsewhere in this book.

Reynafarje et al (1964) present evidence to confirm this falling away of initial erythropoietin production. To detect evidence of erythropoiesis they injected rats with plasma obtained from natives of Morococha (4540 m), from sea-level subjects, and from newcomers to this Andean town 1, 3 and 10 days after their arrival from sea level. They found erythropoietic activity in plasma from newcomers only one day after their arrival. Erythropoietin was not found in plasma taken from newcomers 3 or 10 days after their arrival. The resulting hyperactivity of erythropoietin gives rise to an increase in the iron turnover rate as we have seen and subsequently intestinal iron absorption is stimulated. In fact the effect of erythropoietin on the bone marrow is usually limited by available iron.

Intestinal iron absorption

Iron absorption increases three to four times during early exposure to high altitude and the maximum is reached after one week (Reynafarje and Ramos, 1961). Conversely there is a decrease in iron absorption during descent from high altitude reaching a minimum of one fifth of normal in three weeks. Permanent residents at high altitude do not have a significantly different iron absorption from people born and living at sea-level. Hence bone marrow demands associated with production of red cells is the stimulus for iron absorption rather than a direct action of the oxygen saturation of the blood or of the oxygen tension in the intestinal tissues. This is why high altitude natives with a systemic arterial oxygen saturation of only 80 per cent show only an insignificant elevation of iron absorption.

Haematological changes during physical activity

On exercise there is a slight increase in the haemoglobin and haematocrit which is similar in high altitude natives and in sea-level subjects. The cause is haemoconcentration and probably redistribution of red cells. On recovery the haemoglobin and haematocrit levels return to normal. Hence from a haematological standpoint there are no differences between work performed at sea level and at high altitude (Reynafarje, 1964).

The reversibility of red cell production on loss of the hypoxic stimulus

There is a progressive decrease in red cell iron turnover rate in high altitude natives when they are brought down to sea level reaching its lowest value from the third to the fifth week (Reynafarje et al, 1959). It is then only one third of its initial value indicating great depression of red cell production. Three months after descent to sea level the red cell volume reaches a value lower than normal, indicating a true anaemia. At the same time there is an increase in plasma bilirubin and in fecal urobilinogen, and this has been considered to be consistent with increased red cell destruction as well as inhibition of erythropoietic activity (Merino, 1950). However, Reynafarje (1964) found the red cell life span to be normal, as was also the rate of sequestration in the spleen and liver of autogenous red cells labelled with 51 Cr. Hence there is unlikely to be an accelerated destruction of circulating red cells. On descent to sea level erythropoietin becomes undetectable. There is a progressive decrease in the activity of the bone marrow. The mobilization and utilization of iron decreases progressively, reaching a minimum after two to three weeks. The red cell mass diminishes slowly and after two months is even lower than that of sea-level subjects (Reynafarje, 1966) (Chapter 28). Reynafarje (1966) reported that filtrates of plasma from polycythaemic subjects from the Andes brought to sea level produced a statistically significant depression of erythropoiesis in rats previously exposed to high altitude to make them sensitive to such a depression factor. This allows one to speculate that an inhibitory humoral factor exists in the plasma of high altitude natives brought down to sea level.

Haemoglobin levels in Sherpas

Haemoglobin levels of Sherpas do not reveal the high values found in Quechuas living at 4000 m (Adams and Strang, 1975). These authors studied various haematological parameters in 28 males,

mean age 30 years, and 23 females, mean age 35 years, all being of pure Tibetan ancestry. Their data are summarized in Table 6.1. The lower haemoglo-

Table 6.1 Haemoglobin levels and types in Tibetan Sherpas

Haemoglobin level (g/dl)	17.0±1.25 in males 15.3±0.8 in females
Incidence of Hb A2	2.96± 0.58%
Incidence of Hb F	0.9± 0.3%

bin levels found are not explained by age, systemic disease, haemoglobinopathies or deficiencies of vitamin B12, folate, iron or thyroid (Adams and Strang, 1975). It seems more likely that adaptive evolutionary changes are responsible, as discussed more fully in Chapter 27. Like the llama, Tibetan Sherpas may rely more on greater oxygen extraction than on increased supply by the blood in supplying the body's needs of oxygen at high altitude.

When superimposed on the data of Hurtado and his colleagues (1945) collected from the Quechuas of the Peruvian Andes, the mean haemoglobin level found in Tibetan Sherpas living at an altitude of 4000 m (and an expected oxygen saturation of 84 per cent) corresponds to that expected in Peru at 2300 m and an arterial saturation of 92 per cent (Adams and Strang, 1975). There is no evidence from electrophoresis that the predominant haemoglobin of Sherpas differs from Hb A, and Hb F is not increased (Table 6.1). It is of interest that chronic mountain sickness with greatly elevated levels of haemoglobin is not found in the Himalayas.

Singh and his colleagues (1977) found from an extensive study of Indian soldiers stationed at altitudes between 3692 and 5538 m for prolonged periods that the tendency to polycythaemia persists only during the first ten months stay at high altitude. Subsequently, there is a progressive decline in the haemoglobin and the haematocrit between 10 months and 24 months' stay at high altitude. The cause for this decline is not clear.

Bilirubinaemia at high altitude

Bilirubinaemia occurs in man at high altitude. The serum bilirubin concentration of permanent residents at 4540 m in Peru was found to range from 1.27 to 1.56 mg/100 ml compared with 0.76 mg/

100 ml for men at sea level (Berendsohn, 1962; Merino, 1950; Hurtado et al, 1945). The elevation was due almost entirely to indirect bilirubin. A study was made by Altland and Parker (1977) of the serum bilirubin concentration and the extent of intravascular haemolysis in rats during acclimatization to a simulated altitude of 5500 m. During both continuous and intermittent (4 hours per day) exposure the serum bilirubin was significantly elevated at the end of four to six weeks. The rises occurred only after the development of severe polycythaemia with a haematocrit of 68 per cent, and a haemoglobin level of 21.6 g/dl. An increase in intravascular haemolysis was found after two weeks intermittent exposure and after four weeks continuous exposure to 5500 m. There was no alteration of erythrocyte fragility to account for increased intravascular haemolysis. No liver pathology was observed in the rats exposed to simulated high altitude. Bilirubinaemia in the rat exposed to high altitude may be due to the greatly increased number of erythrocytes and to a proportionate increase in their destruction. Increased intravascular haemolysis associated with the increased blood viscosity and an inability of the liver to handle increased levels of serum bilirubin are other possible factors.

Oxygen release to the tissues

So far in this chapter we have been concerned with the increase in the *quantity* of haemoglobin in man living at high altitude which enables a greater supply of oxygen to be transported to the tissues. The properties of haemoglobin itself must now be considered for they present a seeming paradox. Acclimatization to the hypoxia of high altitude might appear to be better served by an *enhanced affinity* of haemoglobin for oxygen for this will surely carry more oxygen to the tissues. On the other hand it could be considered that acclimatization is better effected by a *decreased affinity* of haemoglobin for oxygen so that it readily yields oxygen to the tissues maintaining an adequate PO_2 for transport to the mitochondria. The answer to this question is complex and depends to some extent on the level of altitude and consequent degree of hypoxia being considered. It also depends to some extent on the species involved. We shall first consider this

problem in relation to man living at high altitude but not extreme altitude (Chapter 30). Central to this matter is the oxygen–haemoglobin dissociation curve.

Oxygen–haemoglobin dissociation curve

As 97 per cent of the oxygen is carried in the erythrocyte and only 3 per cent in the plasma, the increased concentration of haemoglobin in the blood at high altitude means that it transports an increased amount of oxygen to the tissues. However, its availability there depends on the ease with which it will be liberated from haemoglobin. Such affinity is expressed as the familiar oxygen–haemoglobin dissociation curve which was found by Bohr (1904) to have a sigmoid shape (Fig. 6.6). This shape has

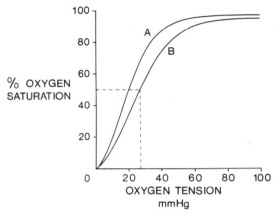

Fig. 6.6 Oxygen–haemoglobin dissociation curves of blood at different (H^+) concentration. A. Normal arterial blood at sea level. B. Normal venous blood at sea level. The value for P_{50} is shown to be 26.6 mmHg.

considerable physiological significance even at sea level because over the flat portion of the curve fluctuations in alveolar oxygen tension do not interfere with the arterial oxygen content. On the other hand, on the steep slope of the curve minimal changes in oxygen tension at the peripheral capillaries allow considerable unloading of oxygen (Finch and Lenfant, 1972). The peculiar sigmoid shape is due to changes in the conformation of the haemoglobin molecule (Finch and Lenfant, 1972).

Crystallographic techniques show that oxyhaemoglobin has a slightly different configuration than deoxyhaemoglobin in that in the former the pair of alpha chains and the pair of beta chains are slightly closer together (Muirhead et al, 1967). The combination of an oxygen molecule with a haem group alters the position of the ferrous ion in the haem ring, changing in turn the position of certain amino acids and changing the affinity for oxygen of the haem group in the neighbouring subunit chain. Hence in effect the uptake of each oxygen molecule in turn enhances the uptake of more (Perutz, 1970). There is a suggestion that upon oxygenation of the third haem the quarternary structure of the haemoglobin changes from the structure of deoxyhaemoglobin to that of oxyhaemoglobin. These molecular events account for the changing of the oxygen affinity of haemoglobin with oxygenation causing the oxygen dissociation curve to have a sigmoid shape.

The position of the oxygen dissociation curve at high altitude

The position of the curve is given by the partial pressure of oxygen in plasma (PO_2) associated with 50 per cent oxygen saturation of blood (that is when half the total haem groups are combined with oxygen) at 37°C and pH 7.4. This particular partial pressure is known as the P_{50}. It is of historical interest that Barcroft et al (1923) suggested that an increased affinity of haemoglobin for oxygen might be an important factor in acclimatization. They found that fetal blood was superior to adult blood in its affinity for oxygen, facilitating the intrauterine transfer of oxygen. Following an expedition to Peru, Barcroft (1925) came to the conclusion that there was a tendency for the dissociation curve to shift to the left. Later studies by Dill and his colleagues in Colorado (1931) and Keys et al (1936) did not confirm this. They thought there was a decrease in the affinity of haemoglobin for oxygen at high altitude. This increase in P_{50} and the shift of the oxygen–haemoglobin dissociation curve to the right is favourable in maintaining an adequate level of oxygen diffusion in the tissues as follows. The rôle of the position of the curve in unloading oxygen is to maintain a relatively high PO_2 when oxygen has been removed from haemoglobin during passage through a systemic capillary. The amount of oxygen per minute which will diffuse from the blood to the mitochondria of tissue cells varies directly with the

difference in PO_2 between these two regions. Since the mitochondrial PO_2 is probably below 1 mmHg the PO_2 in the capillary determines the rate of diffusion of oxygen. A shift of the oxygen–haemoglobin dissociation curve to the right gives a higher PO_2 for every value of saturation and thereby the requisite PO_2 further along the capillary.

This favourable effect of maintaining a higher PO_2 for tissue diffusion of oxygen was recognized by Aste-Salazar and Hurtado (1944) but Lenfant and Sullivan (1971) point out that the advantages of such a shift depend on the altitude. Below an elevation of 3500 m there is a substantial advantage but at higher altitudes the advantage is very small. The reason for this is clearly that the PA_{O_2} decreases greatly so that oxygen loading of the blood in the lungs is impaired. Hence the beneficial effects of increased unloading of oxygen in the tissues is counterbalanced by the decreased oxygen loading in the lungs. In the newcomer to high altitude the hyperventilation which occurs increases his PA_{O_2} enhancing the advantage gained by the progressive shift of the oxygen–haemoglobin dissociation curve to the right.

2, 3-Diphosphoglycerate (DPG)

As we have seen above, events within the haemoglobin molecule determine the shape of the oxygen–haemoglobin dissociation curve. Certain conditions stabilize the deoxy shape of the haemoglobin molecule favouring oxygen release (Finch and Lenfant, 1972). These include hydrogen ions, carbon dioxide, and 2, 3-diphosphoglycerate. Although 2, 3-DPG was described in red cells half a century ago (Greenwald, 1925), it is only since 1967 that its rôle in regulating the oxygen affinity of haemoglobin has been recognized (Benesch and Benesch, 1967; Chanutin and Curnish, 1967). It is generated by the anaerobic glycolytic pathway and each erythrocyte contains some $15\,\mu mol/g$ of haemoglobin. It is able to enter the core of the haemoglobin molecule, when it is in the deoxy form, between the β chains and binding itself to each (Finch and Lenfant, 1972). This stabilization of the deoxy form favours oxygen release so it means in effect that each red cell has its own mechanism for reacting to hypoxia. When deoxyhaemoglobin increases there is an increase in red cell glycolysis leading to an increase in DPG and oxygen availability (Hamasaki et al, 1970). In passing we may note that levels of 2,3-DPG are increased in diseases associated with chronic hypoxia such as chronic lung disease (Oski et al, 1969) and cyanotic heart disease (Woodson et al, 1970). High levels have been found in various forms of anaemia. It seems likely that in these conditions the increased amounts of 2,3-DPG may have some rôle in making more oxygen available to the tissues. It has been shown that 2,3-DPG levels rise in people living at high altitudes (Lenfant et al, 1968; Torrance et al, 1970 a and b). In particular, studies performed on large numbers of residents in Leadville and Climax in Colorado (Eaton et al, 1969) confirmed consistently higher levels of 2,3-DPG in their blood than in controls from the plains.

About half of the change in P_{50} and 2,3-DPG occurs within 15 hours of exposure to high altitude (Lenfant et al, 1971). Mulhausen et al (1968) demonstrated a change in the P_{50} from 26.7 mmHg to 30.6 mmHg within a day of ascent to 3500 m. The intra-erythrocytic organic phosphates appear to affect the affinity of haemoglobin for oxygen by the direct binding action described above and by lowering the intracellular pH (Bohr effect, see below). Lenfant et al (1971) believe that the shift of the oxygen dissociation curve is not so important as hyperventilation as a mechanism of acclimatization.

Alkalosis and 2,3-DPG

Lenfant et al (1971) studied the oxygen dissociation curve and oxygen transport parameters in four normal subjects and in two subjects made acidotic with acetazolamide before, during, and after a stay of four days at an altitude of 4500 m. In the normal subjects on ascent the oxygen dissociation curve shifted rapidly to the right and appeared to be brought about by an increase of 2,3-DPG which was related to an increase in plasma pH above sea-level values. In the acidotic subjects there was no rise in plasma pH, no increase in 2,3-DPG and no shift in the oxygen dissociation curve. Gerlach and his colleagues (1970) found an increased erythrocytic level of 2,3-DPG in rats following exposure to 11 per cent oxygen. However, the addition of 5 per cent carbon dioxide to the inspired gas prevented

alkalosis and the expected DPG changes. Such data are consistent with the view that the altitude-induced increase in 2,3-DPG is the result of alkalosis accompanying exposure (Klocke, 1972).

Age and 2,3-DPG

The same elevation in 2,3-DPG is found in young rats. Thus in the study of Martin and his colleagues (1975) young rats aged 2, 12 and 24 months showed significant increases of respectively 21, 14 and 22 per cent of erythrocyte 2,3-DPG following exposure for four weeks to a simulated altitude of 7010 m. However, rats aged 40 months were not capable of the same exaggerated degree of erythropoiesis and the increase in 2,3-DPG in erythrocytes was much less.

The higher Bohr effect in Quechuas

Morpurgo et al (1970) confirmed the 'shift to the right' of the oxygen dissociation curve to which we have referred above in acclimatized Quechuas of the Peruvian Andes. They were able to show that another feature of acclimatization in Quechua Indians was an increased Bohr effect, that is, a greater decrease in affinity of haemoglobin for oxygen at lower tissue pH. They determined the oxygen dissociation curves of haemoglobin at pH 7.4 and 6.7 for haemolysates from 26 native highlanders in the Andes and from eight subjects of European origin living for different periods at high altitude. In addition the oxygen dissociation curves for haemoglobin were determined on 18 Europeans living at sea level in Rome. The typical 'shift to the right' of the oxygen–haemoglobin dissociation curve was noted at pH 7.4 and pH 6.7. However, in the latter case the mean P_{50} and P_{80} values (the partial pressure of oxygen required for 50 per cent and 80 per cent saturation of haemoglobin) of Europeans at high altitude were lower than those of native highlanders. Thus in native highlanders the Bohr effect is increased at P_{50} and even more at a P_{80} (Fig. 6.7 Table 6.2).

Such an increased Bohr effect might arise from a modification of the haemoglobin molecule. However, Morpurgo et al (1970) failed to detect any abnormal haemoglobin. Hence there must have been some other factor in their haemolysates capable

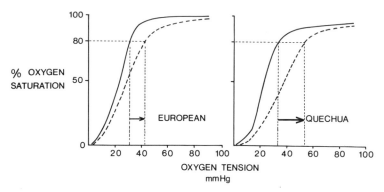

Fig. 6.7 Compared to Europeans the Quechua Indians show an increased Bohr effect, so that there is a greater decrease in affinity of haemoglobin for O_2 at the lower tissue pH of 6.7 (---) compared to a pH of 7.4 (continuous line). Note the greater increase in the value for P_{80} with the fall in pH in the case of the Quechuas. This means that in the hypoxic conditions of life at high altitude there is a higher Po_2 available for use by the tissues where the pH is lower.

Table 6.2   Comparison of partial oxygen pressures required for 50 per cent (P_{50}) and 80 per cent (P_{80}) saturation of haemoglobin in haemolysates diluted in 0.1 M phosphate buffer, at pH 7.4 and pH 6.7, from Europeans and Quechua Indians (with standard deviations of the mean). The Bohr effect is calculated as the difference between the values obtained at pH 6.7 and 7.4. (From the data of Morpurgo et al, 1970).

	No. of individuals	P_{50}			P_{80}		
		pH 7.4	pH 6.7	Bohr effect	pH 7.4	pH 6.7	Bohr effect
Europeans at sea level	18	19.5 ±1.30	26.4 ±1.48	6.9	27.0 ±1.91	37.9 ±2.80	10.9
Europeans above 3500 m	8	24.8 ±2.23	31.3 ±3.21	6.5	35.1 ±2.86	46.3 ±4.14	11.2
Quechuas above 3500 m	26	25.0 ±2.80	33.7 ±2.94	8.7	35.5 ±4.83	52.7 ±5.98	17.2

of decreasing the oxygen affinity of haemoglobin at lower pH. 2,3-DPG influences the Bohr effect when present in approximately equimolar concentration to haemoglobin. The increased Bohr effect facilitates the amount of oxygen made available to the tissues.

The shift of the oxygen–haemoglobin dissociation curve to the left at extreme altitude

In this account of the transport and release of oxygen to the tissues we have seen that in man living permanently at high but not extreme altitude, that is between altitudes of say 3000 m to 5500 m, there is a 'shift of the oxygen-haemoglobin dissociation curve to the right'. This follows an increase in erythrocyte 2,3-DPG and a *decrease* in haemoglobin-oxygen affinity. This increases the unloading of oxygen to

the tissues. However, it is clear that such a shift to the right exacerbates the dangerous degree of arterial desaturation and introduces the possibility of severe hypoxaemia and death. Such considerations reactivate the classical controversy in the early part of the century when Barcroft et al (1923) suggested that an *increased* affinity of haemoglobin for oxygen might be an important factor in acclimatization.

Eaton and his colleagues (1974) sought to establish whether or not artificially increased oxygen-haemoglobin affinity would protect rats when they were subsequently acutely exposed to very low environmental oxygen pressures. Fourteen Sprague-Dawley rats were given drinking water containing 0.5 per cent sodium cyanate which irreversibly carbamoylates haemoglobin amino groups, thereby increasing oxygen-haemoglobin affinity. A fortnight later 80 per cent of the reactive haemoglobin amino groups had been carbamoylated and the P_{50} of the test animals had been reduced from 37.3 to 21 mmHg. The concentration of 2,3-DPG in the erythrocytes had fallen. The control and test rats were then exposed to a simulated altitude of 9180 m. Eight of the ten control animals died but all the test rats, with their artificially-induced 'shift to the left', survived. Such experimental evidence supports the view that *increased affinity* of haemoglobin for oxygen has survival value at *extreme* altitude. It suggests that the characteristic shift to the right of the oxygen-haemoglobin dissociation curve in the highlander and in the newcomer to the mountains is of little value at extreme altitude. On the steep portion of the oxygen–haemoglobin dissociation curve any increase in oxygen delivery to the tissues by decreased affinity of haemoglobin for the gas will be accompanied by an almost equal loss in Pa_{O_2}.

The relative merits of polycythaemia and increased or decreased affinity of haemoglobin for oxygen were also studied by Penney and Thomas (1975). They tested the survival over 90 minutes of acute decompression (228 mmHg) of three groups of rats which had been treated in various ways to bring about changes in the level of 2,3-DPG and the value of P_{50} (Fig. 6.8). The first group was exposed to a simulated altitude of 4570 m for up to nine weeks to give rise to progressive polycythaemia and to elevate 2,3-DPG levels and raise P_{50} values. A second group was exposed to carbon monoxide for nine weeks, bringing no change in 2,3-DPG but a fall in P_{50} (Fig. 6.8). A third group was treated for ten days with sodium cyanate to produce big decreases in 2,3-DPG and P_{50}. Cyanate reacts irreversibly by carbamoylating the alpha amino groups of both the alpha and beta chains of haemoglobin, increasing its affinity for oxygen. When exposed to the acute hypoxia, only 5 per cent of the control animals survived (Fig. 6.8). Some 44 per cent of the polycythaemic rats with a raised P_{50} survived. More of the animals in the other two groups where P_{50} had been artificially diminished lived through the acute hypoxia. No fewer than 75 per cent of the cyanate-treated animals with big falls in 2,3-DPG levels and P_{50} values survived (Fig. 6.8). Clearly such experimental results indicate that survival at extreme altitude is better aided by *increased* oxygen haemoglobin affinity and a shift of

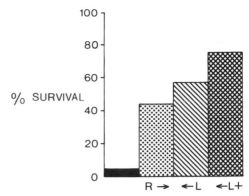

Fig. 6.8 The percentage survival in four groups of rats exposed to a simulated altitude of 8840 m for 90 minutes. The filled column represents control animals. The stippled column represents rats previously acclimatized to an altitude of 4570 m for nine weeks. After such acclimatization there is a rise in P_{50} indicating a fall in the systemic arterial saturation but there is a rise in 2,3-DPG suggesting an increased release of oxygen to the tissues. The hatched column represents rats exposed to carbon monoxide for nine weeks bringing no change in 2,3-DPG but a fall in P_{50} indicating increased oxygen saturation of the blood. The cross-hatched column represents rats treated with sodium cyanate for ten days to bring about big falls in P_{50} and 2,3-DPG indicating increased arterial oxygen saturation but decreased release of oxygen to the tissues. The rise in P_{50} is indicated in the diagram by a shift to the right of the oxygen dissociation curve while a fall in P_{50} is indicated by a shift of the curve to the left. (Based on data from Penney and Thomas, 1975).

the oxygen–haemoglobin dissociation curve *to the left*. Once the level of PA_{O_2} falls below the 'shoulder' of the curve, saturation declines rapidly thereby decreasing total oxygen transport. If the affinity of haemoglobin for oxygen increases, its oxygen saturation is moved back from the mid portion of the sigmoid to the shoulder. Oxygen unloading becomes more difficult but oxygen transport is increased by the higher oxygen saturation of the haemoglobin.

Oxygen–haemoglobin dissociation curve in animals indigenous to high altitude

Monge and Whittembury (1974) contrast the 'shift to the right' of the oxygen–haemoglobin dissociation curve of man at high altitude with the 'shift to the left' of indigenous high altitude animals. Such highland animals include camelids such as the alpaca, vicuña and llama, rodents such as the chinchilla and vizcacha, ruminants such as the yak, and birds such as the hualiata. All of these indigenous high altitude animals have a higher haemoglobin–oxygen affinity than their sea-level relatives such as the camel, rabbit, ox, and a variety of sea-level birds.

The blood of animals indigenous to high altitude

Some animals such as alpacas, llamas, and vicuñas live permanently at altitudes exceeding 4000 m. Reynafarje and Faura (1965) studied the blood of these camelids from Corpaconcha (4200 m) and certain interesting points of difference from haematological findings in man at high altitude emerged. These animals have a considerably lower level of haemoglobin than in the native Quechua Indian (Fig. 6.1), the level being in the region of 14 g/dl rather than 20 g/dl as in man. The haematocrit is also appreciably lower in these animals, lying below 40 per cent in contrast to the 60 per cent found in man (Fig. 6.2). In contrast the animals have considerably larger numbers of red cells (Fig. 6.3). The mean corpuscular volume (Fig. 6.4) and the mean haemoglobin (Fig. 6.5) are much smaller in the high altitude animals.

The erythrocytes are small and ellipsoidal (Fig. 6.9). There is a hyperplastic bone marrow. The iron turnover is greater in alpacas, llamas and vicuñas than in man living at the same altitude. The life span of erythrocytes is about 60 days. Such findings indicate hyperactivity of the erythropoietic tissue in these high altitude animals with a compensated red cell destruction rate. Certainly these animals indigenous to high altitude seem to need less haemoglobin than man to adapt to high altitude (see Chapter 27). This may be related to the small size of the erythrocytes and a high red cell count leading to a greater surface area for a more efficient contact with oxygen in the lung.

Effect of high altitude and strength of flight on the haematocrit of birds

The studies of Carpenter (1975) suggest that the haematocrits in avian species indigenous to high altitude are not greater than those of sea-level birds (Table 6.3). However, birds which characteristically undertake lengthy flights adjust to these conditions of heightened oxygen demand by increasing their haematocrits (Table 6.3).

Haemoglobin concentration and viscosity

The increased haemoglobin concentration heightens the viscosity of the blood (Denning and Watson, 1906). Above a critical haematocrit, which is about 75 per cent, oxygen transport decreases due to a fall in flow rate (Murray et al, 1963; Stone et al, 1968).

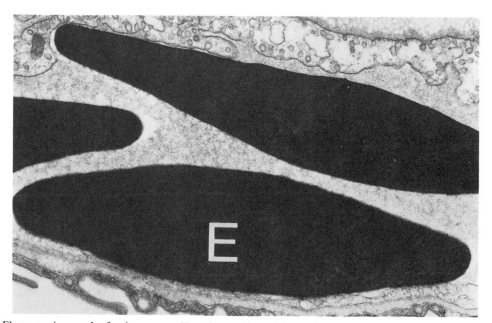

Fig. 6.9 Electron micrograph of pulmonary capillary from a llama. The erythrocytes (E) are ellipsoidal (× 25 000).

Table 6.3   Level of haematocrit, altitude-habitat, and strength of flight in various species of birds.

Species	Haema-tocrit	High or Low Altitude Species	Weak or Strong Flight	Reference
Gallus domesticus (Phasianidae) (Chicken)	29.0	L	W	Sturkie (1965)
Phasianus colchicus (Phasianidae) (Ring-necked pheasant)	34.0	L	W	Sturkie (1965)
Meleagris gallopavo (Meleagrididae) (Turkey)	35.9	L	W	Sturkie (1965)
Rhea americana (Rheidae)	33.8	H	W	Altman and Dittmer (1966)
Colaptes rupicola puna (Picidae)	37.0	H	W	Carpenter (1975)
Columbia livia (Columbidae) (Pigeon)	56.4	L	S	Sturkie (1965)
Metripelia melanoptera melanoptera (Columbidae)	53.5	H and L	S	Carpenter (1975)
Oreotrochilus estella estella (Trochilidae)	55.1	H	S	Carpenter (1975)
Chloëphaga melanoptera (Anatidae) (Bolivian goose)	59.1	H	S	Altman and Dittmer (1966)

Biological significance of shifts in the oxygen–haemoglobin dissociation curve

To summarise, we see that a shift of the oxygen–haemoglobin dissociation curve to the right implies a lowered affinity of haemoglobin for oxygen with elevated levels of 2,3-DPG. This means that oxygen unloading at the tissues is facilitated, maintaining an elevated PO_2 in association with some loss of oxygen saturation of the haemoglobin (Table 6.4). This system is characteristic of acclimatization (Chapter 27), allowing high survival at high altitude but it is not so advantageous under conditions of extreme hypoxia (Table 6.4). It is typical of man at high altitude and is best developed in the native highlander.

On the other hand, a shift of the curve to the left implies increased affinity of haemoglobin for oxygen so that oxygen transport and haemoglobin saturation are enhanced at the expense of oxygen unloading at the tissues (Table 6.4). Such a system is found in indigenous high altitude animals and is characteristic of adaptation in contrast to acclimatization (Fig. 6.10). It may also be induced artifically in laboratory animals as we have already seen (Fig. 6.10). A shift of the oxygen–haemoglobin dissociation curve to the left favours survival at extreme altitudes.

Table 6.4   The biological significance of shifts of the oxygen-haemoglobin dissociation curve at high altitude.

Shift of O₂–Hb dissociation curve	To the Right	To the Left
P_{50}	Raised	Lowered
2,3-DPG	Raised	Lowered or weak interaction between haemoglobin and intra-erythrocytic phosphates
Affinity of Hb for O_2	Lowered	Raised
Survival at high altitude (3000 m to 5500 m)	High	Raised
Survival at extreme altitude (>5800 m)	High compared to sea-level subjects	Higher still
Groups showing features	Quechua Indians of the Peruvian Andes and to less extent newcomers to high altitude	Sherpas of the Himalayas. Indigenous high altitude animals. Rats artificially treated with NaOCN
Typical biological status	Acclimatization	Adaptation

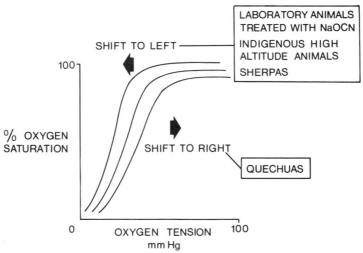

Fig. 6.10   Diagram to illustrate that in the acclimatized Quechua there is a shift to the right of the oxygen-haemoglobin dissociation curve. However, in adapted indigenous high-altitude animals the curve shifts to the left. A leftward shift has also been reported in Sherpas and may be induced experimentally by treatment with sodium cyanate.

REFERENCES

Adams, W. H. & Strang, L. J. (1975) Hemoglobin levels in persons of Tibetan ancestry living at high altitude. *Proceedings of the Society for Experimental Biology and Medicine* **149**, 1036.

Altland, P. D. & Parker, M. G. (1977) Bilirubinaemia and intravascular haemolysis during acclimatization to high altitude. *International Journal of Biometeorology*, **21**, 165.

Altman, P. L. & Dittmer, D. S. (Editors) (1966) *Environmental Biology*. Table 104. Bethesda, Maryland: Federation of American Societies for Experimental Biology.

Aste-Salazar, J. & Hurtado, A. (1944) The affinity of hemoglobin for oxygen at sea level and at high altitude. *American Journal of Physiology*, **142**, 733.

Barcroft, J. (1925) *Respiratory Function of the Blood. Part 1. Lessons from High Altitude.* London and New York: Cambridge University Press.

Barcroft, J., Binger, C. A., Bock, A. V., Doggart, J. H., Forbes, H. S., Harrop, G., Meakins, J. C. & Redfield, C. (1923) Observations upon the effect of high altitude on the physiological processes of the human body carried out in the Peruvian Andes. *Philosophical Transactions of the Royal Society of London*, **211, B.**, 351.

Benesch, R. & Benesch, R. E. (1967) The effect of organic phosphates from the human erythrocyte on the allosteric properties of haemoglobin. *Biochemical and Biophysical Research Communications*, **26**, 162.

Berendsohn, S. (1962) Hepatic function at high altitude. *Archives of Internal Medicine*, **109**, 256.

Berlin, N. H., Reynafarje, C. & Lawrence, J. H. (1954) Red cell life span in the polycythaemia of high altitude. *Journal of Applied Physiology*, **7**, 271.

Bohr, C. (1904) Theoretische Behandlung der Quantitativen Verhältnisse bei der Sauerstoffaufnahme des Hämoglobins. *Zentralblatt für Physiologie*, **17**, 682.

Carpenter, F. L. (1975) Bird hematocrits: effects of high altitude and strength of flight. *Comparative Biochemistry and Physiology*, **50A**, 415.

Chanutin, A. & Curnish, R. R. (1967) Effect of organic and inorganic phosphates on the oxygen equilibrium of human erythrocytes. *Archives of Biochemistry and Biophysics*, **121**, 96.

Denning, A. D. & Watson, J. H. (1906) The viscosity of blood. *Proceedings of the Royal Society of London (Biological Sciences)* **78**, 328.

Dill, D. B., Edwards, H. T., Fölling, A., Oberg, S. A., Pappenheimer, A. M. & Talbott, J. H. (1931) Adaptations of the organism to change in oxygen pressure. *Journal of Physiology, (London)*, **71**, 47.

Eaton, J. W., Brewer, G. J. & Grover, R. F. (1969) Role of red cell 2, 3 diphosphoglycerate in the adaptation of man to altitude. *Journal of Laboratory and Clinical Medicine*, **73**, 603.

Eaton, J. W., Skelton, T. D. & Berger, E. (1974) Survival at extreme altitude; protective effect of increased hemoglobin-oxygen affinity. *Science*, **183**, 743.

Faura, J., Ramos, J., Reynafarje, C., English, E., Finne, P. & Finch, C. A. (1969) Effect of altitude on erythropoiesis. *Blood*, **33**, 668.

Finch, C. A. & Lenfant, C. (1972) Oxygen transport in man. *New England Journal of Medicine*, **286**, 407.

Gerlach, E., Duhm, J. & Deuticke, B. (1970) Metabolism of 2,3-diphosphoglycerate in red blood cells under various experimental conditions. In: *Red cell metabolism and function.* (Edited by G. J. Brewer).

Greenwald, I. (1925) A new type of phosphoric acid compound isolated from blood with some remarks on the effect of substitution and the rotation of L-glyceric acid. *Journal of Biological Chemistry*, **63**, 339.

Hamasaki, N., Asakura, T. & Minakami, S. (1970) Effect of oxygen tension on glycolysis in human erythrocytes. *Journal of Biochemistry, (Tokyo)*, **68**, 157.

Huff, R. L., Lawrence, J. H., Siri, W. E., Wasserman, L. R. & Henessy, T. G. (1951) Effects of changes in altitude on haematopoietic activity. *Medicine*, **30**, 197.

Hurtado, A. (1932) Studies at high altitude. Blood observations on the Indian natives of the Peruvian Andes. *American Journal of Physiology*, **100**, 487.

Hurtado, A. (1964) Some physiologic and clinical aspects of life at high altitude. In: *Aging of the lung* (10th Hahnemann

Symposium). p. 257. (Edited by L. Cander & J. H. Moyer). New York: Grune and Stratton.

Hurtado, A., Merino, C. & Delgado, E. (1945) Influence of anoxaemia on the hemopoietic activity. *Archives of Internal Medicine*, **75**, 284.

Keys, A., Hall, F. G. & Barron, E. S. (1936) The position of the oxygen dissociation curve of human blood at high altitude. *American Journal of Physiology*, **115**, 292.

Klocke, R. A. (1972) Oxygen transport and 2,3-diphosphoglycerate (DPG). *Chest*, **62**, 795.

Lenfant, C., Torrance, J., English, E., Finch, C. A., Reynafarje, C., Ramos, J. & Faura, J. (1968) Effect of altitude on oxygen binding by hemoglobin and on organic phosphate levels. *Journal of Clinical Investigation*, **47**, 2652.

Lenfant, C. & Sullivan, K. (1971) Adaptation to high altitude. *New England Journal of Medicine*, **284**, 1298.

Lenfant, C., Torrance, J. D. & Reynafarje, C. (1971) Shift of the O_2-Hb dissociation curve at altitude: mechanisms and effect. *Journal of Applied Physiology*, **30**, 625.

Martin, L. G., Connors, J. M., McGrath, J. J. & Freeman, J. (1975) Altitude-induced erythrocytic 2,3-DPG and haemoglobin changes in rats of various ages. *Journal of Applied Physiology*, **39**, 258.

Merino, C. (1950) Studies on blood formation and destruction in the polycythaemia of high altitudes. *Blood*, **5**, 1.

Merino, C. & Reynafarje, C. (1949) Bone marrow studies in the polycythaemia of high altitude. *Journal of Laboratory and Clinical Medicine*, **34**, 637.

Monge, M. C. & Monge, C. C. (1966) *High-Altitude Diseases: Mechanism and Management*, pp 58 and 59. Springfield, Illinois: Charles C. Thomas.

Monge, C. C. & Whittembury, J. (1974) Increased hemoglobin-oxygen affinity at extremely high altitudes. *Science*, **186**, 843.

Morpurgo, G., Battaglia, P., Bernini, L., Paolucci, A. M. & Modiano, G. (1970) Higher Bohr effect in Indian natives of Peruvian Highlands as compared with Europeans. *Nature*, **227**, 387.

Muirhead, H., Cox, J. M., Mazzarella, L. & Perutz, M. F. (1967) Structure and function of haemoglobin. III. A three-dimensional former synthesis of human deoxy-haemoglobin at 5.5A resolution. *Journal of Molecular Biology*, **28**, 117.

Mulhausen, R. O., Astrup, P. & Mellemgaard, K. (1968) Oxygen affinity and acid-base status of human blood during exposure to hypoxia and carbon monoxide. *Scandinavian Journal of Clinical Laboratory Investigation, Suppl.* **103**, p. 9.

Murray, J. F., Gold, P. & Johnson, B. I. Jr. (1963) The circulatory effects of haematocrit variations in normovolemic and hypervolemic dogs. *Journal of Clinical Investigation*, **42**, 1150.

Oski, F. A., Gottlieb, A. J., Delivoria-Papadopoulos, M. & Miller, W. W. (1969) Red cell 2,3-diphosphoglycerate levels in subjects with chronic hypoxaemia. *New England Journal of Medicine*, **280**, 1165.

Peñaloza, D., Sime, F. & Ruiz, L. (1971) Cor pulmonale in chronic mountain sickness: present concepts of Monge's disease. In: *High Altitude Physiology: Cardiac and Respiratory Aspects*. Ciba Foundation Symposium. Edited by R. Porter & J. Knight. Edinburgh: Churchill Livingstone.

Penney, D. & Thomas, M. (1975) Hematological alterations and response to acute hypobaric stress. *Journal of Applied Physiology*, **39**, 1034.

Perutz, M. F. (1970) Stereochemistry of co-operative effect in haemoglobin. *Nature (London)*, **228**, 726.

Reynafarje, C. (1964) Hematologic changes during rest and

physical activity in man at high altitude. In: *The Physiological Effects of High Altitude*, p. 73. Edited by W. H. Weihe. Oxford: Pergamon Press.

Reynafarje, C. (1966) Physiological patterns: hematological aspects. In: *Life at High Altitudes*, Scientific Publication No. 140. p. 32. Washington: Pan American Health Organization.

Reynafarje, C., Lozano, R. & Valdivieso, J. (1959) The polycythemia of high altitude. Iron metabolism and related aspects. *Blood*, **14**, 433.

Reynafarje, C., Ramos, J., Faura, J. & Villavicencio, D. (1964) Humoral control of erythropoietic activity in man during and after altitude exposure. *Proceedings of the Society for Experimental Biology and Medicine*, **116**, 649.

Reynafarje, C. & Ramos, J. (1961) The influence of altitude changes on intestinal iron absorption. *Journal of Laboratory and Clinical Medicine*, **57**, 848.

Reynafarje, C. & Faura, J. (1965) Kinetics of red cell formation and destruction in high altitude adapted animals. 23rd International Congress of Physiological Sciences. *Abstract Papers* 182.

Singh, I., Chohan, I. S., Lal, M., Khanna, P. K., Srivastava, M. C., Nanda, R. B., Lamba, J. S., Malhotra, M. S. (1977) Effects of high altitude stay on the incidence of common diseases in man. *International Journal of Biometeorology*, **21**, 93.

Siri, W. E., Van Dyke, D. C., Winchell, H. S., Pollycove, M., Parker, H. G. & Cleveland, A. S. (1966) Early erythropoietin, blood and physiological responses to severe hypoxia in man. *Journal of Applied Physiology*, **21**, 73.

Stone, H. O., Thompson, J. K. Jr. & Schmidt-Nielsen, K. (1968) Influence of erythrocytes on blood viscosity. *American Journal of Physiology*, **214**, 913.

Sturkie, P. D. (1965) *Avian Physiology* p. 5. Ithaca, New York: Cornell University Press.

Surks, M. I., Chinn, K. S. K. & Matoush, L. O. (1966) Alterations in body composition in man after acute exposure to high altitude. *Journal of Applied Physiology*, **21**, 1741.

Thomas, J. M., Lefrak, S. S., Irwin, R. S., Fritts, H. W. & Caldwell, P. R. B. (1974) The oxyhemoglobin dissociation curve in health and disease. *American Journal of Medicine*, **57**, 331.

Torrance, J. D., Jacobs, P., Restrepo, A., Eschbach, J. W., Lenfant, C. & Finch, C. A. (1970 a) Intraerythrocytic adaptation to anemia. *New England Journal of Medicine*, **283**, 165.

Torrance, J. D., Lenfant, C., Cruz, J. & Marticorena, E. (1970 b) Oxygen transport mechanisms in residents at high altitudes. *Respiration Physiology*, **11**, 1.

Tramezzani, J. H., Morita, E. & Chiocchio, S. R. (1971) The carotid body as a neuroendocrine organ involved in the control of erythropoiesis. *Proceedings National Academy of Science, USA*, **68**, 52.

Viault, F. (1891) Sur la quantité d'oxygène contenue dans le sang des animaux des hauts plateaux de l'Amérique du Sud. *Comptes rendus hebdomadaires des séances de l'Académie des sciences. Paris*, **112**, 295.

Whittembury, J. & Monge, C. C. (1972) High altitude, haematocrit and age. *Nature*, **238**, 278.

Woodson, R. D., Torrance, J. D., Shapell, S. D. & Lenfant, C. (1970) The effect of cardiac disease on hemoglobin-oxygen binding. *Journal of Clinical Investigation*, **49**, 1349.

7

Tissue diffusion

The final area in which respiratory acclimatization to high altitude takes place is tissue diffusion. At sea level there is a diffusing pressure of some 30 mmHg from the venous end of the capillary to the immediate vicinity of the mitochondria (Luft, 1972). It has been established in studies on single mitochondria that their critical PO_2 is of the order of 1 to 3 mmHg (Chance et al, 1964; Lübbers and Kessler, 1968; Luft, 1972). It is very difficult to assess what changes occur in PO_2 in tissue and systemic capillaries at high altitude because neither are directly accessible *in vivo*. However, so far as can be judged from measurement of the mixed venous PO_2, the levels of PO_2 in capillaries and tissues must be considerably lower at high altitude than at sea level. The critical PO_2 in the tissues below which the mitochondria can no longer function efficiently is likely to have been reduced in spite of the various mechanisms of respiratory acclimatization to which we have so far referred in the two preceding chapters. A drop of 10 to 15 mmHg in the capillary at its venous end would seriously endanger the survival of cells at the periphery of the surrounding cylinder of tissue being provided with oxygen unless further factors facilitating diffusion to the mitochondria operated. It would appear that two such factors are increased capillary density diminishing the distance through which oxygen has to diffuse in the tissues, and an increased amount of myoglobin in the tissues aiding diffusion of oxygen and constituting a reservoir of oxygen for hypoxic tissues.

Capillary density

One possible microanatomical adjustment to the hypoxaemia resulting from life at high altitude would be a reduction of the distance over which oxygen has to diffuse from blood capillaries to reach the mitochondria. Capillary counts have in fact demonstrated that in acclimatized animals there is an increased number of capillaries per unit of tissue in the cerebral cortex (Diemer and Henn, 1965), skeletal muscle (Valdivia, 1958) and myocardium (Cassin et al, 1966). Detailed histological morphometric studies on the sternothyroid muscle were carried out by Banchero (1975) on three adult mongrel dogs native to Denver (1600 m) before and after exposure for three weeks to a simulated altitude of 4880 m. Subsequently these studies were extended to include five mongrel dogs native to the Andes (4350 m) (Eby and Banchero, 1976). The sternothyroid was chosen in order to minimise the possible effects of exercise on the skeletal muscle. The mean capillary density in cross sections of the muscle (Fig. 7.1a) and the relative surface area of capillaries per unit volume of tissue (Fig. 7.1b) both rose after three weeks at 4880 m but did not reach the levels found in native highland dogs.

In addition to an increase in capillary density an added microanatomical component of acclimatization appears to be a significant decrease in size of muscle fibres, this again diminishing the distance over which the oxygen has to diffuse to reach mitochondria. Thus in the studies of Banchero the average diameter of the muscle fibres fell (Fig. 7.1c) and the relative surface area of muscle fibres increased (Fig. 7.1d) after the animals from 1600 m had been at 4880 m for three weeks but even then the levels did not reach those seen in native highland dogs (Fig. 7.1c and d).

Such findings appear to be significant if one accepts the Krogh model of a cylinder for tissue oxygen diffusion. In this model the partial pressure

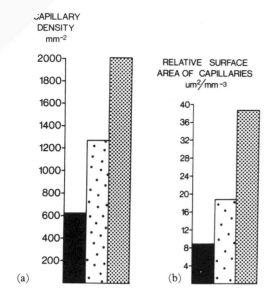

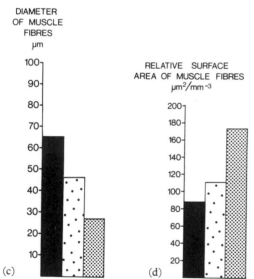

Fig. 7.1 Results of histological morphometric studies by Banchero (1975) and Eby and Banchero (1976) on the sternothyroid muscle of three adult dogs native to Denver (1600 m) (filled column), the same three dogs after residence for three weeks at a simulated altitude of 4880 m (lightly stippled column), and of five dogs native to the Andes (4350 m) (stippled column). (a) Capillary density. (b) Relative surface area of capillaries. (c) Diameter of muscle fibres. (d) Relative surface area of muscle fibres. (For explanation see text).

of oxygen at any point in the muscle fibre is a function of the square of the distance between the capillary and that point. Hence shorter diffusion distances are effective in facilitating oxygenation of

cells. The rôle of the diminution in muscle fibre size seems to be important. In the studies of Banchero (1975) exposure to high altitude for only three weeks led to a reduction of 30 per cent in average fibre diameter, representing a reduction in cross sectional area of about 50 per cent to be served, if a circular or hexagonal fibre shape is assumed. It has to be kept in mind that on acute exposure to hypoxia high capillary counts could indicate capillary recruitment in the general congestion that takes place rather than the development of new capillaries (Fig. 7.2).

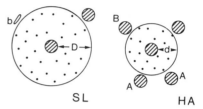

Fig. 7.2 Diagram to illustrate contributory factors in acclimatization in tissue diffusion. Compared to sea level (SL) at high altitude (HA) there is an increased number of new blood capillaries (A) and opening (B) of hitherto closed capillaries (b). The diameter of striated muscle fibres at high altitude (d) is smaller than at sea level (D).

The effectiveness of a combination of increased capillary density and decreased muscle diameter is illustrated by the calculations of Eby and Banchero (1976) and Banchero et al (1976). They estimated that at an atmospheric pressure of 635 mmHg, the P_{O_2} at the remotest extremity of the tissue cylinder at the venous end of a capillary (sometimes picturesquely referred to as 'the lethal corner') was 25 mmHg. At an increased altitude of 4350 m with a P_B of 435 mmHg, if there were no factors for acclimatization in the form of diminished muscle diameter and increased capillary density, the corresponding P_{O_2} would be 16 mmHg. In fact due to such factors it is 31 mmHg, an actual improvement over the level achieved at the lower altitude.

Myoglobin

In spite of the fact that the increased number of capillaries is able to bring the supply of oxygen closer to the cells, in the final analysis oxygen must pass from those capillaries to the mitochondria by the very slow process of diffusion. There is some evidence that the rate of diffusion is enhanced by increased amounts of tissue myoglobin.

Myoglobin is a protein found within cells which has a molecular weight of about 17 500 and consists of a single chain of 152 amino acid residues and one iron-containing haem group. This protein is of great relevance to high altitude studies because it has the property of combining loosely and reversibly with oxygen. Hence from the time of the early studies of Millikan (1939) the general view has been that myoglobin takes up oxygen rapidly at low oxygen tissue tensions and acts as a reserve store of oxygen which is available during periods of activity. As noted by Ward (1975) myoglobin will combine with oxygen even when tissue PO_2 is very low. Thus at a PO_2 of 40 mmHg haemoglobin is 75 per cent saturated, whereas myoglobin is 95 per cent saturated. When PO_2 is 10 mmHg, haemoglobin is only 10 per cent saturated, whereas myoglobin is 70 per cent saturated.

The chemical basis for the combination with oxygen probably lies in the fact that one of the coordination sites of the iron atom situated on one aspect of the tetrapyrrole structure is probably occupied by a water molecule and it is this which is presumably displaced by oxygen when myoglobin is converted to its oxygenated form (Mahler and Cordes, 1966). A survey of aqueous extracts of 8000 samples of human skeletal muscle by Boulton and Huntsman (1972) revealed five variants of myoglobin. These resulted from single amino-acid substitution. Boulton (1973) examined human psoas muscle and found a myoglobin content of about 4.5 mg per gram muscle. If one assumes this to be a representative skeletal muscle, 0.4 to 0.5 g (10 to 12 per cent) of the total body iron of a 70 kg man is present as myoglobin.

There seems little doubt that myoglobin occurs in high concentration in muscles which carry out sustained or periodic heavy work (Millikan, 1939). Thus high concentrations are to be found in mammalian myocardium. In chicken the leg muscles which walk and scratch for food are red and contain much myoglobin but the pectoral muscles which are very intermittently used are white (Wittenberg, 1966). On the other hand the pectoral muscles of flying birds which are constantly used are red. The sustained slow swimming muscles of fish are red but the mass of body musculature used for

sudden spurts of activity is white (Wittenberg, 1966). The amount of myoglobin in muscle depends on how hard it works. Thus the myoglobin content of rats, pigs and racehorses is increased by habitual exercise (Lawrie, 1950, 1953a). Myoglobin increases in muscle with age and this has been ascribed to the need to aid diffusion as the tissue elements increase in bulk (Lawrie, 1950, 1953a). The myoglobin content of cells is closely related to their capacity for oxygen uptake as indicated by the activities of the cytochrome oxidase and succinic dehydrogenase systems (Lawrie, 1953b and c). A cautionary note is sounded by Harris (1971) as to the function of myoglobin as a reserve store of oxygen in the tissues. He points out that while myoglobin combines reversibly with oxygen it gives it up slowly.

However, there is no doubt that increased amounts of myoglobin are present in the tissues of man and animals at high altitude. An increased myoglobin content of skeletal muscle at high altitude has been reported in dogs (Hurtado et al, 1937), guinea pigs (Tappan and Reynafarje, 1957), and rats (Anthony et al, 1959). Reynafarje (1962) found that in the sartorius muscle the mean level of myoglobin (in mg/g fresh tissue) was 6.07 in nine healthy young men of average age 28.8 years at sea level, but was 7.03 in nine healthy young highlanders of average age 24.3 years at 4330 m. Anthony and his colleagues expressed the opinion that in their animals the increment may have been more apparent than real since there was a concomitant loss of body weight. Nevertheless, an absolute increase in the quantity of myoglobin in heart muscle has been found in rats exposed continuously to a simulated altitude of 5490 m (Anthony et al, 1959), and in rats subjected intermittently to a simulated altitude of 7620 m.

An alternative function for this increased amount of myoglobin at high altitude rather than as a reserve store of oxygen is that it facilitates diffusion of the gas (Biörck, 1949). This is a physiochemical phenomenon discovered independently by Wittenberg (1959) and Scholander (1960). Apparently haemoglobin molecules must move to facilitate the diffusion of oxygen. Thus the haemoglobins of earthworms have a molecular weight of three million and do not facilitate the diffusion of oxygen. Proteins of lower molecular weight, such as myoglobin, are capable of facilitating diffusion of the gas. Since haemoglobin molecules are closely packed in an orderly lattice (Wittenberg, 1966), translational movements of the molecules are very limited. It is more likely that facilitation of diffusion of oxygen is mediated by rotation of molecules, particularly if they are in an environment of concentrated proteins. Wittenberg (1966), is of the opinion that oxygen molecules diffuse from the capillaries to the mitochondria by random movement interspersed with larger advances of translational and rotational movements of myoglobin molecules. There is evidence that the myoglobin-facilitated oxygen flux is about six times as great as the ordinary diffusive flux and accounts for most of the oxygen reaching the mitochondria. When a high altitude animal such as the alpaca is brought down to sea level, it shows a progressive diminution in the content of myoglobin in its skeletal muscles (Reynafarje et al, 1975) (Fig. 7.3).

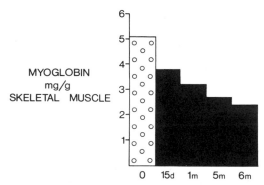

Fig. 7.3 The myoglobin content of skeletal muscle in six alpacas at 4540 m (open circles column) and 15 days, 1, 5 and 6 months after removal to sea level (filled columns). (From data of Reynafarje et al, 1975).

Cytochrome P-450 as a tissue oxygen carrier

Longmuir and Pashko (1977) believe that cytochrome P-450 can act as a tissue oxygen carrier and that it is increased in amount following only three hours exposure to hypoxia. They believe that this cytochrome forms an array of fixed or relatively immobile carrier molecules situated close together (1–3 nm) on the endoplasmic reticulum so that oxygen molecules can rapidly move from one to the other from the plasma membrane to the mitochondria.

Mitochondria

Mitochondria are the subcellular respiratory units that account for the bulk of oxygen consumed by the body. As we have just noted animals at high altitude, even when fully acclimatized, have a mean 'tissue' PO_2 below sea-level standards. It is conceivable that an increase in number of mitochondria would increase the probability of an oxygen molecule finding an enzyme site in a short time and would thus have an apparent effect of increasing intracellular 'diffusion capacity' (Ou and Tenney, 1970). Hearts from eight cattle born and raised at 4250 m were used by Ou and Tenney (1970) to prepare mitochondria which were spun down at high speed. They showed a 40 per cent increase in number of mitochondria compared to hearts from sea-level cattle. The mitochondrial size, however, remained the same. Ou and Tenney (1970) felt that on account of the small dimensions of cells and of the diffusion speed of oxygen in tissues, time delay for a diffusion process is so short that no major advantage is gained by such an increase in number of mitochondria at rest. However, they thought such an increase in mitochondria could assume significance at exercise with increased metabolism.

Kearney (1973), working in our laboratory, was unable to show any quantitative difference in the myocardial mitochondria of rabbits and guinea pigs from Cerro de Pasco (4330 m) and from sea level. Random electron micrographs from both cardiac ventricles of each animal were recorded on plates for later stereological analysis using the methods introduced by Weibel et al (1966, 1969), Weibel (1969). Mitochondrial volume expressed as a percentage of cytoplasmic volume was the same in rabbits, and guinea pigs from sea level and high altitude (Fig. 7.4). The numbers of mitochondria per millilitre of cytoplasm were also the same in these animal species at low and high altitude (Fig. 7.5). There was no significant difference between control and high altitude rabbits and guinea pigs so far as the surface areas of outer and inner mitochondrial membranes, expressed per millilitre of cytoplasm, were concerned (Fig. 7.6 and 7.7). The results of Kearney (1973) differ from those of Ou and Tenney (1970). It may be that this is accounted for by the different species studied. On the other hand it may be an expression of the different techniques used. Ou and Tenney (1970) counted granules in tissue homogenates, as we have noted above, and they were well aware of the fact that not all the granules counted were mitochondria

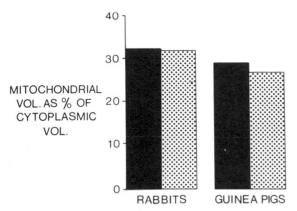

Fig. 7.4 The mitochondrial volume expressed as a percentage of cytoplasmic volume in three rabbits and three guinea pigs born and bred at sea level (filled columns) and the same numbers of the two species born and bred at Cerro de Pasco (4330 m) (stippled columns). There is no significant difference between the mitochondrial volume in high and low altitude representatives of these species. (Figs. 7.4 to 7.7 are based on data from Kearney, 1973).

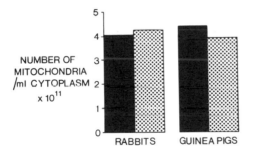

Fig. 7.5 The numbers of mitochondria per ml of cytoplasm in the same animals listed in Fig. 7.4. Identification of groups as in that figure. There is no significant difference between the numbers of mitochondria in high and low altitude representatives of these species.

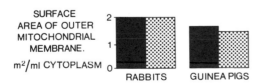

Fig. 7.6 The surface area of the outer mitochondrial membrane in m²/ml cytoplasm in the same animals listed in Fig. 7.4. Identification of groups as in that figure. There is no significant difference between the surface area of the outer mitochondrial membrane in high and low altitude representatives of these species.

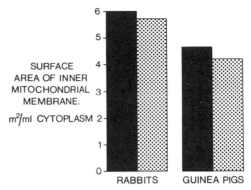

Fig. 7.7 The surface area of the inner mitochondrial membrane in m²/ml cytoplasm in the same animals listed in Fig. 7.4. Identification of groups as in that figure. There is no significant difference between the surface area of the inner mitochondrial membrane in high and low altitude representatives of these species.

but could have been other types of cell organelle such as the lysosome. We have reported an increase in succinic dehydrogenase activity in the myocardium of high altitude rabbits and guinea pigs at Cerro de Pasco (4330 m) (Harris et al, 1970) but it would appear from the results of Kearney (1973) that this heightened activity is not due to an increase in mitochondrial mass or internal surface area.

Oxidative phosphorylation in the mitochondria

Finally by the processes of acclimatization and adaptation described in the preceding chapters and in Chapter 27 oxygen reaches the mitochondria at an adequate partial pressure. Here the oxygen is reduced to water by hydrogen ions and electrons accepted from the respiratory chain (Fig. 7.8). Phosphate is utilised to combine with adenosine diphosphate (ADP) to give rise to the energy-rich triphosphate, ATP. Thus by this process of oxidative phosphorylation in the mitochondria the ultimate function of the provision of energy is achieved.

The mitochondrial respiratory chain is itself driven by hydrogen ions and electrons from the Krebs cycle (Figs. 7.8 and 7.9). This cycle of tricarboxylic acids also liberates to the blood carbon dioxide the transport of which we consider in Chapter 5.

Acetyl coenzyme A which enters this cycle (Fig. 7.10) is produced either from glucose through the

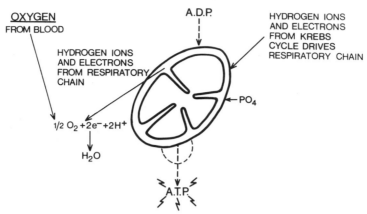

Fig. 7.8 The final destination of oxygen delivered by the processes of acclimatization. Oxygen from the blood diffuses through the tissues to the vicinity of mitochondria where it combines with hydrogen from the respiratory chain to form water. Hydrogen and electrons from the Krebs cycle drive the respiratory chain, and phosphate combines with ADP to form the energy-rich ATP.

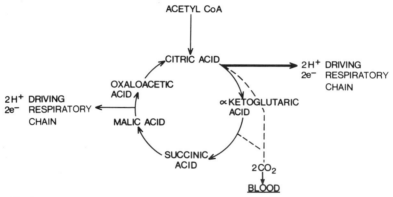

Fig. 7.9 The source of hydrogen driving the respiratory chain and of the carbon dioxide diffusing into the blood is the tricarboxylic-acid cycle.

process of glycolysis or from fatty acids through the process of lipolysis with the formation of the intermediary fatty acyl coenzyme A (Fig. 7.10) (Chapter 19). In the process of glycolysis with the formation of pyruvate there is liberation into the blood stream of lactate (Chapters 19 and 30). Each molecule of glucose passes through the glycolytic pathway (Fig. 7.10), and then the Krebs cycle (Fig. 7.9) and respiratory chain (Fig. 7.8) to liberate 38 molecules of ATP. This is the biochemical significance of the delivery of oxygen at adequate partial pressure by the processes of acclimatization, to allow the formation of energy-rich chemical bonds which are available for the vital functions of the cell and for conversion into physical energy in such phenomena as muscle contraction (Chapter 19).

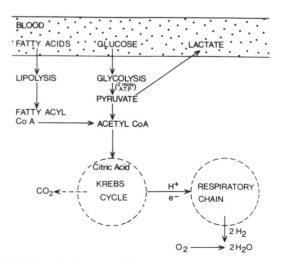

Fig. 7.10 The formation of fatty acyl CoA from fatty acids and of acetyl CoA from glucose. Acetyl CoA enters the Krebs cycle which donates hydrogen to drive the respiratory chain.

REFERENCES

Anthony, A., Ackerman, E. & Strother, G. K. (1959) Effects of altitude acclimatization on rat myoglobin; changes in myoglobin content of skeletal and cardiac muscle. *American Journal of Physiology*, **196**, 512.

Banchero, N. (1975) Capillary density of skeletal muscle in dogs exposed to simulated altitude. *Proceedings of the Society for Experimental Biology and Medicine*, **148**, 435.

Banchero, N., Gimenez, M., Rostami, A. & Eby, S. H. (1976) Effects of simulated altitude on O_2 transport in dogs. *Respiration Physiology*, **27**, 305.

Biörck, G. (1949) On myoglobin and its occurrence in man. *Acta. Medica Scandinavica Supplementa*, **226**, 133.

Boulton, F. E. (1973) The myoglobin content of human skeletal muscle. *British Journal of Haematology*, **25**, 271.

Boulton, F. E. & Huntsman, R. G. (1972) Variants of human myoglobin: their oxygen dissociation curves. *British Journal of Haematology*, **22**, 633.

Cassin, S., Gilbert, R. D. & Johnson, E. M. (1966) 'Capillary development during exposure to chronic hypoxia'. *Technical Report*, **66**, 16. San Antonio, Texas: Brooks, A. F. B.

Chance, B., Schoener, B. & Schindler, F. (1964) The intercellular oxidation-reduction state. In: *Oxygen in the Animal Organism*, p. 367. New York: Macmillan.

Diemer, K. & Henn, R. (1965) Kapillarvermehrung in der Hirnrinde der Ratte unter chronischem Sauerstoffmangel. Die Naturwissenschaften, **52**, 135.

Eby, S. H. & Banchero, N. (1976) Capillary density of skeletal muscle in Andean dogs. *Proceedings of the Society for Experimental Biology and Medicine*, **151**, 795.

Harris, P. (1971) In: *High Altitude Physiology, Cardiac and Respiratory Aspects*, p. 8. Ciba Foundation Symposium. Edited by R. Porter & J. Knight. Edinburgh: Churchill Livingstone.

Harris, P., Castillo, Y., Gibson, D., Heath, D. & Arias-Stella, J. (1970) Succinic and lactic dehydrogenase activity in myocardial homogenates from animals at high and low altitudes. *Journal of Molecular and Cellular Cardiology*, **1**, 189.

Hurtado, A., Rotta, A., Merino, C. & Pons, J. (1937) Studies of myohemoglobin at high altitude. *American Journal of the Medical Sciences*, **194**, 708.

Kearney, M. S. (1973) Ultrastructural changes in the heart at high altitude. *Pathologia et Microbiologia*, **39**, 258.

Lawrie, R. A. (1950) Some observations on factors affecting myoglobin concentrations in muscle. *Journal of Agricultural Science*, **40**, 356.

Lawrie, R. A. (1953a) Effect of enforced exercise on myoglobin concentration in muscle. *Nature*, **171**, 1069.

Lawrie, R. A. (1953b) The activity of the cytochrome system in muscle and its relation to myoglobin. *Biochemical Journal*, **55**, 298.

Lawrie, R. A. (1953c) The relation of energy-rich phosphate in muscle in myoglobin and to cytochrome-oxidase activity. *Biochemical Journal*, **55**, 305.

Longmuir, I. S. & Pashko, L. (1977) The role of facilitated diffusion of oxygen in tissue hypoxia. *International Journal of Biometeorology*, **21**, 179.

Lübbers, D. W. & Kessler, M. (1968) Oxygen supply and rate of tissue respiration. In: *Oxygen Transport in Blood and Tissues*, p. 90. Edited by D. W. Libben et al. Stuttgart: Thieme.

Luft, U. C. (1972) Principles of adaptations to altitude. In: *Physiological Adaptations. Desert and Mountain*, p. 143. Edited by M. K. Yousef, S. M. Horvath & R. W. Bullard. New York and London: Academic Press.

Mahler, H. R. & Cordes, E. H. (1966) *Biological Chemistry*. New York: Harper.

Millikan, G. A. (1939) Muscle hemoglobin. *Physiological Reviews*, **19**, 503.

Ou, L. C. & Tenney, S. M. (1970) Properties of mitochondria from hearts of cattle acclimatized to high altitude. *Respiration Physiology*, **8**, 151.

Reynafarje, B. (1962) Myoglobin content and enzymatic activity of muscle and altitude adaptation. *Journal of Applied Physiology*, **17**, 301.

Reynafarje, C., Faura, J., Villavicencio, D., Curaca, A., Reynafarje, B., Oyola, L., Contreras, L., Vallenas, E. & Faura, A. (1975) Oxygen transport of hemoglobin in high-altitude animals (Camelidae), *Journal of Applied Physiology*, **38**, 806.

Scholander, P. F. (1960) Oxygen transport through hemoglobin solutions. *Science*, **131**, 585.

Tappan, D. V. & Reynafarje, B. (1957) Tissue pigment manifestations of adaptation to high altitude. *American Journal of Physiology*, **190**, 99.

Valdivia, E. (1958) Total capillary bed in striated muscle of guinea pigs native to the Peruvian Mountains. *American Journal of Physiology*, **194**, 585.

Ward, M. (1975) *Mountain Medicine. A Clinical Study of Cold and High Altitude*. London: Crosby Lockwood Staples.

Weibel, E. R. (1969) Stereological principles for morphometry in electron microscopic cytology. *International Review of Cytology*, **26**, 235.

Weibel, E. R., Kistler, G. S. & Scherle, W. F. (1966) Practical stereological methods for morphometric cytology. *Journal of Cell Biology*, **30**, 23.

Weibel, E. R., Stäubli, W., Gnägi, H. R. & Hess, F. A. (1969) I Morphometric model, stereological methods and normal morphometric data for the rat liver. *Journal of Cell Biology*, **42**, 68.

Wittenberg, J. B. (1959) Oxygen transport: a new function proposed for myoglobin. *Biological Bulletin*, **117**, 402.

Wittenberg, J. B. (1966) Myoglobin facilitated diffusion of oxygen. *Journal of General Physiology*, **49**, 57.

8

The carotid bodies

Since the classical studies of Heymans et al (1930), following the suggestions of De Castro (1928), it has generally been accepted that the carotid bodies and the smaller glomera monitor the degree of oxygen saturation of systemic arterial blood. As we have seen in Chapter 5 it is now generally accepted that the initial hyperventilation on first exposure to high altitude follows stimulation of the carotid bodies. Since the major environmental hazard of life at high altitude is chronic hypoxia, one would anticipate that much would be known about the alterations in form and function of the carotid bodies in mountain dwellers. Such is not the case and it was only in 1969 that Arias-Stella reported that the carotid bodies of subjects living at high altitudes in the Andes are larger than those of people dwelling on the coastal plain of Peru.

The human carotid body at high altitudes

Subsequently this early report was extended when Arias-Stella and Valcarcel (1973) described the carotid bodies in two series of necropsies, one from sea level and one at 4330 m. The cases were matched for age and sex and mostly comprised accidental deaths with no significant cardiovascular or pulmonary pathology. The cases were compared in three age groups as shown in Figure 8.1. The carotid bodies of the highlanders were larger in each group and the difference became greater with age (Fig. 8.2).

The combined weights of the carotid bodies at high altitude were correspondingly greater than at sea level (Fig. 8.1). It is of interest that while the mean weights of the carotid bodies varied very little

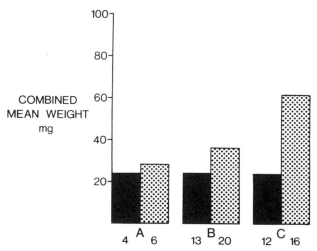

Fig. 8.1 The combined weights of the carotid bodies from two series of necropsies, one from sea level (filled columns) and one from 4330 m (stippled columns). The cases from sea level and high altitude are compared in three age groups: 10–20 years (A), 21–40 years (B), and 41–70 years (C). The numbers of cases in each of the six subgroups is indicated beneath the respective columns. Note that the carotid bodies of the highlanders are heavier in each group. In addition there is a definite progressive increment in weight with age in the high altitude series. (Data from Arias-Stella and Valcarcel, 1973 and 1976).

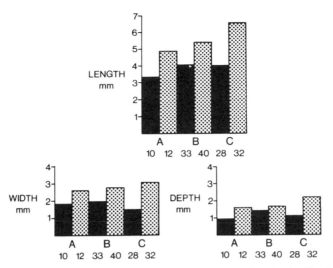

Fig. 8.2 The dimensions of the carotid bodies from the two series of necropsies referred to in Fig. 8.1, one from sea level (filled columns) and one from 4330 m (stippled columns). As in the previous figure the cases from sea level and high altitude are compared in three age groups: 10–20 years (A), 21–40 years (B), and 41–70 years (C). The numbers of cases in each of the six subgroups is indicated beneath the respective columns. Note that the dimensions of the carotid bodies of the highlanders are greater in each group. In addition there is a definite progressive increment in dimensions with age in the high altitude series. (Data from Arias-Stella and Valcarcel, 1976).

in the three age groups from sea level, a definite progressive increment in weight was found with age at high altitude. This finding deserves further investigation. It is of interest to note that at high altitude there is also a relation between age and haematocrit level (Chapter 6). The physiological studies on the sensitivity of the chemoreceptors at high altitude have been made largely in groups of young subjects. It would be interesting to investigate whether differences exist in the degree of ventilatory response to hypoxia in high altitude subjects in relation to their age.

The carotid bodies of animals at high altitude

Following the observations of Arias-Stella we compared the size of the carotid bodies of groups of guinea pigs, rabbits and dogs from high and low altitudes and found the former to be significantly larger (Edwards et al, 1971a) as shown in Figure 8.3. Blessing and Wolff (1973) demonstrated that enlargement of the carotid bodies could be induced in rats by subjecting them to a simulated high altitude of 7500 m for a period of three months. The carotid bodies of their rats increased in volume from 32.8 to 194.5 × $10^6/\mu m^3$. Arias-Stella and Bustos (1976) showed that the carotid bodies of cattle

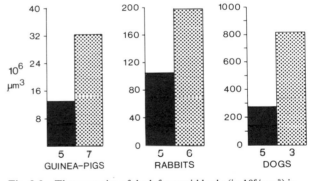

Fig. 8.3 The mean size of the left carotid body (in $10^6/\mu m^3$) in guinea pigs, rabbits and dogs born and bred at low and high altitude respectively. The low altitude animals are represented by filled columns and the high altitude animals by stippled columns. The numbers of animals studied in each of the six groups are indicated beneath the columns. The size of the carotid body is larger at high altitude in each of the three species. (Data from Edwards et al, 1971a).

enlarge in response to the chronic hypoxia of high altitude (Fig. 8.4).

The reversibility of enlargement of the carotid bodies

The increase in size of the carotid bodies in response to hypoxia is rapid and is equally rapidly reversible once the hypoxic stimulus is withdrawn. In one experiment we studied the tissue volume of the

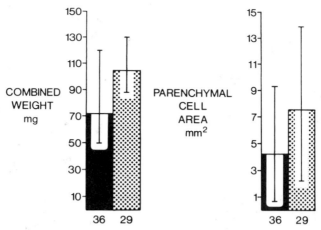

Fig. 8.4 Combined weight and 'functional area' of bovine carotid bodies at sea level and at high altitude (4300 m). The 'functional area' is that occupied by parenchymal cells on sections comprising the entire half of the surface of carotid bodies. The figure illustrates that the carotid bodies of high altitude cattle are heavier and show a hyperplasia of glomic cells. (Data from Arias-Stella and Bustos, 1976).

carotid body by an application of Simpson's rule to histological sections in three groups of 10 adult male Wistar albino rats (Heath et al, 1973). The first group was kept for five weeks in a hypobaric chamber exposed to a barometric pressure of 380 mmHg equivalent to a simulated altitude of 5500 m above sea level. The second was exposed to the same barometric pressure for five weeks and then allowed to recover in room air for a further period of five weeks. The third group acted as controls and was kept at normal barometric pressure throughout. In the control animals the mean carotid body volume (expressed in units of $10^6/\mu m^3$) was 13.45, but after exposure to chronic hypoxia for only five weeks, this volume rose to 47.81. Once the hypoxic stimulus was withdrawn the tissue volume fell after only five weeks to 19.82. This rapid reaction to hypoxia and equally rapid regression on its removal recalls the behaviour of the muscular hypertrophy of the right ventricle, pulmonary trunk and terminal portion of the pulmonary arterial tree referred to elsewhere in this book (Chapters 11 and 12).

The organic basis for enlargement of the carotid bodies

To gain insight into the functional significance of enlargement of the carotid bodies in man and animals at high altitude it is important to know which component of the anatomy of the chemoreceptor tissue is responsible for the increase in size. The carotid body consists of glomic cells, blood vessels and supporting connective tissue and nerves. There are two main types of cell in the carotid body called the chief (Type 1) cell and the sustentacular (Type 2) cell (Heath et al, 1970). Chief cells exist in three distinct forms designated light, dark and pyknotic. The light cell (Fig. 8.5) is the most abundant and is about 13 μm in diameter and has an ill-defined cytoplasmic outline. Its nucleus is 7 μm in diameter and has an open chromatin pattern. The cytoplasm is pale and eosinophilic and contains vacuoles up to 7 μm in diameter. Such vacuoles are more apparent in tissue not fixed immediately after death and are believed to be autolytic in nature, although as we shall see later, some vacuolation may be associated with hypoxia. In the dark variant of the chief cell the chromatin pattern of the nucleus has a denser and more granular pattern (Fig. 8.6). The cytoplasm is more deeply eosinophilic and its cytoplasmic margin is better defined. The pyknotic variant is some 10 μm in diameter with granular, deeply eosinophilic cytoplasm (Fig. 8.7). The nucleus is deeply basophilic, apparently structureless and excentric. The Type 2 or sustentacular cell has an elongated nucleus up to 13 μm long and 4 μm wide. On electron microscopy this Type 2 cell can be shown to be wrapped around small aggregates of Type 1 cells

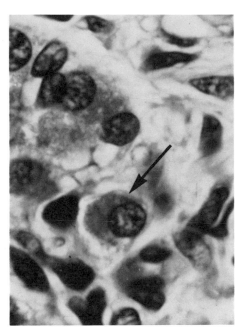

Fig. 8.5 A chief cell of the light variety from a human carotid body (arrow). Its cytoplasm has an ill-defined outline and its nucleus has an open chromatin pattern. To the left and above this is a further collection of light cells. (Haematoxylin and eosin, × 1500).

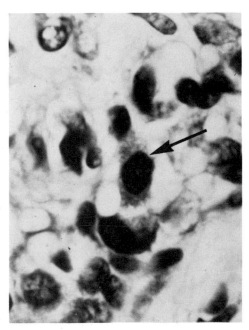

Fig. 8.6 A chief cell of the dark variety from a human carotid body is seen in the centre of the field (arrow). The nuclear chromatin pattern is denser than in the light cell and the cytoplasm is more deeply staining. Note the marked vacuolation in the cytoplasm. (Haematoxylin and eosin, × 1500).

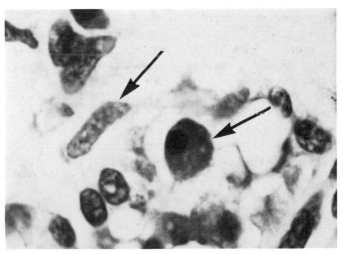

Fig. 8.7 A pyknotic type of chief cell from a human carotid body is situated in the centre of the figure (right arrow). The dark granular cytoplasm and excentric pyknotic nucleus are well seen. To the left of the pyknotic cell is a sustentacular cell with an elongated nucleus (left arrow). (Haematoxylin and eosin, × 1500).

by means of cytoplasmic processes (Grimley and Glenner, 1968).

We have studied the histological appearances of the enlarged carotid bodies of guinea pigs, rabbits and dogs which had been born and lived all their lives at 4330 m. In low altitude animals of the same species, the carotid body is compact and is composed of small clusters of glomic cells of the types described above (Fig. 8.8). High altitude glomera appear larger, even before confirmation by

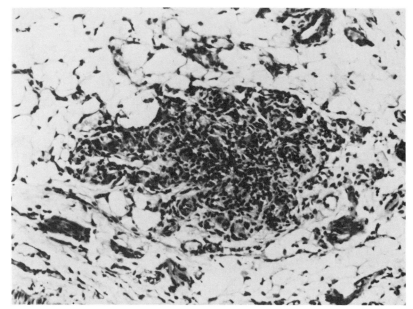

Fig. 8.8 Carotid body from low altitude guinea pig consisting of a compact, fusiform collection of glomic cells situated in the adipose tissue between the internal and external carotid arteries. (Haematoxylin and eosin, × 150).

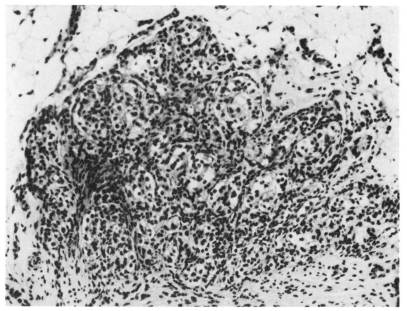

Fig. 8.9 Carotid body from high altitude guinea pig. Note the larger size and obvious clusters of glomic cells. (Haematoxylin and eosin, × 150).

micrometry (Fig. 8.9), and the glomic tissue appears lighter in colour. We found that in guinea pigs and rabbits this difference is due to a relatively greater hyperplasia of the light form of chief cells with associated vacuolation of the cytoplasm (Edwards et al, 1971a) (Fig. 8.5). This suggested to us that this type of glomic cell specifically responded to the stimulus of chronic hypoxia. The constancy of the proportion of light Type 1 cells in the dogs does not negate this suggestion since in this species almost all the Type 1 cells are light-staining even at sea level.

Arias-Stella and Valcarcel (1976) found similar

histological changes in the carotid bodies in native highlanders at 4330 m. The lobes and lobules were larger and more numerous in those living at high altitude. The collagen bands separating the lobules were thinner. The concentration of parenchymal cells was so dense and diffuse as to give the section a homogeneous appearance. The carotid bodies were congested and the glomic cells showed pronounced vacuolation. These authors attempted to measure what they termed 'the functional area' of parenchymal elements. They used one section comprising the entire half surface of the organ from each paraffin block for this purpose. A morphometric technique was employed. They found that the area occupied by parenchymal cells was significantly greater in the native highlanders than in sea-level subjects (Fig. 8.10). The proliferated

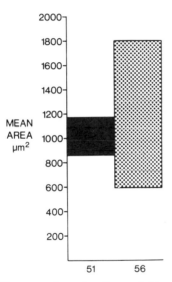

Fig. 8.10 The range of area (μm^2) occupied by parenchymal cells on sections composing the entire half surface of carotid bodies from lowlanders at sea level (filled column) and highlanders at 4330 m (stippled column). The numbers beneath the columns indicate the number of subjects studied in both categories. This 'functional area' is seen to be increased in highlanders. (Data from Arias-Stella and Valcarcel, 1976).

cells also showed depletion of yellow-green, naturally fluorescing, bioamine-containing granules. As in man, guinea-pigs, and rabbits the enlargement of the carotid bodies of cattle is due to hyperplasia of glomic cells with an increase of the 'functional area' (Fig. 8.4).

The severe degree of vacuolation in the glomic cells of animals and man from high altitude is consistent with the view that this cytological appearance may be ascribed to the effects of hypoxia. One must exercise caution in assessing the significance of such vacuolation, however, because it may in part be an expression of autolysis. For this reason any glomic tissue to be studied by light or electron microscopy should be placed in the appropriate fixative immediately. While on a visit to Peru we were shown by Arias-Stella that in cattle living at high altitude the hyperplasia of the chief cells in about a quarter of the animals becomes pronounced and associated with cellular pleomorphism and nuclear irregularities. These histological appearances are reminiscent of chemodectoma.

Chemodectomas

This interesting observation suggests that carotid body tumours in man and animals born and living at high altitude may represent the extreme degree of hyperplastic response of chemoreceptor tissue to prolonged and severe hypoxia. Arias-Stella and Bustos (1976) found that as many as 40 per cent of high altitude cattle develop histological changes suggestive of chemodectoma. Saldaña et al (1973) take the view that chemodectomas represent hyperplasia rather than neoplasia since they have a very slow rate of growth and a benign clinical course. They found that all but two of 25 Peruvian adults with chemodectomas of the head and neck had been born and lived at altitudes between 2105 and 4350 m above sea level, whereas only 38 per cent of the Peruvian population of 13.5 million live at or above an altitude of 3000 m, (Peruvian Census, Monge, 1966). Their studies of the prevalence of this tumour in Peru show that chemodectomas are about 10 times more frequent at high altitude than at sea level. In their series of 23 cases they found that all the carotid body tumours in their patients were benign while one patient had a malignant glomus jugulare tumour. Females predominated over males by about six to one and left sided tumours were three times as common. Histologically the tumour cells were arranged in clusters and comprised chief cells in which the dark variety predominated. It is of considerable interest that a biological environmental factor like the partial pressure of oxygen in the atmosphere can be regarded as a cause of neoplastic proliferation.

Chemodectoma and thyroid carcinoma

Albores-Saavedra and Durán (1968) reported the association of chemodectoma with thyroid carcinoma in two patients in Mexico City (2380 m). In these cases the carcinomas were of papillary and follicular type in contra-distinction to the medullary carcinoma of thyroid which occurs in association with phaeochromocytoma. Both the carotid body and the adrenal medulla share a common origin from neural ectoderm and a common function in secretion of catecholamines. Two of the patients from the Andes with chemodectomas reported by Saldaña and Salem (1970) and Saldaña et al (1973) also died of metastasising thyroid carcinoma.

Clinical application

Chronic hypoxia affects many patients with heart and lung disease as well as subjects living at high altitude. Hence the original observation of Arias-Stella (1969) that the carotid bodies of Quechua Indians were larger than those of Mestizos living on the coastal plain stimulated investigation to see if the same increase in size of chemoreceptor tissue applied to patients with chronic respiratory disease. In a study of the post mortem size and structure of the human carotid body at low altitude Heath et al (1970) found that the average weight of the right carotid body was 12.9 mg while that of the left was 11.3 mg. In two patients who had a post mortem diagnosis of cor pulmonale complicating emphysema, however, the combined carotid body weight was 60.9 mg and 68.8 mg respectively, while the combined weight of the carotid bodies in a case of the Pickwickian syndrome was 45.5 mg. Thus in this small group, selected because of the high likelihood of chronic hypoxia having occurred during life, the mean weight of the carotid bodies (58.4 mg) was significantly higher than that of the rest (21.4 mg). The patient with the carotid body weight of 60.9 mg was known to have a systemic arterial P_{O_2} of 36 mmHg and P_{CO_2} of 54 mmHg during life. These observations were consistent with enlargement of the glomic tissue in response to the chronic hypoxia of disease similar to that brought about by high altitude.

In a further series of 44 successive cases coming to routine necropsy Edwards et al (1971b) point-counted lungs fixed in distension so that the type and severity of any emphysema present could be determined accurately. The mean combined weight of the carotid bodies in 13 cases with no evidence of emphysema was 21.1 mg. In four of 11 cases of emphysema studied, the right ventricular weight was normal and the mean combined carotid body weight was 32.4 mg. In the seven cases of emphysema with right ventricular hypertrophy the mean combined carotid body weight was 56.2 mg. In one case in this latter group the combined carotid body weight was as high as 89.3 mg. There was no relation between the type and severity of emphysema on the one hand and the size of the carotid bodies on the other. Rather there appeared to be a relation between the weight of the carotid bodies and the weight of the right ventricle. Since the latter was probably associated with hypoxic vasoconstriction in the lung elevating pulmonary vascular resistance, it was assumed that the link between enlarged carotid bodies and right ventricular hypertrophy was chronic hypoxia. Thus here we can see a direct application of studies in high altitude to cardiopulmonary pathology.

Chemodectomas and lung disease

In one of the cases of right ventricular hypertrophy complicating pulmonary emphysema referred to above the enlarged right carotid body contained hyperplastic nodules composed of the dark variety of chief cell (Heath et al, 1970). Such histological appearances suggested that, as at high altitude, the chronic hypoxia of some cases of pulmonary emphysema might predispose to chemodectoma formation. Subsequently Chedid and Jao (1974) reported the occurrence of 11 tumours of the carotid body and one chemodectoma of the ganglion nodosum of the right vagus nerve in six members of two consecutive generations of a family. All affected members had bilateral tumours of the carotid bodies with a single exception. No fewer than four of these six patients had associated chronic obstructive pulmonary disease with a persistently high systemic arterial P_{CO_2} and low P_{O_2} levels. One case of bilateral carotid body tumour reported by Saldaña et al (1973) occurred in a 29-year-old Indian male at 3870 m who had worked as a miner for 12 years and developed silicosis. The associated effects of the hypoxia of lung disease and high altitude appeared to have combined to produce the chemodectomas. It is clear that chronic hypoxia occurring as a complication of cardiopulmonary pathology can lead to enlargement of the carotid bodies and chemodectoma formation with or without the associated hypoxia of diminished barometric pressure of high altitude.

Morphometry of the carotid bodies at simulated high altitude

Laidler and Kay (1975a) carried out morphometric studies on the carotid bodies of rats, some of which had been kept in ambient air at sea level and some of which had been maintained in a decompression chamber at a subatmospheric pressure of 460 mmHg for between 25 and 96 days. Point-counting techniques were used to measure the volume proportions of the constituents of the carotid bodies. The constituents studied were glomic cells, capillary blood vessels, connective tissues, other tissues like arterioles and nerve trunks, and an amorphous hyaline material.

These authors were able to confirm that under conditions of hypoxia the mean value of the total volume of the combined left and right carotid bodies rose from $47.16 \times 10^6/\mu m^3$ in the control rats to $187.39 \times 10^6/\mu m^3$ in the test animals. However, their morphometric studies did not show a percentage increase of glomic cells to account for this increase in size (Fig. 8.11). There was, nevertheless, an absolute increase in the total number of glomic cells in the carotid bodies from a mean value of 22 910 in the control rats to a mean value of 40 790 in the hypoxic animals.

It is of interest to note that Laidler and Kay (1975b) found that five of their 10 hypoxic rats had normal numbers of Type 1 cells and in the rat exposed to the longest duration of hypoxia the number had risen only slightly above that found in the controls. The remaining four rats in their series showed a pronounced increase in the number of Type 1 cells in proportion to the duration of hypoxia. The mean diameter of the nuclei of Type 1 cells in the hypoxic rats (5.5 μm) was greater than that of the controls (5.0 μm). These authors are of the opinion that enlargement of the carotid bodies in

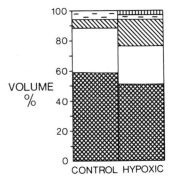

Fig. 8.11 The volume proportions of constituents of the carotid bodies of rats at sea level and simulated high altitude (460 mmHg for 25 to 96 days). (After data from Laidler and Kay, 1975a.) Cross-hatched columns represent glomic tissue, open columns fibrous connective tissue, hatched columns blood capillaries, dash line columns other tissues such as arterioles and nerve trunks and vertically hatched columns, amorphous hyaline.

It will be seen that Kay and Laidler found that in the enlarged carotid bodies of their hypoxic rats, although there was an absolute increase in the total number of glomic cells in the carotid bodies, there was a percentage fall in volume of glomic cells (see text). There was an increase in the percentage of the total volume occupied by blood capillaries. It should be noted that these results were obtained from an acute experiment employing simulated high altitude.

the rat subjected to hypoxia for some time in a hypobaric chamber is largely due to dilatation of capillaries. The functional significance of this increased vascularity is not clear. According to them it may be nothing more than a non-specific reaction designed to increase blood flow and thus oxygen transport to a hypoxic organ with increased metabolic activity. The functional significance of the intracapillary amorphous hyaline material is not yet known.

A careful distinction should be made between the cause of rapid enlargement of the carotid bodies due to acute exposure to hypoxia and chronic enlargement of these peripheral chemoreceptors due to long term exposure. It seems likely that when laboratory animals are acutely exposed to simulated high altitude in a hypobaric chamber their carotid bodies become engorged with blood due to increased blood flow as noted above. However, it is probable that the slow enlargement of the carotid bodies with age over decades as described by Arias-Stella and Valcarcel (1973) has an entirely different basis of cellular hyperplasia. We have personally examined sections of bovine carotid bodies shown to us by Arias-Stella and they show undoubted cellular hyperplastic changes which in a minority of instances mimic chemodectoma. We think this is an example of the distinction that must be made between the examination of animal tissue from acute experiments in a hypobaric chamber and that of tissues from man and animals born and living for many years at high altitude. Experiments in low pressure chambers simulate some of the conditions of natural high altitude but the two situations are not identical. Furthermore the tissues of low altitude animals placed for a short period in hypobaric chambers should not be regarded as identical with those of man and animals native to high altitude. Biochemical factors in the tissues, such as levels of myoglobin and enzymes, are likely to be very different.

Ultrastructural changes

Ultrastructural changes in chemoreceptor tissue exposed to the chronic hypoxia of high altitude also occur. They centre upon the osmiophilic bodies in the chief cells which were first described in the rabbit by Lever and Boyd (1957) and subsequently in the cat by Lever et al (1959). These workers found that the bodies stained positively by the diazonium reaction and concluded that they contained phenolic amines. Blümcke et al (1967) were convinced that the osmiophilic bodies were the source of bioamines and called them 'catecholamine bodies'. Studies have revealed characteristic electron microscopic changes in these osmiophilic bodies in guinea pigs at high altitude (Edwards et al, 1972). In sea-level guinea pigs the average diameter of the bodies varies from 100 to 150 nm (Fig. 8.12). The bodies consist of a central dense osmiophilic core with a very narrow clear halo subjacent to an outer limiting membrane. In some bodies the central core is less dense and appears to merge into the substance of the surrounding cytoplasm. The core does not assume an excentric position in control animals. The chief cells are bounded together by numerous processes and interdigitations which belie the apparently simple ovoid structure seen on light microscopy.

In high altitude guinea pigs there are ultrastructural changes in these osmiophilic bodies (Fig. 8.13). The halo between the central core and the outer limiting membrane becomes much broader. The osmiophilic core tends to be smaller, excentric and less dense, so that in some animals only faint remnants of the core are seen. In others the core is lost so that clear vesicles up to 350 nm replace the osmiophilic bodies. In some cells large clear spaces up to 2 μm in diameter are present and many

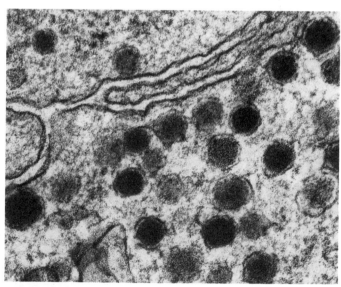

Fig. 8.12 Electron micrograph of carotid body from a low altitude guinea pig. It shows membrane-bounded granules in the cytoplasm of a chief cell. In many granules there is a central dense osmiophilic core with a very narrow clear halo subjacent to an outer limiting membrane. In others there is no limiting membrane and the contents appear to merge into the surrounding cytoplasm. (Electron micrograph, $\times$ 52 500).

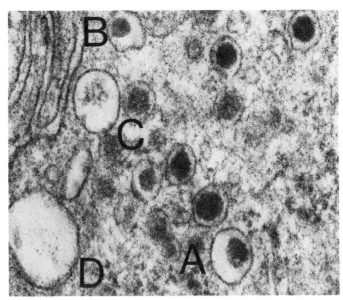

Fig. 8.13 Electron micrograph of carotid body from a high altitude guinea pig. It shows that some of the neurosecretory vesicles are distended by widening of the clear halo, which leaves the hitherto central osmiophilic core in an excentric position (A). In others the core is much less dense than normal (B) or is represented by only faint remains (C). In some granules the core has been lost and the appearances are those of a micro-vacuole (D). (Electron micrograph, × 52 500).

represent enlargement of these vesicles and correspond to the vacuoles seen in the cytoplasm of the chief cells on light microscopy.

In extreme hypoxia the osmiophilic bodies move to the periphery of the cell until their limiting membranes are in contact with the cytoplasmic membrane (Blümcke et al, 1967). The entire contents of the bodies are then discharged into the intercellular space. After 20 minutes of extreme hypoxia there is a total discharge of catecholamines from the receptor cells of the carotid body as confirmed by fluorescence microscopy. It seems likely that the less dramatic change of widening of the halo and the formation of vesicles as described above are those normally found in animals exposed to the chronic hypoxia of high altitude.

The biogenic amines

It is impossible to be certain of the functional significance of enlargement of the carotid bodies at high altitude and of the ultrastructural features that we describe above for the simple reason that the physiology of chemoreception is still not properly understood. What is clear, however, is that in some way the biogenic amines appear to be involved. The four amines concerned are dopamine (3,4-dihydroxyphenylethylamine), serotonin (5,6-hydroxytryptamine), norepinephrine, and epinephrine. In a series of 76 pairs of carotid bodies removed at necropsy the percentage content of these four biogenic amines was respectively 64, 18.8, 14.8, and 2.5 (Steele and Hinterberger, 1972). The concentration of dopamine in the carotid body is 20 to 40 μg g^{-1} while that of norepinephrine is only 1.5 μg g^{-1} (Dearnaley et al, 1968). This predominance of dopamine over norepinephrine occurs in only one other area of the body, namely the corpus striatum and substantia nigra in the brain. While Steele and Hinterberger (1972) were able to demonstrate some association between high levels of dopamine in the carotid bodies and cerebrovascular disease, and to suggest a possible rôle of high carotid body serotonin content in some forms of systemic hypertension, they were unable to find any disturbance in the relative contents of the biogenic amines in subjects dying in respiratory failure with sustained chronic hypoxaemia.

There is some evidence to suggest that in rats there are two classes of Type 1 cell. The first, called 'small vesicle cells', contain neurosecretory vesicles between 47 and 55 nm (Hellström, 1975) and are believed to store norepinephrine, (Hellström and Koslow, 1975). The second, called 'large vesicle cells', contain vesicles between 63 and 78 nm (Hellström, 1975), and are believed to store dopamine (Hellström and Koslow, 1975, 1976). These two classes of cell can be readily demonstrated on electron microscopy by the use of synthetic precursors of catecholamines such as 5-OH-Dopa which accumulates as 5-OH-Dopamine, and less readily by the physiological precursor L-Dopa (Hellström, 1975).

The biogenic amines may be demonstrated histochemically by the chromaffin, Giemsa ferric-ferricyanide, and diazonium reactions (Lever et al, 1959). They also exhibit formaldehyde-induced fluorescence. Hypoxia brings about a pronounced decrease in this fluorescence in the carotid bodies of 9 to 16 day-old-rats (Hervonen et al, 1972). Arias-Stella and Valcarcel (1976) also report a reduction in the amount of formaldehyde-induced fluorescence in the carotid bodies of highlanders. Such observations imply a discharge of amines from the chemoreceptors on exposure to the hypoxaemia of high altitude. Glucocorticoids prevent the loss of fluorescence (Hervonen et al, 1972), indicating storage of catecholamines in previously existing glomic cells. Confirmation that methylprednisole and hydrocortisone will increase fluorescence in rat carotid bodies has been provided by Korkala and his associates (1973). Hence the discharge of the vesicles from the cell in extreme hypoxia (Blümcke et al, 1967) and the gradual dissolution of their osmiophilic cores during life at high altitude as described earlier in the chapter imply increased secretion of biogenic amines in hypoxic conditions. We may now consider how this excess fits in with theories as to chemoreceptor function in hypoxia.

Dopamine as an inhibitory transmitter

One would anticipate that, under the circumstances just referred to, the carotid body has a local humoral function releasing the bioamines to stimulate adjacent nerve terminals. Lever et al (1959) and Grimley and Glenner (1968) advanced this view, visualising a local build up of neurotransmitter substances around sensory afferent nerves leading to depolarisation of sensory fibres with the formation of impulses in the sinus nerve. However, Eyzaguirre

and Koyano (1965) found no increase in the frequency of electric discharges in the sinus nerve after the application of serotonin, epinephrine or norepinephrine. It was clear that the concept of a straightforward stimulatory rôle for released bioamines from chief cells was suspect. Furthermore Blümcke et al (1967) believed that acetylcholine was in some way a transmitter for physiological stimuli in the process.

In 1975 Osborne and Butler suggested that secretion of dopamine from Type 1 cells acts as an inhibitory transmitter (Fig. 8.14). Under eupoxic conditions the dopamine may hyperpolarise the nerve endings by increasing their potassium conductance, thus reducing the spontaneous discharge frequency. Under conditions of hypoxia, such as occur at high altitude, the release of dopamine from Type 1 cells is thought to be reduced allowing the nerve endings to return to their depolarised state (Fig. 8.14). Osborne and Butler (1975) further believe that this depolarisation may cause release of a neurotransmitter, possibly acetyl choline, from the *efferent synapses* of the nerve terminals. This neurotransmitter acts on the Type 1 cells still further reducing the rate of dopamine secretion. This in turn brings about even greater depolarisation of the nerve terminal instigating an even greater increase in discharge frequency. Hence a hypoxaemia initiates a vicious circle of decreased dopamine secretion, increased depolarisation and increased discharge frequency. The studies of Sampson (1971, 1972) also support the hypothesis that catecholamines are released from Type 1 cells in the carotid body on activation of the efferent pathway and thus inhibit chemoreception.

Carlsson (1975) has suggested that the depletion of catecholamines by reserpine is due to the fact that the amines are prevented by reserpine from gaining access to the vesicles and so accumulate in the cell cytoplasm where they are deaminated. Our own studies of the ultrastructure of the carotid body in states of hypoxia have demonstrated the formation of vacuoles and a reduction in size and density of the core of the dense cored vesicles (Fig. 8.13). It is conceivable that this appearance is due to a breakdown of catecholamines such as dopamine so removing the inhibitory effect suggested by Osborne and Butler (1975). Hess (1976) reported that calcium ions inhibit catecholamine depletion by

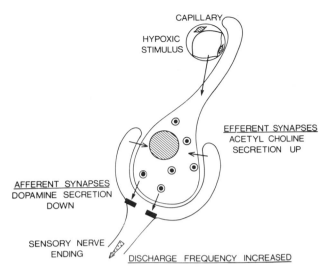

CAPILLARY

HYPOXIC
STIMULUS

EFFERENT SYNAPSES
ACETYL CHOLINE
SECRETION   UP

AFFERENT SYNAPSES
DOPAMINE SECRETION
DOWN

SENSORY NERVE
ENDING

DISCHARGE FREQUENCY INCREASED

Fig. 8.14 Diagram to illustrate the hypothesis of Osborne and Butler (1975) as to the effects of sustained hypoxia, as would be experienced at high altitude, on the chief (Type 1) cells of the carotid body. They believe that the stimulus of sustained hypoxia leads to a diminished secretion of dopamine from the vesicles of the Type 1 cells. This leads to a removal of inhibition of the afferent synapses and thus to increased discharge frequency in the sinus nerve. This provokes secretion of acetyl choline at the efferent synapses which further diminishes secretion of dopamine.

reserpine from chief cells in the carotid bodies. He speculates that calcium might counteract the effects of reserpine by occupying attachment sites of the vesicular membrane.

Blunted ventilatory response to hypoxia

The enlargement of the carotid bodies which occurs at high altitude is not associated with increased ventilatory sensitivity to hypoxia. Indeed it is now firmly established that high altitude natives, both in the Andes and in the Himalayas show a blunted ventilatory response to hypoxia (Chiodi, 1957; Lahiri and Milledge, 1965; Severinghaus et al, 1966; Sørensen and Severinghaus, 1968a; Lefrançois et al, 1968). As we have seen in Chapter 5 sea-level man shows pronounced hyperventilation early on exposure to high altitude. However, although the native highlander hyperventilates compared to man at sea level, he still hypoventilates compared to the newcomer to high altitude. This relative hypoventilation has been found in subjects native to high altitude in the Argentinian Andes and in the Himalayas of Nepal (Chiodi, 1957; Pugh et al, 1964).

Genetic factors are unlikely to be a common denominator in this blunted ventilatory response to hypoxia (Lahiri, 1971). Both the Quechua people of the Andes and the Sherpas of the Himalayas are mongoloid, but genetically unrelated Caucasians living permanently in Kashmir at 3500 m (Ramaswamy, 1962) and in Colorado at 3100 m (Forster et al, 1969) also show a blunted response to hypoxia.

It seems more likely that a blunted ventilatory response to hypoxia is acquired (Chapter 5). Two factors seem to be of importance in the desensitisation to hypoxia, the age of the subject at the time of exposure to the hypoxia and the duration of the exposure. Children who spend only two or three years at high altitude fail to achieve normal sensitivity to hypoxia even after prolonged residence at sea level (Sørensen and Severinghaus, 1968b; Lahiri et al, 1969). On the other hand the offspring of high altitude human natives born at sea level, and who thus have never been exposed to chronic hypoxia, showed the same ventilatory response as subjects who had no altitude ancestry, both with respect to a hypercapnic and hypoxic stimulus. None of these subjects studied by Lahiri (1971) came from stock who were known to have blunted hypoxic sensitivity so that the evidence for ventilatory sensitivity to hypoxia in the offspring of high altitude natives born at sea level is suggestive rather than conclusive.

There is also evidence that attenuation of the ventilatory response occurs during prolonged residence at high altitude (Severinghaus et al, 1966; Sørensen and Severinghaus, 1968c). The reduction in sensitivity to oxygen is often accompanied by a decrease in sensitivity to carbon dioxide (Chiodi, 1957). A characteristic feature of this blunted ventilatory response to the hypoxia of high altitude is its irreversibility. It is not restored even after prolonged residence at sea level.

There appear to be important clinical applications of this diminished ventilatory response at high altitude. Thus a blunted sensitivity also occurs in chronic respiratory disease complicated by hypoxia (Flenley and Millar, 1967; Richards et al, 1968). Such patients also show an insensitivity to carbon dioxide. Blunted ventilatory responses also occur in cyanotic congenital heart disease (Sørensen and Severinghaus, 1968c; Edelman et al, 1970). In these patients there is no decrease in sensitivity to carbon dioxide. There is still controversy as to whether the decreased sensitivity to hypoxia is reversible after surgical correction of the congenital heart disease and improvement of the hypoxia.

All of these findings are consistent with the view that exposure to hypoxia during early life exercises an important effect on the hypoxic control of ventilation in man, a low Pa_{O_2} producing a low hypoxic sensitivity and a high Pa_{O_2} keeping it high.

Lahiri (1971) found that yaks, cattle, sheep and goats born and raised at high altitude in the Himalayas showed hypoxic responses comparable to those of sea-level man and unlike man native to high altitude. Brooks and Tenney (1968) did not find any blunting of hypoxic sensitivity in the llama, and Lefrançois et al (1968) found no blunting of such sensitivity in dogs native to high altitude.

The blunted ventilatory response and the carotid bodies

It is tempting to correlate the enlargement of the carotid bodies that we describe in this chapter with its associated histological and ultrastructural features and the blunted ventilatory response to hypoxia. This enlargement of the carotid bodies occurs in high altitude man, and patients with chronic respiratory disease who are known to show such a diminished ventilatory response. However, this association does not appear to hold in the case of high altitude cattle in which Arias-Stella has demonstrated enlargement and histological changes resembling chemodectoma and yet in which Lahiri (1971) has shown a normal ventilatory response. The same lack of association is found in other animal species at high altitude apparently except in those which are indigenous to mountains such as the yak and llama where the carotid bodies may be of normal size due to evolutionary adaptation. Also it is difficult to reconcile the irreversibility of hypoxic desensitivity reported by Sørensen and Severing-

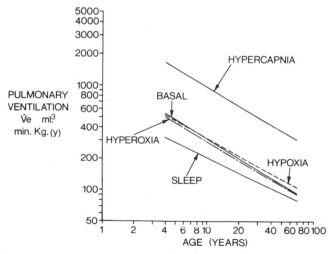

Fig. 8.15   The progression from hyperventilation to hypoventilation with increasing age in native highlanders. Basal and sleeping conditions are shown, together with the effects on the stimuli of hypercapnia, hypoxia and hyperoxia. The diagram is from data collected by Sime (1973) and quoted by Arias-Stella and Valcarcel (1976).

haus (1968a) and Lahiri et al (1969) and the ready reversibility of enlargement of the carotid body in experimental rats in a hypobaric chamber (Heath et al, 1973).

Arias-Stella and Valcarcel (1976) related the augmented carotid body size and weight with increasing age of highlanders to progressive insensitivity of these chemoreceptors. In support of this they refer to the studies of Sime (1973) who was able to demonstrate that there is a progressive fall in pulmonary ventilation with age at high altitude, (Fig. 8.15). He submitted highlanders at 4330 m to steady state exposure to various levels of Pa_{O_2} ranging from hyperoxia to hypoxia, in isocapnic as well as in hypercapnic conditions. Subjects sleeping naturally were also studied. Sime's studies (1973) showed that highlanders progress irreversibly with age from hyperventilation to hypoventilation. This applies equally whether the subjects are tested under basal, acute hypoxic or acute hyperoxic conditions. In comparison to the levels obtained under isocapnia, hypercapnia led to hyperventilation and sleep to hypoventilation.

Effect on ventilatory acclimatization of denervation of carotid bodies

The hypoxic chemoreflex drive of ventilation is reduced by about half in dogs whose carotid bodies have been denervated (Bouverot and Bureau, 1975). These authors studied three conscious dogs in a hypobaric chamber at simulated altitudes of 140 m and at 3550 m before and after chronic bilateral denervation of their carotid bodies. They studied resting ventilation, pulmonary gas exchange, respiratory gases, pH of the arterial blood, acid-base status of the cerebrospinal fluid and ventilatory responses to transient oxygen-inhalation. Bouverot and Bureau (1975) found that at low altitude, denervation of the carotid sinuses and bodies resulted in hypoventilation and respiratory acidosis in the arterial blood and cerebrospinal fluid. At high altitude, initial hypoxic hyperventilation and the related alkalosis in blood and cerebrospinal fluid referred to in Chapter 5 occurred within 30 minutes in intact dogs but was not observed in denervated

dogs. A further increase in ventilation was achieved upon three hours of exposure to simulated high altitude in intact animals while a delayed hyperventilation occurred after 24 hours in denervated dogs. In view of our comments in Chapter 5 on the rôle of the cerebrospinal fluid in the control of ventilation it should be noted that Bouverot and Bureau (1975) found that neither in the intact nor the denervated dogs were the ventilatory changes at simulated altitude related to changes in the pH of cerebrospinal fluid. These authors state that they do not understand the mechanism responsible for the delayed ventilatory response observed in dogs with denervated carotid bodies at high altitude. It may be concluded that the rate of ventilatory acclimatization to altitude is dependent upon the strength of the arterial chemoreceptor drive from the carotid bodies. Integrity of this chemoreflex is essential in determining the eupneic level of ventilation and normal acid-base status of the blood and cerebrospinal fluid at both low and high altitude.

The carotid bodies and systemic hypertension

Earlier in this chapter we have already referred to the fact that Steele and Hinterberger (1972) suggested a possible rôle of high carotid body serotonin content in some forms of systemic hypertension. This suggestion is supported by reports of an association between chemodectomas and systemic hypertension in highlanders and lowlanders in Peru by Saldaña et al (1973). In a case of hyperplasia of glomic tissue around pulmonary veins that we reported in a woman of 63 years, death was due to pontine haemorrhage secondary to systemic hypertension (Edwards and Heath, 1972). Interestingly enough, our studies of the carotid bodies in cases of chronic bronchitis have revealed a statistically significant relation between the weight of the carotid bodies and the weight of the left ventricle (Edwards et al, 1971b). A relation between chemoreception and baroreception is also suggested by the presence in the glomic arteries of elastic tissue reminiscent of the structure of the carotid sinus (Edwards and Heath, 1970; Heath and Edwards, 1971).

REFERENCES

Albores-Saavedra, J. & Durán, M. E. (1968) Association of thyroid carcinoma and chemodectoma. *American Journal of Surgery*, **116**, 887.

Arias-Stella, J. (1969) Human carotid body at high altitudes. Item 150 in the sixty-ninth programme and abstracts of the American Association of Pathologists and Bacteriologists, San Francisco, California.

Arias-Stella, J. & Bustos, F. (1976) Chronic hypoxia and chemodectomas in bovines at high altitudes. *Archives of Pathology*, **100**, 636.

Arias-Stella, J. & Valcarcel, J. (1973) The human carotid body at high altitudes. *Pathologia et Microbiologia*, **39**, 292.

Arias-Stella, J. & Valcarcel, J. (1976) Chief cell hyperplasia in the human carotid body at high altitudes. Physiologic and pathologic significance. *Human Pathology*, **7**, 361.

Blessing, M. H. & Wolff, H. (1973) The carotid bodies at simulated high altitude. *Pathologia et Microbiologia*, **39**, 310.

Blümcke, S., Rode, J. & Niedorf, H. R. (1967) The carotid body after oxygen deficiency. *Zeitschrift für Zellforschung und Mikroskopische Anatomie*, **80**, 52.

Bouverot, P. & Bureau, M. (1975) Ventilatory acclimatization and CSF acid-base balance in carotid chemodenervated dogs at 3550 m. *Pflügers Archiv. (European Journal of Physiology)*, **361**, 17.

Brooks, J. G. III & Tenney, S. M. (1968) Ventilatory response of llama to hypoxia at sea level, and high altitude. *Respiration Physiology*, **5**, 269.

Carlsson, A. (1975) Monoamine-depleting drugs. *Pharmacology and Therapeutics; Part B: General and Systematic Pharmacology*, **1**, 393.

Chedid, A. & Jao, W. (1974) Hereditary tumors of the carotid bodies and chronic obstructive pulmonary disease. *Cancer*, **33**, 1635.

Chiodi, H. (1957) Respiratory adaptations to chronic high altitude hypoxia. *Journal of Applied Physiology*, **10**, 81.

De Castro, F. (1928) Sur la structure et l'innervation du sinus carotidien de l'homme et des mammifères. Nouveaux faits due l'innervation et al function du glomus caroticum. Etudes anatomiques et physiologiques. *Trabajos del Laboratorio de Investigaciones Biologicas de la Universidad de Madrid*, **25**, 331.

Dearnaley, D. P., Fillenz, M. & Woods, R. I. (1968) The identification of dopamine in the rabbit's carotid body. *Proceedings of the Royal Society*, **170**, 195.

Edelman, N. H., Lahiri, S., Braudo, I., Cherniack, N. S. & Fishman, A. P. (1970) The blunted ventilatory response to hypoxia in cyanotic congenital heart disease. *New England Journal of Medicine*, **282**, 405.

Edwards, C. & Heath, D. (1970) Site and blood supply of the intertruncal glomera. *Cardiovascular Research*, **4**, 502.

Edwards, C. & Heath, D. (1972) Pulmonary venous chemoreceptor tissue. *British Journal of Diseases of the Chest*, **66**, 96.

Edwards, C., Heath, D., Harris, P., Castillo, Y., Krüger, H. & Arias-Stella, J. (1971a) The carotid body in animals at high altitude. *Journal of Pathology*, **104**, 231.

Edwards, C., Heath, D. & Harris, P. (1971b) The carotid body in emphysema and left ventricular hypertrophy. *Journal of Pathology*, **104**, 1.

Edwards, C., Heath, D. & Harris, P. (1972) Ultrastructure of the carotid body in high-altitude guinea-pigs. *Journal of Pathology*, **107**, 131.

Eyzaguirre, C. & Koyano, H. (1965) Effects of some pharmacological agents on chemoreceptor discharge. *Journal of Physiology, London* **178**, 410.

Flenley, D. C. & Millar, J. S. (1967) Ventilatory response to oxygen and carbon dioxide in chronic respiratory failure. *Clinical Science*, **33**, 319.

Forster, H. V., Dempsey, J. A., Birnbaum, M. L., Reddan, W. G., Thoden, J. S., Grover, R. F. & Rankin, J. (1969) Comparison of ventilatory responses to hypoxic and hypercapnic stimuli in altitude-sojourning lowlanders, lowlanders residing at altitude and native altitude residents. *Federation Proceedings, Federation of American Society for Experimental Biology*, **28**, 1274.

Grimley, P. M. & Glenner, G. C. (1968) Ultrastructure of the human carotid body. A perspective on the mode of chemoreception. *Circulation*, **37**, 648.

Heath, D. & Edwards, C. (1971) The glomic arteries. *Cardiovascular Research*, **5**, 303.

Heath, D., Edwards, C. & Harris, P. (1970) Post mortem size and structure of the human carotid body. *Thorax*, **25**, 129.

Heath, D., Edwards, C., Winson, M. & Smith, P. (1973) Effects on the right ventricle, pulmonary vasculature, and carotid bodies of the rat on exposure to, and recovery from, simulated high altitude. *Thorax*, **28**, 24.

Hellström, S. (1975) Type 1 cells of carotid body from rats treated with 5-OH-Dopa and L-Dopa: an electron microscopical study. *Journal of Neurocytology*, **4**, 439.

Hellström, S. & Koslow, S. H. (1975) Biogenic amines in carotid body of adult and infant rats—a gas chromatographic—Mass spectrometric assay. *Acta Physiologica Scandinavica*, **93**, 540.

Hellström, S. & Koslow, S. H. (1976) Effects of glucocorticoid treatment on catecholamine content and ultrastructure of adult rat carotid body. *Brain Research*, **102**, 245.

Hervonen, A., Kanerva, L., Korkala, O. & Partenen, S. (1972) Effects of hypoxia and glucocorticoids on the histochemically demonstrable catecholamines of the newborn rat carotid body. *Acta Physiologica Scandinavica*, **86**, 109.

Hess, A. (1976) Calcium inhibits catecholamine depletion by reserpine from carotid body glomus cells. *Brain Research Bulletin*, **1**, 359.

Heymans, C., Bouckaert, J. J. & Dautrebande, L. (1930) Sinus carotidien et réflexes respiratoires: influences respiratoires réflexes de l'acidose, de l'alcalose, de l'anhydride carbonique, de l'ion hydrogène et de l'anoxemie. Sinus carotidiens et échanges respiratoires dans les poumons et au dela des poumons. *Archives Internationales de Pharmacodynamie et de Therapie*, **39**, 400.

Korkala, O., Eränkö, O., Partanen, S., Eränkö, L. & Hervonen, A. (1973) Histochemically demonstrable increase in the catecholamine content of the carotid body in adult rats treated with methylprednisolone or hydrocortisone. *Histochemical Journal*, **5**, 479.

Lahiri, S. & Milledge, J. S. (1965) Sherpa Physiology, *Nature, London*, **207**, 610.

Lahiri, S., Kao, F. F., Velasquez, T., Martinez, C. & Pezzia, W. (1969) Irreversible blunted respiratory sensitivity to hypoxia in high altitude natives. *Respiration Physiology*, **6**, 360.

Lahiri, S. (1971) Genetic aspects of the blunted chemoreflex ventilatory response to hypoxia in high altitude adaptation. In: *High Altitude Physiology: Cardiac and Respiratory Aspects*, p. 103. Ciba Foundation Symposium. Edited by R. Porter & J. Knight. Edinburgh: Churchill Livingstone.

Laidler, P. & Kay, J. M. (1975a) A quantitative morphological study of the carotid bodies of rats living at a simulated altitude of 4300 metres. *Journal of Pathology*, **117**, 183.

Laidler, P. & Kay, J. M. (1975b) The effect of chronic hypoxia on the number and nuclear diameter of type 1 cells in the carotid bodies of rats. *American Journal of Pathology*, **79**, 311.

Lefrançois, R., Gautier, H. & Pasquis, P. (1968) Ventilatory oxygen drive in acute and chronic hypoxia. *Respiration Physiology*, **4**, 217.

Lever, J. D. & Boyd, J. D. (1957) Osmiophile granule in the glomus cells of the rabbit carotid bodies. *Nature, London*, **179**, 1082.

Lever, J. D., Lewis, P. R. & Boyd, J. D. (1959) Observations on the fine structure and histochemistry of the carotid body in the cat and rabbit. *Journal of Anatomy*, **93**, 478.

Monge, C. (1966) Demografia y Altitud en el Peru. In: *Poblacion y Altitud*. Edited by L. Sobrevilla. Lima, Peru: Imprenta Sesator.

Osborne, M. P. & Butler, P. J. (1975) New theory for receptor mechanisms of carotid body chemoreceptors. *Nature*, **254**, 701.

Pugh, L. G. C. E., Gill, M. B., Lahiri, S., Milledge, J. S., Ward, M. P. & West, J. B. (1964) Muscular exercise at great altitudes. *Journal of Applied Physiology*, **19**, 431.

Ramaswamy, S. S. (1962) In: *International Symposium on Problems of High Altitude*. p. 74. Edited by S. P. Bhatia New Delhi: Indian Armed Forces Medical Services.

Richards, D. W., Fritts, H. W. Jr. & Davis, A. L. (1968) Observations on the control of respiration in emphysema: the effects of oxygen on ventilatory response to CO_2 inhalation. *Transactions of the Association of American Physicians*, **71**, 142.

Saldaña, M. J. & Salem, L. E. (1970) High altitude hypoxia and chemodectomas. *American Journal of Pathology*, **59**, 91a.

Saldaña, M. J., Salem, L. E. & Travezan, R. (1973) High altitude hypoxia and chemodectomas. *Human Pathology*, **4**, 251.

Sampson, S. R. (1971) Catecholamines as mediators of efferent inhibition of carotid body chemoreceptors in the cat. *Federation Proceedings*, **30**, 551.

Sampson, S. R. (1972) Mechanism of efferent inhibition of carotid body chemoreceptors in the cat. *Brain Research*, **45**, 266.

Severinghaus, J. W., Bainton, C. R. & Carcelen, A. (1966) Respiratory insensitivity to hypoxia in chronically hypoxic man. *Respiration Physiology*, **1**, 308.

Sime, F. (1973) Ventilacion humana en hipoxia cronica. Etiopatogenia de la Enfermedad de Monge a desadaptacion cronica a la altura. Doctoral Thesis. Universidad Peruana Cayetano Heredia, Lima, Peru, 1973.

Sørensen, S. C. & Severinghaus, J. W. (1968a) Irreversible respiratory insensitivity to acute hypoxia in man born at high altitude. *Journal of Applied Physiology*, **25**, 217.

Sørensen, S. C. & Severinghaus, J. W. (1968b) Respiratory sensitivity to acute hypoxia in man born at sea level living at high altitude. *Journal of Applied Physiology*, **25**, 211.

Sørensen, S. C. & Severinghaus, J. W. (1968c) Respiratory insensitivity to acute hypoxia persisting after correction of tetralogy of Fallot. *Journal of Applied Physiology*, **25**, 221.

Steele, R. H. & Hinterberger, J. (1972) Catecholamines and 5-hydroxytryptamine in the carotid body in vascular, respiratory, and other diseases. *Journal of Laboratory and Clinical Medicine*, **80**, 63.

Abnormal haemoglobins, and leukocytes

Sickle cell anaemia at high altitude

There is a quaint Peruvian saying to the effect that 'Blackbirds do not sing well in the Andes'. This illustrates an awareness of the fact that, when Negroes are exposed to high altitude, many of them are taken ill. Those affected are subjects with the sickle cell trait who are commonly asymptomatic at sea level. Not infrequently this illness takes the form of acute abdominal pain. It represents the classical abdominal crisis of sickle cell disease which frequently has a basis in splenic infarction brought about by blockage of branches of the splenic artery by masses of erythrocytes altered by their exposure to deoxygenation.

Sickle cell anaemia is a hereditary, haemolytic anaemia characterised by an abnormal haemoglobin HbS. It is thus one of the group of diseases now generally termed 'haemoglobinopathies'. Ingram (1956) showed that in haemoglobin S the beta polypeptide differed from the normal adult haemoglobin A by the substitution of only one amino acid among the 287. In one of the repeating eight amino acid groups valine replaced the normally occurring glutamic acid in the sixth position. This substitution alters the configuration of the terminal residues of the beta chain so that on deoxygenation it interacts at a complementary site on an adjacent alpha chain. This interaction results in molecular stacking with the production of tactoids and a rigid deformity of erythrocytes. Such stacking of red cells leads to occlusion of arteries, one of the most frequent sites being the splenic artery and its branches, leading to repeated infarction with fibrous scarring and contraction of the spleen. This presents clinically as the abdominal crisis referred to above. The condition occurs at low altitude and indeed we have previously reported such a case at sea level in which the outstanding feature was widespread pulmonary thrombosis leading to right ventricular hypertrophy and the development of bronchopulmonary anastomoses (Heath and McKim Thompson, 1969). However, such abdominal crises are particularly likely to occur in patients exposed to chronic hypoxia. The tendency to sickling and the severity of the deformity are dependent both upon the amount of haemoglobin S in the erythrocyte and the level of oxygen tension.

The genetic transmission of the condition involves an allele controlling the production of haemoglobin. The genetic defect is virtually limited to Negroes and hence it is this section of the Peruvian population which is at risk. In sickle cell anaemia 80 to 100 per cent of the haemoglobin is in the S form, the remaining being HbA (Robbins, 1967). Patients with the sickle cell trait have only 20 to 40 per cent HbS. Such heterozygous patients are likely to be asymptomatic at sea level but susceptible to abdominal crises at high altitude.

Red cells with 100 per cent HbS will sickle at normal oxygen tensions but as the level of HbS falls there needs to be progressively lower oxygen tension to induce sickling. In Peru many Negroes live on the coast and they are able to reach altitudes exceeding 4000 m in a few hours by car or train. Such hypoxic conditions are ideal for the development of deoxygenation and the abdominal crises of sickle cell anaemia (Monge and Monge, 1966). These authors refer to the clinical observations of Aste-Salazar who reported that the subjects are usually in good health when setting out for high altitude destinations. After two hours' travel by train or car and at altitudes between 3500 m and

4500 m they often develop acute abdominal pain. When they are brought to sea level and investigated, they are found to have haemoglobins of both A and S type, indicative of the sickle cell trait. This disease may be partly responsible for the surprisingly small Negro population inhabiting the mountains of Peru, and this despite the accessibility of the mountains and the long established settlement of Negroes on the coast.

Other clinical features which may develop are anaemia, jaundice, and a wide variety of symptoms and signs depending upon which organ is involved by thrombosis of its nutrient artery. Sometimes the symptoms are bizarre. Pathologically the most characteristic finding is progessive splenic infarction with fibrous scarring and contraction of the organ. Pulmonary thrombosis may lead to cor pulmonale and the formation of bronchopulmonary anastomoses (Heath and McKim Thompson, 1969). There is a hyperplasia of the bone marrow and there may be fatty changes in the heart and liver. It is clear from these considerations that the sickle cell trait is not harmless and those with the condition should be cautioned not to undertake trips to high altitude without an adequate supply of oxygen.

Splenic infarction during air travel

Splenic infarction may occur during flight in unpressurized aircraft in subjects with the sickle cell trait. Cooley et al (1954) reported the case of six Negro soldiers who developed nausea, vomiting, fever and pain in the left upper abdomen during prolonged flights in unpressurized aircraft at altitudes between 3050 m and 4570 m. All were in good health with no previous manifestations of sickle cell disease. Subsequently other cases of splenic infarction were observed in military hospitals and this led Smith and Conley (1955) to obtain, for electrophoretic study, blood from 15 persons who developed splenic infarction during flight. HbS was found to be present in the blood of each of them. Of great interest was the fact that in three cases HbC was also present, indicating that they were cases of sickle cell/HbC disease. The remaining 12 had normal HbA in addition to the S component. In 11 of these subjects with splenic infarction the amount of normal haemoglobin exceeded the amount of sickle haemoglobin which is

the pattern expected in the sickle cell trait. None of the 15 persons was suspected of having sickling prior to the onset of splenic infarction. The flights leading to the infarction were usually taken in unpressurized aircraft between 3050 m and 4570 m, although in two instances splenic infarction occurred on commercial flights as low as between 1220 m and 1830 m. Both of these subjects were found to have sickle cell/HbC disease. Since eight per cent of American Negroes have the sickle cell trait it would seem that the development of splenic infarction during flight is not rare. Sickle cell/HbC disease is, however, extremely uncommon requiring the inheritance of both a gene for sickling and one for HbC. It has now emerged that about a quarter of persons who develop splenic infarcts at high altitude have sickle cell/HbC disease. Air travel presents a considerable hazard to persons with this combination of abnormal haemoglobins. Sickle cell/HbC should disqualify an individual from flying. Routine tests to detect sickling are often unreliable and electrophoresis of the haemoglobin should be used to detect the susceptible. O'Brien et al (1972) described splenic infarction, documented by radioisotopic scan, in a 26-year-old Caucasian man of Sicilian descent while he was hiking at an altitude of only 760 m. Electrophoresis demonstrated a haemoglobin A/S pattern with 42.4 per cent HbS and 0.5 per cent HbF.

Occasionally splenic scans do not confirm the presence of splenic infarction but suggest that symptoms are due to 'splenic sequestration' of sickled cells (Githens et al, 1977). These authors reported five cases of the splenic sequestration syndrome in children with sickle/HbC disease in association with a change in altitude. In four of them it occurred during or immediately following a trip to mountain altitudes greater than 2740 m. In the fifth child the crisis occurred ten days after travel in a pressurized plane from sea level to Denver (1600 m).

Splenic necrosis and pseudocyst formation at high altitude

Rywlin and Benson (1961) reported massive necrosis of the spleen resulting in a pseudocyst in a white Peruvian youth aged 19 years with the sickle cell trait, after a car trip to the Andes at an altitude of

4570 m. A sickle cell preparation was positive. Haemoglobin electrophoresis revealed 61.35 per cent HbA, 38.40 per cent HbS and 0.25 per cent HbF. The patient inherited the HbS from his mother who was an Argentinian and did not manifest any negroid features. The father was born in Belgium and had HbA only.

Methaemoglobinaemia at high altitude

In methaemoglobinaemia the ferrous iron of normal haemoglobin is converted into the ferric form and as such it cannot combine with oxygen. Hypoxaemia ensues and as a result secondary polycythaemia may develop. At sea level the condition is usually produced by drugs like nitrites, sulphonamides, phenacetin, acetanilide, aniline and nitrobenzene (Walters and Israel, 1974). The condition may be reversed by reducing agents such as ascorbic acid and methylene blue.

Methaemoglobinaemia may also be due to congenital defects in the red cells. Thus one form of haemoglobinopathy is due to HbM which is oxidised abnormally easily. In this type of methaemoglobinaemia restorative agents are powerless. Clinically there is a slate-blue cyanosis and spectroscopic examination is necessary to identify methaemoglobin and distinguish it from reduced haemoglobin.

Ferric iron combines so firmly with one atom of oxygen that the gas is not liberated even on exposure to a vacuum. Hence methaemoglobin has no oxygen-carrying capacity and during recent years it has become apparent that one of the functions of red cell metabolism is to provide reducing potential in order to protect the cell against such oxidation.

In view of these considerations it is surprising that methaemoglobin has been reported as being increased at high altitude both in highlanders and indigenous animals such as the llama. Methaemoglobin has been found in the erythrocytes of subjects living permanently above 3500 m. Its presence does not appear to be due to ingested chemicals or drugs. The existence of haemoglobin M has been eliminated as a cause. There is no evidence of NADH-linked methaemoglobin reductase deficiency. Hence in spite of normal red cell metabolism, methaemoglobin levels are abnormally high in low oxygen tension. The methaemo-

globinaemia disappears when the subjects descend to low altitude.

The level of methaemoglobin appears to be inversely related to the red cell count so that it is much increased in relatively anaemic subjects. Gourdin et al (1975) studied two groups of highlanders. The first consisted of 208 adult Quechuas living at 3500 m in central Peru. The mean methaemoglobin percentage was 5.3 in those with a haemoglobin level between 14 and 21 g/dl but was 10.9 in anaemic subjects with a haemoglobin level below 14 g/dl (Fig. 9.1). The second group

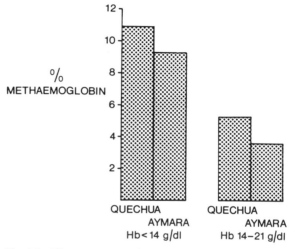

Fig. 9.1 The percentages of methaemoglobin in the blood of Quechuas and Aymaras at 3500 m with haemoglobin levels respectively, less than 14 g/dl, and between 14 and 21 g/dl. (Based on data from Gourdin et al, 1975).

comprised 71 Aymaras and in them the mean methaemoglobin percentage was 2.1 in those with a haemoglobin level exceeding 21 g/dl, 3.7 in those with haemoglobin concentrations between 14 and 21 g/dl, and 9.3 in anaemic subjects with a haemoglobin level beneath 14 g/dl (Fig. 9.1). Methaemoglobin levels of 15 to 20 per cent have been found in the llama which as we have seen (Chapter 6) has a low haematocrit. At sea level in some areas where there are such chemicals as nitrites in the soil, the level of methaemoglobin may rise to 1.5 per cent.

The presence of methaemoglobin in man and animals at high altitude is unexpected and its inverse proportion to the haemoglobin level even more so (Gourdin et al, 1975). As we have noted above, methaemoglobin has no oxygen-carrying capacity and its presence would appear to impose an additional disadvantage to highlanders. Gourdin and his colleagues speculate that at very high altitudes the not altogether advantageous shift of the oxygen–haemoglobin dissociation curve to the right may be too great so that there is a decrease in oxygen uptake (Chapters 6 and 27). They speculate that small amounts of methaemoglobin promote a shift of the curve to the left thus providing some regulation. The llama has much methaemoglobin and its oxygen dissociation curve is well to the left of that of man. Smith and Ou (1975) on the contrary believe that the phenomenon represents an adverse effect of altitude and is secondary to an increased rate of haemoglobin autoxidation. It is known that this rate increases with decreasing oxygen tensions until it reaches a maximum at oxygen tensions corresponding to haemoglobin half-saturation. Hence at high altitude in the absence of a compensatory increase in methHb reductase activity, methaemoglobin would accumulate. Conceivably this effect might be more pronounced in anaemic than polycythaemic subjects because of the lower venous oxygen tension in the former.

Haemoglobin F

According to Reynafarje et al (1975), alkali denaturation of blood reveals a persistence of fetal

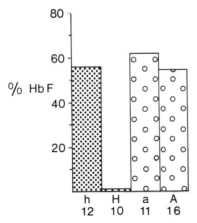

Fig. 9.2 The percentage of haemoglobin F in human newborns (at one week of age) (h), adult human beings (H), newborn alpacas (at one week of age) (a), and adult alpacas (A). The number at the foot of each column represents the number of subjects studied in that category. (From data of Reynafarje et al, 1975).

type haemoglobin, HbF, in 55 per cent of adult alpacas in comparison with 0.7 per cent of adult native highlanders (Fig. 9.2). These authors relate this persistence of haemoglobin F with the increased affinity of alpaca blood for oxygen with the consequent shift of the oxygen–haemoglobin dissociation curve to the left, the significance of which we discuss in Chapter 27.

Carboxyhaemoglobin. Smoking at high altitude

On a recent visit to Bolivia we were interested to hear anecdotal evidence to the fact that amongst the Indians of the Altiplano around La Paz (3800 m) it is the subjects who smoke cigarettes who develop the highest levels of haemoglobin. It has certainly been established at sea level that pregnant women who smoke have higher haemoglobin levels than those who do not (Davies et al, 1979). Furthermore, the smokers have higher levels of carboxyhaemoglobin and a greater affinity of haemoglobin for oxygen with a significantly lower P_{50} and hence a lower availability of oxygen for the tissues per gram of haemoglobin (Davies et al, 1979). This lowered oxygen availability is, however, compensated for by the higher level of haemoglobin. In pregnant women who stopped smoking there was an immediate reduction in CO Hb and a decrease in HbO_2 affinity leading to a significant increase of eight per cent of oxygen available for the tissues within 48 hours (Davies et al, 1979). Applying these results to life at high altitude it is likely that the habitual smoker at such elevations would in the same way have a haemoglobin level even higher than his fellow native highlanders but that this increased amount of haemoglobin would not result in a greater availability of oxygen to the tissues in view of the presence of carboxyhaemoglobin and the resultant shift to the left of the oxygen–haemoglobin dissociation curve (Chapter 6).

MacLean (1979) felt that these studies in pregnant women could be applied to the problem of smoking and acclimatization to high altitude. As a smoker he had observed that during skiing and mountaineering he suffered less from the symptoms of acute mountain sickness than his non-smoking companions. He noted that he and several smoking mountaineers either smoked less or stopped smoking completely during these expeditions and all had suffered significantly less from altitude sickness than their non-smoking companions. Such anecdotal evidence is certainly in keeping with a fall in carboxyhaemoglobin levels, a shift of the oxygen–haemoglobin dissociation curve to the right with an elevation of P_{50} and an increase in the availability of oxygen to the tissues.

Leukocytes

The native highlander of the Peruvian Andes has a leukocyte count and a differential count considered normal at sea level, according to Hurtado (1964a), (Table 9.1). Under conditions of complete physical

lymphopenia and eosinophilia are said to develop (Verzar, 1952).

Hurtado (1964b) has studied the effect of exercise on leukocyte count at high altitude. It has been known for a long time that a pronounced leukocytosis of segmented neutrophils and lymphocytes occurs regularly with strenuous exercise. Counts as high as $22.0 \times 10^9/l$ have been recorded for a runner after making a 100 yard dash in 11 seconds and $35.0 \times 10^9/l$ on completing a quarter-mile run in less than a minute (Garrey and Bryan, 1935). Leukocyte counts in excess of $20 \times 10^9/l$ are regularly recorded for marathon runners covering 26 miles in up to three hours and these

Table 9.1 Leukocyte and differential count at sea level and at high altitude. (After Hurtado, 1964a)

	Sea level (Lima) 140 subjects		4540 m (Morococha) 72 subjects	
	Mean	SD	Mean	SD
Leukocytes ($\times 10^9/l$)	6.68	1.21	7.04	1.62
Neutrophils, stab, %	4.0	2.6	4.1	2.4
Neutrophils, segmented, %	52.7	7.5	51.3	10.6
Neutrophils, total, %	56.5	8.1	55.4	10.7
Eosinophils, %	4.1	3.0	3.3	2.4
Basophils, %	0.4		0.3	
Monocytes, %	6.4	2.6	5.7	2.6
Lymphocytes, %	32.4	10.4	35.8	10.4

and mental relaxation the leukocyte count lies between 5.0 to $7.0 \times 10^9/l$ (Wintrobe et al, 1974), so that the normal sea-level counts given by Hurtado are somewhat high (Table 9.1). Nevertheless, the counts given by him for subjects at 4540 m were presumably arrived at by the same techniques so that his conclusion that there is no difference in leukocyte count at high altitude and sea level appears to be valid.

Leukocyte concentration fluctuates during the day, from day to day, and in response to a wide range of physiological stimuli, some of which are increased at high altitude. Thus heat and solar radiation, whose influence at high altitude we describe in Chapter 22, is said to cause a leukocytosis (Kennedy and MacKay, 1936). Ultraviolet radiation (Chapter 2) is said to cause a lymphocytosis (Wintrobe et al, 1974). Acute hypoxia is said to cause a leukocytosis (Cress et al, 1943). It has also been reported that in the first few days after an individual has arrived at high altitude, some leukocytosis accompanied by

values may take several hours to return to normal (Wintrobe et al, 1974). Thus leukocytosis which occurs on exercise is dependent more on the intensity of activity rather than its duration (Farris, 1943). Hurtado (1964b) studied this phenomenon of exercise leukocytosis in 12 athletes and 11 non-athletes at sea level and in 12 native highlanders at high altitude (Fig. 9.3). He found that the leukocytosis exhibited by highlanders on exercise to exhaustion ($2.45 \times 10^9/l$) was only half of that shown at sea level by athletes ($5.23 \times 10^9/l$) or non-athletes ($5.09 \times 10^9/l$). Finally we may note that nausea and vomiting, so common in acute mountain sickness, may be associated with a leukocytosis (Wintrobe et al, 1974).

Leukaemia at high altitude

It has been the impression of some authorities on high altitude studies that leukaemia is rare in man in a high environment (Hurtado, 1966). This

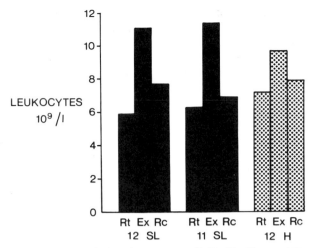

Fig. 9.3 Leukocyte counts at rest (Rt), on exercise (Ex), and on recovery (Rc) in 12 athletes and 11 non-athletes at sea level and in 12 highlanders native to 4520 m. (Based on data from Hurtado, 1964b).

impression has recently been subjected to statistical investigation by Eckhoff et al (1974), who correlated leukaemia mortality rates with altitude in the United States. Radiation has long been associated with an increase in the risk of contracting leukaemia. However, this increased risk is well established only for high doses of radiation. No positive experimental evidence has been obtained for low doses and low dose rates. According to Eckhoff et al (1974) the International Commission on Radiologicial Protection, in evaluating the consequences of radiation exposure, estimates that total leukaemia risk per rad of dose is 20 cases per million persons (ICRP, 1966). Background radiation dose from all sources at sea level has been estimated to be about 0.1 rad per year with cosmic radiation contributing about one third of this value (Eisenbud, 1963). At an altitude of 1520 m the background radiation dose from all sources is some 0.2 rad per year, with cosmic radiation accounting for one quarter of this value. At 3050 m the cosmic radiation is roughly double this level. At such levels of increased cosmic ray dose one would expect 0.14 more deaths from leukaemia per 10^5 people per year at 3050 m than at sea level (Eckhoff et al, 1974). These authors correlated a topographical distribution of the population of the

U.S.A. with the mortality rate from leukaemia. In over five thousand geographical areas they expressed the mortality rate for leukaemia per 10^5 persons. The leukaemia mortality rates were summed for 60 m increments for altitudes up to 1980 m and for 120 m increments between 1980 m and 2260 m. Above the latter altitude the data were expressed for a mean elevation of 2710 m. This statistical study suggested that the mortality rate increases with an altitude up to 610 m and at a much higher rate than would be expected to result from the general increase in background radiation. At altitudes above 610 m this trend seems to be reversed so that the mortality rate seems to decrease with increasing altitude. Such data illustrate that, if background radiation is a contributory factor to mortality from leukaemia, it is certainly much smaller than other leukemogenic agents. Surprisingly, it seems that there are factors involved in the aetiology of leukaemia which actually *decrease* with increasing altitude and increasing dose of cosmic background radiation. This concept, derived from statistical studies, agrees with the generally accepted view expressed by Hurtado (1966) that leukaemia is rare at high altitude.

REFERENCES

Cooley, J. C., Peterson, W. L., Engel, C. E. & Jernigan, J.P. (1954) Clinical triad of massive splenic infarction, sicklemia trait, and high altitude flying. *Journal of the American Medical Association*, **154**, 111.

Cress, C. H., Clare F. B. & Gellhorn, E. (1943) The effect of anoxic and anemic anoxia on the leukocyte count. *American Journal of Physiology*, **140**, 299.

Davies, J. M., Latto, I. P., Jones, J. G., Veale, A. & Wardrop, C. A. J. (1979) Effects of stopping smoking for 48 hours on oxygen availability from the blood: a study on pregnant women. *British Medical Journal*, ii, 355.

Eckhoff, N. D., Shultis, J. K., Clack, R. W. & Ramer, E. R. (1974) Correlation of leukaemia mortality rates with altitude in the United States. *Health Physics*, 27, 377.

Eisenbud, M. (1963) *Environmental Radioactivity*. New York: McGraw-Hill.

Farris, E. J. (1943) The blood picture of athletes as affected by intercollegiate sports. *American Journal of Anatomy*, 72, 223.

Garrey, W. E. & Bryan, W. R. (1935) Variations in white cell blood cell counts. *Physiology Reviews*, 15, 597.

Githens, J. H., Gross, G. P., Eife, R. F. & Wallner, S. F. (1977) Splenic sequestration syndrome at mountain altitudes in sickle/haemoglobin C disease. *Journal of Pediatrics*, 90, 203.

Gourdin, D., Vergnes, H. & Gutierez, N. (1975) Methaemoglobin in man living at high altitude. *British Journal of Haematology*, 29, 243.

Heath, D. & McKim Thompson, I. (1969) Bronchopulmonary anastomoses in sickle-cell anaemia. *Thorax*, 24, 232.

Hurtado, A. (1964a) Animals in high altitudes: resident man. In: *Handbook of Physiology*, p. 843. (Section 4: 'Adaptation to the Environment') Washington: American Physiological Society.

Hurtado, A. (1964b) Some physiologic and clinical aspects of life at high altitudes. In: *Aging of the Lung*, p. 257. Edited by L. Cander & J. H. Moyer. New York: Grune and Stratton.

Hurtado, A. (1966) Needs for further research. In: *Life at High Altitudes*, p. 80. Washington: Pan American Health Organization Scientific Publication No. 140.

Ingram, V. M. (1956) A specific chemical difference between the globins of normal and human sickle anaemia haemoglobin. *Nature*, 178, 792.

Internation Commission of Radiological Protection (1966) The evaluation of risks from radiation. *Health Physics*, 12, 239.

Kennedy, W. P. & MacKay, L. (1936) The normal leucocyte picture in a hot climate. *Journal of Physiology*, 87, 336.

Maclean, N. (1979) Smoking and acclimatisation to altitude. *British Medical Journal*, ii, 799.

Monge, M. C. & Monge, C. C. (1966) *High-Altitude Diseases. Mechanism and Management*. Springfield, Illinois: Charles C. Thomas.

O'Brien, R. T., Pearson, H. A., Godley, J. A. Spencer, R. P. (1972) Splenic infarct and sickle-cell trait. *New England Journal of Medicine*, 287, 720.

Reynafarje, C., Faura, J., Villavicencio, D., Curaca, A., Reynafarje, B., Oyola, L., Contreras, L., Vallenas, E. & Faura A. (1975) Oxygen transport of hemoglobin in high-altitude Camelidae. *Journal of Applied Physiology*, 38, 806.

Robbins, S. L. (1967) In: *Pathology*, p. 626. Philadelphia, London: W. B. Saunders.

Rywlin, A. M. & Benson, J. (1961) Massive necrosis of the spleen, with formation of a pseudocyst. *American Journal of Clinical Pathology*, 36, 142.

Smith, E. W. & Conley, C. L. (1955) Sicklemia and infarction of the spleen during aerial flight. Electrophoresis of the hemoglobin in 15 cases. *Bulletin of the Johns Hopkins Hospital*, 96, 35.

Smith, R. P. & Ou, L. C. (1975) Methaemoglobinaemia at high altitude. *British Journal of Haematology*, 31, 411.

Verzar, F. (1952) Die Zahl der Lymphocyten und eosinophilen Leukocyten in 1800 und 3450 m Höhe. *Schweizerische Medizinische Wochenschrift*, 82, 324.

Walters, J. B. & Israel, M. S. (1974) *General Pathology*, 4th Edition, Edinburgh: Churchill Livingstone.

Wintrobe, M. M., Lee, G. R., Boggs, D. R., Bithell, T. C., Athens, J. W. & Foerster, J. (1974) *Clinical Hematology*, 7th Edition. Philadelphia: Lea and Febiger.

10

Platelets and disorders of blood coagulation

Thrombocytes at high altitude

The blood platelet count falls on ascent to high altitude. Fourteen men studied by Gray et al (1975) showed a fall of seven per cent in the count after two days at 2990 m. After a further two days at 5370 m there was a reduction of 25 per cent from control values, but after another eight days the platelet count rose again to a point only seven per cent below the baseline at sea level (Fig. 10.1). This drop in the

platelet count was not striking, dropping only from 643 to 594 10^9/l. However, a much more severe and persistent fall in platelet count was produced in mice exposed to the same reduction in barometric pressure (Birks et al, 1975). After this degree of decompression for two days there was an initial small rise in the platelet count, probably due to haemoconcentration (Fig. 10.2). This was followed

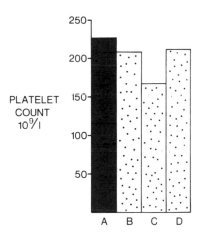

Fig. 10.1 The platelet count at various natural altitudes in 14 young men. (A) Control values at 790 m; (B) After airlift to 2990 m for five days; (C) In eight subjects after airlift to 5370 m for two days; (D) After ten days at 5370 m. (Based on data from Gray et al, 1975).

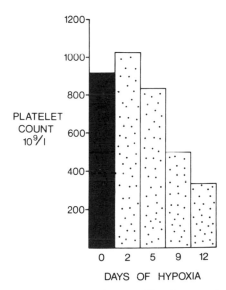

Fig. 10.2 Effect of hypoxia (produced by a barometric pressure of 380 mmHg) on the platelet count in groups of six mice after, 0, 2, 5, 9 and 12 days. An initial slight thrombocytosis, probably due to haemoconcentration, is followed by a severe and persistent thrombocytopenia. (Based on data from Birks et al, 1975).

number of circulating thrombocytes has been confirmed in rats subjected to simulated high altitude for 22 days by exposing them to a reduced barometric pressure of 380 mmHg (De Gabriele and Penington, 1967). These experimental conditions produced gross polycythaemia with a rise of packed cell volume from 45 to 65 per cent but the fall in

by a severe and persistent thrombocytopenia with a rapid decline in the platelet count between the fifth and ninth days of hypoxia. By the twelfth day the platelet count in the mice was only 36 per cent of that in the control group (Fig. 10.2).

Such a rapid and severe decline in the platelet count could be brought about by an increased destruction or decreased production of platelets; it could also be brought about by changes in blood volume or by sequestration of platelets. Since Birks and his colleagues (1975) found only a slight decline in platelet count during the first five days of hypoxia, and since the life-span of platelets in rodents is four to five days, one can assume that the loss of thrombocytes is not due to increased destruction. They excluded as causes of the diminished platelet count, dilution by an expanded blood volume or sequestration in an associated enlarged spleen. Furthermore, mice made polycythaemic by daily injections of six units of erythropoietin showed no significant decline in their platelet counts so thrombopoiesis does not appear to be depressed by excess erythropoietin as such. Rather the thrombocytopenia appears to be due to either a decreased rate of production of platelets in hypoxia or a defect in their structure or metabolism under such conditions.

Decompression and platelet count

Decompression *per se* may lead to a sharp fall in platelet count. Sixteen men were decompressed to simulate a rate of ascent of 1520 m per minute to an altitude of 6100 m where they remained for two hours (Gray et al, 1975). During this period they breathed oxygen at a partial pressure of 150 mmHg but in spite of this they showed a significant fall in platelet count from control levels of 232 to 207 10^9/l, a fall of approximately 10 per cent which persisted for three days. This thrombocytopenia was regarded in part as being due to a reduction in platelet half-life for autologous platelet survival was diminished in four of five men. However, the fall in platelet count was also considered to be brought about to some extent by sequestration of thrombocytes in the lung. Platelets labelled with ^{51}Cr sequester in the pulmonary vascular bed of rabbits which are decompressed in normoxic and hypoxic conditions (Gray et al, 1975). It will be apparent that the explanation advanced for the thrombocytopenia in man at high altitude by Gray and his colleagues (1975) differs from that put forward for mice by Birks et al (1975).

The thrombocytopenia of high altitude is in some

ways analogous to that which occurs during decompression in divers. In Caisson disease the fall in platelet count is associated with the formation of asymptomatic bubbles which cause thrombocytes to adhere to them. Strong electrokinetic forces develop at gas-liquid interfaces which disrupt the secondary and tertiary structure of protein molecules leading to partial unfolding and subsequent formation of macromolecules and large molecular-weight complexes which form the nidus for platelet adhesion. The fate of the postulated platelet-bubble aggregates has not been established, although it seems possible that they may form microemboli which are filtered by the pulmonary vascular bed. However, there are considerable differences between Caisson disease and high altitude decompression. In the first place the pressure changes employed in simulated high altitude studies are of the order of 0.5 atmosphere and these are small compared with those of 8.0 atmospheres dealt with in diving studies. Furthermore, after diving, full equilibration following decompression may take many hours instead of the rapid decompression which occurs after exposure to simulated high altitude. The magnitude of the fall in platelets on decompression at high altitude is only half of that reported in most diving studies.

Ultrastructure of platelets

In subjects with significant pulmonary hypertension at high altitude and in cases of high altitude pulmonary oedema the electrophoretic mobility of platelets is reduced. No ultrastructural abnormality of the thrombocytes has been reported in these two conditions. The integrity of the plasma membrane, the capacity for pseudopodia formation, and the ability for degranulation are all intact.

Hypercoagulability at high altitude

Animal studies reveal that acute exposure to high altitude brings about an undeniable hypercoagulability. Genton and his colleagues (1970) established baseline data on blood coagulation in calves four to six weeks of age at Denver (1600 m) and subsequently when the animals were taken to Mount Evans (4310 m) for 10 days prior to their return to the lower altitude. They demonstrated in their animals the same reduction in the number of circulating platelets to which we have referred above as occurring in man and rodents exposed to high altitude. At the same time the number of circulating young adhesive platelets rose greatly to an average figure some 142 per cent greater than baseline values. There was a striking decrease of 63 per cent in platelet half life. Platelet count and adhesiveness returned rapidly to normal within one week of returning to the lower altitude.

These changes in the platelets were associated with a considerable shortening of the first stage of clotting or thromboplastin formation which occurred by the second day of exposure to high altitude and persisted for the remainder of the period of hypoxia. The first stage of clotting returned to normal within one week of returning to 1600 m. The second stage of clotting, as measured by the one-stage prothrombin time, showed significant prolongation averaging 50 per cent above the baseline. The third stage of clotting, fibrinogen conversion, did not change significantly. During their first week at high altitude the calves showed a fall in plasma fibrinogen levels until they were about a third below baseline values. Plasminogen levels showed a rapid fall to an average of 50 per cent below baseline levels. There was a decrease of some 35 per cent in fibrinogen half life. Fibrinolytic activity was not found to be altered (Genton et al, 1970).

Disorders of blood coagulation in man in high altitude pulmonary hypertension

Disturbances of the coagulation of blood also occur in man. Most of the information we have on this subject comes from Indian studies and particularly from the work of Singh and Chohan. Consequently confirmation of these findings awaits parallel studies by other groups. Singh and Chohan (1972b) studied the coagulability of 38 soldiers who were resident at altitudes between 3690 m and 5490 m for two years. Six of them were thought to have developed pulmonary hypertension on clinical, radiological and electrocardiographic grounds, although this was not confirmed by cardiac catheterization. Sixteen men at sea level were also investigated. In all those at high altitude there was a significant increase of plasma fibrinogen levels and fibrinolytic activity, and both were thought to be induced by hypoxic

stress. However, in soldiers with clinical evidence of pulmonary hypertension plasma fibrinogen levels were lower suggesting that in them there is a constant depletion of fibrinogen by its conversion into fibrin. This mirrors the finding of diminished plasma fibrinogen levels in calves referred to above. This conversion seems to be facilitated by platelet adhesiveness in the presence of factors V and VIII for all these components were increased. Platelet factor 3 was also increased and this probably implies platelet aggregation.

An abrupt fall in the plasma fibrinogen level has also been reported in eight men exposed for two days to a simulated high altitude of 4400 m (Maher et al, 1976). However, in this study no alterations were found in the platelet count, prothrombin and thrombin times, and platelet factor 3 availability. There was a shortening of partial thromboplastin time and an abrupt fall of factor VIII. All the abnormalities found returned to normal values within one to two days of acute exposure to simulated altitude.

Disorders of blood coagulation in high altitude pulmonary oedema

In high altitude pulmonary oedema (Chapter 15) the disturbances of blood coagulation are said to be similar to those found in Indian soldiers with pulmonary hypertension (Singh and Chohan, 1973). However, in cases with pulmonary oedema there appears to be a breakdown of the fibrinolytic system resulting in increase in plasma fibrinogen levels and in the time required for lysis of clot in venous blood (Singh et al, 1969; Singh and Chohan, 1972a). Both fibrinogen levels and venous clot lysis time were significantly higher during the acute phase of oedema than during convalescence and may be regarded as characteristic of the disease according to Singh et al (1969).

These disorders of blood coagulation are thought by Singh and Chohan (1973) to be responsible for intravascular sludging of erythrocytes and the formation of fibrin thrombi within the pulmonary vasculature in cases of high altitude pulmonary oedema. Certainly in this disease small fibrin thrombi are common in the pulmonary capillaries, venules and arterioles. Nayak et al (1964) studied necropsy specimens of lung from 13 cases of high altitude pulmonary oedema and found hyaline fibrin thrombi in no fewer than six of them. Usually they appeared as clear homogeneous masses but sometimes they had an indistinct laminated appearance which always gave a strongly positive staining reaction with phosphotungstic acid haematoxylin. In one case multiple fibrin thrombi were not confined to the pulmonary capillary bed but also occurred in the kidney where they plugged the glomerular and peritubular capillaries. Similar thrombi and small plugs of fibrin strands were found in the hepatic sinusoids. It is these minute fibrin thrombi which have been considered by Singh et al (1965b) to be involved in the pathogenesis of high altitude pulmonary oedema.

The chain of events in the formation of fibrin thrombi would appear to be as follows. Increase of platelet factor 3 and increased platelet adhesiveness stimulates their aggregation. The plasma membrane of the thrombocyte provides an active catalytic surface for the interaction of plasma coagulation factors leading to thrombus formation. The resulting consolidation of platelets leads to their degranulation with the release of more factor 3 and fibrin formation. Singh and Chohan (1972b) believe such a self-perpetuating system of fibrin thrombus formation is maintained until the high altitude pulmonary oedema clears.

Reversibility of disorders of coagulation

Disorders of blood coagulation induced by exposure to high altitude revert to normal on descent. Platelet adhesiveness, clot lysis time, and factors V and VIII returned to normal values within one week of Indian soldiers descending to sea level (Singh and Chohan 1972b). Platelet factor 3 and plasma fibrinogen returned to normal in one to three weeks.

The action of frusemide

In Chapter 15 we discuss the use of frusemide for prophylaxis against, and treatment of, high altitude pulmonary oedema. Its beneficial effects have been ascribed to its potency and rapidity of action in inducing diuresis and its effect on reducing pulmonary blood volume. However, Singh and Chohan (1973) found that frusemide also reverses adverse changes in fibrinolytic activity, blood coagulation and platelet function. Their studies

were carried out on four patients with high altitude pulmonary oedema at 3690 m on the first day of their illness before and 30 minutes after the administration of 40 mg intravenous frusemide. It is possible that at least part of the efficacious action of frusemide depends on its prevention of the widespread formation of fibrin thrombi in the pulmonary capillary bed.

Hypercoagulability, pulmonary thrombosis and pulmonary hypertension at high altitude

One could anticipate that in subjects exposed to high altitude the resulting hypercoagulability of the blood and sequestration of platelets in the pulmonary vascular bed might provoke pulmonary thrombosis. If this were extensive enough, it might in turn be expected to lead to an elevation of pulmonary arterial pressure. In fact Singh (1973) describes just such a clinical picture in some of the Indian soldiers arriving at high altitude. He reports that symptoms of significant pulmonary hypertension develop in the affected subjects after a stay of five to 42 months at altitudes between 3660 m and 5490 m. The pathology of these cases is that of a thrombotic, occlusive hypertensive pulmonary vascular disease. Singh (1973) says that once this vascular condition has developed periodic returns to sea level for two to three months once a year do not alter the picture. The pulmonary hypertension either continues to persist at sea level, or, if it abates, it reappears within two to three weeks after the individual returns to high altitude. Unless the affected individuals are evacuated to sea level in time, on return to sea level they may not recover and may die from progressive right ventricular failure. Singh and Chohan (1972a) record that such troops may develop thrombosis in the peripheral and splenic veins and in the coronary, cerebral and mesenteric arteries usually when the subject has been at high altitude for several weeks or months. Thrombosis of such magnitude may also occur in high altitude climbers and it is in this context that we consider the entity at greater length in Chapter 30. In our experience thrombosis of this degree of severity at high altitude is rare but Singh and Chohan (1972b) believe that pulmonary thrombi may form the organic basis for the pulmonary hypertension characteristic of the highlander (Chapter 11).

We are unfamiliar with such a clinical syndrome in the Andes and it is certainly a different entity from the clinicopathological picture of highly reversible hypoxic pulmonary hypertension and hypertensive pulmonary vascular disease with which we are familiar in Peruvian highlanders or laboratory animals in decompression chambers. Singh and Chohan (1972b) believe that the basis for the pulmonary hypertension of high altitude in general cannot rest on muscularization of the terminal portions of the pulmonary arterial tree such as we describe at length in Chapter 11. They fail to see why administration of oxygen does not give total and immediate reversibility of high altitude pulmonary hypertension if it is based on pulmonary vasoconstriction and muscularization. In contradistinction to this view, we think that such lack of total immediate reversibility is entirely explicable on the basis of muscularization of the pulmonary arterioles. As we explain at length in Chapter 28 and demonstrate in Figure 28.1, the immediate partial reversibility of pulmonary hypertension on the inhalation of oxygen is due to a relaxation of pulmonary vasoconstriction, but the complete reversibility requires the regression of the muscular coat around the pulmonary arterioles and this takes months to achieve.

Singh and Chohan (1972b) support their concept of the raised pulmonary arterial pressure of high altitude having an occlusive basis of pulmonary thrombosis by the findings at necropsy in one of their patients. He developed pulmonary hypertension at high altitude and then died 35 months after descent to sea level. At post mortem there was extensive occlusive thrombosis in the pulmonary trunk extending from near the pulmonary valve to the hila of the lungs (Singh et al, 1965a). The widespread occlusion of many small pulmonary arteries may have been autochthonous or thromboembolic in origin. It is of interest to note that underlying the thrombus, the media of the pulmonary trunk showed patchy destruction, with an inflammatory arteritis and a cellular infiltration around the vasa vasorum. These findings are consistent with a primary arteritis involving the pulmonary trunk with secondary thrombosis. Alternatively, they may represent one of the major episodes of thrombosis to which we refer above and which we describe in Chapter 30. While we must accept the thrombotic clinicopathological syndrome occurring in Indian troops described by Singh and his colleagues (1965a) we do not agree with them that in all or even most cases the 'histopathological evidence suggests that high altitude pulmonary hypertension is of obstructive origin and results from thrombosis of the smaller pulmonary arteries'. We believe, as stated in Chapter 11, that the organic basis for increased pulmonary vascular resistance at high altitude is usually entirely muscular in nature. This accounts for its reversibility which is in striking contrast to the natural history of recurrent pulmonary thrombosis or pulmonary thromboembolism which Singh et al (1965a) put forward as the underlying pathology. Furthermore the pulmonary vascular alterations characteristic of exposure to high altitude can be readily induced in experimental animals by subjecting them to simulated high altitude. When this is done, the vascular lesions so produced in the lungs are entirely muscular and characteristic of 'hypoxic pulmonary vascular disease'. They are not occlusive in nature.

Post-operative hypoxaemia and thrombosis

Nunn and Payne (1962) pointed out that during the first day after a surgical operation, the systemic arterial oxygen tension frequently encountered was equivalent to that which would be found at an altitude of 3050 m to 5180 m. Atkins and Lempke (1970) also noted this similarity and advocated that vigorous efforts to prevent hypoxaemia in the post-operative period should figure in any prophylactic regime to prevent systemic venous thrombosis and pulmonary thromboembolism.

Initiating factors

Singh and Chohan (1972b) speculate that ADP may play some part in initiating pulmonary thrombosis and pulmonary hypertension in subjects living at high altitude. Thrombus formation in the pulmonary arteries was induced in calves aged two to four months living at 3350 m by daily intravenous injections of 100 μg ADP for one to four weeks (Reeves et al, 1968). Singh and Chohan (1972a) also consider it possible that platelet adhesiveness may be related to increased levels of catecholamines in the blood at high altitude.

REFERENCES

Atkins, P. & Lempke, R. E. (1970) The effect of hypoxia on the platelet count. *British Journal of Surgery*, 57, 583.

Birks, J. W., Klassen, L. W. & Gurney, C. W. (1975) Hypoxia-induced thrombocytopenia in mice. *Journal of Laboratory and Clinical Medicine*, 86, 230.

De Gabriele, G. & Penington, D. G. (1967) Physiology of the regulation of platelet production. *British Journal of Haematology*, 13, 202.

Genton, E., Ross, A. M., Takeda, Y. A. & Vogel, J. H. K. (1970) Alterations in blood coagulation at high altitude. In: *Hypoxia, High Altitude and the Heart*. Edited by J. H. K. Vogel. Aspen. Colorado. Advances in Cardiology, 5, 32. Basel: S. Karger.

Gray, G. W., Bryan, A. C., Freedman, M. H., Houston, C. S., Lewis, W. F., McFadden, D. M. & Newell, G. (1975) Effect of altitude exposure on platelets. *Journal of Applied Physiology*, 39, 648.

Maher, J. T., Levine, P. H. & Cymerman, A. (1976) Human coagulation abnormalities during acute exposure to hypobaric hypoxia. *Journal of Applied Physiology*, 41, 702.

Nayak, N. C., Roy, S. & Narayanan, D. C. P. (1964) Pathologic features of altitude sickness. *American Journal of Pathology*, 45, 381.

Nunn, J. F. & Payne, J. P. (1962) Hypoxaemia after general anaesthesia. *Lancet*, ii, 631.

Reeves, J. T., Jokl, P., Merida, J. & Leathers, J. E. (1968) Pulmonary vascular obstruction following administration of high-energy nucleotides. *Exercise and Altitude*, p. 122. Edited by E. Jokl, & P. Jokl. Basel: S. Karger.

Singh, I. (1973) Pulmonary hypertension in new arrivals at high altitude. World Health Organization meeting on Primary Pulmonary Hypertension. Geneva, October, 1973.

Singh, I., Chohan, I. S. & Mathew, N. T. (1969) Fibrinolytic activity in high altitude pulmonary oedema. *Indian Journal of Medical Research*, 57, 210.

Singh, I. & Chohan, I. S. (1972a) Abnormalities of blood coagulation at high altitude. *International Journal of Biometeorology*, 16, 283.

Singh, I. & Chohan, I. S. (1972b) Blood coagulation changes at high altitude predisposing to pulmonary hypertension. *British Heart Journal*, 34, 611.

Singh, I. & Chohan, I. S. (1973) Reversal of abnormal fibrinolytic activity, blood coagulation factors and platelet function in high altitude pulmonary oedema with frusemide. *International Journal of Biometeorology*, 17, 73.

Singh, I., Khanna, P. K., Lal, M., Hoon, R. S. & Rao, B. D. P. (1965a) High altitude pulmonary hypertension. *Lancet*, ii, 146.

Singh, I., Kapila, C. C., Khanna, P. K., Nanda, R. B. & Rao, B. D. P. (1965b) High altitude pulmonary oedema. *Lancet*, i, 229.

Pulmonary hypertension

Pulmonary hypertension at high altitude

Healthy man born and living at high altitude has some degree of pulmonary arterial hypertension at rest. Peñaloza et al (1962, 1963) and Sime and his colleagues (1963) have studied the pulmonary haemodynamics of 38 healthy adults and 32 healthy children at Morococha (4540 m) and Cerro de Pasco (4330 m) in the Peruvian Andes. In the former town the mean barometric pressure is 446 mmHg and the atmospheric PO_2 80 mmHg, while in the latter settlement these values are respectively 455 and 90 mmHg. We have expressed some of their findings diagrammatically in Figures 11.1 and 11.2.

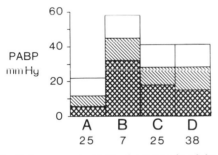

Fig. 11.1 Pulmonary arterial blood pressure in adults at sea level and in children and adults at high altitude. Cross-hatched column = Diastolic blood pressure; Hatched column = Mean blood pressure; Open column = Systolic blood pressure; Group A (25 subjects) = Adults at sea level; Group B (7 subjects) = Children 1 to 5 years of age at 4540 m; Group C (25 subjects) = Children 6 to 14 years of age at 4540 m; Group D (38 subjects) = Adults at 4540 m. (Compiled from data in Peñaloza et al, 1962 and Sime et al, 1963).

Compared to an average level of 22/6 mmHg (mean 12 mmHg) for sea-level residents, the adult at high altitude has a mild pulmonary arterial hypertension of 41/15 mmHg (mean 28 mmHg). The pulmonary

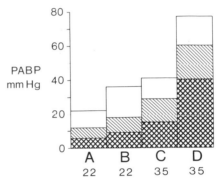

Fig. 11.2 Pulmonary arterial blood pressure at rest and on exercise in residents at sea level and at high altitude. Cross-hatched column = Diastolic blood pressure; Hatched column = Mean blood pressure; Open column = Systolic blood pressure; Group A (22 subjects) = Sea-level residents at rest; Group B (22 subjects) = Sea-level residents during exercise; Group C (35 subjects) = High altitude residents at rest; Group D (35 subjects) = High altitude residents during exercise. (Compiled from data in Peñaloza et al, 1962).

wedge pressure is not elevated. The average total pulmonary vascular resistance in the subjects studied by Peñaloza et al (1962) was found to be increased from a value of 159 dynes s cm^{-5} in residents at sea level to 401 dynes s cm^{-5} in adult subjects at high altitude.

In young children between the age of one to five years the level of pulmonary arterial pressure is considerably greater. Thus Sime et al (1963) found the average pulmonary arterial mean pressure in seven young children to be no less than 45 mmHg with a systolic pressure as high as 58 mmHg (Fig. 11.1). However, after the age of five years the pulmonary arterial pressure falls to adult levels (Fig. 11.1). It is of interest to note that the same degree of pulmonary hypertension that persists in high altitude children to the age of five years is to be found in infants at sea level only for the first week of life (James and Rowe, 1957; Rudolph and Cayler, 1958; Rudolph et al, 1961). After only a fortnight of postnatal life at sea level the pulmonary arterial pressure has fallen to adult levels (Keith et al, 1958). As we shall see later in this chapter these differences in the pulmonary haemodynamics of adults and children at high altitude are related to the fact that the pulmonary vasculature of the young is more muscular.

The effects of high altitude on pulmonary haemodynamics are shown in a more pronounced manner on exercise. Peñaloza et al (1962) found that the average pulmonary arterial mean pressure rose as high as 60 mmHg on exercise in subjects at high altitude, the systolic level being 77 mmHg (Fig. 11.2). Sea-level subjects who go to reside for a long time in the mountains will also develop pulmonary hypertension but its degree will depend on whether the individual is resistant or susceptible to its development as discussed later in this chapter.

Its reversibility

The pulmonary hypertension of high altitude can be partially reversed immediately on the administration of oxygen but total reversal by removal of the subject to sea level takes a much longer time. Thus administration of 35 per cent oxygen to produce an oxygen tension similar to that found at sea level will immediately reduce the pulmonary arterial pressure

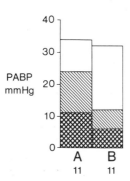

Fig. 11.3 Pulmonary arterial blood pressure in high altitude subjects in their native habitat and after two years residence at sea level. Cross-hatched column = Diastolic blood pressure; Hatched column = Mean blood pressure; Open column = Systolic blood pressure; Group A (11 subjects) = Subjects at 4330 m; Group B (11 subjects) = The same subjects after 2 years residence at sea level. (Compiled from data in Peñaloza et al, 1962).

of highlanders by 15 to 20 per cent (Peñaloza et al, 1962). On the other hand, considering long term reversibility, when subjects born and bred at high altitude are taken down to sea level, they show a considerable reduction in pulmonary arterial pressure (Fig. 11.3). Thus in 11 inhabitants of Cerro de Pasco (4330 m) the average pulmonary arterial mean pressure halved from 24 to 12 mmHg after two years' residence at sea level (Peñaloza et al, 1962). In this chapter we consider the physiological and anatomical basis for this pulmonary hypertension which is moderate in childhood and persists into adult life at a reduced level but which is also readily reversible.

Pulmonary hypertension in the fetus

Physiologically the fetus has much in common with the high altitude dweller and hence consideration of the pulmonary circulation *in utero* is helpful in understanding the pathogenesis of high altitude pulmonary hypertension. The intrauterine umbilical arterial oxygen tension is about 20 mmHg which corresponds to an atmospheric oxygen tension of 61 mmHg which would be found at an elevation of about 7500 m (Mithoefer, 1966). In the fetus there is physiological pulmonary hypertension and the interlobular and intralobular pulmonary arteries of the solid lung are thick-walled and muscular (Civin and Edwards, 1951). At birth the lungs expand and begin to move rhythmically in respiration and as a consequence the pulmonary arteries dilate. There is at the same time a sudden relief from the severe degree of hypoxaemia to which the fetal lung is subjected.

The extent of oxygenation of the air which suddenly inflates the alveolar spaces depends upon whether the child is born at sea level or at high altitude. At sea level PA_{O_2} is high and the muscular tone of the pulmonary arteries freshly mechanically dilated by the expansion of the lung is kept low by the high pressure of oxygen in the alveolar air. This leads to the development of a thin-walled pulmonary arterial tree associated with a low intravascular pressure and resistance in the lung throughout the whole of adult life. In contrast, in the infant born at high altitude PB_{O_2} is lower than at sea level. As a result, although the pulmonary arteries are opened by the mechanical expansion of the lungs at birth, they are subjected to chronic alveolar hypoxia which induces pulmonary vasoconstriction and a higher level of pulmonary arterial tone. It is salutary to realise that even at altitudes exceeding 4000 m the newborn baby is in effect suddenly transported to low altitudes (Arias-Stella, 1966).

The mechanism of hypoxic pulmonary hypertension

The lowered barometric pressure at high altitude and the consequent diminution of the partial pressure of oxygen in the alveolar spaces are responsible for the maintenance of increased tone in the peripheral portion of the pulmonary arterial tree of highlanders at birth and throughout infancy and childhood into adult life. Since the early studies of von Euler and Liljestrand (1946) on the pulmonary circulation of the cat it has become clear that hypoxia is one of the most powerful constrictors of pulmonary arterial blood vessels. Hence in those who live in high mountains the chronic hypoxia inherent in the environment stimulates throughout life a persistent constriction of the pulmonary arteries which in the case of the highlanders have never lost their muscular coat. The reduction of the arteriolar lumen elevates pulmonary vascular resistance producing mild pulmonary hypertension.

The mechanism by which a diminution in oxygen tension brings about pulmonary vasoconstriction has always been controversial and it remains so today. It has generally been accepted in the past that alveolar *hypoxia* is a more potent vasoconstrictor than vascular *hypoxaemia* but even this concept is now challenged (Fishman, 1976). If one does accept the traditional premise, it is likely that pulmonary vasoconstriction is mediated by an indirect mechanism (Reeves et al, 1962) some agent lying between alveolar space and pulmonary arteriole, itself receiving the hypoxic stimulus and converting this into a chemical messenger acting on the vascular smooth muscle. It was postulated that this chemical mediator was histamine (Hauge, 1968; Hauge and Melman, 1968). Since most histamine in the lung is stored in mast cells they came to be championed as the sought-for mediator (Bergofsky, 1969).

Lung mast cells at simulated high altitude

Mast cells in the lung are to be found in the pleura,

around the bronchial tree, and around pulmonary blood vessels. A glance at Figure 11.4 will show that the perivascular mast cell is in an ideal anatomical situation to act as a mediator, monitoring alveolar oxygen tension and responding to hypoxia by releasing histamine which could act on arteriolar smooth muscle inducing constriction. There is certainly evidence that perivascular mast cells are degranulated *in vivo* during acute alveolar hypoxia (Haas and Bergofsky, 1972). Hypoxia also releases histamine from mast cells isolated from the peritoneal cavity without injuring them (Haas and Bergofsky, 1972).

An increased histamine-forming capacity in the lung might be brought about by specific enzyme induction of histidine carboxylase in individual mast cells without an increase in their number but Kay et al (1974) found that the mean pulmonary histamine-forming capacity does not increase in hypoxic rats. There is rather a striking increase in the number of mast cells in the lungs of rats subjected to a simulated high altitude. Thus Kay et al (1974) exposed eight rats to a barometric pressure of 380 mmHg for 20 days and found that the lung mast cell density more than doubled. Our own experiments on rats (Williams et al, 1977) have confirmed these observations but have extended them to reveal that the appearance of large numbers of mast cells in the lungs is highly reversible. As soon as the hypoxia is relieved the mast cells disappear. The calling forth of mast cells into the lung by hypoxia is more pronounced in adult than young rats but its magnitude is independent of the sex of the animals studied (Williams et al, 1977). A hyperplasia of mast cells in the lungs of hypoxic rats has also been observed by Mungall and Barer (1975).

While it is easy to demonstrate an accumulation of mast cells in the lungs under conditions of simulated high altitude it is far more difficult to come to any conclusion as to what function they are fulfilling. There have been reports of experimental work which supports the concept that mast cells have a rôle as a mediator in bringing about hypoxic pulmonary vasoconstriction. Thus dogs premedicated with disodium cromoglycate, a substance which inhibits the release of histamine from mast cells, have been reported as failing to increase pulmonary vascular resistance while breathing a

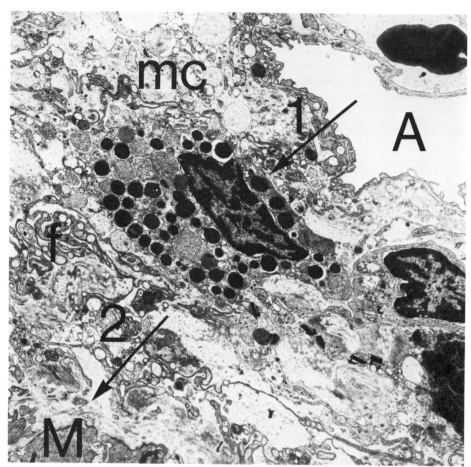

Fig. 11.4 Perivascular mast cell, mc, in a Wistar albino rat exposed to a reduced barometric pressure of 263 mmHg for 8 hours. It lies amongst fibrocytes, f, and is situated between the alveolar space, A, and the smooth muscle, M, of the media of a pulmonary artery. Its cytoplasm contains compact granules, which appear round and black, and granules in the phase of dissolution. It has been speculated that the diminished partial pressure of oxygen diffusing from the hypoxic alveolar air stimulates the mast cell (arrow 1). According to one school of thought this cell then liberates histamine which diffuses to affect the smooth muscle cells of the arterial media (arrow 2). As noted in the text, however, this hypothesis is by no means universally accepted. (Electron micrograph × 7500).

hypoxic gas mixture. However, one disturbing fact which emerged from the studies of Mungall (1976) was that in chronically hypoxic rats, the mast cells did not proliferate until after 21 days of exposure, whereas right ventricular hypertrophy was apparent after only 14 days, a rather unlikely event if the mast cells were instrumental in raising pulmonary vascular resistance leading to right ventricular hypertrophy. We shall return to this point below. Another noteworthy fact is that during a review of the world literature on the pharmacology of the pulmonary circulation we found that in man histamine has always brought about pulmonary vasodilatation rather than vasoconstriction (Harris and Heath, 1977). This raises the interesting possibility that the accumulation of mast cells around small muscular pulmonary arterial vessels at high altitude may represent an attempt to bring about a relief rather than an exaggeration of elevated pulmonary vascular resistance.

Lung mast cells at natural high altitude

Tucker et al (1977) studied the changes in the density and distribution of lung mast cells in six mammalian species exposed to an altitude of 4500 m (P_B = 435 mmHg) for 19 to 48 days. Controls were studied at 1600 m (P_B = 635 mmHg). Calves, pigs, rats, sheep, guinea pigs and dogs were investigated. None of these species showed a general hyperplasia of mast cells throughout the body so that their density in kidney, liver, myocardium and skin was normal. Increased mast cell-activity in mammals at high altitude is confined to the lung and it is here that its functional significance is to be found. In calves the increase in mast cells is found in all tissues of the lung so that they accumulate in alveolar walls, in the pleura, around bronchi and around pulmonary blood vessels. As we shall see in Chapter 13 the bovine pulmonary vasculature is naturally very muscular and Tucker and his colleagues (1977) found accumulations of mast cells around pulmonary arteries of all diameters in this species. The same finding was encountered in the pig which also has muscular blood vessels in the lung. In pigs, rats and sheep at high altitude there were significant increases in mast cell density around muscular pulmonary arterioles 20 to 130 μm in diameter. In all these four species significant positive correlations were found between perivascular mast cell density and right venticular hypertrophy. It should, however, be pointed out that species with the thickest pulmonary arteries at high altitude do not necessarily have the highest perivascular mast cell counts. In fact high densities of mast cells have been reported around the thin-walled pulmonary arteries of the dog at high altitude.

Tucker and his colleagues (1977) believe their findings indicate that at high altitude and in other states of hypoxia the proliferation of mast cells around pulmonary arteries may be more closely related to the muscular changes in those blood vessels than to the existing hypoxia. This leads them to consider the idea referred to above that mast cells increase in number in *response* to the pulmonary hypertension rather than to induce it. The histamine or other agents liberated by such mast cells may dilate the pulmonary arteries in an attempt to restrain the hypoxic pressor response of high altitude rather than constrict them. This is a remarkable example of how the interpretation of facts in biological science may alter so rapidly as to be diametrically opposite to what they were a decade ago.

Direct action of hypoxia on pulmonary vascular smooth muscle

Hypoxia may exert its constricting effect directly on the smooth muscle cells of the terminal portions of the pulmonary arterial tree (Harris and Heath, 1977). Muscle cells in this situation are exposed to partial pressures of oxygen which far exceed those required for the maintenance of function of mitochondria. This is true even at high altitude under conditions of chronic alveolar hypoxia. Hence, if vascular smooth muscle cells are to be sensitive to any fall in alveolar oxygen tension and respond to it, some biochemical system of amplification of the stimulus would have to be involved. Such an arrangement could involve the production or utilization of ATP. Disturbance of production could involve such processes as high rates of hydrolysis of ATP (Chapter 19).

A system of biochemical amplification involving a disturbance of utilization of ATP could conceivably involve some enzyme activity which is dependent on a high concentration of the phosphate. Nothing is known of the relative dependence on ATP of those enzyme activities in pulmonary arterial smooth muscle which might be involved in the process of contraction. Hence we must look to the myocardium for an analogous situation that may act as a model for a possible mode of constriction of vascular smooth muscle in response to hypoxia.

In the myocardium the ATP concentration at half-maximal activity (the Michaelis constant) is about a thousand times higher for the calcium-stimulated ATPase of the sarcolemma, and for sarcolemmal sodium:potassium ATPase, than it is for myofibrillar ATPase. In this tissue the two sarcolemmal activities appear normally to be operating near their Michaelis constants under which conditions they would be particularly susceptible to small changes in ATP concentration and a mechanism of amplification would exist. An inhibition of the calcium pump would directly increase the cytoplasmic concentration of calcium ions, while an inhibition of the sodium pump would indirectly have the same effect. Thus it is not difficult to imagine mechanisms whereby small decreases in cytoplasmic ATP concentration increase the nett transport of calcium ions across the cell membrane or increase membrane permeability to calcium while remaining sufficient to sustain contractile activity. Should that happen in pulmonary arterial smooth muscle, contraction could follow the direct action' of hypoxia on the muscle cells (Harris and Heath, 1977).

The possible rôle of the renin-angiotensin system in the constriction of the terminal portion of the pulmonary arterial tree in states of chronic hypoxia is discussed in Chapter 25.

Peripheral muscularization of the pulmonary arterial tree

The persistent pulmonary vasoconstriction induced by the chronic hypoxia leads to muscularization of the terminal portion of the pulmonary arterial tree (Figs. 11.5 and 11.6). Quantitative studies show

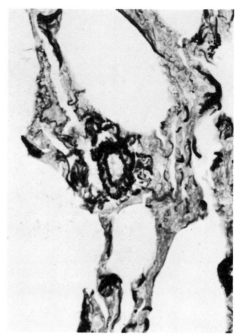

Fig. 11.6 Transverse section of a muscularized pulmonary arteriole from a male Indian native of La Paz (3800 m) aged 35 years. This arteriole is only 33 μm in diameter and it will be readily appreciated that such minute muscular vessels are associated with an elevation of pulmonary vascular resistance. (Elastic Van Gieson × 375).

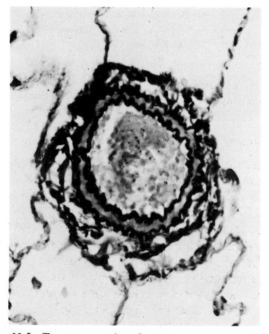

Fig. 11.5 Transverse section of a pulmonary arteriole from a Quechua Indian living at high altitude in the Peruvian Andes. There is a distinct muscular media sandwiched between internal and external elastic laminae. This muscularization of the terminal portion of the pulmonary arterial tree is in striking contrast to the normal pulmonary arteriole at sea level which has a wall consisting of a single elastic lamina. (Elastic Van Gieson × 375).

that this is the result of muscularization of a considerable number of the peripherally situated pulmonary arterioles rather than by medial hypertrophy in parent muscular arteries (Arias-

Stella and Saldaña, 1963). These small muscular vessels come to resemble systemic arterioles (Figs. 11.5 and 11.6) and increase pulmonary vascular resistance (Sime et al, 1963). This in turn leads to right ventricular hypertrophy and enlargement of the main pulmonary arteries which may be detected on radiological examination (Fig. 11.7). There is no occlusive intimal fibroelastosis such as occurs in the pulmonary vasculature in association with the pulmonary hypertension induced by congenital cardiac shunts and hence high altitude pulmonary hypertension is characteristically reversible as observed by Peñaloza et al (1962) (Fig. 11.3) and referred to above. This is not to say, however, that the pulmonary arteries of adult highlanders do not show the customary age-change of intimal fibrosis. Through the courtesy of Dr Rios Dalenz we have been able to confirm this ourselves in specimens of lung from residents of La Paz, Bolivia (3800 m) (Fig. 11.8). Hypervolaemia and polycythaemia appear to be only of minor importance in the pathogenesis of the raised pressure in the pulmonary

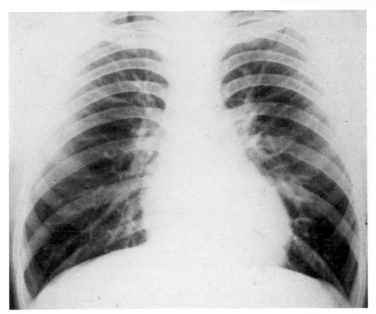

Fig. 11.7 Antero-posterior radiograph of the chest from a Quechua Indian aged 42 years living at 4330 m. There is evidence of right ventricular hypertrophy and enlargement of the main pulmonary artery consistent with an elevation of pulmonary arterial pressure due to chronic hypoxia.

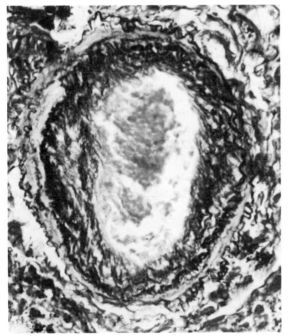

Fig. 11.8 Transverse section of a muscular pulmonary artery from an Indian native of La Paz (3800 m) aged 83 years. It shows prounced age-change intimal fibroelastosis. (Elastic Van Gieson × 397).

circulation. When a native of high altitude is taken to sea level hypervolaemia and polycythaemia

disappear quickly but pulmonary hypertension regresses slowly (Fig. 28.1) (Chapter 28).

Ultrastructure

We have carried out electron microscopic studies on the pulmonary vasculature in rats exposed to a simulated high altitude of 5500 m in a decompression chamber (Smith and Heath, 1977). Hypoxia causes constriction of the smooth muscle cells of the muscular pulmonary arteries, pulmonary arterioles and pulmonary veins, and as we shall see in the following chapter, of the pulmonary trunk. This constriction gives rise to characteristic ultrastructural changes since the constricted or contracted smooth muscle cell does not merely become shorter and thicker but becomes covered by small protuberances. These have been seen directly in scanning electron micrographs of the contracted isolated smooth muscle cell from the stomach wall of *Bufo marinus* (Fay and Delise, 1973). These protuberances are formed by evaginations of smooth muscle cytoplasm and the mode of their formation is shown diagrammatically in Figure 11.9. When the muscle cell constricts, cytoplasmic evaginations are squeezed out between the attachment points for myofilaments. These extrusions are devoid of

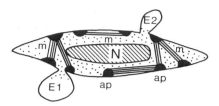

Fig. 11.9 Diagram to illustrate the mode of formation of muscular evaginations. On contraction of the myofilaments, m, which connect to 'attachment points', ap, muscle cell cytoplasm forms evaginations, shown here as E1 and E2. The cytoplasm of these evaginations is clear, being devoid both of myofilaments and organelles, and may be mistaken for cysts in electron micrographs. The nucleus, N, holds a central position in the cell.

myofilaments and organelles and as a result have a clear cytoplasm that can easily be mistaken on electron microscopy for a cyst. In rats exposed to the hypoxia of simulated high altitude such muscular evaginations protrude through deficiencies in the inner elastic lamina of the pulmonary trunk to press

Fig. 11.10 Transverse section of a muscularized pulmonary arteriole from a rat subjected to a simulated altitude of 5500 m for two weeks. There is a distinct media of smooth muscle cells, M, bounded by inner (arrow 1) and outer (arrow 2) elastic laminae. Endothelial cells, E, project into the lumen. The lumen contains erythrocytes. (Electron micrograph, × 3750).

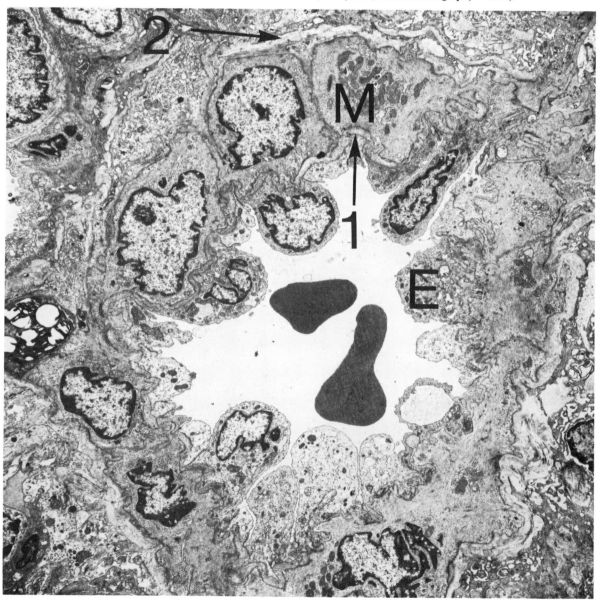

on the undersurface of the endothelial cells of the intima (Chapter 12, Fig. 12.1). They also protrude beneath the endothelial cells of the intima of pulmonary veins and we illustrate them in Figure 15.5 and consider their possible rôle in the causation of high altitude pulmonary oedema in Chapter 15. Muscular evaginations also protrude into the adventitia of muscular pulmonary arteries in conditions of simulated high altitude. Such are the ultrastructural features of the initial stage of constriction of smooth muscle cells in response to hypoxia.

Electron microscopy also reveals the nature of the muscularization of the terminal portions of the pulmonary arterial tree to which we have referred above. In the pulmonary arterioles of the control rat individual smooth muscle cells are seen immediately beneath the endothelium. On exposure to the chronic hypoxia of simulated high altitude there is a hyperplasia of immature smooth muscle cells to form a distinct media between external and internal elastic laminae (Fig. 11.10). The external lamina is thick and represents the original single lamina of the normal pulmonary arteriole (Fig. 11.11). The inner lamina is thin and is newly formed internal to the

smooth muscle cells (Fig. 11.11). Thus the histological features of the muscularised pulmonary arteriole in the rat subjected to simulated high altitude are the same as those seen in the highlander (Fig. 11.5).

Hypoxic hypertensive pulmonary vascular disease

Similar muscularization of the terminal portions of the pulmonary arterial tree is seen in a group of diseases of man characterized by chronic hypoxia (Heath, 1973). These include chronic bronchitis and emphysema, kyphoscoliosis, the Pickwickian syndrome, enlarged adenoids (Levy et al, 1967), and anomalous aortic rings compressing the trachea. In pathological conditions we have termed this 'hypoxic hypertensive pulmonary vascular disease' (Hasleton et al, 1968). On account of the lack of occlusive intimal fibrosis both the arterial changes and the associated pulmonary hypertension are readily reversible. The reversibility of hypoxic pulmonary hypertension is supported in patients with chronic bronchitis and emphysema by the long term administration of oxygen.

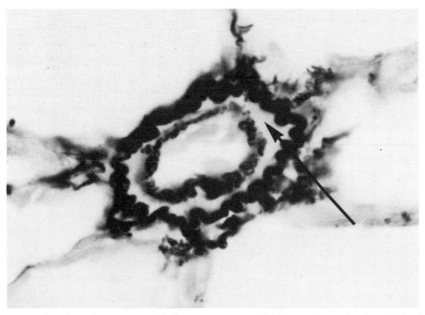

Fig. 11.11 Transverse section of a pulmonary arteriole from a rat exposed for five weeks to a simulated altitude of 5500 m. The normal pulmonary arteriole in the rat, as in man, has a wall consisting of a single elastic lamina. The pulmonary arteriole shown here is abnormally muscularized. A distinct media of circularly orientated smooth muscle (arrow) has formed internal to the original thick elastic lamina. On the inner aspect of the muscle layer a new thin internal elastic lamina has been laid down. The vessel now resembles a systemic arteriole and is capable of elevating pulmonary vascular resistance to give rise to pulmonary arterial hypertension and right ventricular hypertrophy. (Elastic Van Gieson × 1125).

Intimal longitudinal muscle

Longitudinal muscle fibres develop in the intima of pulmonary arterioles in highlanders (Wagenvoort and Wagenvoort, 1973). They are embedded in a fine network of elastic tissue and tend to occur in individuals living above 3300 m even in the absence of medial hypertrophy of 'muscular pulmonary arteries'. This development of intimal longitudinal muscle is also characteristic of hypoxic hypertensive pulmonary vascular disease referred to above.

Right ventricular hypertrophy

The pulmonary vasoconstriction and muscularization of the terminal portion of the pulmonary arterial tree elevate pulmonary vascular resistance leading to pulmonary hypertension and right ventricular hypertrophy. Arias-Stella and Recavarren (1962) demonstrated this by comparing the weights of the left and right ventricles of the heart in necropsy in 59 children who lived between 3700 m and 4260 m and in 70 children who lived at sea level. The results of their studies are summarized diagrammatically in Figure 11.12. At sea level, although the right ventricle weighs as much as the left at birth, by the age of four months the

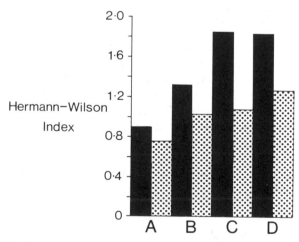

Fig. 11.12   The Hermann-Wilson Index in newborns and children at sea level and at high altitude; this index is the ratio of left to right ventricular weight. Stippled columns = 59 subjects at high altitude (3700 m to 4260 m). Filled columns = 70 subjects at sea level; Group A = Newborns; Group B = 1 day to 3 months; Group C = 4 months to 23 months; Group D = 2 years to 10 years. (Compiled from data in Arias-Stella and Recavarren, 1962).

predominance of the left ventricle seen in the adult is already established. In contrast, at high altitude right ventricular hypertrophy persists.

The speed of development and regression of right ventricular hypertrophy in rats subjected to a simulated high altitude

Some idea of the speed with which hypoxia may induce right ventricular hypertrophy and with which oxygenation may reverse it may be gained from experiments in which rats are subjected to a simulated high altitude in a decompression chamber. Exposed to a simulated altitude of 5500 m, they rapidly develop right ventricular hypertrophy over a period of only five weeks (Heath et al, 1973). After a period of relief from hypoxia lasting a further five weeks there is an equally remarkable regression to a virtually normal right ventricle. The rapid development and regression of right ventricular hypertension and hypertrophy in rats under similar conditions have also been reported by Abraham et al (1971). Leach et al (1977) studied the speed of regression of right ventricular hypertrophy, muscularization of pulmonary arterioles and polycythaemia during *intermittent* correction of hypoxia. After inducing these changes in rats, kept in a hypobaric chamber for three or more weeks, they studied their resolution in a normal environment and in an intermittently normal and hypoxic environment. In both situations right ventricular hypertrophy was resolved in six weeks and in both, pulmonary vascular changes remained unresolved after twelve weeks. The only difference noted between the two methods of recovery was that polycythaemia resolved in six weeks in the normal environment but had still not regressed after twelve weeks in the intermittently normoxic environment.

Genetic factors influencing the development of pulmonary hypertension at high altitude

So far in this account we have considered the development of pulmonary hypertension in a somewhat mechanistic fashion, perhaps suggesting that hypoxia has a stereotyped, uniform action on vascular smooth muscle. This is not true. There is much evidence to indicate that different groups and different individuals may show considerable variation in their response to the pressor effect of hypoxia on the pulmonary circulation.

Studies from the field of veterinary medicine indicate that susceptibility or resistance to the development of pulmonary hypertension at high altitude in cattle is influenced by genetic factors. Weir and his colleagues (1974) measured the pulmonary arterial pressure in 49 apparently healthy animals residing at an elevation of 3000 m in Colorado. Two groups of cattle were selected from this herd. One consisted of the 15 animals with the lowest pulmonary arterial pressure; they were designated as 'resistant' to the development of pulmonary hypertension. A second group consisted of 10 animals with clinical evidence of brisket disease (Chapter 13); they were regarded as 'susceptible' to pulmonary hypertension. Both groups were taken down to an altitude of 1500 m where the susceptible animals recovered. The animals within each group were bred and the first generation offspring were studied at low and high altitudes. Pulmonary vascular resistance increased to a much greater extent in the offspring of the susceptible group at high altitude. This suggested that the susceptible and resistant groups were genetically different. The second generation offspring of the original susceptible and resistant groups were then studied similarly. After two weeks at 3400 m the mean pulmonary arterial pressure in the susceptible animals virtually doubled from 23 to 44 mmHg with a corresponding rise in pulmonary vascular resistance. On the other hand the increase in pulmonary arterial pressure in the resistant animals from 28 to 32 mmHg was minimal. The difference increased still further after a further two weeks. These observations support the hypothesis that genetic factors are important in determining the magnitude of the tendency towards pulmonary hypertension at high altitude.

The mechanism of susceptibility is obscure. Pulmonary vascular hyper-reactivity does not seem to be the distinguishing feature since prostaglandin $F_{2\alpha}$, a potent pulmonary vasoconstrictor, exerts the same effect in susceptible and resistant cattle. Cattle prone to develop pulmonary hypertension appeared to hypoventilate compared with their resistant fellows and hence could have been subjected to a more severe hypoxic stimulus (Weir et al, 1974).

Further evidence to support this view comes from the fact that susceptible cattle had a significantly greater increase in haematocrit than did the resistant animals implying a greater hypoxic stimulus to erythropoiesis in the former. On the other hand earlier studies by members of this group (Will et al, 1962) had revealed no difference in Pa_{O_2} and Pa_{CO_2} between cattle with severe pulmonary hypertension and those with only moderate pulmonary hypertension after five months at 3030 m.

The mild pulmonary hypertension of Quechuas

At the outset of this chapter we described the characteristic development of pulmonary hypertension in the Quechua Indian of the Peruvian Andes. We did not emphasise at that point, however, that the elevated pulmonary arterial pressure of the native highlander is mild. This is without doubt a reflection of the fact that the Quechua has lived in his mountain home for countless generations over a period approaching 35 000 years (Grover, 1965).

Pulmonary hypertension in new communities at high altitude

In contrast to the highlanders of the Andes there are mountain communities like Leadville, Colorado (3100 m), which were settled barely a century ago. In such communities the population probably still contains individuals who are prone to develop pulmonary hypertension, as well as others who are not, in a manner analogous to susceptible and resistant traits in cattle. Vogel et al (1962) identified one family in which five children had pulmonary arterial pressures well above the mean values for that population. Such families in the community account for the disproportionately high mean pulmonary arterial pressure at only 3100 m when compared with the populations in the Andes.

Individual hyper-reactivity to hypoxia

There is histological evidence to support the concept of individual variation to the reaction of vascular smooth muscle to hypoxia. Wagenvoort and Wagenvoort (1973) carried out morphometric studies of the pulmonary vasculature of groups of highlanders from 21 to 58 years of age from Denver

(1600 m), Johannesburg (1800 m), Leadville, Colorado (3100 m) and from the Peruvian Andes (over 4000 m). They found the mean medial thickness of 'muscular pulmonary arteries' at sea level to be 5.1 per cent, at moderate elevation (Denver and Johannesburg) to be 4.9 per cent, and at high altitude (Leadville and Peru) to be 6.6 per cent. However, of the eight subjects living at high altitude three showed unequivocal medial hypertrophy of the parent 'muscular pulmonary arteries', with percentage medial thickness of 8.4, 8.6 and 9.8 per cent suggesting that there is individual variation in the response of the pulmonary vasculature to hypoxia at high altitude. Even at sea level there are subjects who have thick-walled pulmonary arteries for no apparent reason; such hyper-reactivity usually remains latent at sea level but it may become manifest with increasing elevation.

We have recently been able to confirm this variation in the response of the pulmonary arteries of the individual highlander to the hypoxia of high altitude. Through the courtesy of Dr Rios Dalenz we have recently been able to carry out a histological examination of specimens of lung from 20 long-term residents of La Paz, Bolivia (3800 m). The study, as yet unpublished, showed that whereas four of six Indians showed unequivocal muscularization of the terminal portions of the pulmonary arterial tree (Fig. 11.13) this change was found in only one of six

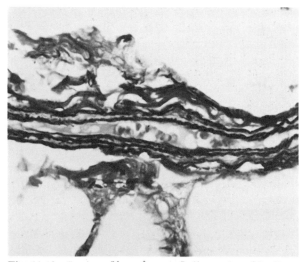

Fig. 11.13 Section of lung from an Indian native of La Paz (3800 m) aged 35 years showing muscularization of the terminal portions of the pulmonary arterial tree. A minute muscular vessel, only some 25 μm in diameter, has been cut in longitudinal section and shows a distinct media. (Elastic Van Gieson × 397).

citizens of Caucasian extraction (Fig. 11.14) and in only one of four mestizos examined. Results of this type from this small series suggest that hypoxic hypertensive pulmonary vascular disease is not invariably present in highlanders.

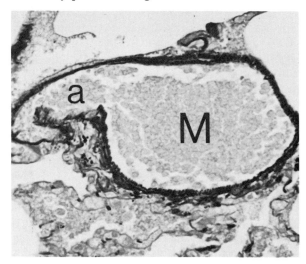

Fig. 11.14 This figure illustrates that not all highlanders have a muscular pulmonary vasculature. It shows a section of lung from an Indian native of La Paz (3800 m) aged 33 years. The muscular pulmonary artery, M, cut in transverse section, has a very thin media. An arteriolar branch, a, arises from it; this vessel has no muscular coat and its wall comprises only a single elastic lamina. (Elastic Van Gieson × 397).

Primary pulmonary hypertension at high altitude

One might be led to think that the persistent constriction of the terminal portions of the pulmonary arterial tree brought about by the stimulus of chronic hypoxia at high altitude would result in a high incidence of primary pulmonary hypertension in mountain peoples. Such is not the case and the answer probably lies in the fact that natural selection over countless generations has left a population in which the hyper-reactors to hypoxia have been removed.

However, in recently established communities at high altitude such as in the area of Leadville (3100 m) and Climax, Colorado, a few children develop primary pulmonary hypertension. Khoury and Hawes (1963) described two such cases and considered the clinical and haemodynamic data from 11 others. The affected children were usually between six weeks and six years of age and became cyanosed and breathless. They developed clinical and electrocardiographic evidence of right ventricular hypertrophy and severe pulmonary hypertension. In the patients who died, the pulmonary vasculature showed intimal proliferation and medial necrosis in the pulmonary arteries and arterioles (Khoury and Hawes, 1963). Such vascular changes are in keeping with the plexogenic pulmonary arteriopathy of primary pulmonary hypertension and quite unlike the purely muscular changes in the pulmonary arteries and arterioles which characterize people who live at high altitude. One rather odd finding reported by Khoury and Hawes (1963) is that children with primary pulmonary hypertension at high altitude improve on removal to a lower altitude. Such reversibility of pulmonary hypertension and its clinical effects are not characteristic of this condition at sea level.

The influence of age and sex on the response of the murine pulmonary circulation to simulated high altitude

Age and sex appear to influence the effect of chronic hypoxia on the pulmonary vasculature of rats at simulated high altitude. Female weanling rats will readily develop muscularization of pulmonary arterioles when kept in a decompression chamber (Abraham et al, 1971) but exposure of adult male rats to hypoxia of identical severity and duration does not appear to bring about the same degree of change in the pulmonary vasculature (Heath et al, 1973). Subsequent studies in our laboratory were carried out on young and old male and female rats subjected for 39 days to a diminished barometric pressure of 380 mmHg simulating an altitude of 5500 m above sea level (Smith et al, 1973). They confirmed that there is a greater tendency to muscularization of the pulmonary arterial tree in females.

Patent ductus arteriosus

An interesting complication of the effects of chronic hypoxia and pulmonary arterial hypertension of infants born at high altitude is the higher incidence in them of patient ductus arteriosus. It is the most frequent form of congenital cardiac anomaly in Mexico City which is at an altitude of 2380 m (Chavez et al, 1953). Natives of the Peruvian Andes show a higher incidence of patent ductus (Alzamora et al, 1953; Alzamora-Castro et al, 1960). This has been confirmed by a screening cardiovascular examination for this condition of 5000 school children of both sexes, born and living in towns between 3500 m and 5000 m above sea level (Marticorena et al, 1962). The incidence of patent ductus arteriosus at these altitudes is eighteen times that reported at sea level by different authors (Gardiner and Keith, 1951; Richards et al, 1955). Between 3500 and 5000 m the incidence is thirty times greater than at sea level (Peñaloza et al, 1964).

One can appreciate the reasons for the high incidence of patent ductus at high altitude by considering the normal mechanism of closure of this channel. In the fetus, when the lungs are not yet expanded, the blood pressure in the pulmonary arteries exceeds that in the aorta so blood flows from right to left through the ductus. Once the umbilical cord has been cut and the lungs expand, the blood pressure falls sharply in the pulmonary arteries and rises in the systemic circulation. As a result richly oxygenated blood now flows from left to right through the ductus constricting it and leading to its functional closure. Functional closure of the ductus is hindered by hypoxia. Furthermore, if the ductus has already closed functionally, the breathing of hypoxic air or gas mixture leads to pulmonary hypertension and causes the ductus to reopen (James and Rowe, 1957; Rowe and James, 1957; Eldridge and Hultgren, 1955). On the basis of such experimental studies it is readily understood how the breathing of hypoxic ambient air at high altitude will hinder closure of the ductus and lead to the complication of this congenital cardiac anomaly by pulmonary hypertension. Peñaloza and his colleagues (1964) note that the clinical picture of patent ductus arteriosus at high altitude is frequently atypical with the complication of pulmonary hypertension.

Through the kindness of clinicians and pathologists at La Paz (3800 m) we have had the opportunity to study the histology of the pulmonary blood vessels in lung biopsy specimens from cases of patent ductus arteriosus at high altitude. We find evidence of hypertensive pulmonary vascular disease of severity corresponding to grades 1 to 3 (Figs. 11.15 and 11.16) on the criteria we suggested some years ago (Heath and Edwards, 1958).

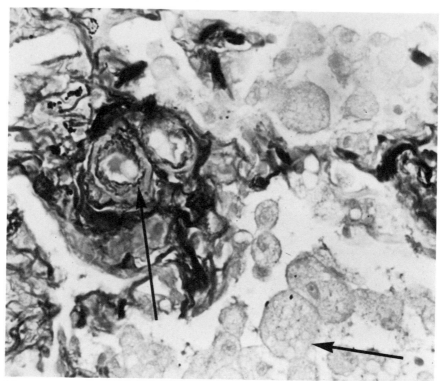

Fig. 11.15 Grade 1 hypertensive pulmonary vascular disease in a case of patent ductus arteriosus occurring in a young woman of 20 years who was born and lived all her life in Potosi (4000 m). There is muscularization of the terminal portions of the pulmonary arterial tree. This change extends distally to a remarkable extent. Thus the muscularized pulmonary arteriole indicated by the upper arrow has a diameter comparable with that of the larger macrophages in the surrounding alveolar spaces, indicated by the lower arrow. (Elastic Van Gieson × 600).

REFERENCES

Abraham, A. S., Kay, J. M., Cole, R. B. & Pincock, A. C. (1971) Haemodynamic and pathological study of the effect of chronic hypoxia and subsequent recovery of the heart and pulmonary vasculature of the rat. *Cardiovascular Research*, 5, 95.

Alzamora, V., Rotta, A., Battilana, G., Abugattas, R., Rubio, C., Bouroncle, J., Zapata, C., Santa-Maria, E., Binder, T., Subira, R., Paredes, D., Pando, B. & Graham, G. G. (1953) On the possible influence of great altitudes on the determination of certain cardiovascular anomalies. *Pediatrics*, 12, 259.

Alzamora-Castro, V., Battilana, G., Abugattas, R. & Sialer, S. (1960) Patent ductus arteriosus and high altitude. *American Journal of Cardiology*, 5, 761.

Arias-Stella, J. (1966) Mechanism of pulmonary arterial hypertension. In: *Life at High Altitudes*, p. 9. Scientific Publication No. 140. Washington: Pan American Health Organization.

Arias-Stella, J. & Recavarren, S. (1962) Right ventricular hypertrophy in native children living at high altitude. *American Journal of Pathology*, 41, 55.

Arias-Stella, J. & Saldaña, M. (1963) The terminal portion of the pulmonary arterial tree in people native to high altitudes. *Circulation*, 28, 915.

Bergofsky, E. H. (1969) Ions and membrane permeability in the regulation of the pulmonary circulation. In: *The Pulmonary Circulation and Interstitial Space*, p. 289. Edited by A. P. Fishman & H. H. Hecht. Chicago: University of Chicago Press.

Chavez, I., Espino-Vela, J., Limon, R. & Dorbecker, N. (1953) La persistencia del conducto arterial. Estudio de 200 casos. *Archivos del Instituto de cardiologia de Mexico*, 23, 687.

Civin, W. H. & Edwards, J. E. (1951) The post-natal structural changes in the intrapulmonary arteries and arterioles. *Archives of Pathology*, 51, 192.

Eldridge, D. L. & Hultgren, H. N. (1955) The physiologic closure of the ductus arteriosus in the newborn infant. *Journal of Clinical Investigation*, 34, 987.

Euler, U. S. von & Liljestrand, G. (1946) Observations on the pulmonary arterial blood pressure in the cat. *Acta physiologica Scandinavica*, 12, 301.

Fay, F. S. & Delise, C. M. (1973) Contraction of isolated smooth-muscle cells—structural changes. *Proceedings of the National Academy of Sciences U.S.A.*, 70, 641.

Fishman, A. P. (1976) Hypoxia on the pulmonary circulation. How and where it acts. *Circulation Research*, 38, 221.

Gardiner, J. H. & Keith, J. D. (1951) Prevalence of heart disease in Toronto children 1948–1949, Cardiac registry. *Pediatrics*, 7, 713.

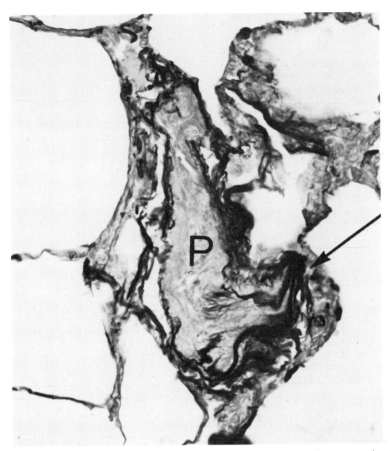

Fig. 11.16   Grade 3 hypertensive pulmonary vascular disease in a case of patent ductus arteriosus occurring in a young woman of 22 years who was born and lived all her life in La Paz (3800 m). There is muscularization of the pulmonary arteriole (arrow) which also shows intimal fibrosis. The proliferation of fibrous tissue extends into the thin-walled pre-capillary vessel (P) into which the arteriole leads. (Elastic Van Gieson × 600).

Grover, R. F. (1965) Pulmonary circulation in animals and man at high altitude. *Annals of the New York Academy of Science*, **127**, 632.

Haas, F. & Bergofsky, E. H. (1972) Role of the mast cell in the pulmonary pressor response to hypoxia. *Journal of Clinical Investigation*, **51**, 3154.

Harris, P. & Heath, D. (1977) *The Human Pulmonary Circulation. Its Form and Function in Health and Disease.* Second Edition. London: Churchill Livingstone.

Hasleton, P. S., Heath, D. & Brewer, D. B. (1968) Hypertensive pulmonary vascular disease in states of chronic hypoxia. *Journal of Pathology and Bacteriology*, **95**, 431.

Hauge, A. (1968) Role of histamine in hypoxic pulmonary hypertension in the rat. I. Blockade or potentiation of endogenous amines, kinins and ATP. *Circulation Research*, **22**, 371.

Hauge, A. & Melman, K. L. (1968) Role of histamine in hypoxic pulmonary hypertension in the rat. II. Depletion of histamine, serotonin and catecholamines. *Circulation Research*, **22**, 385.

Heath, D. (1973) The cardiopulmonary system at high altitude. *Medikon*, **9**, 10.

Heath, D. & Edwards, J. E. (1958) The pathology of hypertensive pulmonary vascular disease: a description of six grades of structural changes in the pulmonary arteries with special reference to congenital cardiac septal defects. *Circulation*, **18**, 533.

Heath, D., Edwards, C., Winson, M. & Smith, P. (1973) Effects on the right ventricle, pulmonary vasculature, and carotid bodies of the rat of exposure to, and recovery from, simulated high altitude. *Thorax*, **28**, 24.

Kay, J. M., Waymire, J. C. & Grover, R. F. (1974) Lung mast cell hyperplasia and pulmonary histamine-forming capacity in hypoxic rats. *American Journal of Physiology*, **226**, 178.

James, L. S. & Rowe, R. D. (1957) The pattern of response of pulmonary and systemic arterial pressures in newborn and older infants to short periods of hypoxia. *Journal of Pediatrics*, **51**, 5.

Keith, J. D., Rowe, R. D. & Vlad, P. (1958) *Heart diseases in infancy and childhood.* New York: Macmillan.

Khoury, G. H. & Hawes, C. R. (1963) Primary pulmonary hypertension in children living at high altitude. *Journal of Pediatrics*, **62**, 177.

Leach, E., Howard, P. & Barer, G. R. (1977) Resolution of hypoxic changes in the heart and pulmonary arterioles of rats during intermittent correction of hypoxia. *Clinical Science and Molecular Medicine*, **52**, 153.

Levy, A. M., Tabakin, B. S., Hanson, J. S. & Narkewicz, R. M. (1967) Hypertrophied adenoids causing pulmonary hypertension and severe congestive heart failure. *New England Journal of Medicine*, **227**, 506.

Marticorena, E., Peñaloza, D., Severino, J. & Hellriegel, K. (1962) Incidencia de la persistencia del conducto arterioso en las grandes alturas. *Memorias del IV Congreso Mundial de Cardiologie, Mexico*, **1-A**; 155.

Mithoefer, J. C. (1966) The respiration of Andean natives, In: *Life at High Altitudes*, p. 21. Pan American Health Organization. Washington: Scientific Publication No. 140.

Mungall, I. P. F. (1976) Hypoxia and lung mast cells: influence of disodium cromoglycate. *Thorax*, **31**, 94.

Mungall, I. P. F. & Barer, G. F. (1975) Lung vessels and mast cells in chronically hypoxic rats. In: *Progress in Respiration Research*, Vol. 9, *Pulmonary Hypertension*, p. 144. Basle: Karger.

Peñaloza, D., Sime, F., Banchero, N. & Gamboa, R. (1962) Pulmonary hypertension in healthy man born and living at high altitudes. *Medicina Thoracalis*, **19**, 449.

Peñaloza, D., Sime, F., Banchero, N., Gamboa, R., Cruz, J. & Marticorena, E. (1963) Pulmonary hypertension in healthy men born and living at high altitudes. *The American Journal of Cardiology*, **11**, 150.

Peñaloza, D., Arias-Stella, J., Sime, F., Recavarren, S. & Marticorena, E. (1964) The heart and pulmonary circulation in children at high altitudes. Physiological, anatomical and clinical observations. *Pediatrics*, **34**, 568.

Reeves, J. T., Leathers, J. E., Eisemann, B. & Spencer, F. C. (1962) Alveolar hypoxia versus hypoxaemia in the development of pulmonary hypertension. *Medicina Thoracalis*, **19**, 369.

Richards, M. R., Merritt, K. K., Samuels, M. H. & Longmann, A. G. (1955) Congenital malformations of the cardiovascular system in a series of 6053 infants. *Pediatrics*, **15**, 12.

Rowe, R. D. & James, L. S. (1957) The normal pulmonary arterial pressures during the first year of life. *Journal of Pediatrics*, **51**, 1.

Rudolph, A. M. & Cayler, G. G. (1958) Cardiac catheterization in infants and children. *Pediatric Clinics of North America*, **5**, 907.

Rudolph, A. M., Drorbaugh, J. E., Auld, P. A. M., Rudolph, A. J., Nadas, A. S., Smith, C. A. & Hubbell, J. P. (1961) Studies on the circulation in the neonatal period. The circulation in the respiratory distress syndrome. *Pediatrics*, **27**, 551.

Sime, F., Banchero, N., Peñaloza, D., Gamboa, R., Cruz, J. & Marticorena, E. (1963) Pulmonary hypertension in children born and living at high altitudes. *American Journal of Cardiology*, **11**, 143.

Smith, P. & Heath, D. (1977) Ultrastructure of hypoxic hypertensive pulmonary vascular disease. *Journal of Pathology*, **121**, 93.

Smith, P., Moosavi, H., Winson, M. & Heath, D. (1973) The influence of age and sex on the response of the right ventricle, pulmonary vasculature, and carotid bodies to hypoxia in rats. *Journal of Pathology*, **112**, 11.

Tucker, A., McMurtry, I. F., Alexander, A. F., Reeves, J. T. & Grover, R. F. (1977) Lung mast cell density and distribution in chronically hypoxic animals. *Journal of Applied Physiology*, **42**, 174.

Vogel, J. H. K., Weaver, W. F., Rose, R. L., Blount, S. G. & Grover, R. F. (1962) Pulmonary hypertension and exertion in normal man living at 10 150 feet. In: *Normal and Abnormal Pulmonary Circulation*, p. 269. Edited by R. F. Grover. Basle: Karger.

Wagenvoort, C. A. & Wagenvoort, N. (1973) Hypoxic pulmonary vascular lesions in man at high altitude and in patients with chronic respiratory disease. *Pathologica et Microbiologia*, **39**, 276.

Weir, E. K., Tucker, A., Reeves, J. T., Will, D. H. & Grover, R. F. (1974) The genetic factor influencing pulmonary hypertension in cattle at high altitude. *Cardiovascular Research*, **8**, 745.

Will, D. H., Alexander, A. F., Reeves, J. T. & Grover, R. F. (1962) High altitude-induced pulmonary hypertension in normal cattle. *Circulation Research*, **10**, 172.

Williams, A., Heath, D., Kay, J. M. & Smith, P. (1977) Lung mast cells in rats exposed to acute hypoxia, and chronic hypoxia with recovery. *Thorax*, **32**, 287.

12

The pulmonary trunk

The pulmonary trunk and its branches, the right and left main pulmonary arteries, conduct the flow of blood from the right ventricle into the lungs and the structure of these vessels is modified by the pressure within them. In particular the thickness of the media is related to the presence of pulmonary arterial hypertension and the pattern of the elastic tissue within it to the time of onset of the raised presssure. As a result the microanatomy of the pulmonary trunk found in natives at high altitude differs from that seen in lowlanders. To understand the form of the pulmonary trunk at high altitude it is necessary to relate it to what is found at sea level.

Medial thickness of the pulmonary trunk

At sea level. In the fetus the media of the pulmonary trunk is as thick as that of the aorta (Heath et al, 1959) since it is subjected to physiological pulmonary hypertension at this stage of life. During the first few days of life the mean pulmonary arterial pressure averages about 30 mmHg, although there is a wide variation in this figure. For purposes of comparison it must be remembered that the mean systemic arterial pressure is low at this time of life, averaging about 50 mmHg (Harris and Heath, 1962). At birth at sea level with the sudden expansion of the lungs there is a precipitate fall in pulmonary arterial pressure. In the ensuing first months of life this lowered intravascular pressure in the pulmonary trunk is associated with thinning of its media. As a result, in infants between 6 and 24 months the ratio of the thickness of the media of the pulmonary trunk to that of the aorta (PT/A ratio) falls to a range of 0.4 to 0.8 from the ratio of unity characteristic of fetal life. At sea level the pulmonary arterial pressure remains low throughout the rest of

life and as a consequence of this PT/A is in the range of 0.4 to 0.7.

At high altitude. In natives at high altitude a mild degree of pulmonary arterial hypertension persists from birth as we have seen in Chapter 11. Hence the stimulus for the relative thinning of the media of the pulmonary trunk never occurs. Thus Saldaña and Arias-Stella (1963c) studied necropsy specimens of pulmonary trunk in 100 persons who were born and lived permanently in places between 3490 and 4540 m above sea level. The ages of the cases varied from birth to 78 years. A hundred control cases from sea level were also studied. These authors found that the absolute thickness of the media of the pulmonary trunk is greater at high altitude than at sea level throughout life. PT/A is also elevated in high altitude subjects. It should be noted, however, that the elevation of this ratio is brought about not only by an absolute increase in the medial thickness of the pulmonary trunk, but by an associated absolute decrease in the thickness of the aortic media. This aortic thinning appears to be related to the existence of a lower systemic blood pressure in high altitude subjects: this subject is discussed further in Chapter 17.

The speed of development and regression of hypertrophy of the pulmonary trunk at simulated high altitude

Experiments in which animals are exposed to simulated high altitude in a hypobaric chamber illustrate how rapidly the media of the pulmonary trunk hypertrophies in response to the chronic hypoxia of diminished barometric pressure. In one such investigation adult male Wistar albino rats were exposed for five weeks to a reduced barometric pressure of 380 mmHg, equivalent to an altitude of 5500 m above sea level (Heath et al, 1973). In that time the PT/A ratio rose from the control value of 0.45 to 0.95. This remarkably rapid hypertrophy regressed almost back to normal (PT/A of 0.51) in another group of rats, exposed to hypoxia of the same length and severity, which was then allowed to recover in room air for the same short period of five weeks. This equally remarkable rapid regression depends upon the largely reversible nature of the muscular changes in the terminal portions of the pulmonary arterial tree that characterize hypoxic hypertensive pulmonary vascular disease (Chapter 11).

Muscular evaginations

When smooth muscle cells contract, evaginations of their cytoplasm protrude from their surface so that it no longer remains smooth but becomes covered by bulbous excrescences. These evaginations occur between the points of sarcolemmal thickening to which are attached the intracytoplasmic actin filaments. The cytoplasm of the evaginations is free of myofilaments and organelles, thus appearing in electron micrographs paler and clearer than that of the parent muscle cell from which it takes origin (Smith et al, 1978). Such evaginations are to be found in control material especially when the lung has been allowed to collapse after death, but they are more pronounced in constricting muscle, including the vascular muscle of the elastic and muscular pulmonary arteries and especially of the pulmonary veins. We have studied such evaginations in the pulmonary trunks of rats subjected to a simulated altitude of 5500 m (Smith et al, 1978). The evaginations insinuate through deficiencies in the internal elastic lamina to press on to the outer aspects of the endothelial cells of the intima, distorting them (Fig. 12.1). In this situation the evaginations are connected to their parent cell by only a narrow isthmus and have so clear a cytoplasm as to resemble a fluid-filled cyst rather than a continuation of a muscle cell (Fig. 12.1). A search for such evaginations of smooth muscle in the pulmonary blood vessels of highlanders has not been made but would probably prove positive.

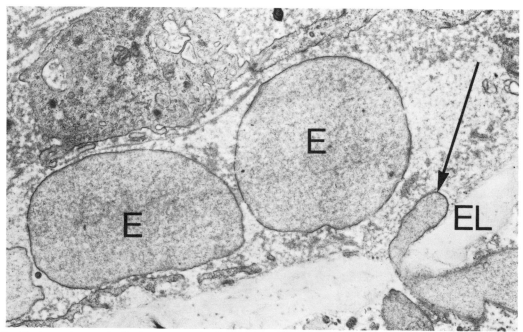

Fig. 12.1 Electron micrograph of pulmonary trunk of Wistar albino rat subjected for four weeks to a barometric pressure of 380 mmHg which simulates an altitude of 5500 m. Evaginations (E) of vascular smooth muscle cells are seen. Their substance is electron-lucent with total loss of cellular detail. They bear no resemblance structurally to the muscle from which they are derived. In the lower right-hand corner, a small muscular evagination (arrow) has extended through a gap in the internal elastic lamina (EL). (Electron micrograph × 12 500).

The pattern of elastic tissue in the media of the pulmonary trunk

At sea level. In the fetus the pulmonary trunk is subjected to a similar pressure as that found in the aorta. As a result not only is it as thick as the aorta but the pattern of elastic tissue in its media is reminiscent of that of the aorta. The configuration of the elastic fibrils in the fetal pulmonary trunk is so similar to that in the aorta, that at a glance the two may be confused, but differences do exist which make identification of elastic tissue from the fetal pulmonary trunk clear after careful examination (Fig. 12.1). The fibrils in the pulmonary trunk are

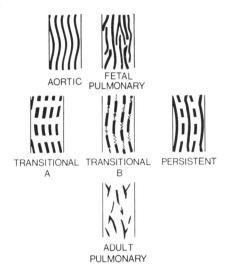

Fig. 12.2 The elastic tissue patterns of the pulmonary trunk at high altitude. The elastic tissue of the aortic media is in the form of long elastic fibrils which are parallel to one another and packed together compactly. The elastic tissue of the media of the fetal pulmonary trunk is also packed tightly but the fibrils are shorter and branched. On the following line are the transitional forms which are found in the evolution to the adult pulmonary configuration in which the elastic fibrils from an open network of branched, irregularly-shaped fibrils. At sea level transitional forms A and B are found. In Form A the long elastic fibrils fragment into rods. In Form B the rods are still connected by very thin elastic fibrils. At high altitude the 'persistent configuration' is found. In this, long continuous elastic fibrils are found in association with short stick-like rods. These different elastic tissue patterns in the pulmonary trunk are referred to in the text (Based on data from Saldaña and Arias-Stella, 1963a, b).

fewer and, although mostly parallel to each other, are less regular. They are somewhat coarser and they branch in places; they are shorter than the aortic fibrils, and even very short ones may be found (Fig.

12.2). The thickness of the fibrils is less uniform along their length, the ends of some showing small club-like expansions. These differences are not superficially evident, however, and the configuration of fibrils in the fetal pulmonary trunk appears generally parallel and compact like that in the aorta. Since the similarities of the media of the fetal pulmonary trunk to that of the fetal aorta outweigh the differences, the fetal pulmonary trunk may be said to have an *aortic* configuration (Heath et al, 1959).

This pattern of elastic tissue is seen in the major pulmonary arteries at birth and until the age of about six months when changes which no doubt started at birth become evident. The elastic fibrils still tend to be parallel but divide transversely into much more numerous, short, stick-like structures (Transitional A pattern in Fig. 12.2) some of which show an accentuation of the clubbed terminations referred to above. As a result of this fragmentation compactness is lost and the elastic tissue pattern becomes more open and loose. Hence at sea level a glance at a section of the pulmonary trunk in infancy is already sufficient for recognising it as pulmonary: at the same time the tendency for the fragments of elastic fibrils to be parallel to one another and to retain a fair uniformity of width demonstrates their derivation from an aortic type of elastica.

When we originally described this elastic tissue pattern we designated it the *transitional* pattern. Subsequently Saldaña and Arias-Stella (1963a) termed it the '*Transitional A*' pattern (Fig. 12.2). They described in addition a '*Transitional B*' pattern in which the zones of apparent fragmentation really correspond to segments of variable length where the compact elastic fibrils have been replaced by numerous delicate elastic fibrils that maintain the continuity of the long fibres (Fig. 12.2).

Saldaña and Arias-Stella (1963a) also described a '*persistent*' type of elastic pattern in the pulmonary trunk. In this the pulmonary artery shows such a high content of elastic fibres that at first sight it resembles the aorta. However, there is a greater degree of fragmentation. There is a combination of long thick fibres with others which are markedly fragmented, the long ones being more numerous (Fig. 12.2). Both of these elastic tissue patterns are seen in the pulmonary trunk at sea level. There is evidence to suggest that the physiological basis for a transitional A pattern of elastica is a pronounced postnatal fall of pulmonary pressure which reaches the lowest levels of normality. On the other hand the high content of long elastic fibres in cases of 'persistent pattern' could be related to a postnatal fall in pulmonary pressure detained at the highest levels of normality. Levels of pulmonary arterial pressure with the transitional B pattern would be intermediate between those found with transitional A and persistent patterns.

By the end of the second year of life at sea level the adult pulmonary type of elastic pattern of the media has become established. In this the elastic tissue is irregular and more sparse than in the aorta. Widely spread, the fibrils assume grotesque shapes with many club-like terminal expansions (Fig. 12.2). Individual laminae are short and can be traced for only short distances, while numerous slender fibrils intervene. Branching of individual fibrils is common, and the laminae run in all directions. It will be seen that this pattern is distinct from that of the aorta for the elastic laminae are not long, parallel or uniform. Fenestrations are numerous and occasionally clumps of amorphous elastic tissue are found. A thick internal elastic lamina forms and there are zones of compressed elastic fibrils in the innermost and outermost parts of the media. There is much fragmentation of the elastic fibrils.

At high altitude. In people born and living permanently at high altitudes, the evolution of the elastic configuration of the pulmonary trunk differs notably from that observed at sea level. An abnormal maintenance of the 'aortic' type of pulmonary trunk and a high incidence of the 'persistent' configuration, even in adult life, are the main features in high altitude dwellers compared with those at sea level (Fig. 12.3). Saldaña and Arias-Stella (1963b) studied the pulmonary trunk of 267 persons native to the central Andean region of Peru. They found that in places between 4040 m and 4540 m above sea level there is a maintenance of the 'aortic' type of elastic pattern up to the age of nine years (Fig. 12.3). This elastic type evolves exclusively into the 'persistent' configuration which is commonly observed for the rest of life (Fig. 12.4). This is clearly related to the persistence of a mild pulmonary arterial hypertension from birth. However, according to Saldaña and Arias-Stella (1963b) there may be a late conversion of the 'persistent' type into the 'adult' type late in middle age, especially after 60 years of age (Fig. 12.3). The explanation for this is obscure at present.

In places located between 3440 m and 3840 m the 'aortic' type of pulmonary trunk is retained up to three years (Fig. 12.3). Cases with the 'transitional' configuration occur but they rapidly evolve into the 'persistent' type. This pattern is found mainly in childhood and adolescence and is also present in a great proportion of young adults. By the age of 55 years the pattern of elastic tissue is of the adult

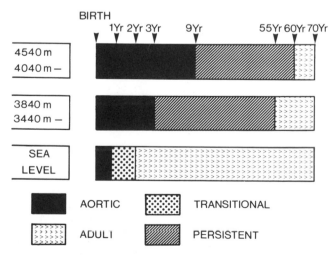

Fig. 12.3   The elastic tissue patterns in the media of the pulmonary trunk throughout life at sea level and at high altitude. The two ranges of elevation studied were 3440 m to 3840 m, and 4040 m to 4540 m. Note that at high altitude the 'persistent configuration' of elastic tissue pattern extends into middle age and beyond. (Based on data from Saldaña and Arias-Stella 1963b).

pulmonary type, but once again the reason for this late transition is not clear.

Chemical composition of the pulmonary trunk in the highlander

As would be expected, the different histological appearance of the pulmonary trunk of the highlander referred to above is reflected in its chemical composition (Heath, 1966; Castillo et al, 1967). There is a higher proportion of elastin in the pulmonary trunk with a persistent configuration of elastic tissue. Thus in 41 specimens of that artery showing a normal adult pulmonary configuration of elastic tissue in subjects from Lima at sea level (Fig. 12.4 right) the average elastin content was 27.2 per cent with a tendency for the content to increase with age. In contrast, the elastin contents of the pulmonary trunks showing a persistent configura-

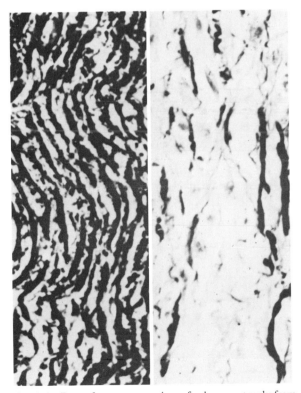

Fig. 12.4 Parts of transverse sections of pulmonary trunks from a native highlander born and bred at an altitude of 4330 m in the Peruvian Andes (left) and a sea-level mestizo (right). The coastal dweller shows the open network of branched, irregularly-shaped fibrils typical of the adult pulmonary trunk. The highlander shows the more compact, persistent configuration of elastic tissue described in the text. (Both Elastic Van Gieson, × 150).

tion of elastic tissue in four highlanders native to 4330 m (Fig. 12.4 left) were 31.0, 31.4, 37.9 and 40.0 per cent. Moreover, it should be noted that the average age of these four highland subjects was only 22.5 years compared to an average age of 45.5 years for the low-altitude group (Heath, 1966; Castillo et al, 1967).

In contrast the collagen content of the pulmonary trunk in three native highlanders from 4330 m did not differ appreciably from that in 28 sea-level subjects. The average collagen content in those from low altitude was 31.0 per cent whereas that in the three highlanders was respectively 28.3, 31.7 and 37.9 per cent (Heath, 1966; Castillo et al, 1967). When collagen and elastin were both estimated in 27 specimens with an adult pulmonary pattern of elastic tissue, there proved to be a positive relation between the ratio of elastin to collagen and the age of the subject. When this ratio was calculated in specimens from three highlanders native to 4330 m, it was found to be inappropriately high for the age (Heath, 1966; Castillo et al, 1967).

Extensibility of the pulmonary trunk in the highlander

The relation between length and tension in any artery is a curved one, so that the greater the tension applied the less extensible is the vessel (Harris and Heath, 1977) (Fig. 12.5). This state of affairs is precisely the reverse of that which occurs in an

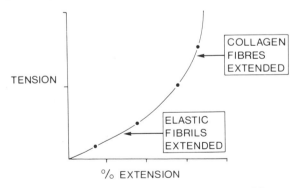

Fig. 12.5 Idealized extensibility curve of an artery. The more the artery is stretched the less extensible it becomes. This is because, when low tension is applied to it, the highly extensible elastic fibrils are stretched. On the other hand, when greater tension is applied, the much less extensible collagen fibres are stretched. Thus it will be seen that the characteristic extensibility curve of an artery is produced by the extension of two sets of fibres.

inorganic material such as a wire. The basis for this remarkable phenomenon was explained by the studies of Roach and Burton (1957) who demonstrated that the initial more horizontal part of the arterial extensibility curve registers the extension of elastic fibrils, whose relaxed length is shorter relative to that of collagen, while the later more vertical portion registers the extension of collagen fibres. Hence the arterial extensibility curve is in reality two curves, one for elastic fibrils and one for collagen fibres (Fig. 12.5). We have previously found that the extensibility of the pulmonary trunk in low-level subjects conforms to that of the idealised arterial extensibility curve the degree of extension falling with age (Harris et al, 1965) (Fig. 12.6). This finding is consistent with the increasing proportion of elastin with age.

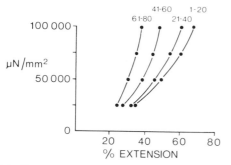

Fig. 12.6 Extensibility curves of the pulmonary trunk for different age groups indicated over the curves. It is apparent that this vessel becomes progressively less extensible with increasing age.

As we have just seen above, the pulmonary trunk of the native highlander is abnormal in that it contains an excess of elastic tissue, as demonstrated visibly by an 'aortic' or 'persistent' configuration of elastic tissue, and chemically by an increased content of elastin. At the same time the collagen content of the pulmonary trunk of the highlander is no different from that of the sea-level dweller. This being the case, one would anticipate that the extensibility of the pulmonary trunk of the highlander would be abnormally low at a *low* extensile force. This is precisely what one finds. Subjects indigenous to high altitude show an abnormally low extensibility of their pulmonary trunk at a low extensile force of 25 000 μN mm^2 (Heath, 1966). In one of our investigations (Castillo

et al, 1967) we found that in Quechuas in the Andes there was an abnormally high ratio of extension at 100 000 μN mm^2 to that at 50 000 μN mm^2 and it seems likely that this is due to the normal proportion of collagen in the pulmonary trunk of the highlander referred to above, coupled with the excess elastic tissue so that the ratio of elastin to collagen was abnormally high.

When the degree of extension is expressed as a percentage of that predicted, it is found that the pulmonary trunk of the native highlander with an aortic or persistent configuration of elastic tissue shows a maximal decrease in extensibility at a low degree of extension (Fig. 12.7). The extensibility of

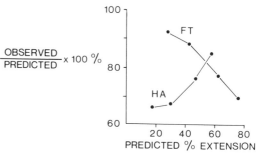

Fig. 12.7 The relation between the degree of extension of the pulmonary trunk and the degree of extension expressed as a percentage of that predicted. In high altitude subjects (HA) the extensibility of this vessel on slight extension is less than would be anticipated and this is because the pulmonary trunk contains excessive amount of elastic tissue giving rise to the 'persistent pattern' of elastica in the media. In contrast in patients with Fallot's tetrad (FT) the extensibility of this vessel is normal on slight extension but it is less than would be expected on further extension because the media is composed of excessive amounts of collagen with a sparsity of elastic tissue. The greater the tension applied to the pulmonary trunk of the highlander the more normal becomes the extensibility of the vessel and this is because it contains normal amounts of collagen.

the pulmonary trunk of the native highlander makes an interesting contrast with that of the patient with Fallot's tetrad. In the former, as we have seen, there is an excess of elastic fibrils and no increase in collagen, so that the observed extensibility is lower than anticipated with low levels of tension but rises with increasing tension (Curve HA in Fig. 12.7). In Fallot's tetrad, however, there is a paucity of elastic tissue and increased amounts of collagen, so that the observed extensibility falls with increasing tension (Curve FT in Fig. 12.7).

The pulmonary trunk of the llama

We may at this point briefly consider the pulmonary trunk of an indigenous high altitude animal, the llama. In contrast to man the media of the pulmonary trunk in this species is highly muscular with thick bundles of smooth muscle in the outer media separated from one another by dense bands of elastic fibrils (Heath et al, 1968, 1974). The mean percentage of dry weight of the vessel is composed of 24.0 per cent of elastin and 20.2 per cent of collagen. The extensibility of the tissue of the pulmonary trunk of the llama is greater than the average found in the first two decades of life in man (Castillo et al, 1967) and it seems that this is because such extensibility is simply a direct function of time rather than a proportional funtion of the average longevity of this species. The proportion of collagen in the pulmonary trunk of the llama is lower than the figure of 31 per cent found in man (Castillo et al, 1967). The histology, extensibility and chemical composition of the pulmonary trunk of the llama do not differ significantly from those of the pulmonary trunk of other animals at high altitude such as cattle and sheep.

An application to human disease

The alteration of the structure of the media of the pulmonary trunk of highlanders by a persistence of pulmonary arterial hypertension from birth has its counterpart in congenital heart disease. In a child with a large post-tricuspid shunt such as a ventricular septal defect or a patent ductus arteriosus the pulmonary circulation is similarly exposed to raised levels of pulmonary arterial pressure and flow from birth. Under these circumstances an aortic pattern of elastic tissue is found in the media of the pulmonary trunk and the normal sea-level transition to an adult pulmonary configuration of elastica does not take place (Heath et al, 1959).

The endothelium of the pulmonary trunk at simulated high altitude

The endothelial cells of the pulmonary trunk have a different size and shape from those of the aorta. These differences are well demonstrated when the cell borders are stained by silver salts. The endothelial cells of the pulmonary trunk are rounded or rectangular (Fig. 12.8) whereas those of

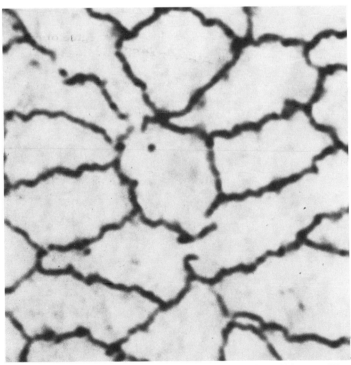

Fig. 12.8 Endothelial pavement pattern of the pulmonary trunk of a control Wistar albino rat. The cells are rounded or rectangular. (Florey silver method, × 1000).

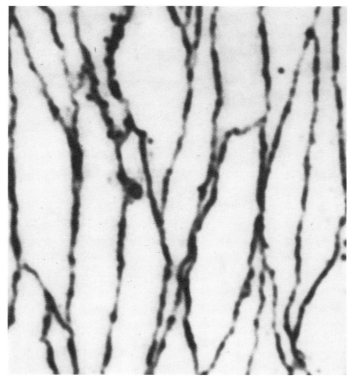

Fig. 12.9 Endothelial pavement pattern of the ascending aorta of a control Wistar albino rat. The cells are fusiform with the long axis pointed in the direction of blood flow. (Florey silver method, × 1000).

the aorta are smaller and fusiform being pointed in the direction of blood flow (Fig. 12.9). When rats are subjected to simulated high altitude in a decompression chamber, they develop pulmonary hypertension as a consequence of constriction and muscularization of the terminal portions of the pulmonary arterial tree as described in Chapter 11. At the same time their pulmonary trunk becomes dilated and hypertrophied. The endothelial pavement pattern becomes disrupted with many of the cells assuming a fusiform shape reminiscent of aortic endothelium (Fig. 12.10) (Kombe et al, 1980). Many small endothelial cells are found in the

pulmonary trunk suggesting division of cells to line the enlarging blood vessel. In contrast the endothelial cells of the inferior vena cava merely increase in size to cope with the dilatation of the vein. These changes in the endothelial lining of the pulmonary trunk at simulated high altitude appear to be related to pulmonary hypertension rather than to hypoxaemia since the same alterations in the endothelial cell pavement pattern can be induced in rats by the administration to them of *Crotalaria spectabilis* seeds which lead to an elevated pulmonary arterial pressure (Kibria et al, 1980).

REFERENCES

Castillo, Y., Krüger, H., Arias-Stella, J., Hurtado, A., Harris, P. & Heath, D. (1967) Histology, extensibility and chemical composition of pulmonary trunk in persons living at sea level and high altitude in Peru. *British Heart Journal*, **29**, 120.

Harris, P. & Heath, D. (1962) In: *The Human Pulmonary Circulation. Its Form and Function in Health and Disease*, p. 145. Edinburgh: Livingstone.

Harris, P. & Heath, D. (1977) *The Human Pulmonary Circulation.*

Its Form and Function in Health and Disease. (Second Edition). Edinburgh: Churchill Livingstone.

Harris, P., Heath, D. & Apostolopoulos, A. (1965) The extensibility of the human pulmonary trunk. *British Heart Journal*, **27**, 651.

Heath, D. A. (1966) Morphological patterns: the structure, composition, and extensibility of the pulmonary trunk at sea level and high altitude in Peru. In: *Life at High Altitudes*, p. 13.

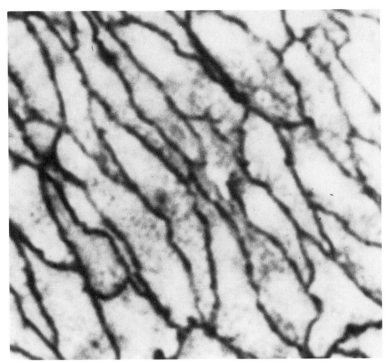

Fig. 12.10 Endothelial pavement pattern of the pulmonary trunk of a Wistar albino rat subjected to a reduced barometric pressure of 380 mmHg (simulating an altitude of 5500 m) for one month. The normally large and rounded endothelial cells have changed into smaller, fusiform cells pointed in the direction of blood flow and hence reminiscent of aortic endothelium. (Florey silver method, × 1000).

Washington: Pan American Health Organization Scientific Publication No. 140.

Heath, D., Edwards, C., Winson, M. & Smith, P. (1973) Effects on the right ventricle, pulmonary vasculature, and carotid bodies of the rat on exposure to, and recovery from, simulated high altitude. *Thorax*, **28**, 24.

Heath, D., Harris, P., Castillo, Y. & Arias-Stella, J. (1968) Histology, extensibility and chemical composition of the pulmonary trunk of dogs, sheep, cattle and llamas living at high altitude. *Journal of Pathology and Bacteriology*, **96**, 161.

Heath, D., Smith, P., Williams, D., Harris, P., Arias-Stella, J. & Krüger, H. (1974) The heart and pulmonary vasculature of the llama (*Lama glama*). *Thorax*, **29**, 463.

Heath, D., Wood, E. H., DuShane, J. W. & Edwards, J. E. (1959) The structure of the pulmonary trunk at different ages and in cases of pulmonary hypertension and pulmonary stenosis. *Journal of Pathology and Bacteriology*, **77**, 443.

Kibria, G., Heath, D., Smith, P. & Biggar, R. (1980) Pulmonary endothelial pavement patterns. *Thorax*, **35**, 186.

Kombe, A. H., Smith, P., Heath, D. & Biggar, R. (1980)

Endothelial cell pavement patterns in the pulmonary trunk in rats in chronic hypoxia. *British Journal of Diseases of the Chest*. In press.

Roach, M. R. & Burton, A. C. (1957) The reason for the shape of the distensibility curves of arteries. *Canadian Journal of Biochemistry and Physiology*, **35**, 681.

Saldaña, M. & Arias-Stella, J. (1963a) Studies on the structure of the pulmonary trunk. I. Normal changes in the elastic configuration of the human pulmonary trunk at different ages. *Circulation*, **27**, 1086.

Saldaña, M. & Arias-Stella, J. (1963b) Studies on the structure of the pulmonary trunk. II. The evolution of the elastic configuration of the pulmonary trunk in people native to high altitudes. *Circulation*, **27**, 1094.

Saldaña, M. & Arias-Stella, J. (1963c) Studies on the structure of the pulmonary trunk. III. The thickness of the media of the pulmonary trunk and ascending aorta in high altitude natives. *Circulation*, **27**, 1101.

Smith, P., Heath, D. & Padula, F. (1978) Evaginations of smooth muscle cells in the hypoxic pulmonary trunk. *Thorax*, **33**, 31.

13

Brisket disease and pulmonary arterial pressor response in animals

In Chapter 16 we shall consider the clinical and pathological features of a syndrome in Andean natives which has been succinctly termed Monge's disease. This is widely regarded as a result of loss of acclimatization to high altitude and it usually necessitates removal of the patient to sea level. Cattle may also suffer from a disease brought about by exposure to a mountain environment and it too may prove fatal unless the affected animal is taken to lower pastures. This condition is termed 'brisket disease' or 'high mountain disease'. In describing its features in this chapter we shall show that it should not be regarded as a bovine form of Monge's disease.

Geographical location

The disease occurs in certain areas in the intermountain region of the United States notably in Utah and Colorado (Hecht et al, 1959). For over a century it has been the custom of ranchers in this area to graze livestock during the summer months from July to October at altitudes ranging from 2500 to 3700 m (Hecht et al, 1962a). In Utah the disease is most prevalent in mountain ranges in two main regions. One is the southern slopes of the Unitah range and the other is along the western slopes of the Wasatch mountains (Hecht et al, 1959). It is of interest to note that the disease is generally regarded as being restricted to this region of the United States. Monge and Monge (1966) refer to the reports of Cuba-Caparó (1949, 1950) who described cases of high altitude disease in lambs born in Peru. It should be noted, however, that in these cases polycythaemia was prominent and as we shall see below this haematological finding is not characteristic of brisket disease. Furthermore oedema was absent.

Species involved

Apart from these isolated reports from Peru the disease seems to occur only in the European type of cattle (*Bos taurus*). It has been reported in Hereford cattle and also in the Holstein, Ayrshire, Angus, Shorthorn, Jersey and Swiss breeds. The disease has not been seen in French Charolais cattle. As might be anticipated species indigenous to high altitude areas do not develop brisket disease. Thus Indian and African bovine species living respectively in the Himalayas and the highlands of Ethiopia tolerate high altitude well. The yak (*Bos grunniens*), of the Himalayas and the Tibetan plateau, is also resistant to the disease. Although the American buffalo is an animal of the plains, a herd of this species lives without difficulty at 3000 m in Utah (Hecht et al, 1962a).

Clinical features

In 1915 reports began to appear in the literature describing the occurrence of a fatal disease in cattle characterized by systemic oedema (Newsom, 1915; Glover and Newsom, 1915, 1917, 1918). The oedema found in the disease is so severe that Hecht et al (1962a) believe it unlikely that cases occurring before 1915 would have been overlooked. They consider that it may be a new disease. According to them the incidence of the disease in cattle has varied from herd to herd, from year to year, and from location to location.

High mountain disease develops mostly in calves brought to high altitude for the first time but it may occur in young animals born in the highlands. The condition may occur in adult cattle in which it tends to run a protracted course (Hecht et al, 1959); 1 to 5 per cent of the animals in affected herds may be involved.

There is intense dyspnoea so that the animal is unable to tolerate even mild exertion. Even if the animal survives grazing, the excitement of round-up may lead to death. The mucous membranes are cyanosed. The affected calf has a rough coat and its head is lowered with droopy ears (Hecht et al, 1962a). A foul-smelling diarrhoea termed 'scours' is common. In the late stages of the disease there is oedema of the dependent parts of the trunk. Such oedema occurs particularly in the region between the forelegs and the neck, the 'brisket' of commerce, and hence the condition is commonly referred to as 'brisket disease' (Fig. 13.1). Associated odema of

Fig. 13.1   Brisket disease. There is oedema between the forelegs and the neck (the 'brisket' of commerce) and associated oedema of the lower jaw.

the lower jaw may give the affected animal an appearance reminiscent of mumps. There is no oedema of the legs. Commonly the sytemic oedema is accompanied by pleural effusion and ascites. The jugular veins are distended and there is commonly a pronounced systolic wave indicative of tricuspid incompetence. The breath sounds are vesicular. On auscultation of the heart there is accentuation of the second sound in the pulmonary area. There is commonly a loud, low-pitched systolic murmur in the apical region consistent with tricuspid incompetence. Frequently there is a presystolic apical gallop. These are the features of right ventricular failure with tricuspid incompetence secondary to pulmonary arterial hypertension. The essence of the problem is to elucidate the basis for the raised pulmonary arterial pressure.

Pathogenesis of the pulmonary hypertension

Three factors have to be considered in respect of the pathogenesis of the elevated arterial pressure. They are the vegetation eaten by the grazing cattle, the daily salt intake and the high altitude at which the cattle live.

In the areas in which the disease is enzootic the vegetation comprises salt grass, marsh marigolds, ranunculus and groundsell. According to Hecht et al (1959) analysis of the flora has not revealed any toxic alkaloids, although it has to be borne in mind that some species of *Crotalaria* (Kay and Heath, 1969) and *Senecio* (Burns, 1972) contain pyrrolizidine alkaloids that may give rise to pulmonary hypertension.

The areas in which the disease occurs are studded with small water pools and creeks and in addition large troughs containing salt are spaced at irregular intervals (Hecht et al, 1962a). These authors point out that affected cattle are likely to have ingested an excess of salt and it is conceivable that this may play some part in the development of the condition.

However, the important factor in the aetiology of the pulmonary hypertension is almost certainly the effects of high altitude. As early as 1918 Glover and Newsom stated: 'We, therefore, have no hesitancy in concluding that the malady is due to failure of acclimatization at high altitudes'. Hecht et al (1962a) and Reeves et al (1960) also accept that influences inherent in high altitude bring about pulmonary arterial hypertension and right ventricular failure.

Reversibility of the pulmonary hypertension

When calves with acute brisket disease are brought down from their summer grazing ranges in the mountains to Salt Lake City at an altitude of 1370 m their clinical abnormalities disappear in four to six weeks (Kuida et al, 1963). At the same time there is a pronounced fall in their pulmonary arterial pressure. Thus, when haemodynamic studies were made on 14 calves that recovered from brisket disease, the mean pulmonary arterial pressure was found to have halved from 63 to 32 mmHg. Pulmonary vascular resistance, right atrial pressure and pulmonary wedge pressure all fell on recovery at the lower altitude. These studies make it clear that the cause of brisket disease is pulmonary hypertension induced by high altitude and that the basis for this is rapidly and completely reversible.

Pulmonary hypertension in normal cattle at high altitude

An understanding of the nature of brisket disease comes from an appreciation of the fact that normal cattle taken to high altitude develop pulmonary hypertension. Will et al (1962) showed that when 10 steers born at an altitude of 1100 m were taken to 3050 m, six showed an elevation of pulmonary arterial pressure from 27 to 45 mmHg. The remaining four animals developed more severe pulmonary hypertension and in two instances the mean pressure exceeded 100 mmHg. In such steers there is no significant increase in cardiac output or blood volume and there is no elevation of pulmonary wedge pressure. It is thus clear that pulmonary hypertension is the result of increased pulmonary vascular resistance and the relief afforded by oxygen administration confirms this. The data of Will et al (1962) and Reeves et al (1962) make it clear that the range of pulmonary arterial and wedge pressure in normal cattle is comparable to normal values in man. However, the data of Will and his associates (1962) indicate that cattle have a most active pulmonary vasomotor system with a tendency for pulmonary arterial pressure to rise on the slightest provocation.

Pulmonary vascular morphology in cattle

This hyper-reactivity of the bovine pulmonary vasculature is related to its muscular nature. In 1961 one of us carried out an extensive histological study of the pulmonary blood vessels in a wide variety of species (Best and Heath, 1961). In that paper we published a figure of a transverse section of a muscular pulmonary artery from a cow. It was pointed out that the media was thick and muscular and that this degree of medial thickness would be taken as evidence of hypertensive pulmonary vascular disease in man. We found that the percentage medial thickness of some muscular pulmonary arteries in normal cattle sometimes exceeded 20 per cent, compared to 5 per cent in man. Some years later we had the opportunity to

study the medial thickness of the muscular pulmonary arteries in two bulls and a cow from Cerro de Pasco at an altitude of 4330 m in the Peruvian Andes. We found it to be in the range of 10 to 13 per cent (Heath et al, 1969). We felt unable to conclude that there was muscularization of the bovine pulmonary vasculature at high altitude since that in sea-level cattle was already so muscular. Alexander (1962) also studied the structure of the pulmonary blood vessels in healthy cattle at sea level. He found that the pulmonary arteries were of elastic type down to an external diameter of 200 to 300 μm. At that level a transition into a muscular type of vessel occurred. He found that the pulmonary arterioles were also muscular there being a distinct muscular media in arterioles as small as 20 μm in diameter. He was also able to confirm our findings of a year earlier (Best and Heath, 1961) that there is discontinuous muscular beading in the walls of pulmonary veins. Alexander (1962) suggested that the muscle was arranged in a spiral manner around the vein.

Wagenvoort and Wagenvoort (1969) carried out detailed quantitative studies of the pulmonary arteries of calves and cattle at sea level, expressing the medial thickness as a percentage diameter of the arteries. Their results are shown in Figure 13.2. They found that in contrast to the range of percentage medial thickness of the muscular pulmonary arteries of 2.8 to 6.8 in the normal adult

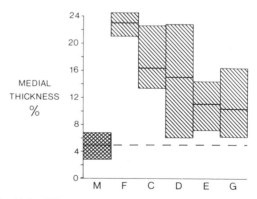

Fig. 13.2 The percentage medial thickness of pulmonary arteries in man and cattle. (Based on data from Heath and Best, 1958; and Wagenvoort and Wagenvoort, 1969.) M = adult human pulmonary arteries. F = fetal cattle. C = calves: 1 day to 3 months. D = calves: 9 months. E = calves: 1 year. G = cattle: more than 1 year. In each column the central bar is the mean value. The horizontal interrupted line is the mean value for the normal adult lung.

human lung (Heath and Best, 1958) the values for cattle were much greater. Thus the range for fetuses was 21.1 to 24.5 per cent, for calves 1 day to 3 months of age was 13.4 to 22.6 per cent and for cattle over one year of age was 6.2 to 16.4 per cent. Clearly the very muscular nature of pulmonary arteries, arterioles and veins reveals a great potentiality for vasoconstriction.

Influence of altitude and age on pulmonary arterial pressure in cattle

It has been known for many years that less than 2 per cent of the native cattle living in the mountains of Colorado develop brisket disease, while at the same altitudes up to 40 per cent or more of the cattle native to low altitude develop the disease. Will et al (1975) found from studies of 195 cattle that there was a tendency for pulmonary arterial pressure to increase with age and altitude but cattle native to high altitudes are more resistant to hypoxic pulmonary hypertension than those from sea level. Resistant and susceptible traits are probably genetically determined (Chapter 11). Calves susceptible to pulmonary hypertension die or are transported to low altitude before producing offspring. Hence the cattle remaining at high altitude have a tendency to have lower pulmonary arterial pressure. Pulmonary arterial blood pressure in cattle increases with age at high altitude but not at sea level. The hypoxia of high altitude elevates the pulmonary arterial blood pressure in all cattle but while this takes weeks or months in newcomers to the mountains it requires years in native highland cattle.

The pulmonary vasculature in brisket disease

Alexander (1962) carried out postmortem angiography on the lungs of calves which had died from brisket disease and found a 'tree-pruning effect' due to restriction of the peripheral pulmonary arterial tree due to vasoconstriction. He carried out histological studies on the pulmonary vasculature in 24 cases of brisket disease and found hypertrophy of the already muscular pulmonary arteries and pulmonary arterioles in most. The changes were entirely muscular in nature and there was no occlusive disease of the intima. This would account

for the reversibility of the pulmonary hypertension in brisket disease. The small amount of intimal disease present was dystrophic in nature consisting of intimal mineralization and thrombotic change.

Alexander (1962) was also able to study the pulmonary blood vessels in 10 yearling Hereford steers which were taken to 3050 m. All 10 developed pulmonary hypertension. In six the pulmonary hypertension was moderate and there proved to be no medial hypertrophy of the muscular pulmonary arteries. In the remaining four, the pulmonary hypertension was severe and in three of them there was hypertrophy of the pulmonary arteries so that even the normal muscularization of the bovine pulmonary arterial tree was exceeded.

Right ventricular hypertophy follows the pulmonary vascular changes (Alexander and Jensen, 1959). Hecht and his associates (1959) used the following formula to assess the relative weights of the right and left ventricles.

$$T = R + \left[\frac{R}{L + R} \right] I$$

where T = the total weight of the right ventricle

R = the weight of the free wall of the right ventricle

L = the weight of the free wall of the left ventricle

I = the weight of the interventricular septum.

In normal cattle the mean value for this ratio is 0.36 but in animals who died of brisket disease it is in the range of 0.45 to 0.65.

Brisket disease and Monge's disease

It is thus clear that brisket disease is not a bovine form of Monge's disease (Table 13.1). As we shall see in Chapter 16 Monge's disease is based on alveolar hypoventilation giving rise to pronounced haematological changes like further exaggeration of polycythaemia and an already increased haematocrit. There is substantial desaturation of sytemic arterial blood and hypercapnia develops. There is a relationship with brisket disease only in the sense that alveolar hypoventilation leads to chronic hypoxia elevating somewhat the pulmonary arterial

Table 13.1   Differences between Monge's disease and brisket disease

	Monge's Disease	Brisket Disease
Unsaturation of systemic arterial blood	Pronounced	Slight
Hypercapnia	Pronounced	Absent
Haematocrit	Greatly raised	Normal
Polycythaemia	Pronounced	None
Pulmonary arterial hypertension	Moderate	Severe
Right ventricular failure	Rare	Very common
Basic mechanism	Alveolar hypoventilation	Hyper-reactivity of muscular pulmonary vasculature
Form of syndrome	Respiratory	Cardiovascular

pressure. However, it should be noted that the degree of pulmonary hypertension is not great and right ventricular failure is not a feature of Monge's disease. Monge's disease is essentially *respiratory* in nature.

In contrast brisket disease is *cardiovascular* in nature (Table 13.1). Arterial oxygen saturation is not greatly impaired, Hecht et al (1959) quoting a range of 78 to 94 per cent and a mean value of 87 per cent. The haematocrit is not raised, being in the range of 30 to 45 per cent (Hecht et al, 1959). Hypercapnia is not present. However, while the respiratory and haematological disorders are slight the vascular disturbance is profound. There is severe pulmonary arterial hypertension due to vasoconstriction of a hyper-reactive and muscular pulmonary vasculature normal for the species. The stimulus for this sustained pulmonary vasoconstriction is the diminished ambient oxygen pressure consequent upon high altitude. The resulting increased pulmonary vascular resistance leads to right ventricular failure.

Left ventricular function in brisket disease

Left ventricular function is impaired in brisket disease (Hecht et al, 1962b). Calves with the condition frequently show elevated pulmonary arterial wedge, left atrial and left ventricular and diastolic pressures. Hecht and his colleagues consider a number of possible explanation for this. They reject the idea that in brisket disease the expansion of the left ventricle might be impaired by the confining restrictions of the pericardium or pericardial fluid. Arterial hypoxaemia is not a striking feature of brisket disease and in itself is unlikely to be a cause of left ventricular failure. No gross pulmonary disease leading to the development of bronchopulmonary anastomoses can be implicated as a cause of enlargement of the left ventricle. There is excessive fluid retention in brisket disease and Hecht and his colleagues (1962b) believe that further consideration should be given to the concept of interstitial oedema of the myocardium leading to impairment of left ventricular function. Finally, impairment and overstretching of myocardial fibres in the left ventricle due to anatomical continuity between the right and left ventricular muscle has been suggested.

It is of interest to recall that there is sometimes impairment of left ventricular function and left ventricular hypertrophy in some cases of pulmonary emphysema in man, a condition sharing the development of right ventricular hypertrophy and muscularization of the distal portion of the pulmonary arterial tree. Edwards (1974) found that in some instances the left ventricular hypertrophy was clearly related to the development of systemic hypertension since there was medial thickening of the internal mammary artery. In others the aetiology of the left ventricular hypertrophy in emphysema is obscure and the hypotheses advanced to explain it (Rao et al, 1968) are the same as those put forward to explain the left ventricular enlargement in brisket disease.

Training effect of high altitude on cattle

Although, as we have seen in this Chapter, cattle are susceptible to high altitude on account of the constricting effect of hypoxia on their unusually muscular pulmonary vasculature, there is some evidence that residence in mountainous regions may induce exercise conditioning in them. Hays (1976) studied two groups of cattle, one kept in an alpine pasture (1700–2600 m) for five months, the other at

400 m. Subsequently both were subjected to treadmill exercise at a simulated altitude of 3500 m. After exercise the heart rates in the alpine cattle were 10 to 20 beats per minute slower than in the valley cattle. This was interpreted as indicating that the five months' residence in the alpine zone had exerted a training effect on the animals.

Pulmonary hypertension in high altitude chicken

The lung structure and respiratory system of birds are considerably different from those of mammals. The greatest difference is the mechanism to maintain the distended state of the lungs. In the mammalian lung this is brought about by a sub-atmospheric pressure in the pleural cavity but in birds it is due to normal lung tissue adhesions in the rib cage. Hence the pulmonary circulation is independent of changes in the intrathoracic pressure (Burton et al, 1968). Nevertheless the chicken responds to high altitude by a consistent pulmonary arterial hypertension. It is quantitatively less than that found in cattle but the pressor response is twice that found in rabbits (Fig. 13.3). The right

contrasted to 15.7 mmHg at sea level). There was associated right ventricular hypertophy. Inhalation of 95 per cent oxygen and 5 per cent carbon dioxide for 10 minutes had no effect on the pulmonary hypertension. Acute hypoxia produces pulmonary hypertension in sea-level chicken but this regresses within two days after return to sea level (Burton et al, 1968). These data are more consistent with the view that hypoxia exerts a direct effect on the vascular smooth muscle of the terminal portions of the pulmonary arterial tree than with the concept that intrathoracic and alveolar pressure may affect the patency of pulmonary blood vessels. The effect of altitude on Pa_{O_2} in various animal species is shown in Figure 13.4 for comparison with the pressor response seen in the same species.

Pulmonary arterial pressor response in various species at high altitude

Tucker and his associates (1975) found that there were remarkable differences in the pulmonary arterial pressor response, as indicated by the

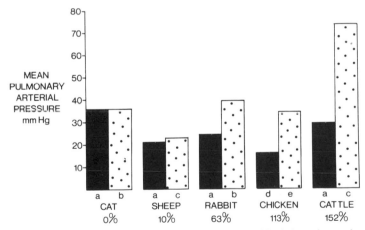

Fig. 13.3 The pressor response of the pulmonary arterial tree to hypoxia at high altitude in various animal species. Altitudes: a = 1585 m b = 4315 m c = 3870 m d = 0 m e = 3810 m. Data on cat and rabbit are systolic right ventricular blood pressures and from Reeves et al (1963a). Data on cattle from Grover et al (1963). Data on sheep from Reeves et al (1963b). Data on chicken from Burton et al (1968).

ventricular hypertrophy in healthy chicken at high altitude may lead to congestive cardiac failure (Olander et al, 1967). Burton et al (1968) found that White Leghorn chickens hatched and living at 3810 m had pulmonary arterial pressure about twice that found at sea level (34.2 mmHg for males as

percentage increase in right ventricular systolic pressure, in seven species exposed to a barometric pressure of 435 mmHg for periods ranging from 19 to 42 days (Fig. 13.5). Calves and pigs were very responsive to hypoxia but dogs, guinea pigs and sheep showed little or no response.

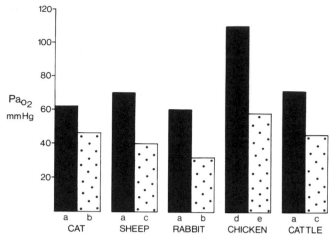

Fig. 13.4   The effect of altitude on Pa_{O_2} in various animal species. The altitudes and sources of data are as in Figure 13.3.

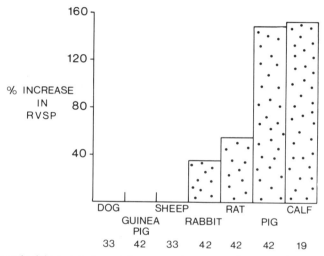

Fig. 13.5 The percentage increase in right ventricular systolic pressure in seven species exposed to a barometric pressure of 435 mmHg for periods ranging from 19 to 42 days (indicated in each instance at the foot of the column). (Based on data from Tucker et al, 1975).

REFERENCES

Alexander, A. F & Jensen, R. (1959) Gross cardiac changes in cattle with high mountain (brisket) disease and in experimental cattle maintained at high altitudes. *American Journal of Veterinary Research*, **20**, 680.

Alexander, A. F. (1962) The bovine lung: Normal vascular histology and vascular lesions in high mountain disease. *Medicina Thoracalis*, **19**, 528.

Best, P. V. & Heath, D. (1961) Interpretation of the appearances of the small pulmonary blood vessels in animals. *Circulation Research*, **9**, 288.

Burns, J. (1972) The heart and pulmonary arteries in rats fed on *Senecio jacobaea*. *Journal of Pathology*, **106**, 187.

Burton, R. R., Besch, E. L. & Smith, A. H. (1968) Effect of chronic hypoxia on the pulmonary arterial blood pressure of the chicken. *American Journal of Physiology*, **214**, 1438.

Cuba-Caparó, A. (1949) Policitemia de la Altura en corderos. *Revista De La Facultad de Medicina Veterinaria. Universidad Nacional Mayor De San Marcos*, **4**, 4.

Cuba-Caparó, A. (1950) Policitemia y mal de montaña en corderos. Tesis doctoral, Facultad de Medicina, Lima.

Edwards, C. W. (1974) Left ventricular hypertrophy in emphysema. *Thorax*, **29**, 75.

Glover, G. H. & Newsom, I. E. (1915) Brisket disease. (Dropsy at high altitude.) *Colorado Agricultural Experimental Station Bulletin 204*.

Glover, G. H. & Newsom, I. E. (1917) Brisket disease. *Colorado Agricultural Experimental Station Bulletin 229*.

Glover, G. H. & Newsom, I. E. (1918) Further studies on brisket disease. *Journal of Agricultural Research*, **15**, 409.

Grover, R. F., Reeves, J. T., Will, D. H. & Blount, S. G. Jnr.

(1963) Pulmonary vasoconstriction in steers at high altitude. *Journal of Applied Physiology*, **18**, 567.

Hays, F. L. (1976) Alp and valley cattle: Exercise in cold, hot and high environments. *Pflügers Archiv.*, **362**, 185.

Heath, D. & Best, P. V. (1958) The tunica media of the arteries of the lung in pulmonary hypertension. *Journal of Pathology and Bacteriology*, **76**, 165.

Heath, D., Castillo, Y., Arias-Stella, J. & Harris, P. (1969) The small pulmonary arteries of the llama and other domestic animals native to high altitudes. *Cardiovascular Research*, **3**, 75.

Hecht, H. H., Lange, R. L., Carnes, W. H. Kuida, H. & Blake, J. T. (1959) Brisket disease. I. General aspects of pulmonary hypertensive heart disease in cattle. *Transactions of Association of American Physicians*, **72**, 157.

Hecht, H. H., Kuida, H., Lange, R. L., Thorne, J. L. & Brown, A. M. (1962a) Brisket disease. II. Clinical features and hemodynamic observations in altitude-dependent right heart failure of cattle. *American Journal of Medicine*, **32**, 171.

Hecht, H. H., Kuida, H. & Tsagaris, T. J. (1962b) Brisket disease. IV. Impairment of left ventricular function in a form of cor pulmonale. *Transactions of the Association of American Physicians*, **75**, 263.

Kay, J. M. & Heath, D. (1969) *Crotalaria Spectabilis, The Pulmonary Hypertension Plant.* Springfield, Illinois: Charles C. Thomas.

Kuida, H., Hecht, H. H., Lange, R. L., Brown, A. M., Tsagaris, T. J. & Thorne, J. L. (1963) Brisket disease. III. Spontaneous remission of pulmonary hypertension and recovery from heart failure. *Journal of Clinical Investigation*, **42**, 589.

Monge, M. C. & Monge, C. C. (1966) In: *High Altitude Diseases. Mechanism and Management*, p. 70. Springfield, Illinois: Charles C. Thomas.

Newsom, I. E. (1915) Cardiac insufficiency at high altitude. *American Journal of Veterinary Medicine*, **10**, 890.

Olander, H. J., Burton, R. R. & Adler, H. E. (1967) The pathophysiology of chronic hypoxia in chickens. *Avian Diseases*, **11**, 609.

Rao, B. S., Cohn, K. E., Eldridge, F. L. & Hancock, E. W. (1968) Left ventricular failure secondary to chronic pulmonary disease. *American Journal of Medicine*, **45**, 229.

Reeves, J. T., Grover, E. B. & Grover, R. F. (1963a) Circulatory responses to high altitude in the cat and rabbit. *Journal of Applied Physiology*, **18**, 575.

Reeves, J. T., Grover, E. B. & Grover, R. F. (1963b) Pulmonary circulation and oxygen transport in lambs at high altitude. *Journal of Applied Physiology*, **18**, 560.

Reeves, J. T., Grover, R. F., Blount, S. G., Alexander, A. F. & Gill, D. (1960) Altitude as a stress to the pulmonary circulation of steers. *Clinical Research*, **8**, 138.

Reeves, J. T., Grover, R. F., Will, D. H. & Alexander, A. F. (1962) Hemodynamics in normal cattle. *Circulation Research*, **10**, 166.

Tucker, A., McMurtry, I. F., Reeves, J. T., Alexander, A. F., Will, D. H. & Grover, R. F. (1975). Lung vascular smooth muscle as a determinant of pulmonary hypertension at high altitude. *American Journal of Physiology*, **228**, 762.

Wagenvoort, C. A. & Wagenvoort, N. (1969) The pulmonary vasculature in normal cattle at sea level at different ages. *Pathologia Europaea*, **4**, 265.

Will, D. H., Alexander, A. F., Reeves, J. T. & Grover, R. F. (1962) High-altitude-induced pulmonary hypertension in normal cattle. *Circulation Research*, **10**, 172.

Will, D. H., Horrell, J. F., Reeves, J. T. & Alexander, A. F. (1975) Influence of altitude and age on pulmonary arterial pressure in cattle. *Proceedings of the Society for Experimental Biology and Medicine*, **150**, 564.

14

Acute mountain sickness

On acute exposure to high altitude the process of acclimatization begins and some of its physiological components give rise to unusual bodily sensations and symptoms. Thus the hyperventilation which helps to diminish the gradient of partial pressure of oxygen in the ambient air to that in the alveolar spaces (Chapter 5), leads to a feeling of breathlessness. The tachycardia on exercise at high altitude may give rise to palpitation. These are features of normal acclimatization to high altitude and they appear to be directly related to hypoxia.

Such sensations should be distinguished from the symptom-complex which develops in susceptible subjects after a time lag of 6 to 96 hours and which constitutes 'acute mountain sickness'. The clinical features of this condition do not seem to be brought about by hypoxia *per se* but by redistribution of body water.

They have been familiar to travellers to the mountains for many years being known as 'soroche' in Peru and 'puna' in Bolivia. As early as 1913 Ravenhill gave an account of the clinical picture of acute mountain sickness which was so clear as to distinguish between pulmonary and cerebral forms, a valid distinction which has assumed great importance subsequently. What he called 'puna of a cardiac type' we should call today 'high altitude pulmonary oedema' (Chapter 15) and his 'puna of a nervous type' would today be termed cerebral oedema which we consider later in this Chapter. In recent years Singh and his colleagues (1969) have gained considerable experience of the condition since it affected large numbers of Indian troops stationed at altitudes between 3350 m and 5500 m in the Himalayas during the border dispute with China.

Clinical features

The commonest symptoms of acute mountain sickness are shown in Figure 14.1 based on a study of 840 cases studied by Singh et al (1969) and in Table 14.1 based on a study of 146 trekkers with the condition in the Himalayas (Hackett et al, 1976).

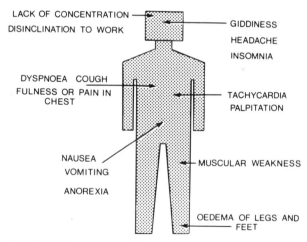

Fig. 14.1 The commonest symptoms in acute mountain sickness. (Based on data from 840 untreated cases studied by Singh et al, 1969).

Table 14.1 The percentage incidence of symptoms in 146 trekkers with acute mountain sickness in the Himalayas (after Hackett et al, 1976).

Headache	96
Insomnia	70
Anorexia	38
Nausea	35
Dizziness	27
Headache unrelieved by analgesics	26
Excessive dyspnoea on exercise or at rest	25
Reduced urine	19
Vomiting	14
Lassitude	13
Incoordination	11

Incidence and importance

Probably the best indication we have of the incidence of acute mountain sickness in lowlanders becoming acutely exposed to high altitude comes from the investigation of Hackett and his colleagues (1976). Their data were obtained from 278 questionnaires completed by unacclimatized hikers at Pheriche (4240 m) in the Himalayas of Nepal. The climbers were ascending on the main trekking trail to visit the Mount Everest Base Camp at 5500 m. The overall incidence of acute mountain sickness in these climbers was 53 per cent, occurring in 53 per cent of men and 51 per cent of women. The equal incidence in the two sexes suggests that the greater number of cases of high altitude pulmonary oedema in men, merely means that they climb high mountains more commonly than women. The incidence of acute mountain sickness was found to be greater in the young.

Duration

Incapacitating illness is likely to be short. In observations on 1925 soldiers between 18 and 53 years of age stationed at altitudes of 3350 m to 5500 m, Singh et al (1969) found the early phase of severe symptoms to last from only two to five days but complete recovery frequently took a long time (Fig. 14.2). One cannot say dogmatically at which altitude symptoms may be expected for individual tolerance to hypoxia is so variable. Singh and his colleagues (1969) believe that oliguria and redis-

tribution of blood in the body are of importance in the development of acute mountain sickness and bring about oedema of the brain and lungs which is responsible for many of the symptoms.

Oliguria

There is experimental evidence to suggest that mild hypoxia induces polyuria and severe hypoxia, oliguria. If exposure to high altitude is tolerated well, there may be a diuresis that lasts for days (Burrill et al, 1945) (Fig. 14.3). This experience of passing copious urine is well known to mountaineers. On the other hand some subjects may become oliguric during the first few hours of their exposure to high altitude (Fig. 14.3). This is the group likely to develop subsequently acute mountain sickness according to Singh et al (1969).

There is no evidence to suggest that urinary output is affected by chemical changes in the blood, alterations of ventilatory volume, respiratory rate, sodium excretion or urinary pH, all of which are modified by exposure to hypoxia (Currie and Ullman, 1961). It seems more likely that the effect of high altitude on the volume of urine passed depends on the activation of the pituitary-adrenal system by hypoxia, so that the onset of antidiuresis or polyuria depends on the level and balance of secretion of anti-diuretic hormone and the adrenal steroids. We consider the endocrinology of this at greater length in Chapter 25. The oliguria and severe hypoxic stress are accompanied by profound depression of

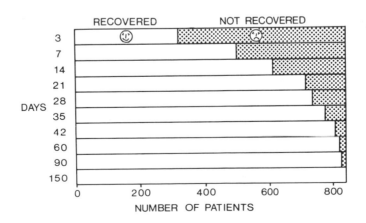

Fig. 14.2 The duration of acute mountain sickness in 840 untreated cases studied by Singh et al (1969).

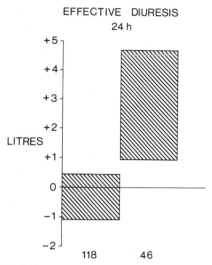

Fig. 14.3 Effective diuresis in 118 men predisposed to acute mountain sickness and in 46 subjects immune to the condition on the first day of their arrival at high altitude. Subjects developing acute mountain sickness are oliguric. (Based on data from Singh et al, 1969).

sodium excretion (Franklin et al, 1951) and this aggravates the antidiuresis and pulmonary congestion.

Ventilation

On acute exposure to the hypoxia of high altitude the subject undergoing successful acclimatization has a surprisingly normal respiratory rate. Thus in one of us (DRW) the resting respiratory rate was 12 per minute at 150 m, 12 at 2500 m, 12 at 3000 m and 12 to 16 at 4330 m. However, the minute volume does increase with altitude even when the subject is at rest. Thus its mean value under basal conditions at sea level was 4.9 l/min, at 3000 m was 5.6 l/min and at 4330 m was 6.6 l/min. The lack of increase in respiratory rate associated with increased minute volume implies an increase in tidal volume. Such hyperventilation has been considered in Chapter 5 to be an important feature of acclimatization; it becomes even more pronounced on exercise.

Significant elevation of the *respiratory rate* at rest at high altitude is likely to be a sign of the development of acute mountain sickness. Thus Singh et al (1969) found that, whereas there was an insignificant difference between the mean respiratory rate of symptomatic and asymptomatic patients on the first two days of exposure to high altitude, there was a significant elevation to a rate of 24 per minute on the third day in symptomatic subjects. This rapidity in breathing was not apparent in persons treated with frusemide so the implication is that the tachypnoea is brought about by pulmonary congestion and exaggerated pulmonary blood volume which increases by 80 per cent between 48 and 72 hours after arrival at high altitude. In a minority of subjects the congestion of the lung may deteriorate into frank oedema. In his early paper Ravenhill (1913) pointed out that this was one of the serious and potentially fatal developments of acute mountain sickness the other being a nervous form which we should now term cerebral oedema.

The pulmonary congestion and increase in lung blood volume are in part related to the reduction in peripheral blood flow which is abrupt during the first day at high altitude and which continues to diminish for two or three days. The peripheral veins are constricted from the very beginning and remain so for five or six days returning to normal on the tenth day. Breathlessness increases on lying down and during sleep it has a tendency to become irregular or frankly Cheyne-Stokes in type. Subjects often have sensations of fullness, discomfort or pain in the chest (Singh et al, 1969) and may wake up suddenly with a feeling of suffocation as we describe below.

Cough

Many people who go to high altitude develop a persistent non-productive cough. It is particularly likely to occur on ascent to 'extreme altitude' (Chapter 30) and is due to mouth-breathing and the inhalation of cold, dry mountain air. Clarke and Duff (1976) report that high altitude climbers who use oxygen are particularly affected, especially if no humidifier is included with the oxygen cylinder. These authors refer to the somewhat homely, but apparently effective, remedy of an inhaler filled with hot water, 5 ml of rum and Mac® lozenges.

Heart rate

The heart rate is surprisingly normal at rest on acute exposure to high altitude. Thus in one of us (DRW) at Lima (150 m) the radial pulse rate was 72 per minute, at Arequipa (2500 m) it was in the range of 76 to 80, at Tarma (3000 m) it was between 72 and 80 and at Cerro de Pasco (4330 m) it was in the range of 68 to 84. However, with slight exercise on acute exposure to 4330 m the pulse rate increased to 106 per minute. Under these conditions the subject has pronounced palpitation and tends to be apprehensive as he senses a general instability of his heart action and cardiovascular system. On moderate exercise the heart rate may increase suddenly to more than double its sea-level value (Fig. 14.4) and it may not return to normal levels for a considerable time. Under such conditions ectopic beats are not infrequent.

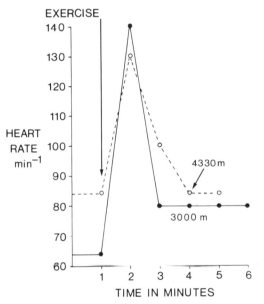

Fig. 14.4 Radial pulse rate immediately before and after a burst of moderate exercise at the point indicated by an arrow in one of the authors (DRW) at 3000 m and 4330 m. At the time of the test the subject was 27 years of age. The heart rate increases rapidly to a high value but soon subsides to a lower value which at 3000 m is higher than the resting value.

Such tachycardia probably has a rôle as an early feature of acclimatization to high altitude. It has been so regarded by various observers for many years. Thus Douglas et al (1912) found a transient rise in heart rate in subjects at 4200 m which was maximal by the ninth day and then gradually returned towards normal but it was still somewhat elevated at the end of a stay of 33 days. Grollman (1930) found the maximal heart rate to occur by the

second day of exposure to high altitude but also found that the rate did not fall to sea-level values. Jackson and Davies (1960) found that the resting heart rate in English climbers ascending Ama Dablam in the Himalayas rose from 60 to 65 beats per minute to 85 to 105 beats per minute at 5790 m.

With the onset of acute mountain sickness two distinct types of reaction of pulse rate emerge, according to Singh et al (1969). Usually there is a slow pulse rate under 70 per minute but in a minority the pulse rate is rapid, exceeding 100 per minute or more. On recovery the pulse rate returns to normal. Singh and his colleagues found that a combination of a slow pulse rate and systemic hypotension led to giddiness. When Cheyne-Stokes respiration became associated as well it led to a loss of consciousness. Kellogg (1964) also finds two distinct reactions of pulse rate at high altitude but he relates this to the degree of fitness achieved at sea level. He finds that the unfit with sedentary jobs at sea level may exhibit a striking tachycardia at high altitude. On the other hand fit subjects appear to be restricted to a lower ceiling of heart rate despite their ability to continue increasing their work performance far beyond the former group. In this connection Kellogg (1964) speculates that in some subjects hypoxia may limit the maximum rate at which the cardiac pacemaker can fire.

Systemic blood pressure

After prolonged residence at high altitude there is a significant fall in systemic blood pressure (Chapter 17). However, even on acute exposure to these conditions there is a slight fall in systemic pressure. In our personal experience the fall is about 5 mmHg after 36 hours exposure at 3000 m and Sime (1975) informs us that he has recorded a fall of up to 15 mmHg in subjects taken to 4540 m.

Headache

Frontal headache is one of the commonest symptoms experienced by the newcomer to high altitude. Few can ascend high mountains without suffering a mild headache and a fuzziness in the head. However, bouts of intense, incapacitating headache may occur and they are commonly associated with giddiness. When such severe headache is present, there is not infrequently some disturbance of vision but ophthalmoscopic examination during attacks does not reveal any abnormality. In our experience the headache is not relieved by oxygen. During attacks there is no elevation of systemic blood pressure. It seems likely that most headache at high altitude is due to mild cerebral oedema. Usually headache is nothing more than an inconvenience to those ascending to high altitude but occasionally its dramatic exacerbation heralds the development of the second major complication of acute mountain sickness, namely cerebral oedema.

Cerebral oedema

The pulmonary congestion referred to above sometimes becomes associated with cerebral oedema and this has been confirmed by examination of biopsy and necropsy specimens of brain. In one 38-year-old German who died with clinical features of the cerebral form of acute mountain sickness Houston and Dickinson (1975) report evidence of widespread and severe brain damage in the form of multiple petechiae, focal degenerations from previous petechiae, small intra-cerebral haemorrhages and subarachnoid haemorrhages. Increasingly severe headache is the likeliest symptom to herald the onset of cerebral oedema in acute mountain sickness but other serious neurological manifestations such as stupor, paralysis or coma may develop. Houston (1976) refers to the onset of mental confusion, hallucinations, ataxia, specific and localized motor weakness such as amblyopia, facial asymmetry and dysarthria. Houston and Dickinson (1975) describe an emotional lability which occasionally becomes frankly psychotic, with disorientation in time and space. Ravenhill (1913) recognised 'puna of nervous type' referring to cases which were characterised by vertigo or convulsions. Exceptionally cranial nerve palsies have been reported and such severe clinical signs may mimic a cerebral tumour (Singh et al, 1969).

The increased water content of the brain is possibly the result of increased capillary permeability secondary to hypoxic damage. Cerebral blood flow increases by 40 per cent within 12 to 36 hours of exposure to high altitude but approaches normal values by the fifth day. The cerebrospinal fluid

pressure rises and embarrasses venous and lymphatic return commonly causing engorgement of the retinal veins and retinal haemorrhages as we describe in Chapter 31. Much rarer is the blurring of the margins of the optic disc with the development of frank papilloedema (Singh et al, 1969). The protein, sugar, chloride and cell count of the cerebrospinal fluid are normal. Queckenstedt's test is negative.

Sleep

The newcomer to high altitude commonly feels tired and sleepy to an extent which is quite disproportionate to the degree of physical activity he has undertaken. Rapid ascent from sea level to over 4000 m may precipitate several hours of sleep. Sleep would seem to be the answer to the persistent tiredness and feeling of exhaustion common to the newcomer to high altitude but in fact it brings its own problems. Night is often the most difficult time for the newcomer. Breathing tends to become difficult on lying flat and the subject not uncommonly experiences symptoms which may be exaggerated by the psychologically susceptible into feelings of suffocation. A characteristic feature of sleep at high altitude is its periodic nature. In Figure 14.5 the periodic breathing of one of the authors (DH) at 4330 m is illustrated diagrammatically. Periods of heavy breathing lasting about ten seconds are followed by long periods of apnoea in which breath sounds cannot be detected by a stethoscope. This disturbed respiration leads to restless nights,

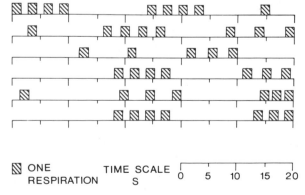

Fig. 14.5 The breathing pattern of one of the authors (DH) during the night of July 18th 1975 at Cerro de Pasco (4330 m). The breathing tends to be periodic. During the silent periods breathing could not be detected by stethoscope.

lack of sleep, and further tiredness seeking to be relieved by more sleep of this disturbed nature. An unfortunate vicious circle is thus created. A tendency for vivid dreaming accompanies this disturbed sleep pattern. In our experience it is advantageous to keep the head raised as much as possible while sleeping, and, if woken up by sensations of suffocation, sitting upright in bed is advantageous. These disturbances of sleep may persist for many days or even weeks at high altitude.

It is difficult to assess accurately the effects of high altitude on sleep and wakefulness because of the psychological aspects. Williams (1959) kept records on the sleeping habits of four subjects during two months of their normal life and during an expedition to the Karakoram mountains of Pakistan. The elevations reached ranged from 3050 m to 6100 m the greater part of the time being spent over 4570 m. Williams found that at high altitude all four subjects tended to sleep longer although there was great variation from day to day. He was well aware that the new sleep patterns reflected the exigencies of mountaineering as much as the effects of high altitude. Thus the hours of sleep tended to move forward to stretch from the early hours of the evening to very early the following morning. This alteration in the sleep pattern was far more likely to be related to the necessity for early morning starts and to the sudden evening drop in temperature than to any effects of high altitude *per se*. There was an increased tendency for interruption of the night's sleep. This characteristic interruption of sleep at high altitude is commonly misinterpreted as indicating that climbers sleep less than at low levels. There is a tendency, imposed by the environment, for the diurnal sleep-wakefulness pattern to coincide with the diurnal light-darkness cycle.

The nature of sleep

In order to assess the significance of this increased desire to sleep on acute exposure to high altitude we must give some consideration to the nature of sleep. The reticular formation of the brain stem appears to be excited by impulses from sense organs via collateral branches from the main, transitional pathways to the cerebral cortex. The excited reticular formation gives off streams of non-specific impulses which ascend to the cortex to increase its efficiency (Oswald, 1974). When the intensity of the upflow from the reticular formation falls off, drowsiness and sleep supervene. The onset of sleep is characterized by a change in the electroencephalograph from alpha rhythm of about 10 cycles per second to larger, slower waves of one to 3 per second, with brief bursts of faster waves termed 'sleep spindles'.

Oswald (1974) notes that the baroreceptors in the walls of the carotid sinus are stimulated by an elevation of systemic arterial pressure and send a powerful stream of nerve impulses to the brain-stem which quieten excitation of the reticular formation. He speculates that there may be a relation between atmospheric pressure and the desire to sleep. He postulates that a fall in barometric pressure would cause a very slight expansion of the carotid sinus, enhancing the tendency to sleep. It is of interest to note that Oswald's speculation is consistent with the strong tendency to sleep on acute exposure to a diminished barometric pressure to which we refer above.

It is now established that there are two varieties of sleep. Orthodox sleep is characterized by big slow waves in the electroencephalograph and quiescent eyes. The latter feature of non-rapid eye movement sleep sometimes gives the alternative term of NREM sleep. In orthodox sleep the throat muscles are tense and the heart is regular. In paradoxical sleep the e.e.g. shows low voltage waves while there is rapid eye movement, hence this form is sometimes called REM sleep. In paradoxical sleep the throat muscles are relaxed and the heart irregular. Paradoxical sleep is deeper than the orthodox variety and its function is believed by Oswald (1974) to allow protein synthesis in the brain. It predominates in infancy when the brain is growing, falls off in old age, or in mentally-retarded children and may be important in the synthesis of durable memory proteins. Orthodox sleep, especially in its third and fourth stages, is believed to be concerned with general body synthesis. It is required when there has been unusual physical activity, weight loss and starvation. Hence from the data presented in Chapter 21 we may suspect that a period of exposure to high altitude might require a compensatory period of increased orthodox sleep. However, this is conjecture and there would appear to be a fruitful area of research here for determining the relative importance and frequency of orthodox and paradoxical sleep in highlanders and those undergoing acclimatization to high altitude.

Reite and his colleagues (1975) studied the sleep of six young men between 19 and 23 years of age during two nights at 50 m and four non-consecutive nights at 4300 m. They employed electroencephalography, electrocardiography and monitoring of the heart and respiratory rate. They found that sleep at high altitude is characterized initially by a significant decrease in Stages 3 and 4, a significant increase in the number of arousals, a trend towards more time spent awake, and less time spent in REM sleep. In terms of actual time spent asleep, however, a relatively good night's sleep was obtained, which suggests that the objective sleep disturbance was not commensurate with the loud subjective complaints of sleeplessness. Periodic respiration during sleep was frequent at high altitude and was quickly terminated by oxygen administration.

Pulmonary gas exchange in acute mountain sickness

The extent to which changes in blood gases are related to the clinical features of acute mountain sickness was studied by Sutton and his colleagues (1976). They assessed the severity of acute mountain sickness by grading each of six clinical features of the condition on a 0 to 4 point scale, four of the features being objective and observable. Such assessment was made on seven soldiers, aged 18 to 30 years, at a staging camp at 2990 m and subsequently at 5360 m and at both altitudes the blood gases were measured. The clinical severity of acute mountain sickness was found to be directly related to Pa_{CO_2} and inversely to pH but it was unrelated to Pa_{O_2} on arrival at high altitude. In other words the subjects who develop the most severe acute mountain sickness are those who have the lowest levels of alveolar ventilation on ascent into the mountains. It is they who show the lowest levels of Pa_{O_2} two days after arrival at high altitude and an increase in A–a difference. The findings of the investigation of Sutton and his colleagues suggest that in acute mountain sickness the severity of cerebral symptoms and pulmonary dysfunction go

hand in hand and this in turn is indicative of simultaneous oedema of the brain and lungs. Thus in acute mountain sickness there seems to be a vicious circle of abnormality of pulmonary gas exchange and cerebral oedema.

Causation

Various theories have been put forward to explain the pathogenesis of acute mountain sickness. Some authorities believe it follows the respiratory alkalosis caused by hyperventilation which is later said to be compensated by the excretion of an alkaline urine. In this connection it is of interest to note that in the subjects studied by Waterlow and Bunjé (1966) at high altitude the samples of urine were almost invariably acid even during the early stages of acclimatization. Of over 500 samples collected during three expeditions only two were alkaline. We have carried out studies of the pH of the urine during early acclimatization at 3000 m and 4330 m. They showed some elevation of the pH on ascent from sea level to 3000 m although the urine never became alkaline. However, on further ascent to 4330 m there was a rapid reversal of the pH of the urine to acidic, sea-level values. It seems likely to us that respiratory alkalosis should be regarded merely as an early stage of acclimatization and not as the cause of acute mountain sickness.

The alternative views of Singh and his colleagues (1969) are set out in an extended diagrammatic form in Figure 14.6. According to their hypothesis, on going to high altitude hypoxia brings about a redistribution of fluid in the body with a decreased flow in the periphery and increased blood volume and congestion in the lung. As we have seen above, Sutton and his associates (1976) support the importance of the rôle of diminished pulmonary gas exchange in the development or exaggeration of acute mountain sickness. Singh et al (1969) noted the increased secretion of adrenal corticosteroids. Sutton (1971) later confirmed an increased serum cortisol in acute mountain sickness. Increased output of anti-diuretic hormone also occurs and predominates so that oliguria ensues. Cerebral oedema may also develop leading to the symptoms to which we have already referred above. Finally there is elevation of cerebrospinal fluid pressure which commonly leads to retinal haemorrhages

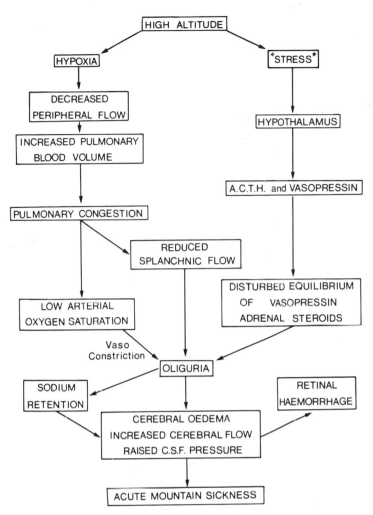

Fig. 14.6 Scheme to illustrate the pathogenesis of acute mountain sickness according to Singh et al (1969).

which we describe in detail in Chapter 31. It has been suggested that after a week most of the redistribution of body water is complete and symptoms of acute mountain sickness diminish (Carson et al, 1969).

Weightlessness in astronauts

It is of interest to note that the same redistribution of blood occurs during the phenomenon of weightlessness which affects astronauts (BBC, 1978). Normally blood pools in the lower limits as a response to gravity. In the weightlessness of space-travel the blood is redistributed to the thorax and head. The pooling of blood in the lungs leads to a sharp depression in the level of secretion of antidiuretic hormone by the pituitary so that astronauts experience a free diuresis as occurs in high altitude climbers. It is a reversal of this depression of ADH secretion which Singh et al (1969) believe to be the basis of the onset of acute mountain sickness. The fall-off in secretion of ADH in astronauts prevents pulmonary oedema in Zero G but can be troublesome on return to conditions of normal terrestial gravity because of depletion of blood supply to the brain. In astronauts pooling of blood in the legs can be stimulated by placing the legs in a vacuum tank (BBC, 1978).

Accommodation to high altitude and the nervous system

The immediate responses of the body to exposure to high altitude are thought to be regulatory reactions by von Muralt (1966) and referred to by him as 'accommodation' to hypoxia. He believes that such reactions represent a defence of the body against imminent danger and are heralded by an increased excitability and responsiveness of the autonomic and central nervous system to oxygen lack. He states that on arrival of subjects at Jungfraujoch (3450 m) there is an immediate lowering between 25 to 40 per cent of the thresholds for all external stimuli such as touch, pain-producing stimuli, carbon dioxide, light and stimuli affecting taste and smell (Fleisch and von Muralt, 1948; Grandjean, 1948). Studies on the patellar and pupillary reflexes showed a drop in threshold, a shortening of reflex time and an increased response. All these effects disappear when oxygen is administered, according to von Muralt (1966). He is of the opinion that this rapid increase in response of all these nervous functions is the primary process of accommodation of the body to the hypoxia of high altitude and all other changes may be secondary.

Treatment

Since oliguria and the retention and redistribution of body water into the pulmonary and cerebral circulation are believed to be of importance in the pathogenesis of acute mountain sickness, it is not surprising that diuretics are widely advocated for the prophylaxis and treatment of the condition.

Acetazolamide

Acetazolamide (Diamox) has mainly been used as a prophylactic against acute mountain sickness. It is one of the aromatic and heterocyclic sulphonamides which are potent carbonic anhydrase inhibitors (Fig. 14.7). Normally in the distal renal tubule and collecting ducts there is reabsorption of sodium with simultaneous excretion of an equivalent amount of hydrogen ions and potassium. There is a secondary reabsorption of bicarbonate as the hydrogen ions react with bicarbonate in the lumen of the tubule converting it into carbonic acid and dissolved

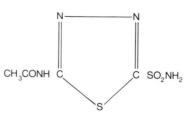

Fig. 14.7 Structural formula of acetazolamide.

carbon dioxide, both of which can diffuse across the renal tubule with ease. The hydrogen ions required for exchange with sodium are generated in the renal tubule cells by the hydration of carbon dioxide to carbonic acid followed by ionization into hydrogen ions and bicarbonate. This process is catalysed by the enzyme carbonic anhydrase. The supply of hydrogen ions for exchange with sodium ions is thus reduced by inhibitors of carbonic anhydrase (Burgen and Mitchell, 1972). Thus the prophylactic administration of acetazolamide for acute mountain sickness leads to a brisk diuresis in which the urine is alkaline and rich in bicarbonate and potassium. An alternative explanation for the action of acetazolamide is that it produces a metabolic acidosis through enhancing excretion of bicarbonate by the kidney by inhibiting carbonic anhydrase. This offsets the respiratory alkalosis of high altitude. It has further been suggested that acetazolamide increases the hydrogen ion concentration of the cerebrospinal fluid thus reversing the fall which occurs on acute exposure to high altitude allowing a greater degree of hyperventilation (Gray et al, 1971) (Chapter 5).

Hackett and his colleagues (1976) recommend acetazolamide as a useful prophylactic for acute mountain sickness. They used it in a double-blind study and found it reduced the incidence and severity of the condition in those who flew to 2800 m but not in those who hiked up to that altitude. The dose used was 250 mg twice daily for four days. The drug was started when the climbers were at an altitude of 3440 m. Hackett et al (1976) insist that acetazolamide must be given at the start of exposure to high altitude, if it is to be successful. In another group of subjects ascending to 4300 m the drug was given in a dose of 250 mg four times daily for two days before ascent and for three days afterwards.

Carson et al (1969) were not so enthusiastic about its use, finding it to bring about only a minimal reduction in symptoms with no improvement in performance.

Frusemide

Frusemide (Lasix) is another diuretic which has been used in the treatment of acute mountain sickness. However, it is a more powerful and more rapidly acting diuretic than acetazolamide and its use might be considered to be more appropriate for the treatment of high altitude pulmonary oedema. However, as we shall see in Chapter 15, its administration in that condition is controversial, for it is not without its discomforts and even dangers. Frusemide is a more powerful diuretic because being related to the thiazides it exerts a powerful depressant action on distal tubular reabsorption (Burgen and Mitchell, 1972). In addition it contains a sulphonamide group (Fig. 14.8) and is thus, like acetazolamide, a carbonic anhydrase inhibitor.

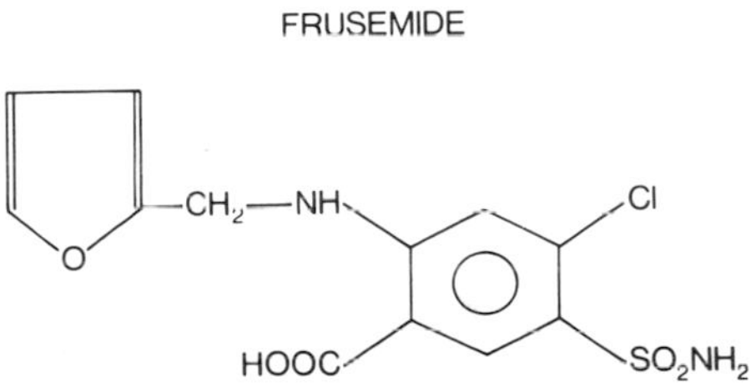

FRUSEMIDE

Fig. 14.8 Structural formula of frusemide.

When given orally, it leads to copious outpouring of an alkaline urine with potassium loss. Frusemide is, however, more rapid in action and produces a greater peak effect. While this may produce very rapid effects in clearing pulmonary oedema, it has to be kept in mind that on the mountainside the production of torrents of urine can be very uncomfortable if it requires constant moving throughout the night from a tent into surrounding sub-zero temperatures. It may also lead to an undesirable contraction of the plasma and extracellular fluid space and dehydration is already one of the drawbacks of sojourn at high altitude (Chapter 30).

Singh et al (1969) used frusemide in a heavy dosage of 80 mg every 12 hours for two days and claimed that on this regime symptoms and physical signs associated with acute mountain sickness were relieved within 6 to 48 hours. When there were associated neurological manifestations, betamethasone was given with the frusemide. Such therapy relieved headache and vomiting within three days, although papilloedema took up to a month to resolve. Singh and his associates point out that in their cases before the combination of frusemide and betamethasone was used for patients with neurological manifestations three died on routine methods of treatment and two survived with optic atrophy. They believe that frusemide is effective as a prophylactic measure against the development of acute mountain sickness in a dosage of 80 mg every 12 hours. Houston and Dickinson (1975) advocate the use of frusemide in combating the cerebral oedema of acute mountain sickness and also rapid injection of hyperosmolar solutions of urea, 50 per cent saline, mannitol or 50 per cent sucrose. They point out that in cases of cerebral oedema the longer the coma persists before descent the poorer the prognosis.

The dosage of frusemide advocated by Singh and his colleagues is heavy and the efficacy of the drug in acute mountain sickness is by no means universally accepted. In particular Wilson (1973) casts doubt on the use of the drug for the prevention and treatment of acute mountain sickness. He thinks it illogical to provoke diuresis in individuals already dehydrated from the effort of climbing, by the low humidity (Chapter 30), and from vomiting. He thinks the good effects noted by Singh were more likely to have been due to betamethasone since similar corticosteroids are used to reduce cerebral oedema. Gray et al (1971) noted that severe ataxia developed in five persons given frusemide in a study to test the efficacy of the agent for preventing acute mountain sickness. Wilson (1973) notes that in one case of soroche occuring in a man of 22 years 80 mg of frusemide produced copious incontinent urination but did not relieve his headache. Gray et al (1971) suggest that frusemide may be dangerous and doses over 200 mg should be avoided. Aoki and Robinson (1971) also found frusemide to be ineffective in relieving symptoms on ascent to high altitude. Bumetanide (Burinex) is a metanilamide derivative and is a diuretic of similar action and efficacy to frusemide. A standard dose is 1 mg daily.

Potassium supplements

Waterlow and Bunjé (1966) believe that potassium depletion may be an important factor in inducing the symptoms of acute mountain sickness. They studied this problem during four expeditions to the Sierra Nevada de Santa Marta in Colombia (3660 m to 4270 m). In one expedition volunteers were given either potassium supplements or a lactose placebo. The subjects receiving lactose had lower serum potassium levels but there was no confirmation that their losses of body potassium were higher. The symptoms of acute mountain sickness tended to be worse and more prolonged in the volunteers on low potassium intake who tended to retain sodium until they were given potassium. Waterlow and Bunjé (1966) suggest that where men are acutely exposed to high altitude a high intake of potassium should be assured. It is of interest to note that elsewhere Waterlow (1966) refers to the fact that Whymper, who first climbed the Matterhorn in 1865 and opened the way to the conquest of peaks throughout the world, mentions in his book *Travels Amongst the Great Andes of the Equator* published in 1892 (Chapter 1), that his doctor recommended potassium chloride to him as an antidote to mountain sickness from which Whymper suffered severely.

Oxygen

Singh et al (1969) believe that there is no direct relation between hypoxia and acute mountain sickness and hence the rôle of oxygen in treatment is not as straightforward as it might at first seem. Houston and Dickinson (1975) also found that the benefit of oxygen is less striking than might be anticipated. Sometimes the headache of acute mountain sickness is dramatically relieved by oxygen but in other cases it gets worse for a while before being slowly relieved. Breathing oxygen at night may prove effective in preventing morning headaches. Frequent intermittent use of small amounts of oxygen should be avoided for this will delay rather than aid acclimatization. However, in our experience, when breathlessness is pronounced, oxygen is helpful. Should basal crepitations increase and incipient pulmonary oedema be apparent, more intensive oxygen therapy should be given as described in Chapter 15, where we consider the treatment of high altitude pulmonary oedema. Progression of this condition requires early descent to lower altitudes.

Diet

According to Ward (1975) a high carbohydrate and low fat diet may ameliorate the symptoms of acute mountain sickness. He recommends a minimal carbohydrate consumption of 320 g per day. So far as diet is concerned it is of historical interest that in 1913 Ravenhill recorded that the Indians of Peru and Bolivia had great faith in several herbs for the relief of *soroche* or *puna*. Of these 'Chacha Como' and 'Flor de Puna' were most used in the form of an infusion. Another herb employed was 'Huaman-ripu'.

The psychological aspect

While it is clear that acute mountain sickness is based on the redistribution of water in the body and that diuretics have thus a place in its treatment, the psychological aspects must not be forgotten. Many of those about to ascend to high altitudes expect to become ill, especially if they are familiar with the dreaded reputation of *soroche* as is the Peruvian coastal dweller. In a delightful aside Ravenhill (1913) noted that the railway on the Bolivian border of Chile rose to its greatest height at Ascotan (3960 m), a fact noticed at the side of the line which immediately induced 'puna' in many travellers. The difference in psychological approach to a journey to high altitude may in part account for why some individuals appear to be more susceptible than others to the development of acute mountain sickness. Singh (1964) expressed the opinion that individuals of nervous temperament and psychological instability are not likely to do well at great elevation. On the other hand, subjects like mining engineers who wish to get on with their duties as soon as possible will tend to accept acute mountain sickness as a necessary but transient discomfiture. However, it is not possible to predict the behaviour of a sea-level subject at high altitude and there is no sure method for the selection of individuals suitable for travel or employment at heights.

In considering the treatment of acute mountain sickness it should be kept in mind that most subjects who experience symptoms on travelling to high altitude need little more than the reassurance of their companions and the time to allow themselves to accommodate to the new environment, and the novel and unpleasant bodily sensations it induces. After a few days in most people the symptoms will lessen and disappear, although in a few the symptoms may persist for many weeks (Fig. 14.2). Acclimatization gradually takes place and a successful transition from sea level to high altitude life will have been made. Some of the symptoms in this early period at high altitude, like the feeling of breathlessness accompanying hyperventilation, are features of acclimatization rather than of acute mountain sickness and they will persist as the process of acclimatization takes place. As noted by Singh et al (1969) and indicated in Figure 14.2, some subjects suffer a persistence of acute mountain sickness. Under such conditions continuing psychological apprehension and instability at high altitudes may indicate a return to sea level. Monge and Monge (1966) point out that in a minority of subjects acclimatization may be only temporary and that after a variable period of time symptoms of acute mountain sickness may return forcing the affected person to seek a lower environment. We have had the experience of having successfully passed through a period of *soroche* at an altitude of 4300 m only to experience a second attack of acute mountain sickness following a period of physical activity at a slightly higher elevation (4600 m).

The avoidance of acute mountain sickness

The golden rule is 'don't go too high too fast'. The traveller should not attempt to ascend from sea level to high altitude rapidly as one of the authors (DH) did in 1965, ascending in a few hours to 4330 m only to regret this action six hours after arrival with the onset of severe acute mountain sickness (A in Fig. 14.9). It is much better to break the journey and spend a few days at an intermediate altitude. When the author mentioned above took the same journey with a period of acclimatization at 3000 m the symptoms of acute mountain sickness at 4330 m proved to be minimal (B in Fig. 14.9). Singh et al (1969) recommended stops of one week each at 2440 m, 3350 m and 4270 m to ascend to an altitude of 5500 m. Acclimatization-stops of this type are

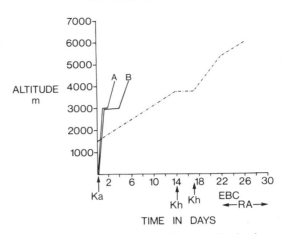

Fig. 14.9 Speed of ascent and its influence on the development of acute mountain sickness. Journey A from sea level to 4330 m in Peru was taken by one of the authors (DH) in six hours and ended with a severe attack of acute mountain sickness with early pulmonary oedema. The same journey, B, with acclimatization for three days at 3000 m led to only mild symptoms. The ascent indicated by the dotted line was that undertaken by the British Everest expedition in 1975. This group underwent successful acclimatization without acute mountain sickness. For this journey, Ka = Katmandu, Kh = Khumde, EBC = Everest Base Camp. RA = rate of ascent (600 m every fourth day).

commonly employed by high altitude climbers. Thus Clarke and Duff (1976) note that the members of the successful British Everest Expedition in 1975 acclimatized by marching from an altitude of 1500 m to 3800 m in two weeks (dotted line in Fig. 14.9). A rest of three days was taken at this elevation and thence a climb of 1600 m to the Base Camp at 5400 m. Thereafter, 'day-return' journeys to the next higher camp before sleeping there, representing an ascent of only 600 m every fourth day. No diuretics were used prophylactically. Supplementary oxygen was used above 7600 m when climbing and sleeping.

Hackett et al (1976) found that acute mountain sickness amongst 278 trekkers in the Himalayas was commoner in those who did not allow time for acclimatization. Thus it occurred in 60 per cent of those who flew from sea level to 2800 m before starting further high climbing but in only 42 per cent of those who walked up from Katmandu (1200 m). Acute mountain sickness was also commoner in those who climbed fast or who spent fewer nights acclimatizing on the route. On arrival at high altitude it is wise not to indulge in too much physical activity in the first few days. Hackett et al

(1976) found there was no relationship between the tendency to develop acute mountain sickness and recent respiratory infections or the load carried up the mountain. Houston and Dickinson (1975) point out that once it was thought safe to allow in the course of slow ascent, one day per 300 m of ascent from 3000 to 4270 m and two days per 300 m thereafter. However, pulmonary and cerebral oedema can still occur with this rate of ascent and they believe it more prudent to allow one day per 150 m of ascent from 2750 m upwards, interrupted by a day of partial rest at 4270 and 5500 m. They quote part of the leaflet prepared for high altitude trekkers by the Himalayan Rescue Association (Fig. 14.10). The basic principle advocated throughout this leaflet is to take one's time and ascend slowly and, should symptoms of cerebral oedema develop, descent is indicated immediately—even it if means travelling by night. One leaflet says 'Beware of the 'do or die' attitude in the Himalayas. All too often it has meant more 'die' than 'do'.'

HIMALAYAN RESCUE ASSOCIATION

ENJOY   YOUR   TREK !

ADVICE ON HIGH ALTITUDE SICKNESS
FOR TREKKERS PROCEEDING
ABOVE 12,000 FEET.

THE GOLDEN RULE:

DO'NT GO "TOO FAST TOO HIGH"

Fig. 14.10 The first page of the leaflet issued by the Himalayan Rescue Association to trekkers climbing above an altitude of 3660 m.

Effect of previous acclimatization

From their experience as members of the successful British Everest Team in 1975 that conquered the South-West face of the mountain, Clarke and Duff (1976) came to the conclusion that the effects of previous acclimatization are long-lasting and aid further acclimatization. It seemed to them that this applied particularly to retinal haemorrhages which occurred rarely in those accustomed to altitudes above 6000 m. Hackett et al (1976) were unable to support this impression from their findings in 278 trekkers in the Himalayas. In fact two of their subjects who developed cerebral oedema were experienced European mountain guides who had visited the Himalayas before.

Age, sex and tolerance to high altitude

There is evidence to suggest that the response of rats to acute hypoxia is influenced by their age and sex. Altland and Highman (1964) studied the percentage survival of rats of various ages to a simulated altitude of 10 360 m in a decompression chamber for six hours. The ages selected were 25 days as representing pre-puberty, 110 and 150 days ('young adults'), 330 days ('middle age') and 720 days ('old age'). They found that tolerance to altitude in these animals declined from birth to puberty. There was then a pronounced improvement in tolerance from 110 to 150 days and this improvement was sustained in rats to 330 days of age. There is a great reduction in altitude tolerance in rats which are two years of age. Altland and Highman (1964) found that tolerance to high altitude was greater in females of all ages. Hence the young female rat is best able to tolerate a simulated high altitude whereas the aged male is at the greatest risk. It is of interest to relate these findings to the results of our studies of the influence of age and sex on the development of hypoxic hypertensive pulmonary vascular disease (Smith et al, 1974), referred to in Chapter 11. We found that the old male rat exposed to simulated high altitude had the greatest tendency to develop cardiomegaly and the least tendency to show muscularization of its pulmonary arterial tree. On the other hand the young female rat showed the most marked propensity for the development of muscularization of the terminal portion of its pulmonary arterial tree. Relating these microanatomical findings we arrive at the somewhat surprising conclusion that in some ways hypoxic pulmonary vasoconstriction is beneficial in aiding survival at high altitude. It is conceivable that the pulmonary vasoconstriction protects against the development of oedema in the lung. There is a clinical application of this hypothesis. Heath et al (1958) reported the cases of two infants with widely patent ductus arteriosi who died on the second day of life from pulmonary oedema. The onset of oedema in both patients appears to have been related to the state of the muscular pulmonary arteries

which had undergone an unusually rapid transition from the fetal to the adult form. Thus in an infant with a defect allowing communication between the systemic and pulmonary circulations the arteries appeared much thinner than comparable vessels seen in the normal infant of this age. According to Klein (1964) the resistance of healthy pilots to acute oxygen deficiency in a decompression chamber simulating 7500 m increases between the ages of 18 and 40, the steepest rise being between 25 and 36 years. Beyond the age of 40 the tolerance seems slowly to decrease again.

REFERENCES

Altland, P. D. & Highman, B. (1964) Effects of age and exercise on altitude tolerance in rats. In: *The Physiological Effects of High Altitude*, p. 301. Edited by W. H. Weihe, Oxford: Pergamon Press.

Aoki, V. S. & Robinson, S. M. (1971) Body hydration and the incidence and severity of acute mountain sickness. *Journal of Applied Physiology*, **31**, 363.

British Broadcasting Corporation In: BBC 2 TV Programme: 'Zero G' February 1978.

Burgen, A. S. V. & Mitchell, J. F. (1972) *Gaddum's Pharmacology*, Seventh Edition. London: Oxford University Press.

Burrill, M. W., Freeman, S. & Ivy, A. C. (1945) Sodium, potassium and chloride excretion of human subjects exposed to simulated altitude of 18,000 feet. *Journal of Biological Chemistry*, **157**, 297.

Carson, R. P., Evans, W. D., Shields, J. L. & Hannon, J. P. (1969) Symptomatology, pathophysiology, and treatment of acute mountain sickness. *Federation Proceedings. Federation of American Societies for Experimental Biology*, **28**, 1085.

Clarke, C. & Duff, J. (1976) Mountain sickness, retinal haemorrhages, and acclimatization on Mount Everest in 1975. *British Medical Journal*, ii, 495.

Currie, J. C. M. & Ullmann, E. (1961) Polyuria during experimental modifications of breathing. *Journal of Physiology*, **155**, 438.

Douglas, C. G., Haldane, J. S., Henderson, Y. & Schneider, E. C. (1912) Physiological observations made on Pike's Peak, Colorado, with special reference to adaptation to low barometric pressure. *Philosophical Transactions of the Royal Society of London*, **203B**, 185.

Fleisch, A. & von Muralt, A. (1948) *Klimaphysiologische untersuchungen in der Schweiz*. Part 1 1944; Part 2 1948. Basel: Berno Schwabe.

Franklin, K. J., McGee, L. E. & Ullmann, E. A. (1951) Effects of severe asphyxia on kidney and urine flow. *Journal of Physiology*, **112**, 43.

Grandjean, E. (1948) Physiologie du climat de la montagne. *Journal de Physiologie*, (Paris), **40**, 1A.

Gray, G. W., Bryan, A. C., Frayser, R., Houston, C. S. & Rennie, I. D. B. (1971) Control of acute mountain sickness. *Aerospace Medicine*, **42**, 81.

Grollman, A. (1930) Physiological variations of the cardiac output of man. VII The effect of high altitude on the cardiac output and its related functions: an account of experiments conducted on the summit of Pike's Peak, Colorado. *American Journal of Physiology*, **93**, 19.

Hackett, P. H., Rennie, D. & Levine, H. D. (1976) The incidence, importance and prophylaxis of acute mountain sickness. *Lancet*, ii, 1149.

Heath, D., Swan, H. J. C., DuShane, J. W. & Edwards, J. E. (1958) The relation of medial thickness of small muscular pulmonary arteries to immediate postnatal survival in patients with ventricular septal defect or patent ductus arteriosus. *Thorax*, **13**, 267.

Houston, C. S. (1976) High Altitude Illness. Disease with Protean Manifestations. *Journal of the American Medical Association*, **236**, 2193.

Houston, C. S. & Dickinson, J. (1975) Cerebral form of high-altitude illness. *Lancet*, ii, 758.

Jackson, F. & Davies, H. (1960) The electrocardiogram of the mountaineer at high altitude. *British Heart Journal*, **22**, 671.

Kellogg, R. H. (1964) Heart rate and alveolar gas in exercise during acclimatization to altitude. In: *The Physiological Effects of High Altitude*, p. 191. Edited by W. H. Weihe. Oxford: Pergamon Press.

Klein, K. E. (1964) In discussion following the paper of Altland and Highman, (1964).

Monge, M. C. & Monge, C. C. (1966) In: *High-altitude Diseases. Mechanism and Management*. Springfield, Illinois: Charles C. Thomas.

von Muralt, A. (1966) Acquired acclimatization: To high altitude. In: *Life at High Altitudes*, p. 53. Pan American Health Organization. Washington. Scientific Publication No. 140.

Oswald, I. (1974) *Sleep*. 3rd Edition. Harmondsworth: Penguin Books.

Ravenhill, T. H. (1913) Some experience of mountain sickness in the Andes. *Journal of Tropical Medicine and Hygiene*, **16**, 313.

Reite, M., Jackson, D., Cahoon, R. L. & Weil, J. V. (1975) Sleep physiology at high altitude. *Electroencephalography and Clinical Neurophysiology*, **38**, 463.

Sime, F. (1975) Personal Communication.

Singh, I. (1964) Medical problems during acclimatization to high altitude. In: *The Physiological Effects of High Altitude*, p. 333. Edited by W. H. Weihe, Oxford: Pergamon Press.

Singh, I., Khanna, P. L., Srivastava, M. C., Lal, M., Roy, S. B. & Subramanyam, C. S. V. (1969) Acute mountain sickness. *New England Journal of Medicine*, **280**, 175.

Smith, P., Moosavi, H., Winson, M. & Heath, D. (1974) The influence of age and sex on the response of the right ventricle, pulmonary vasculature and carotid bodies to hypoxia in rats. *Journal of Pathology*, **112**, 11.

Sutton, J. (1971) Acute mountain sickness. An historical review with some experience from the Peruvian Andes. *Medical Journal of Australia*, **2**, 243.

Sutton, J. R., Bryan, A. C., Gray, G. W., Horton, E. S., Rebuck, A. S., Woodley, W., Rennie, I. D. & Houston, C. S. (1976) Pulmonary gas exchange in acute mountain sickness. *Aviation, Space and Environmental Medicine*, **47**, 1032.

Ward, M. (1975) *Mountain Medicine. A Clinical Study of Cold and High Altitude*. London: Crosby, Lockwood, Staples.

Waterlow, J. C. (1966) Discussion In: *Life at High Altitudes*, p. 70. Washington: Pan American Health Organization. Scientific Publication. No. 140.

Waterlow, J. C. & Bunjé, H. W. (1966) Observations on mountain sickness in the Colombian Andes. *Lancet*, **ii**, 655.

Whymper, E. (1892) *Travels Amongst the Great Andes of the Equator*. London: Murray.

Williams, E. S. (1959) Sleep and wakefulness at high altitudes. *British Medical Journal*, **i**, 197.

Wilson, R. (1973) Acute high-altitude illness in mountaineers and problems of rescue. *Annals of Internal Medicine*, **78**, 421.

15

High altitude pulmonary oedema

As we have seen in the preceding chapter about half of the lowlanders who ascend rapidly to altitudes exceeding about 3000 m develop symptoms and signs of acute mountain sickness. In most people this condition is more in the nature of an inconvenience than an illness but in a minority it rapidly escalates into an acute and rapidly progressive pulmonary oedema which may prove fatal if prompt treatment is not given.

The recognition of high altitude pulmonary oedema as a distinct entity

Until about 1950 the acute onset of severe breathlessness far beyond what one would expect to occur in acute mountain sickness was regarded as due to pneumonia. As early as 1937, however, Hurtado had suspected that the dyspnoea was due to the sudden onset of lung oedema. Then early reports by Peruvian doctors began to appear with increasing recognition of the true nature of the condition (Lundberg, 1952; Bardales, 1955; Lizarraga, 1955). The first clear account of high altitude pulmonary oedema in the English language was given by Houston (1960) who described its onset in an athletic student of 21 years carrying a heavy pack in deep snow and cold weather at an altitude of 3600 m. He also referred briefly to four other cases but the brevity of these reports does not detract from the dramatic nature of the condition that he described as a distinct clinical entity. Thus he reports the companion of one of his subjects, who died from pulmonary oedema, as saying 'He sounded as though he were literally drowning in his own fluid with an almost continuous loud bubbling sound as if breathing through liquid. I noticed that a white froth resembling cotton candy had appeared to well up out of his mouth.' Subsequently with other colleagues (Hultgren et al, 1961) he presented data from 15 patients with high altitude pulmonary oedema at La Oroya in Peru and added information on 14 patients collected by Lizarraga (1955) and on 12 cases studied by Bardales (1955). This review and analysis of 41 cases formed the first major account of high altitude pulmonary oedema in the English language. Thirty-six cases were reported soon after by Marticorena et al (1964) and since that time an extensive literature on the disease has accumulated.

Predisposing factors

The altitude at which the risk of developing oedema of the lung begins has been reported as about 3350 m in the Himalayas (Singh et al, 1965), 3660 m in the Andes and somewhat lower (2590 m) in the Rockies. High altitude pulmonary oedema is a disease affecting young healthy people. Children and teenagers are especially at risk but this is not to say that the condition does not occur in the middle-aged, especially those who venture too quickly in to the high mountainous regions on 'adventure holidays'. The susceptibility of young people in the second or third decades to the disease has been reported by many investigators (Marticorena et al, 1964; Menon, 1965; Scoggin et al, 1977). The age range of patients in the series of 15 cases collected by Hultgren and his colleagues was four to 42 years, the mean age being as low as 16.1 years. The larger reviewed series of 41 cases was composed of 35 males and six females. The person most at risk is the young male and many are of athletic disposition and totally free of heart and lung disease (Houston, 1960). There is nothing to suggest that any preceding respiratory infection is involved.

This is an important point to bear in mind for it indicates that the most meticulous physical examination of persons planning to go to high altitude will not pick out those who will subsequently develop high altitude pulmonary oedema. A clean bill of health at sea level in no way guarantees exemption from the development of lung oedema.

The condition commonly afflicts the unacclimatized subject exposed to diminished barometric pressure who engages too quickly in strenuous physical exercise on arrival. It is thus a hazard for the mountain climber and skier (Fred et al, 1962). However, exercise is not an essential predisposing factor and many cases of high altitude pulmonary oedema occur in people arriving for the first time in mountainous areas by plane. The condition may occur while the subject is at rest (Singh et al, 1965) or even asleep (Marticorena et al, 1964; Peñaloza and Sime, 1969).

There appears to be a distinct individual predisposition to the development of the condition since some patients have repeated attacks of soroche and pulmonary oedema occurring each time they return to the mountains after a sojourn at lower altitudes (Hultgren et al, 1961). One of the patients reported by these authors had four attacks of pulmonary oedema in six years. There may be inherited familial susceptibility as well as individual susceptibility (Hultgren et al, 1961; Fred et al, 1962). There do not appear to be racial differences in predisposition to lung oedema.

A most important predisposing factor is re-entry to hypoxia, the lung oedema occurring in the highlander returning to his mountain home after a period spent at a lower altitude. Thus, although two-thirds of the 332 cases reported in India by Singh et al (1965) were in fresh inductees to high altitude, over eighty were in hill people who had been at lower altitudes for from one day to six months. Hultgren et al (1961) found that only one of the 41 cases they reviewed had not recently been to a lower altitude prior to his illness. Furthermore, only nine had developed pulmonary oedema during their first visit to the mountains. Menon (1965) also found that 65 of his 101 cases occurred in people re-entering high altitude. Marticorena et al (1964) elicited a history of a recent visit to low altitude in 33 of their 36 cases of lung oedema. The length of stay

at a lower elevation prior to subsequent development of the disease was five to 21 days in one series of cases (Hultgren et al, 1961). Previous acclimatization will not necessarily guard against development of the condition (Singh et al, 1965) and ascent for as little as an extra 300 m may induce it. In Peru the condition is commonest in January. At first one might be tempted to relate this to the rain and snow in the mountains at that time but this is also the season when highlanders tend to take their vacation at the coast. Hence January is the month when the number of re-entries to the hypoxia of high altitude is greatest. A recent study from Colorado emphasises once again the importance of re-entry into high altitude in the development of lung oedema and reveals in a striking manner just how brief needs to be the visit to low altitude to put the highlander at risk on his return to the mountains.

The importance of re-entry to hypoxia in pathogenesis

Leadville, Colorado (3100 m) is of unusual interest in illustrating the importance of re-entry to the hypoxia of high altitude in the pathogenesis of high altitude pulmonary oedema. It has a resident population of only 8300 but is visited annually by no fewer than 100 000 tourists. This offers a unique opportunity of contrasting the incidence of lung oedema in newcomers to the area with that in long-term residents re-entering their mountain home after a brief visit to low altitude. During a period of seven years and three months there were 58 suspected cases and of these 39 episodes in 32 patients met strict diagnostic criteria for high altitude pulmonary oedema (Scoggin et al, 1977). This group consisted of 18 males and 14 females ranging in age from 3 to 41 years with an average age of only 12 years. Nineteen patients with shortness of breath were excluded only because of lack of radiographic evidence of oedema.

All but two cases occurred in long-term residents at high altitude and represented 0.9 per cent of the population one to 14 years of age. All but one of the 30 cases occurring in residents of Leadville developed after they had returned from a visit to low altitude. These visits needed only to be brief and in three instances lasted for only one day (Fig. 15.1). In five patients there was more than one attack of lung

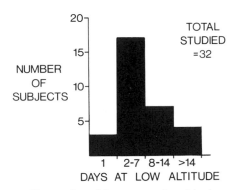

Fig. 15.1 The number of days spent at low altitude preceding the development of high altitude pulmonary oedema at Leadville, Colorado (3100 m) in 32 subjects.

The conclusions from this clinical study are striking. The disease is not rare. It occurs on re-entry to high altitude of those already living at such elevations, rather than in newcomers to the environment. It affects the young predominantly and is a special risk for children. There is likely to be some as yet undefined factor of the pulmonary vasculature of youth that accounts for this predisposition. Even brief exposure to low altitude puts the susceptible subject at risk. The period of exposure to low level before re-entry to high altitude adequate for this is usually less than a week and may be only one day. Subjects who fall prey to this dangerous illness are otherwise in excellent health. Scoggin and his colleagues (1977) conclude that 'high altitude pulmonary edema is a serious health problem for high altitude dwellers in North America, and re-emphasizes the point that acute changes in oxygen tension have profound effects on lung function'.

oedema. In four instances, the recurrences followed a return from a low-altitude visit and in the fifth it followed still higher ascent from the normal altitude of residence. Some residents at high altitude may make repeated trips to low altitude without developing pulmonary oedema in their re-ascent. However, many of the patients studied volunteered that they often had symptoms of respiratory distress similar to their recognised episodes of high altitude pulmonary oedema but that these symptoms resolved spontaneously in a few hours and did not require medical attention. A possible familial tendency for this condition was suggested by the occurrence of high altitude pulmonary oedema in a 21-year-old woman from Leadville, and, three years later, in her three-year-old daughter.

Incidence

From reported series of cases it is now proving possible to give an estimate of the risk of developing oedema of the lung on going to high altitude. This appears to be as high as 1 in 200 ascents. The percentage incidence of clinically significant cases reported from various mountainous areas of the world are shown in Table 15.1. The overall

Table 15.1 The incidence of severe episodes of high altitude pulmonary oedema in adults in various mountainous regions of the world (after Hultgren and Marticorena, 1978).

Area	Altitude (m)	Subjects	Incidence of HAPO (percentage)	Reference
Himalayas	3050–5490	Indian troops	0.57	Menon (1965)
Mt. Kenya	5200	7500 climbers	0.44	Houston quoted by Hultgren and Marticorena (1978)
Himalayas	2800–5500	522 trekkers	1.50	Hackett and Rennie (1976)
Mt. Rainier	4400	141 climbers	0.50	Houston (1976)
Mt. McKinley	6195	587 climbers	1.00	Rodman Wilson quoted by Hultgren and Marticorena (1978)
La Oroya, Central Peru	3750	97 residents in mining community	0.60	Hultgren and Marticorena (1978)

incidence of high altitude pulmonary oedema, including mild cases, was reckoned to be as high as 6.1 per cent by Hultgren and Marticorena (1978) who investigated 97 residents of La Oroya (3750 m) in Central Peru by means of a questionnaire. Although the accuracy of diagnosis by such means must be somewhat inexact, the data of these clinical investigators confirm the view of Scoggin et al (1977) that this disease is commoner than hitherto thought. Furthermore, it is likely that subclinical pulmonary oedema is at least three times as common as the well-defined fully developed form. The study of Hultgren and Marticorena (1978) also confirms the view of Scoggin et al (1977) that high altitude pulmonary oedema is a particular hazard for the young.

Clinical features

Most of the cases which have been reported from the Peruvian Andes have occurred acutely within three to 48 hours of the ascent (Arias-Stella and Krüger, 1963; Marticorena et al, 1964). However, in the Indian cases reported by Singh et al (1965), whereas two-thirds of the cases admittedly occurred within three days of arrival at high altitude, many were delayed to up to ten days. There are initial symptoms of a dry cough associated with breathlessness, palpitation and precordial discomfort. Headache, nausea and vomiting occur commonly. Pronounced weakness and fatigue are common early symptoms. The patient then becomes extremely breathless and begins to cough up foamy pink sputum. Haemoptysis may occur (Hultgren et al, 1961). There is pronounced peripheral vasoconstriction with pallor of the face and coldness and clamminess of the skin. On clinical examination widespread crepitations are heard throughout the lungs. Singh et al (1965) recognise several clinical variants. Thus in some cases the onset is insidious with premonitory malaise, headache, insomnia and anxiety. Only later are these symptoms followed by dry cough and dyspnoea. In others the condition starts more dramatically with acute respiratory symptoms. Bonington (1971) gives a graphic account of the onset of what appears to have been a case of high altitude pulmonary oedema in one of his fellow climbers during the ascent of Annapurna: 'Mike's condition worsened. His breathing seemed

to have got out of control, rising to a crescendo of raucous pants. He was convinced he was dying; it was as if his heart and lungs were exploding.' Such patients develop progressive cough, productive of large quanitities of frothy pink sputum, with dyspnoea and cyanosis and may become moribund. Haemoptysis, sometimes involving fairly large quantities of blood are seen in about 20 per cent of severe episodes. Some patients first show signs of the condition by developing oliguria. Some cases are complicated by cerebral oedema (Chapter 14). They may be giddy and become increasingly apprehensive, fear death, and may become incoherent or irrational or experience hallucinations. Coma may follow and death is then likely to follow in very few hours.

On clinical examination patients with high altitude pulmonary oedema commonly have tachycardia and low grade fever. Tachypnoea up to 40 respirations a minute may develop. Cyanosis of the lips and extremities may be pronounced. The systemic blood pressure may be low. Early harshness of expiration in the interscapular region is soon followed by crepitations and bubbling râles. In severe cases gurgling sounds can be heard without a stethoscope. The second sound in the pulmonary area is often loud and even palpable (Singh et al, 1965). There are no murmurs and there is no clinical evidence of cardiac failure. There is commonly an elevated haematocrit and an acid urine with a high specific gravity compatible with haemoconcentration. Signs of infection are absent, with only minor rise in temperature. There is a normal white cell count and a normal sedimentation rate.

Radiological features

Radiological examination demonstrates the pulmonary oedema as a coarse mottling which is commonly confluent and prominent in the parahilar regions (Marticorena et al, 1964). Singh et al (1965) and Menon (1965) reported that the shadowing was at first pronounced in the upper and middle lobes especially on the right side. However, Hultgren et al (1961) found there was no predilection of the process for either lung or any area of the lung. This is certainly our experience. In some cases we have seen, the oedema presents as a coarse mottling which is bilateral and most pronounced at the apices

(Fig. 15.2a). In others the oedema is confined to one lung and most pronounced at the base (Fig. 15.2c). The typical symmetrical butterfly distribution of pulmonary oedema as seen frequently in uraemia and left ventricular failure was not seen in the cases of Lizarraga (1955) and Bardales (1955), and in only

two of the cases of Hultgren et al (1961). Basal horizontal lines (Kerley B lines) and pleural effusion are rarely seen.

The pulmonary conus is usually prominent and the hilar and major pulmonary arteries are dilated in most cases, such dilatation being consistent with

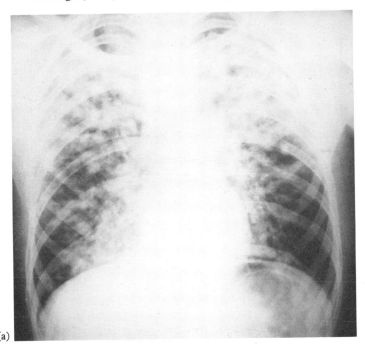

(a)

Fig. 15.2 (a) Radiograph of the chest from a Peruvian youth of 17 years returning to an altitude of 4330 m after one week's residence at sea level. There is coarse mottling of the lung fields indicative of high altitude pulmonary oedema. The changes are bilateral and most pronounced at the apices. (From Peñaloza and Sime, 1969).

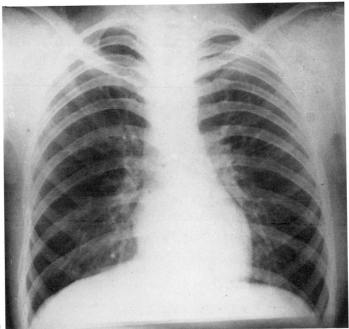

(b)

(b) Radiograph of the chest from the same case illustrated in Fig. 15.2a taken three days later. There has been clearing of the pulmonary oedema.

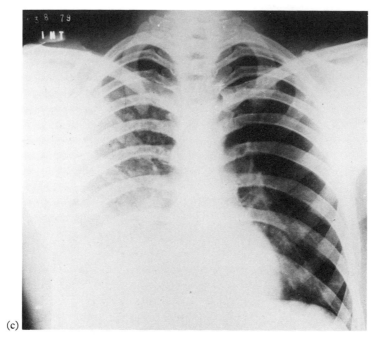

(c)

Fig. 15.2 (continued)   (c) Radiograph of the chest from a French student of 24 years who developed high altitude pulmonary oedema whilst on a trekking holiday in the Bolivian Andes at an altitude of around 3800 m. In this instance the oedema is unilateral, being confined to the right lung, and most pronounced at the right base.

elevation of pulmonary arterial pressure, With clinical improvement of the oedema the shadowing in the lung fields disappears within 6 to 48 hours (Singh et al, 1965) (Fig. 15.2b). At the same time the fullness of the hilar vessels recedes. After clinical recovery Hultgren and his associates found no consistent change in heart size which was often unchanged by the onset of pulmonary oedema. In some cases recovery from the attack was followed by a decrease in the cardio-thoracic ratio; in others it was increased. The right atrial shadow tends to get smaller after clinical recovery from the oedema. Chest radiographs may reveal exudates persisting for days or even a fortnight after clinical recovery. Hence radiological examination is important to ensure a return to normal in patients who are symptom-free.

Electrocardiographic features

The majority of subjects who develop high altitude pulmonary oedema have electrocardiographic evidence suggestive or characteristic of right ventricular hypertrophy since they are mountain dwellers re-entering high altitude. Peñaloza and Sime (1969)

recorded in two cases sinus tachycardia, right deviation of ÂQRS, and tall peaked T waves over the right precordial leads. Marticorena et al (1964) reported sinus tachycardia, slight peaked P waves, tall R waves in aVR, Rs complexes in V3R and deep S waves in V6. They also recorded positive displacement of the RS–T segment and tall, peaked T waves. In one severe case reported by Singh et al (1965) there were similar electrocardiographic findings with right axis deviation, clockwise rotation, T wave inversion in leads V1 to V5, a prominent R in leads aVR and V1 and peaked P waves. Some patients, however, have a normal electrocardiogram during an attack of high altitude pulmonary oedema. On recovery there is a decrease in heart rate, a lowering of the P wave and a decrease in the degree of right ventricular strain (Hultgren et al, 1961). After clinical recovery the electrocardiogram may take three to six weeeks to return to normal (Singh et al, 1965).

Pathology

In cases of high altitude pulmonary oedema occurring in native highlanders there is pronounced right ventricular hypertrophy indicative of significant pulmonary hypertension (Arias-Stella and Krüger, 1963). The lungs do not collapse and are congested on pressure yielding a foamy pink fluid. Histological examination shows dilatation of pulmonary capillaries leading to thickening of the alveolar septa. Oedema coagulum is found in the alveolar ducts and spaces (Figs. 15.3a, b). Hyaline membranes are in close contact with the alveolar walls and their histochemistry is identical with that of membranes found in the respiratory distress syndrome of the newborn. There may be haemorrhage into alveolar spaces. There is dilatation of pulmonary lymphatics and interstitial pulmonary oedema.

The small muscular pulmonary arteries show medial hypertrophy and crenation of elastic laminae consistent with vasoconstriction (Fig. 15.3c). The pulmonary arterioles are muscularized. There is sludging of the red cells in the pulmonary capillaries (Singh et al, 1965) and recent thrombi are occasionally seen in the pulmonary arteries and septal capillaries (Fred et al, 1962; Marticorena et al, 1964; Arias-Stella and Krüger, 1963). Singh et al (1965) refer to perivascular seepage of blood in the areas of vascular dilatation; they also report haemorrhages in the congested brain.

The haemodynamics of high altitude pulmonary oedema

Some insight into the mechanism of production of high altitude pulmonary oedema may be obtained from the results of cardiac catheterization carried out during an attack. Fred et al (1962) studied the pulmonary haemodynamics of a 48-year-old doctor who had an attack of pulmonary oedema after two days of vigorous skiing at altitudes between 2600 and 3125 m. They found a raised pulmonary arterial mean pressure of 46 mmHg with normal left atrial and pulmonary venous pressures, the latter being obtained by advancing the catheter through a probe-patent foramen ovale into the left atrium and thence into a pulmonary vein. Peñaloza and Sime (1969) reported findings at cardiac catheterization in two young men, aged 17 and 21 years, who developed high altitude pulmonary oedema on returning to their homes at 4300 m after a brief visit to sea level. During the attack both were found to have severe hypoxaemia, and pulmonary arterial

mean blood pressures of 62 and 63 mmHg respectively. The levels of pulmonary vascular resistance were elevated to 1301 and 1118 dynes s cm^{-5} respectively. These findings were associated with low cardiac output and pulmonary wedge pressure. The degree of pulmonary hypertension was significantly reduced after inhalation of 100 per cent oxygen. Following recovery, physiological observations were similar to those seen in healthy residents well acclimatized to high altitude. The data were regarded by Peñaloza and Sime (1969) as indicating pulmonary arteriolar constriction at the precapillary level due to severe hypoxia. As this constriction occurs proximal to the pulmonary capillary bed it would not explain the development of pronounced pulmonary oedema. Fred et al argue that to explain the coexistence of acute pulmonary arterial hypertension and acute pulmonary oedema in the presence of normal left atrial and pulmonary venous pressure one must postulate some increased vascular resistance situated in either the pulmonary venous capillaries or venules. Roy et al (1969) studied six subjects with high altitude pulmonary oedema and also found pulmonary arterial hypertension with normal left atrial and pulmonary capillary pressure.

Pathogenesis

The origins of high altitude pulmonary oedema are not known and, as might be anticipated in such a situation, there are a number of hypotheses to explain it. As we have seen the condition can develop rapidly within as little as six hours after arrival at high altitude. Furthermore, it is commonly reversed very speedily on treatment with oxygen. Hence the functional or organic basis for the condition must be rapid both in formation and reversibility.

It is clear from the series of cases referred to in this chapter, from the haemodynamic studies of Fred et al (1962) and from the necropsy findings of Arias-Stella and Krüger (1963) that many subjects who develop high altitude pulmonary oedema have had chronic and significant pulmonary hypertension. It is also clear, however, from the development of the condition in air travellers arriving for the first time at high altitude (Menon, 1965) and in lowlanders ascending to the mountains that chronic

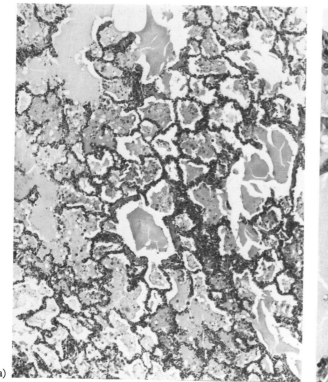

(a)

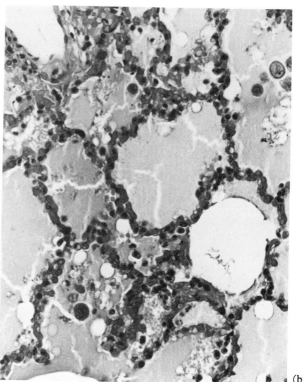

(b)

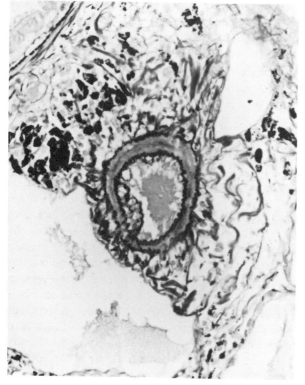

(c)

Fig. 15.3 (a: above left) Section of lung from a case of high altitude pulmonary oedema occurring in a Mestizo male aged 29 years. A native of Huanuco (3410 m), he lived for one month at sea level before returning to an altitude of 4330 m. There is extensive filling of the alveolar spaces by oedema coagulum. (Haematoxylin and eosin, × 55). (b: above right) Section of lung from the same case illustrated in Fig. 15.3a. The alveoli are filled with oedema coagulum and the pulmonary capillaries are congested. (Haematoxylin and eosin, × 240). (c: below left) Transverse section of small muscular pulmonary artery from the same case illustrated in Fig. 15.3a. There is medial hypertrophy and crenation of elastic laminae consistent with pulmonary vasoconstriction. These are the histological features forming the anatomical basis for the pulmonary hypertension which characterises cases of high altitude pulmonary oedema. (Elastic Van Gieson, × 335).

pulmonary hypertension is not necessary as a predisposing factor. However, what does seem undeniable is the acute onset, or worsening, of pre-existing pulmonary arterial hypertension probably under the hypoxic stimulation of re-entry to high altitude. As we have seen above this is not associated with an elevation of left atrial or pulmonary venous pressure but with pulmonary capillary hypertension due to some increase in vascular resistance situated in either the pulmonary venous capillaries or small venules. Such haemodynamic associations are found in pulmonary veno-occlusive disease (Thadani et al, 1975), and here there are organic fibrous occlusions in the venous capillaries and venules of the lung. Hence one might anticipate that in high altitude pulmonary oedema there are lesions in the same anatomical situation but of such a nature as to be consistent with rapid formation and reversibility. Ultrastructural studies of the lungs of rats subjected to severe simulated high altitude show that organic lesions of a highly reversible nature may develop but it is not known if these are of any significance in the causation of the condition.

Ultrastructure of pulmonary capillaries and veins

We have studied the ultrastructure of pulmonary capillaries in rats exposed for 12 hours in a hypobaric chamber to a subatmospheric pressure of 265 mmHg which simulates a high altitude roughly corresponding to the summit of Mount Everest (Heath et al, 1973). Under these conditions there is a formation of multiple endothelial vesicles which form and protrude into the pulmonary capillaries (Fig. 15.4). When seen in a longitudinal section these vesicles have an elongated shape which accommodates itself to the confines of the capillary into which it projects. These extrusions are large enough to occlude pulmonary capillaries into which they project. They arise by pedicles from localized widened areas of the fused basement membranes of the alveolar wall where there seems to be an accumulation of fluid. The oedema vesicles are covered by an exceedingly thin layer of cytoplasm which consists of part of the overlying endothelial cell of the pulmonary capillary stretched by the localized accumulation of fluid. When an oedema vesicle is cut in transverse section without fortuitously including its pedicle, it appears as a round body covered by a double membrane and gives the spurious appearance of lying free in the pulmonary capillary (Fig. 15.4). In conditions of simulated high altitude many capillaries in the lung contain oedema vesicles which reduce the diameter of the capillaries distorting contained erythrocytes. The membranous pneumocytes show micropinocytosis with hydropic degeneration.

In passing it should be noted that the oedema vesicles described here are in no sense specific. Indeed such vesicles in the thinner portions of the alveolar wall over the convexities of the pulmonary capillaries have in the past been regarded as characterisic of pulmonary oedema produced by toxic substances. Thus they have been produced by ammonium sulphate (Hayes and Shiga, 1970), alpha naphthyl urea (Meyrick et al, 1972) and the seeds of *Crotalaria spectabilis* containing monocrotaline (Kay et al, 1969). Clearly one must exercise considerable caution before ascribing functional significance to structural change in the absence of physiological data. However, the lesions described by Heath et al (1973) are composed largely of oedema fluid and could form rapidly and thus account for the very sudden onset of the symptoms of high altitude pulmonary oedema. Likewise they could shrink equally rapidly on return to low altitude or on the administration of oxygen. These oedema vesicles could produce a haemodynamic effect by protruding into pulmonary venous capillaries. This would meet the requirements laid down by Fred et al (1962) to explain the coexistence of acute pulmonary arterial hypertension and acute pulmonary oedema in the presence of normal left atrial and pulmonary venous pressure. It would be possible to object to the view that these oedema vesicles are of functional significance on the grounds that the pulmonary capillary bed has an enormous reserve capacity but we are unable to assess the validity of this objection without knowing what percentage of capillaries are involved by these intraluminal projections.

Another ultrastructural change that occurs rapidly on the exposure of rats to simulated high altitude is the development of evaginations of smooth muscle cells in the walls of pulmonary veins and venules. These protrude towards the intima pressing on the undersurface of endothelial cells (Fig. 15.5). Such evaginations are produced by constriction of the muscle cells as we have already seen in Chapter 12. Thus the ultrastructural changes we describe here clearly indicate that hypoxia induces constriction of pulmonary veins and it is possible that they accompany or even precede the development of high altitude pulmonary oedema. It is of interest to recall that constriction of small pulmonary veins is thought to occur in centrogenic pulmonary oedema as found in cases of head injury.

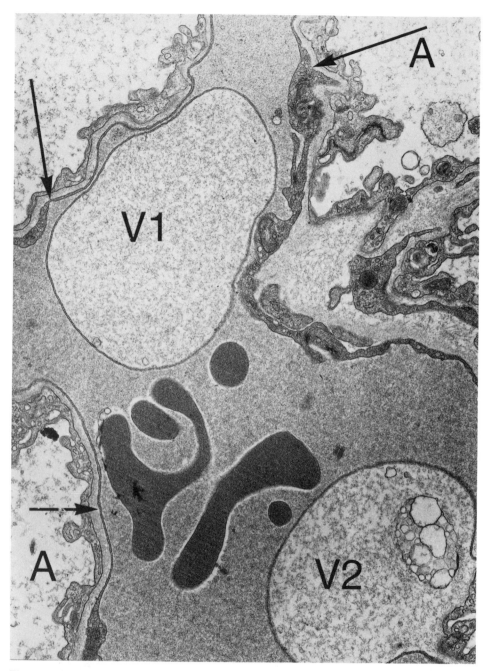

Fig. 15.4 Electron micrograph of pulmonary capillary with surrounding aleveolar spaces, A, from Wistar albino rat subjected for 12 hours to a barometric pressure of 265 mmHg simulating the altitude of the summit of Mount Everest. Endothelial vesicles, V1 and V2, project into the pulmonary capillaries from the alveolar-capillary wall. Vesicle V2 shows the mode of formation of these intracapillary projections. It arises by a pedicle from an oedematous area of the fused basement membrane (arrow) of the pulmonary endothelium and the overlying flattened membranous pneumocytes. When this pedicle is not fortuitously included in the section, the endothelial vesicle gives the spurious appearance of lying free in the lumen of the capillary (as in V1). The vesicles are covered by an attentuated cytoplasmic lining derived from the pulmonary endothelial cell (× 12 500).

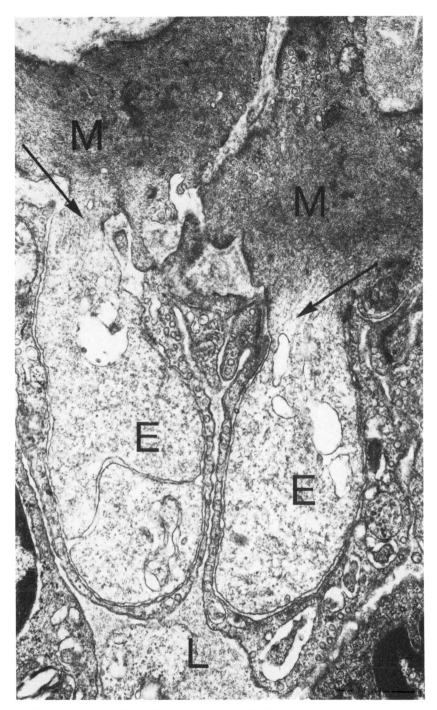

Fig. 15.5 Electron micrograph of part of the wall of a pulmonary vein from a rat exposed to severe, acute hypoxia in a hypobaric chamber mimicking conditions at the summit of Mount Everest. Vacoconstriction has occurred in response to the hypoxic stimulus with the formation of muscular evaginations, E, the cytoplasm of which is clear being devoid of myofilaments and organelles. The sites of junction of the evaginations with the cytoplasm of the parent smooth muscle cells, M, are indicated by arrows. The edge of the lumen of the vein, L, is seen ($\times$ 17 500).

There is a parallel here for cerebral oedema may develop in acute mountain sickness and become accompanied by pulmonary oedema.

Elevated pulmonary capillary pressure

Since pulmonary oedema implies exudation of fluid through the pulmonary capillaries it is clear that there are likely to be two basic mechanisms for its production at high altitude, namely increased pulmonary capillary pressure and/or increased pulmonary capillary permeability. Hultgren et al (1961) believe that several processes may occur in man during initial exposure to high altitude which could be conducive to the development of an elevated pulmonary capillary pressure. They are a redistribution of blood from the systemic to pulmonary circulation, changes in total blood volume as induced by high altitude, constriction of pulmonary veins or venules, and acute left ventricular failure.

Menon (1965) is in a minority in believing that acute myocardial failure secondary to hypoxia is of some significance in the pathogenesis of high altitude pulmonary oedema and he thinks this to be of importance in the treatment of the condition as we note below. While Hultgren et al (1961) recognize that in experimental animals severe hypoxia may give rise to left ventricular failure (Rivera-Estrada et al, 1958), they believe that acute left ventricular failure is not the sole basis for the pulmonary oedema. Most observers agree that in high altitude pulmonary oedema there is no consistent cardiac enlargement, or clinical evidence of left ventricular failure.

Some believe that the condition is a form of neurogenic pulmonary oedema (Theodore and Robin, 1975). A massive central sympathetic discharge might shift blood from the high-resistance systemic to the low-resistance pulmonary circulation with resultant pulmonary hypertension, lung haemorrhage and increased capillary permeability, malperfusion and maldistribution of ventilation (*Lancet* Annotation, 1976). Certainly Weil et al (1971) have detected a decrease in systemic venous compliance, mediated through the sympathetic system. Cruz et al (1976) confirmed the venoconstriction but concluded that hypoxia was responsible for the decreased venous compliance and hypocapnia for the increased resistance and decreased flow.

Total blood volume and high altitude pulmonary oedema

As we shall see in Chapter 17 prolonged exposure to high altitude results in an increase in blood volume as much as 30 per cent. This level is reached after about six weeks continuous exposure and may increase slowly for up to a year. Much of it is due to an increase in the red cell volume. On return to sea level there is a rapid decrease in red cell volume as described in Chapter 6, and this is accompanied by a compensatory rise in plasma volume. Hultgren et al (1961) believe this may be of importance in the pathogenesis of high altitude pulmonary oedema. They emphasise that subjects who have lived at a high elevation and then returned to sea level may have, beginning about a week after arrival, a higher than normal plasma volume for an unknown period of time. If such a person returns to high altitude during this time he may well be more susceptible to the development of pulmonary oedema than the total newcomer to the mountains who will have a lower plasma volume. This may be a contributory factor in explaining why many subjects with lung oedema at high altitude develop the condition after spending some time at sea level.

Increased pulmonary capillary permeability

It has been known for thirty years that dogs exposed to simulated altitudes up to 6400 m show an increase in the flow of lymph from the right lymph duct which ceases immediately on the inhalation of 100 per cent oxygen (Warren and Drinker, 1942; Warren et al, 1942). Such evidence suggests that hypoxia may increase pulmonary capillary permeability.

J receptors and pulmonary oedema at high altitude

Paintal (1970) believes that at high altitude the stimulation of J receptors in the alveolar walls has a physiological rôle in the avoidance of pulmonary oedema. He postulates that lying in the interstitial tissue of the alveolar walls are minute sensory nerve fibres with a diameter of the order of 0.1 to 0.3 μm (Paintal, 1970). Due to the position of the pulmonary capillaries lying on a connective tissue scaffolding they have two distinct microanatomical aspects (Fig. 15.6). One is bordered by a very thin wall composed of the attenuated cytoplasmic extensions of membranous pneumocytes, the fused basement membrane of alveolar epithelium and pulmonary capillary, and the ultrathin cytoplasm of the endothelium of the pulmonary capillary (Fig. 15.6). This part of the alveolar-capillary membrane is concerned with the exchange of respiratory gases. The other aspect of the alveolar-capillary membrane is composed of the pulmonary capillary endothelium, its basement membrane, interstitial tissues of collagen and reticulin, the basement membrane of the alveolar epithelium and the epithelium itself (Fig. 15.6). This thicker part of the alveolar-capillary wall is concerned with the movement of the interstitial fluid of the lung. The J receptors of the alveolar wall are believed to lie in the position shown in Figure 15.6. Their designation is

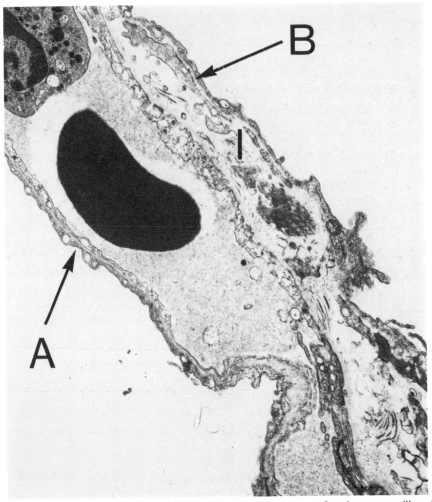

Fig. 15.6   Electron micrograph of lung from a Wistar albino rat. It shows the two aspects of a pulmonary capillary. The thin wall of the capillary (arrow A) is concerned with the exchange of respiratory gases. From the alveolar space inwards it consists of the attenuated cytoplasm of the membranous pneumocytes, the fused basement membrane of the alveolar capillary wall, and the thinned cytoplasm of the pulmonary endothelial cells. The thick wall of the capillary (arrow B) is concerned with the movement of interstitial fluid in the lungs. It comprises interstitial tissue, I, which includes collagen fibres embedded in ground substance. This is bounded internally by the pulmonary endothelium with its basement membrane, and externally by the attenuated cytoplasm of the membranous pneumocytes and their basement membrane. The J receptors are thought to lie in the interstitial layer, I. ($\times$ 12 500).

an abbreviation of 'juxta-pulmonary capillary receptors' which refers to their anatomical position. These receptors are rapidly stimulated by injection of phenyl diguanide into the right atrium or ventricle, or insufflation of volatile anaesthetics into the lungs, consistent with their postulated sitation in the alveolar wall.

J receptors lie in the interstitial tissue and are perhaps connected to collagen fibres. They are stimulated when more fluid enters the interstitial tissue which acts like a sponge. Even a small increase in pulmonary capillary pressure might be expected to cause a slight increase in interstitial volume, stimulating the endings. Paintal (1970) is of the opinion that the increased pulmony arterial pressure and plasma volume which occur in subjects at high altitude especially on exercise, produce congestion and hypertension in the pulmonary capillaries which in a minority of cases progress to pulmonary oedema. He cites the experience of Vogel et al (1963) who found that during exercise in 28 healthy residents between 13 and 17 years of age at Leadville, Colorado (3100 m) the pulmonary arterial pressure rose from 25 mmHg to 54 mmHg. Paintal (1970) believes that exercise at high altitude leads to a rise in pulmonary capillary pressure with an increase in interstitial pulmonary volume. The J receptors in the alveolar walls are stimulated and a feeling of breathlessness ensures (Paintal, 1955, 1968). He believes there is a reflex response (the 'J reflex') in which excitation of cerebral pathways leads to inhibition of skeletal muscles and a decrease in bodily exercise. There is experimental evidence to support this concept of a 'J reflex'. Deshpande and Devanandan (1970) studied the reflex effect of stimulating J receptors on the monosynaptic reflexes of hind-limb muscles and found that stimulating the endings produced inhibition of monosynaptic reflexes of both flexor and extensor muscles of the hind limb. It is post-synaptic as it is abolished by strychnine. The J reflex is abolished by decerebration thereby indicating that higher centres are involved.

Transarterial leakage

Severinghaus believes that there is no direct hypoxic damage to the lung in high altitude pulmonary oedema and does not refer to the oedema vesicles described above. He is of the opinion that the condition results from transarterial leakage giving rise to perivascular oedema, basing this hypothesis on the results of experiments studying the relationship between pulmonary arterial pressure and transarterial leakage in excised rat lungs. The terminal pulmonary arterial bed was blocked by infusing polystyrene microspheres 12 to 35 μm intravenously as a bolus. A catheter was tied into the pulmonary artery and the blood from the animal's inferior vena cava was then infused into it at a pressure of 100 mmHg for 10 minutes. This maneouvre induced perivascular oedema cuffs without haemorrhage in seven out of eight lungs. Whayne and Severinghaus (1968) note that the perivascular cuffs observed in their study contained protein and occasional erythrocytes. It seemed to them that this leakage was not a general transarterial migration of water down a hydrostatic gradient but rather a focal rupture or separation of elements of the arterial wall that permits the passage of plasma and blood, or both, into the extra-vascular spaces. Severinghaus (1971) refers to this as 'discrete focal damage' but he does not describe the nature of this pulmonary vascular pathology. We are not familiar with a focal arterial lesion of this type in states of chronic hypoxia. Neither, apparently, is Arias-Stella (1971) who, in discussing the hypothesis of Severinghaus, notes that perivascular oedema and haemorrhage are not features of high altitude pulmonary oedema. However, one is aware that, when material is impacted in a pulmonary artery for any length of time, it may become surrounded by a granulomatous reaction and the entire granuloma is extruded through the wall of the vessel. Such a reaction occurs around impacted fragments of cotton wool (Heath and Mackinnon, 1962) but 'transarterial leakage' around particulate matter of this type must be very different from the process conceived by Severinghaus as occurring in the hypoxia of high altitude.

Fibrin thrombi in pulmonary capillaries

Platelet and fibrin agglutination and microthrombi have been found in the pulmonary capillaries in individuals dying from high altitude pulmonary oedema (Nayak et al, 1964; Hultgren et al, 1962;

Arias-Stella and Krüger, 1963). Disorders of blood coagulation in high altitude pulmonary oedema predisposing to such thrombi are considered at length in Chapter 10.

High altitude pulmonary oedema in Kenya

According to Bulstrode (1975) high altitude pulmonary oedema has not been reported as occurring on Mount Kilimanjaro (5960 m) whereas it is commonplace on Mount Kenya (5200 m). This is in spite of the fact that the mountain is higher and attracts as many tourists each year. Bulstrode (1975) believes that a possible explanation may be that water is scarce on Mount Kilimanjaro and must be carried up from 3050 m while it is available on Mount Kenya as high as 4880 m above which tourists do not climb. Hence rehydration on Mount Kilimanjaro may not occur until after the descent when hypoxia is no longer significant, while rehydration on Mount Kenya may occur at 4880 m where the subjects may still be hypoxic.

Treatment

Familiarity with the clinical features of high altitude pulmonary oedema and with the means of avoiding and treating it is becoming of increasing importance. There have always been small mining communities at high altitude, such as at La Oroya (3750 m) in Peru, where the physician has had to be aware of the dangers of too rapid an ascent into the mountains or of a return to great heights after a period at lower elevations. In recent years, however, the growing popularity of adventure and trekking holidays in such areas as the Himalayas has made it very desirable for family doctors to be able to advise their patients about to take such vacations as to how to avoid developing the condition. Occasionally special circumstances arise where medical advice is sought as to a safe code of practice for groups of sea-level subjects being sent to work at high altitude; an example of this is the manning of the new infrared telescope on the summit of Mauna Kea in Hawaii referred to in Chapter 1. For all these differing groups of subjects the clear message should be that avoidance is better than cure.

Prophylaxis

Most cases of high altitude pulmonary oedema can be avoided if the subject does not attempt to ascend too rapidly and if he is not tempted to undertake too much exertion soon after arrival at high altitude. Above an elevation of 2130 m a greater rate of ascent than 300 m a day should be avoided (*Lancet* annotation, 1976). Estimates of a safe rate of ascent become ever more conservative with growing experience of this dangerous disease and two such experienced authorities as Houston and Dickinson (1975) now recommend an ascent of only 150 m a day above 2750 m. Over-exertion should be avoided during the early hours and days of arrival at high altitude. It should be kept in mind that children and adolescents are peculiarly prone to the condition. Acclimatized subjects returning to the mountains after a few days at sea level are more at risk than newcomers to high altitude. Those who have experienced a previous attack of high altitude pulmonary oedema should be especially careful.

Early diagnosis

Prompt recognition of the condition is vital. The onset of repeated coughing and the early symptoms described above should be viewed seriously. If anyone looks uncharacteristically weak or fatigued he should be brought down the mountain immediately (*Lancet* annotation, 1976). One should not wait for obvious râles or bubbling to indicate the onset of undeniable and severe pulmonary oedema. The subject with early suggestive symptoms should be taken down the mountain quickly and intellectual debate on the validity of the diagnosis can then take place at a lower and safer altitude.

Descent and bed rest

There is no doubt that absolute bed rest is important in treatment since physical exercise will further aggravate existing oedema. However, it must be borne in mind that delaying descent for anything but the mildest cases is likely to be highly dangerous. Clearly the patient must be moved down as soon as possible but the conditions of descent must be made as comfortable as can be managed, avoiding all unnecessary exertion. A litter may be necessary for an affected climber. Prompt evacuation to levels below 2400 m usually brings about rapid recovery (Wilson, 1973) but this cannot be relied on. Dickinson (1979) speaking at a Symposium on Acute Mountain Sickness at the University of Birmingham reported the case of a middle-aged doctor who was diagnosed immediately on the appearance of pulmonary oedema and urgently taken down the mountain to 'safe' levels but he died in spite of such prompt recognition and treatment.

Oxygen

Oxygen is a vital component in treatment. Its administration lowers pulmonary arterial pressure dramatically. High flow rates are necessary. Thus Menon (1965) gave the gas continuously at 6 to 8 l/min. finding that an intermittent flow below 4 l per minute was ineffective. He found that there was a response to oxygen therapy within 30 minute to 2 hours with relief of cough, cyanosis, chest pain, and a diminution of pulse rate. Oxygen treatment was tapered off after 8 to 12 hours. Some patients may show no improvement even after inhalation of the gas for 12 to 48 hours and cyanosis may persist. Wilson (1973) recommends that oxygen should be delivered through a tight-fitting mask for at least 24 hours. He also recommends that the rate of flow should be 6 to 8 l/min but is of the opinion that if supplies of oxygen are small, lower flows may still prove beneficial. Hultgren et al (1961) recommended that all parties climbing above 4570 m should have available emergency oxygen sufficient to provide a minimum of four l/min for several days. They should also have appropriate tools and spare parts to utilise additional oxygen if dropped by plane. Wilson (1973) points out that efforts to arrange descent should not await oxygen therapy which can be continued *en route*.

Frusemide

Frusemide has been used as a prophylactic agent to prevent the development of high altitude pulmonary oedema but its use is controversial as described in Chapter 14. The dehydrating effect of frusemide under mountain conditions, themselves predisposing to dehydration (Chapter 30), may be dangerous.

Bonington (1971) gives an account of the effects of such prophylaxis by dehydration on one of his companions during the climbing of Annapurna. 'The pills in Dave's box had the effect of dehydrating the patient by making him want to urinate, and hence reducing the likelihood of fluids forming in the lung. Mick decided to try out the treatment and took the prescribed dose. As a result he had to get out of bed every half hour or so through the night, a grim and exhausting experience in sub-zero temperatures.'

The use of frusemide may cause a further decrease of the already low blood volume, lowering systemic blood pressure and cardiac output. This will lead to the conversion of an ambulatory case into a litter patient.

Indian physicians have claimed that frusemide in a dosage of 80 mg daily for two or more days may prevent high altitude pulmonary oedema but others have found that such a dose gave rise to headache, vomiting, ataxia, and even coma (Wilson, 1973). More warranted is the cautious use of the drug in a dosage of 20 mg every 12 to 24 hours. However, the general consensus of opinion is that diuretics cannot be relied upon either to prevent or correct high altitude pulmonary oedema. They offer no clear benefit and may actually exert harmful effects.

Bumetanide

Bumetanide is a derivative of metanilamide which is as effective a diuretic as frusemide at only one-fortieth the molar dose (Seth et al, 1975). When the drug was given intravenously in a dose 1 to 3 mg to 35 patients with pulmonary oedema of different aetiology, it had a diuretic action within 20 minutes and continued to do so for about five hours. Bumetanide is more effective in acute than chronic pulmonary oedema and hence might be expected to be of value in the treatment of high altitude pulmonary oedema. Pines (1974) reported the effects of the drug in two cases of acute oedema developing in subjects at high altitude. He prefers the term 'oedema of mountains' since this makes the point that the fluid retention is generalized involving the kidneys and brain as well as the lungs. The cases he treated occurred amongst 30 climbers scaling peaks up to 7450 m in the Hindu Kush range of Afghanistan (Chapter 3). The first was in a man of

45 years who climbed to 5470 m and returned to base camp at 4860 m with swelling of the face, hands and ankles with severe oliguria. An oral dose of 2 mg of bumetanide caused clearance of the predominantly renal oedema which, however, recurred twice on further ascents. The second occurred in a man of 41 years who developed cerebral oedema at 6080 m. Although this subject had developed coma with signs of paraplegia, he improved rapidly after an intravenous dose of 1 mg of bumetanide.

Other drugs have been employed in the treatment of high altitude pulmonary oedema but their use is controversial. Thus, as noted above, some authors such as Menon (1965) believe left ventricular failure to be of importance in the aetiology of the condition and he advocates the use of intravenous digoxin for which excellent results are claimed even in the absence of oxygen therapy. Most authorities believe that digoxin has no place in the treatment of high altitude pulmonary oedema.

Morphia has a traditional use in the treatment of pulmonary oedema with a cardiac basis. However, in high altitude pulmonary oedema unwanted depression of respiration can occur particularly if cerebral oedema is present. The use of morphine has met with favour with Indian physicians but there

has been less confidence in it elsewhere. Some authors (Wilson, 1973) believe that parenteral morphine, in a dose of 10 to 15 mg, may be helpful because it calms the victim and dilates peripheral veins, pooling blood there. Others, like Hultgren, point out that there have been no controlled studies of its use in high altitude pulmonary oedema and believe that its use should be avoided.

Antibiotics have been given to prevent superadded respiratory infection but Menon (1965) notes that the white cell count is returned to normal equally well by oxygen therapy alone.

Prognosis

If high altitude pulmonary oedema is not treated promptly and effectively, it may prove fatal. On the other hand, if it is treated adequately, the prognosis is excellent since the condition is totally and rapidly reversible. Clinical improvement should follow in 30 minutes to 2 hours. Singh et al (1965) found that the lungs were free of oedema in four days and the chest radiograph was clear in five days. In his series of 101 cases Menon (1965) found that two patients developed high altitude pulmonary oedema for a second time.

REFERENCES

Arias-Stella, J. & Krüger, H. (1963) Pathology of high altitude pulmonary edema. *Archives of Pathology*, 76, 147.

Arias-Stella, J. (1971) Discussion of paper by Severinghaus, J. W. quoted below.

Bardales, A. (1955) Algunos casos de edema pulmonar aguda por soroche grave. *Anales de la Facultad de Medicina, Universidad Nacional Mayor de San Marcos*, 38, 232.

Bonington, C. (1971) *Annapurna South Face*, p. 172. London: Cassell.

Bulstrode, C. J. K. (1975) A preliminary study into factors predisposing mountaineers to high altitude pulmonary oedema. *Journal of the Royal Naval Medical Service*, 61, 101.

Cruz, J. C., Grover, R. F., Reeves, J. T., Maher, J. T., Cymerman, A. & Denniston, J. C. (1976) Sustained venoconstriction in man supplemented with CO_2 at high altitude. *Journal of Applied Physiology*, 40, 96.

Deshpande, S. S. & Devenandan, M. (1970) Reflex inhibition of monosynaptic reflexes by stimulation of type J pulmonary endings. *Journal of Physiology*, 206, 345.

Dickinson, J. G. (1979) Severe acute mountain sickness. *Postgraduate Medical Journal*, 55, 454.

Fred, H. L., Schmidt, A. M., Bates, T. & Hecht, H. H. (1962) Acute pulmonary edema of altitude. Clinical and physiologic observations. *Circulation*, 25, 929.

Hackett, P. & Rennie, D. (1976) The incidence, importance and prophylaxis of acute mountain sickness. *Lancet*, ii, 1149.

Hayes, J. A. & Shiga, A. (1970) Ultrastructural changes in pulmonary oedema produced experimentally with ammonium sulphate. *Journal of Pathology*, 100, 281.

Heath, D. & Mackinnon, J. (1962) Cotton-wool granuloma of pulmonary artery. *British Heart Journal*, 24, 518.

Heath, D., Moosavi, H. & Smith, P. (1973) Ultrastructure of high altitude pulmonary oedema. *Thorax*, 28, 694.

Houston, C. S. (1960) Acute pulmonary edema of high altitude. *New England Journal of Medicine*, 263, 478.

Houston, C. (1976) High altitude illness: Disease with protean manifestations. *Journal of the American Medical Association*, 236, 2193.

Houston, C. S. & Dickinson, J. (1975) Cerebral forms of high altitude illness. *Lancet*, ii, 758.

Hultgren, H. N. & Marticorena, E. A. (1978) High altitude pulmonary edema. *Chest*, 74, 372.

Hultgren, H. N., Spickard, W. B., Hellriegel, J. & Houston, C. S. (1961) High altitude pulmonary edema. *Medicine*, 40, 289.

Hultgren, H., Spickard, W. & Lopez, C. (1962) Further studies of high altitude pulmonary oedema. *British Heart Journal*, 24, 95.

Hurtado, A. (1937) *Aspectos Fisiologicos y Patologicos de la Vida en la Altura*. Imp. Edit. Rimac, SA. Lima.

Kay, J. M., Smith, P. & Heath, D. (1969) Electron microscopy of *Crotalaria* pulmonary hypertension. *Thorax*, 24, 511.

Lancet Annotation (1976) See Nuptse and die. *Lancet*, **ii**, 1177.

Lizarraga, L. (1955) Soroche agudo: Edema agudo del pulmón. *Anales de la Facultad de medicina, Universidad Nacional Mayor de San Marcos*, **38**, 244.

Lundberg, E. (1952) Edema agudo del pulmón en el soroche. *Conferencia sustenada en la Asociación Médica de Yauli, Oroya.*

Marticorena, E., Tapia, F. A., Dyer, J., Severino, J., Banchero, N., Gamboa, R., Krüger, H. & Peñaloza, D. (1964) Pulmonary edema by ascending to high altitudes. *Diseases of the Chest*, **45**, 273.

Meyrick, B., Miller, J. & Reid, L. (1972) Pulmonary oedema induced by ANTU, or by high or low oxygen concentrations in rat—an electron microscopic study. *British Journal of Experimental Pathology*, **53**, 347.

Menon, N. D. (1965) High-altitude pulmonary edema. *New England Journal of Medicine*, **273**, 66.

Nayak, N. C., Roy, S. & Narayanan, T. K. (1964) Pathologic features of altitude sickness. *American Journal of Pathology*, **45**, 381.

Paintal, A. S. (1955) Impulses in vagal afferent fibres from specific pulmonary deflation receptors. The response of these receptors to phenyl diguanide, potato starch, 5-hydroxytryptamine and nicotine, and their rôle in respiratory and cardiovascular reflexes. *Quarterly Journal of Experimental Physiology and Cognate Medical Sciences*, **40**, 89.

Paintal, A. S. (1968) Respiratory reflex mechanisms and respiratory sensations. *Indian Journal of Medical Research*, **56**, 1.

Paintal, A. S. (1970) The mechanisms of excitation of type J receptors, and the J reflex. In: *Breathing*, p. 59. Hering-Bruer Centenary Symposium. London:Churchill.

Peñaloza, D. & Sime, F. (1969) Circulatory dynamics during high altitude pulmonary edema. *American Journal of Cardiology*, **23**, 369.

Pines, A. (1974) Oedema of mountains, *British Medical Journal*, **4**, 233.

Rivera-Estrada, C., Saltzman, P., Singer, D. & Katz, L. (1958) Action of hypoxia on the pulmonary vasculature. *Circulation Research*, **6**, 10.

Roy, S. B., Guleria, J. S., Khanna, P. K., Manchanda, S. C., Pande, J. N. & Subba, P. S. (1969) Haemodynamic studies in high altitude pulmonary oedema. *British Heart Journal*, **31**, 52.

Scoggin, C. H., Hyers, T. M., Reeves, J. T. & Grover, R. F.(1977) High-altitude pulmonary edema in the children and young adults of Leadville, Colorado. *New England Journal of Medicine*, **297**, 1269.

Seth, H. C., Coulshed, N. & Epstein, E. J. (1975) Intravenous bumetanide in the treatment of acute and chronic pulmonary oedema. *British Journal of Clinical Practice*, **29**, 7.

Severinghaus, J. W. (1971) Transarterial leakage: a possible mechanism of high altitude pulmonary oedema. *High Altitude Physiology: Cardiac and Respiratory Aspects*, p.61. A Ciba Foundation Symposium. Edited by R. Porter & J. Knight Edinburgh: Churchill Livingstone.

Singh, I., Kapila, C. C., Khanna, P. K., Nanda, R. B. & Rao, B. D.P. (1965) High-altitude pulmonary oedema. *Lancet*, **i**, 229.

Thadani, U., Burrow, C., Whitaker, W. & Heath, D. (1975) Pulmonary veno-occlusive disease. *Quarterly Journal of Medicine*, **43**, 133.

Theodore, J. & Robin, E. D. (1975) Pathogenesis of neurogenic pulmonary oedema. *Lancet*, **ii**, 749.

Vogel, J. H. K., Weaver, W. F., Rose, R. L., Blount, S. G. & Grover, R. F. (1963) In: *Progress in Research in Emphysema and Chronic Bronchitis*, I, 269, 285. Edited by R. F. Grover & H. Herzog New York: Karger.

Warren, M. & Drinker, C. (1942) The flow of lymph from the lungs of the dog. *American Journal of Physiology*, **136**, 212.

Warren, M., Peterson, D. & Drinker, C. (1942) The effects of heightened negative pressure in the chest together with further experiments upon anoxia in increasing the flow of lung lymph. *American Journal of Physiology*, **137**, 641.

Weil, J.W., Byrne-Quinn, E., Battock, D. J., Grover, R. F. & Chidsey, C. A. (1971) Forearm circulation in man at high altitude. *Clinical Science*, **40**, 235.

Whayne, T. F. Jnr. & Severinghaus, J. W. (1968) Experimental hypoxic pulmonary edema in the rat. *Journal of Applied Physiology*, **25**, 729.

Wilson, R. (1973) Acute high-altitude illness in mountaineers and problems of rescue. *Annals of Internal Medicine*, **78**, 421.

16

Monge's disease

A small number of people living at altitudes exceeding 3000 m in the Andes develop a complex clinical picture composed of neuropsychic, haematological, cardiovascular and respiratory elements. This clinical syndrome was first described in 1928 by Monge as 'la enfermedad de los Andes' and it has since come to be termed 'chronic mountain sickness' or 'Monge's disease' in his honour (Fig. 16.1). In this chapter we shall consider its clinical and pathological features and the concept of the condition as a distinct disease entity. Peñaloza and his colleagues (1971) have made a special study of Monge's disease and in this account we shall draw heavily on their unique experience of the clinical features.

Symptoms

Most patients with the disease are young or middle aged men but the mean age approaches 40 years (Sime et al, 1975) (Fig. 16.2a). Thus in the study described by Peñaloza and Sime (1971) and Peñaloza et al (1971), all 10 subjects were male and between 22 and 51 years of age and the mean age was 38 years. The cases of Hurtado (1942) were all men between 24 and 44 years. The disease occurs but rarely in women and their exemption may be ascribed to menstrual loss. It is not found in children.

Fig. 16.1 Professor Carlos Monge, M. (right) with his son Professor Carlos Monge, C., also a distinguished worker in the field of high altitude medicine, in the garden of his home in Lima in July 1965.

Fig. 16.2 (a: right). A young male Quechua Indian, aged 35 years, who resides at 4330 m and has developed Monge's disease.

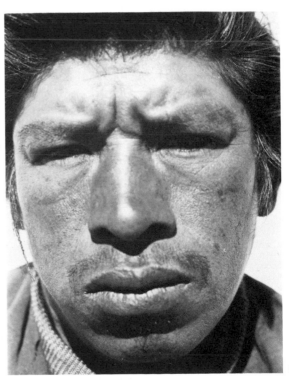

(a)

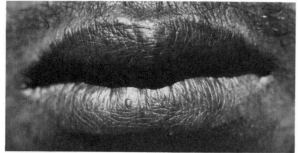

(b)

(c)

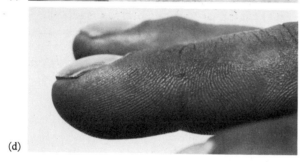

(d)

Fig. 16.2 (continued) (b: top) The lips appear almost black due to a combination of greatly elevated haematocrit and diminished arterial oxygen saturation, this giving the designation 'cardiac negro'. (c: centre) The conjunctival vessels are suffused and congested and were dark wine-red in colour, this being the consequence of the greatly elevated haematocrit and diminished arterial oxygen saturation. (d: bottom) The fingers are clubbed.

Neuropsychic symptoms are common and frequently take the form of headaches, dizziness, paraesthesiae and somnolence (Peñaloza and Sime, 1971). The earliest reports stressed such symptomatology and Monge and Monge (1966) present an imposing list of neuropsychic symptoms ranging from depression and irritability to hallucinations. Untreated early cases sometimes presented with cerebral crises which over a period of hours or even minutes deteriorated through drowsiness to coma which sometimes proved fatal. In present times,

when the effective treatment is known to be removal to sea level and when transport is readily available, such serious developments are virtually unknown. Patients with Monge's disease tire easily and have a decreased exercise tolerance. Breathlessness is not a common symptom. Haemoptysis is said to occur (Hurtado, 1942; Monge and Monge, 1966).

Signs

Physical examination must take into account the altitude at which the patient lives for the higher the location the more erythraemic and cyanosed will the normal healthy residents look. Only when this comparison is made will it be appreciated that in Monge's disease there is a change from the ruddy erythraemic colour of the normal healthy residents to a frankly cyanotic appearance. The combination of the virtually black lips and wine-red mucosal surfaces against the olive green pigmentation of the Indian skin give the patient with Monge's disease a striking appearance (Fig. 16.2a). The greatly elevated haematocrit and diminished arterial oxygen saturation described below give the lips a deeply cyanosed, almost black colour (Fig. 16.2b). The ear lobes and facial skin may also appear very dark. The conjunctivae are dark red in colour and look very suffused and congested (Fig. 16.2c). The fingers are commonly clubbed (Fig. 16.2d). In our experience haemorrhages in the finger nails are characteristic of the condition (see Chapter 23).

When cyanosis is severe, there may be signs of mild congestive cardiac failure. On examination of the chest there may be increased loudness of the second sound in the pulmonary area suggestive of an exaggeration of the degree of pulmonary hypertension normal for highlanders at the altitude in question. An associated mid-systolic murmur may be present. There is a slight but significant elevation of systemic diastolic and mean pressure according to Peñaloza and his colleagues (1971) (Fig. 16.3). Other authors have commented that the systemic blood pressure is normal or low (Monge and Monge, 1966). Further observations on the systemic blood pressure in chronic mountain sickness are made in Chapter 17. The fundi show tortuosity and dilatation of venous vessels. It is reported that the barrel-shaped chest of the Quechua Indian is

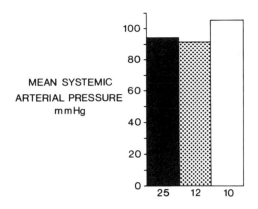

Fig. 16.3 Mean systemic arterial pressure (mmHg) in 25 healthy residents at sea-level (filled column), in 12 healthy highlanders (stippled column), and in 10 cases of Monge's disease (open column). The levels are significantly higher in the cases of chronic mountain sickness. (The data presented diagrammatically in Figs. 16.3, 16.5, 16.6, 16.7 and 16.8 are derived from Peñaloza et al, (1971). Their study was carried out at an altitude of 4375 m with a mean barometric pressure of 446 mmHg and an atmospheric P_{O_2} of 90 mmHg).

exaggerated when he develops symptoms and signs of chronic mountain sickness (Monge and Monge, 1966) but this seems unlikely to us.

Radiological changes

Radiological examination of the chest in chronic mountain sickness reveals cardiac enlargement due to increase in size of the right cardiac chambers (Fig. 16.4). In a minority of cases in which the heart is greatly enlarged the left ventricle is also involved (Fig. 16.4). The mean values per square metre of body surface of the transverse diameter of the heart, frontal area and heart volume in patients with Monge's disease compared with healthy highlanders are shown in Figure 16.5. Prominence of the pulmonary artery and accentuation of the pulmonary vascular markings is found in all patients with Monge's disease but they are especially pronounced when the cardiac enlargement is great. On descent to sea level the cardiac size and the pulmonary vascular markings decrease with the passage of time.

Electrovectorcardiography

Peñaloza and Sime (1971) have described the electrocardiograms and vectorcardiograms in 10 cases of chronic mountain sickness. Peaked P waves with increased voltage were often observed in leads II, III and aVF as well as in the right precordial

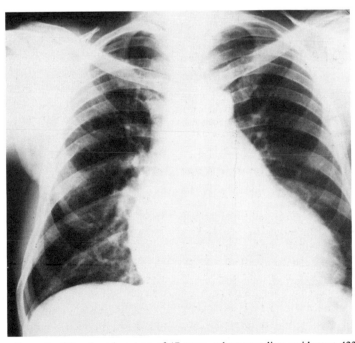

Fig. 16.4 Anteroposterior radiograph of the chest in a man of 47 years, a long standing resident at 4330 m, who developed Monge's disease. There is increased size of the right ventricle, prominence of the pulmonary artery and accentuation of the pulmonary vascular markings. There is also distinct enlargement of the left ventricle.

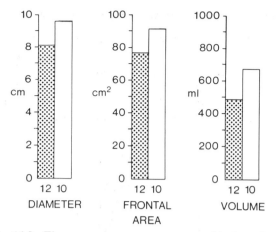

Fig. 16.5 The mean values per square metre of body surface area of the transverse diameter of the heart, frontal area and heart volume in 10 patients with Monge's disease compared with 12 healthy highlanders.

leads. The mean voltage of the P wave in lead II was 2.9 mm compared with 1.0 mm obtained in a group of 12 healthy residents at high altitude. The P loop of the vectorcardiograms was increased in area and voltage and was directed forward and inferiorly. An accentuated degree of right ÂQRS deviation was present in most patients, and the mean value of ÂQRS was +152° compared with +95° in healthy highlanders. The rS pattern in the right precordial leads and complexes of RS or rS type in the left precordial leads were common findings. Tall R waves were found over the left precordial leads in two of three patients in whom radiographic examination showed left cardiac enlargement. The horizontal QRS loop was orientated to the right and backward in most cases with counter-clockwise rotation or a figure-of-eight configuration and enlarged final QRS vectors. Forward orientation and clockwise rotation was seen in only two cases. The vectorcardiographic pattern of biventricular hypertrophy was seen in one patient. Negative T waves over the right precordial leads were found in five patients and in most of them the negative T waves were peaked, deep and symmetrical reproducing the 'ischaemic' T wave pattern. As in the case of the radiological changes, the electrovector-cardiographic alterations slowly return to normal on descent to sea level.

Haematological values

Patients with Monge's disease have higher haemo-

globin and haematocrit values than those seen in healthy residents at high altitudes (Fig. 16.6). A value of 23 g/dl has been taken arbitrarily as the sole diagnostic criterion of chronic mountain sickness according to Monge (1966), the son of the distinguished Peruvian physician (Fig. 16.1). Thus Hurtado (1942) gave for eight cases of Monge's disease a range of 22.9 to 27.2 g/dl for the haemoglobin level and 73 to 83 per cent for the haematocrit. The arbitrary value of 23 g/dl of

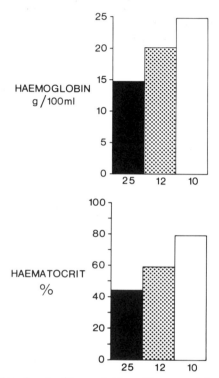

Fig. 16.6 Levels of haemoglobin and haematocrit in the same subjects as in Fig. 16.3. Both levels are higher in healthy highlanders than in sea-level residents but higher still in patients with Monge's disease.

haemoglobin certainly exceeds the figure corresponding to two standard deviations above the mean for healthy acclimatized people living between 4000 m and 4400 m. However, it is overlapped by two standard deviations above the mean for people living above 4400 m (Monge, 1966). This illustrates that there is some difficulty in distinguishing between normal high altitude physiology from the pathophysiology of 'Monge's disease'. The morphology of the circulating red cells is normal. There is no increase in the number of leukocytes in the blood thus distinguishing the condition from polycythaemia vera. Hurtado (1942) gives a range of 3800 to 7200 × 10⁹/l. There is no abnormality in the differential count.

Pulmonary haemodynamics

In Monge's disease the degree of pulmonary arterial hypertension is higher than that observed in healthy highlanders (Peñaloza and Sime, 1971; Peñaloza et al, 1971) (Fig. 16.7a). So too is the mean total pulmonary resistance (Fig. 16.7b). The cardiac index and pulmonary wedge pressure are not significantly different from those found in healthy highlanders.

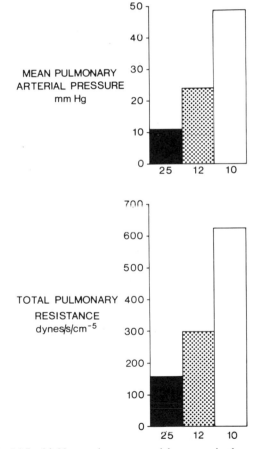

Fig. 16.7 (a) Mean pulmonary arterial pressure in the same subjects as in Fig. 16.3. There is slight pulmonary hypertension in healthy highlanders but moderate elevation of pulmonary arterial pressure in patients with Monge's disease. (b) Total pulmonary vascular resistance in the same subjects as in Fig. 16.3. The resistance is higher in healthy highlanders than in sea-level residents but it is higher still in patients with Monge's disease.

Arterial blood gases and acid-base equilibrium

The arterial oxygen saturation in cases of Monge's disease is even lower than that found in normal, healthy residents at high altitude (Hurtado, 1942, 1955, 1960; Peñaloza and Sime, 1971; Peñaloza et al, 1971) (Fig. 16.8). This suggests that the increased level of polycythaemia is a consequence of

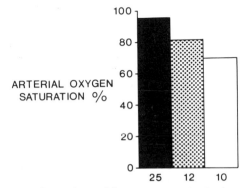

Fig. 16.8 Systemic arterial oxygen saturation in the same subjects as in Fig. 16.3. There is a fall in arterial oxygen saturation in healthy highlanders compared to sea-level residents but the fall becomes even more pronounced in cases of chronic mountain sickness.

the increased hypoxic stimulus. There is also a higher arterial PCO_2 than that normally present in the native population (Monge et al, 1964). The pH remains unaltered as the consequence of an elevation of plasma bicarbonate concentration.

Blood volume

Hurtado (1942) found that in Monge's disease the total blood volume is elevated. He found the range in eight cases to be 149.6 to 211.9 ml/kg with a range of plasma volume of 27.7 to 46.8 ml/kg.

Alveolar hypoventilation

There are two ways in which even more inadequate oxygenation of the blood in the pulmonary capillaries could develop even beyond that due to the lowered partial pressure of oxygen in the alveolar spaces inherent in life at high altitude. One is decreased permeability in the alveolar capillary membrane. The other is deficient ventilation of the alveoli. There is no great diminution of vital capacity in Monge's disease to explain the further desaturation of systemic arterial blood. Hurtado found the range of vital capacity to be 2.24 to 4.46 l in seven subjects. Hence the basis of Monge's disease appears to be alveolar hypoventilation. This leads to chronic alveolar hypoxia, exceeding that inherent in residence at high altitude. This in turn brings about pulmonary vasoconstriction by one of the mechanisms discussed in Chapter 11. An exaggeration of pulmonary arterial hypertension and increased blood viscosity follows. The haematological and neuropsychic features of the condition are also secondary to hypoxaemia.

The concept of loss of acclimatization

The alveolar hypoventilation has been ascribed to a loss of sensitivity of the respiratory centre to carbon dioxide (Hurtado, 1960) or to an irreversible insensitivity of the peripheral chemoreceptors to hypoxia. This loss of respiratory drive is considered by the Peruvian school to represent a loss of acclimatization to high altitude under the persistent stimulus of chronic hypoxia. Monge's disease is clearly a syndrome of alveolar hypoventilation (Fishman et al, 1957; Bergofsky, 1967) occurring at high altitude (Severinghaus et al, 1966). It seems likely that the disease corresponding to Monge's disease at sea level is the 'primary hypoventilation syndrome' (Richter et al, 1957; Rodman and Close, 1959). In this condition there is a diminished ventilatory response to hypoxaemia and hypercapnia so that polycythaemia and cyanosis develop.

Geographical pathology

While Monge's disease is seen not infrequently in the Peruvian Andes a case does not appear to have been reported from the Himalayas at comparable altitudes. One isolated case has been reported from the United States (Hecht and McClement, 1958). This virtual confinement of cases of Monge's disease to the Andes suggests the operation there of a factor peculiar to the region. Since mining is a major activity in the area in such towns as Cerro de Pasco, Morococha, and La Oroya it is a possibility that pollution of the air might predispose to chronic bronchitis and perhaps centrilobular emphysema which could exaggerate the alveolar hypoxia of high altitude. The development of such pulmonary disease associated with the hypoxia of the mountains

might account for one of the clinico-pathological types of Monge's disease referred to below. However, it is much more likely that the special local factor is simply the great altitude at which people live permanently in the Andes. Although the Himalayas are of comparable altitude to the Andes, the permanent settlements of the Sherpas are nowhere as high as those of the Quechuas.

Clinico-pathological types

Arias-Stella and his colleagues (1973) now recognise three clinico-pathological types of Monge's disease.

The first is designated 'chronic soroche' and occurs in people who have to move from sea level to live at high altitude and never adjust to this change. Monge (1943) recognised that some examples of the syndrome to which his name has been coupled were cases of chronic lack of adjustment to high altitude but he did not give details of individual cases.

The second form is seen among people from sea level who have already acclimatized and have been living in good health at high altitude, or among natives of the Andes, in whom organic diseases develop and exaggerate the hypoxaemia. These are diseases which are themselves capable, even at sea level, of producing chronic hypoxia and characteristic changes in the pulmonary arterial tree which we have previously called 'hypoxic hypertensive pulmonary vascular disease', (Hasleton et al, 1968). Such diseases include gross obesity (the Pickwickian syndrome), kyphoscoliosis and pulmonary emphysema. Other conditions referred to as occurring in this clinicopathological type of Monge's disease by Arias-Stella et al (1973) are neuromuscular disorders affecting the thoracic cage, pulmonary tuberculosis and pneumoconiosis. This type may be referred to as 'secondary chronic mountain sickness' (Monge and Monge, 1966) or Monge's syndrome (Arias-Stella et al, 1973).

According to Arias-Stella and his colleagues the third type occurs in persons who are native highlanders or who have acclimatized successfully to life at high altitude and then who subsequently develop the features of chronic mountain sickness, although no organic disease is found to explain their increased hypoxaemia. This is an attractive and logical idea but it should be noted that, as there have been no necropsy studies on unequivocal cases of

Monge's disease, it is impossible at this time to prove or disprove that the condition occurs in the absence of demonstrable morbid anatomical change. Should such cases subsequently be proven to occur they are what Arias-Stella and his colleagues would call 'true Monge's disease'. Patients of this type have been described by Monge (1928), Hurtado (1942) and Peñaloza and Sime (1971).

Age and Monge's disease

There is a correlation between ventilatory rate and age both in healthy highlanders and in cases of chronic mountain sickness (Sime et al, 1975). The relation is hyperbolic and Monge's disease represents the extreme situation in older people with a low ventilatory rate. Cases of the disease in other words are merely extreme examples of a normal trend, the results from such patients lying on the regression line obtained from data on control subjects. The same is true of correlations between haematocrit levels and ventilatory rate, cases of Monge's disease once again lying on the normal regression line, albeit at one extreme edge. Such findings lead Sime et al (1975) to believe that chronic mountain sickness is a clinical manifestation of ageing at high altitude, being the result of an excessive polycythaemia secondary to the fall in ventilatory rate occurring with age. Patients with Monge's disease living at 4500 m and referred to in the literature are commonly in their forties whereas 'controls' for these cases are in their twenties. For example, in the report of Peñaloza et al (1971) the mean age of the cases of Monge's disease was 38 years whereas that of control subjects was 24 years. This all-important age factor must not be overlooked for chronic mountain sickness may evolve as a natural extension of what occurs in younger natives.

At sea level the ventilatory rate does not change with age from 16 to 69 years (Baldwin et al, 1948) but Pa_{O_2} diminishes linearly with age. This drop has been attributed to an increased inequality of the ventilation–perfusion ratio with age (Sorbini et al, 1968). At sea level the Pa_{O_2} falls on the *flat* part of the oxygen–haemoglobin dissociation curve so that a moderate fall in Pa_{O_2} will not result in a noticeable polycythaemic response. However, at high altitude this fall in Pa_{O_2} occurs on the *steep* slope of the curve

so that significant polycythaemia results. As we have already noted in Chapter 6, the haematocrit rises as a function of age in addition to altitude (Whittembury and Monge, 1972).

Sleep and Monge's disease

It has not escaped the attention of some workers that the hypoventilation of sleep mimics the physiological basis of Monge's disease. At high altitude the shallowness of breathing during the hours of sleep induces severe hypoxaemia and appears to be a factor in the pathogenesis of what some American authors term 'chronic mountain polycythaemia' (Kryger et al, 1978a). The combination of ambient hypoxia of high altitude and relative hypoventilation of sleep lowers Pa_{O_2} to the inflection point of the oxygen–haemoglobin dissociation curve so that pronounced oxygen desaturation occurs.

Morbid anatomy

A few years after Monge described the clinical syndrome which has come to bear his name, Hurtado (1942) came to the conclusion that an understanding of the nature of the disease would be helped considerably by establishing the morbid anatomy of the disease.

He said: 'The study of a greater number of cases and the anatomic investigation of the lungs and other organs after death will be necessary for the final understanding of the etiologic mechanism responsible for the hematologic alterations found in some cases of chronic mountain sickness. In this field, as in many others, the morphologic aspects have been unduly neglected in favor of the functional and chemical approximation.' Such data have been slow in coming and in fact it is salutary to realise that since Monge's disease was first described in 1928 there are only three reports of the histopathology in fatal cases of the condition (Fernan-Zegarra and Lazo-Taboada, 1961; Reategui-López, 1969; Arias-Stella et al, 1973). Furthermore, not one of them is a *bona fide* case of Monge's disease since each had some condition like obesity, lordoscoliosis or kyphoscoliosis that might on its own predispose to chronic hypoxia and hypoxic hypertensive pulmonary vascular disease. Hence all the cases so far described in the literature as illustrating the pathological features of the condition are in fact examples of 'secondary Monge's disease'.

Analysis of the macroscopic and histological features in these reports reveals that they are the result of pulmonary hypertension secondary to the alveolar hypoxia. Thus all three reports refer to right ventricular hypertrophy and thickening of peripheral pulmonary arteries, thickening of the media being specified in two instances. An important point emerges from the most satisfactory report on a case of Monge's disease by Arias-Stella et al (1973) although we may note in passing that the patient reported also suffered from dorsal kyphoscoliosis, chronic bronchitis, and slight centrilobular emphysema. They note that the degree of right ventricular hypertrophy in their case is more severe than in normal cases at high altitude (Recavarren

and Arias-Stella, 1964). Similarly the area of muscle in the media of the peripheral pulmonary arteries in their case exceeded that found in the muscularization of the normal pulmonary arterial tree at high altitude (Fig. 16.9a). Clearly such features are consistent with the greater elevation of pulmonary arterial pressure induced by an exaggeration of alveolar hypoxia over that normally experienced at the altitude in question. Arias-Stella and his colleagues (1973) also reported the presence of fresh and partially organized thrombi (Fig. 16.9b and c) in the pulmonary arteries and arterioles. This is consistent with the greater elevation of haematocrit induced by the worsening of the alveolar hypoxia. In passing we may note that their patient also had a

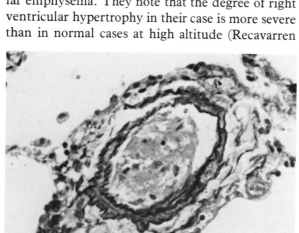

(a)

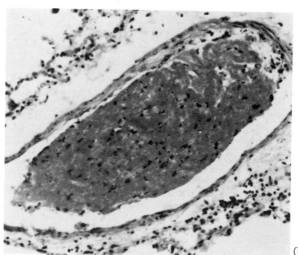

(b)

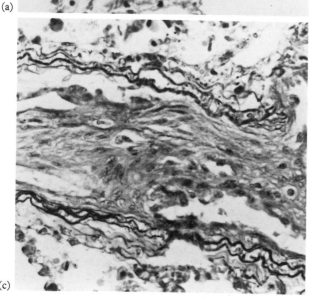

(c)

Fig. 16.9   (a: above left) Transverse section of a muscularized pulmonary arteriole from a case of Monge's disease in a woman aged 48 years. There is a distinct media of circularly-oriented smooth muscle bounded by internal and external elastic laminae. While these changes are in essence precisely those found in the lungs of healthy highlanders they are said to be quantitatively more severe in Monge's disease. Such morbid anatomical changes are associated with the further elevation of pulmonary arterial pressure in this condition. Within the muscularized pulmonary arteriole is recent thrombus. (Elastic Van Gieson × 375). From the case reported by Arias-Stella et al (1973). (b: above right and c: below left) Oblique and longitudinal sections of muscular pulmonary arteries from the same case. They contain recent and organizing thrombus respectively. (b) Haematoxylin and eosin (× 177). (c) Elastic Van Gieson (× 375).

colloid goitre and hyperplasia of the zona glomerulosa of the adrenal cortex (Fig. 16.10). The functional significance of these endocrine changes may be assessed from Chapter 25.

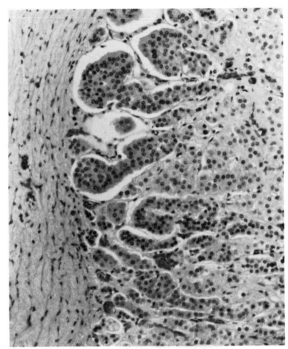

Fig. 16.10 Section of the adrenal cortex from the same case of Monge's disease as illustrated in Fig. 16.9. It shows hyperplasia of the zona glomerulosa. (Haematoxylin and eosin × 150).

The nature of Monge's disease

Consideration of the clinical, physiological, and morbid anatomical features of Monge's disease leads one to accept that the basis for the development of the condition is hypoventilation at high altitude. This exaggerates all the effects of the chronic alveolar hypoxia inherent in a life in the mountains. It seems very likely that Monge's disease may be an expression of ageing at high altitude since increasing age is associated with progressive hypoventilation. The degree of altitude at which the subject lives is also of importance since the greater the elevation the younger the age at which alveolar hypoventilation may bring about the conditions consistent with the development of chronic mountain sickness. The physiological and pathological features of Monge's disease are but exaggerations of those found in healthy native highlanders of the Andes. However,

this exaggeration is of great clinical importance for, whereas the healthy Quechua Indian can in no sense be regarded as suffering from cor pulmonale, accentuation of his normal haematological and haemodynamic characteristics can lead to a potentially fatal condition.

Treatment

The simple and effective treatment for Monge's disease is removal to a lower altitude. When a patient with the condition descends to sea level, his symptoms regress rapidly and there are immediate improvements in the haematological and haemodynamic disturbances. After a stay of two months at sea level the improvement is usually so pronounced that the abnormalities have returned virtually to normal (Table 16.1). Peñaloza et al

Table 16.1 Clinical, haematological and haemodynamic factors in a patient with Monge's disease at high altitude and after two months' residence at sea level (Peñaloza et al, 1971).

	At high altitude (4375 m)	After 2 months at sea level
Heart rate (beats/min)	112	65
Haematocrit (%)	86.0	50.0
Haemoglobin (g/dl)	23.2	17.0
Arterial oxygen saturation (%)	66.6	98.0
Pulmonary arterial mean pressure (mmHg)	62	24

(1971) described the changes which occurred in three patients with Monge's disease after 3, 11, and 60 days sojourn at lower altitude. Cyanosis, the tendency to fatigue, and cerebral symptoms all improve rapidly. The heart decreases in size (Figs. 16.11a and b) and the electrocardiographic indicators of right ventricular hypertrophy diminish. After residence at sea level for two months the heart rate, haematocrit, and haemoglobin level fall substantially; the arterial oxygen saturation increases and there is a sharp fall in the pulmonary

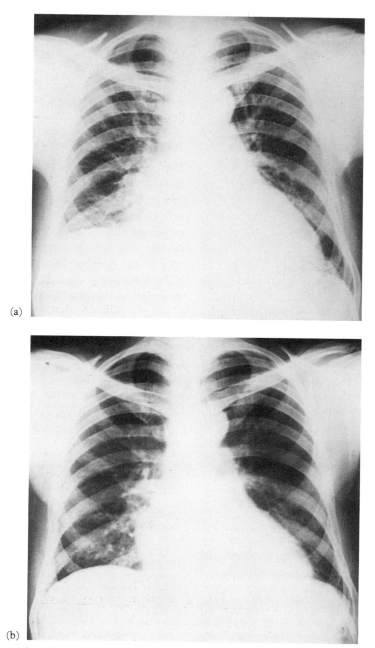

(a)

(b)

Fig. 16.11   (a) Anteroposterior radiograph of the chest in a man of 40 years with Monge's disease who resided at an altitude of 4330 m. There is cardiac enlargement and prominence of the pulmonary artery and vascular markings in the lung. (b) Antero posterior radiograph of the chest from the same patient illustrated in Fig. 16.11a, after he had resided at sea level for nine months. There is diminution in the size of the cardiac shadow.

arterial mean pressure (Table 16.1). There is a considerable decrease of circulating blood volume from a mean value of 175 to 120 ml/kg body weight (Hurtado, 1942). This is largely due to lowering of the red cell volume. At the same time there is an increase in the plasma volume from just over 40 to 60 ml/kg body weight. This increase in plasma volume may be of aetiological significance in the development of high altitude pulmonary oedema in native highlanders returning to the mountains after a period of residence at sea level as we describe in Chapter 15.

Phlebotomy will lead to some improvement in the systemic arterial saturation and a fall in pulmonary arterial pressure. Oxygen has no place in the treatment of Monge's disease for it will achieve little more than a slight fall in pulmonary hypertension. Medroxyprogesterone acetate (MPA) has been reported as an effective form of therapy in excessive polycythaemia of high altitude because of its stimulant effects on ventilation and tidal volume, and the resultant increase in arterial oxygen

saturation (Kryger et al, 1978b). In another clinical trial MPA led to a remission of chronic mountain polycythaemia with reduction of the haematocrit level from 60.1 to 52.1 per cent after 10 weeks in 17 patients (Kryger et al, 1978a).

An interesting sociological slant on the treatment of chronic mountain sickness is that patients who have had the condition relieved by residence at sea level not infrequently insist on returning to their mountain homes because of family reasons. When they do, symptomatology usually returns (Monge and Monge, 1966).

REFERENCES

Arias-Stella, J., Krüger, H. & Recavarren, S. (1973) Pathology of chronic mountain sickness. *Thorax*, **28**, 701.

Baldwin, F., Cournand, A. & Richards, D. W. (1948) Pulmonary insufficiency. I Physiological classification, clinical methods of analysis, standard values in normal subjects. *Medicine*, **27**, 243.

Bergofsky, E. H. (1967) Cor pulmonale in the syndrome of alveolar hypoventilation. *Progress in Cardiovascular Diseases*, **9**, 414.

Fernan-Zegarra, L. & Lazo-Taboada, F. (1961) Mal de Montana crónico. Consideraciones anatomopatológicas y referencias clinicas de un caso. *Revista Peruana de Patologia*, **6**, 49.

Fishman, A. P., Turino, G. M. & Bergofsky, E. H. (1957) The syndrome of alveolar hypoventilation. *American Journal of Medicine*, **23**, 333.

Hasleton, P. S., Heath, D. & Brewer, D. B. (1968) Hypertensive pulmonary vascular disease in states of chronic hypoxia. *Journal of Pathology and Bacteriology*, **95**, 431.

Hecht, H. H. & McClement, J. H. (1958) A case of chronic mountain sickness in the United States. Clinical, physiologic and electrocardiographic observations. *American Journal of Medicine*, **25**, 470.

Hurtado, A. (1942) Chronic mountain sickness. *Journal of the American Medical Association*, **120**, 1278.

Hurtado, A. (1955) Pathological aspects of life at high altitude. *Military Medicine*, **117**, 272.

Hurtado, A. (1960) Some clinical aspects of life at high altitude. *Annals of Internal Medicine*, **53**, 247.

Kryger, M., Weil, J. & Grover, R. (1978a) Chronic mountain polycythemia: A disorder of the regulation of breathing during sleep? *Chest* (Supplement), **73**, 304.

Kryger, M., McCullough, R. E., Collins, D., Scoggin, C. H., Weil, J. V. & Grover, R. F. (1978b) Treatment of excessive polycythemia of high altitude with respiratory stimulant drugs. *American Review of Respiratory Disease*, **117**, 455.

Monge, C. C., Lozano, R. & Carcelen, A. (1964) Renal excretion of bicarbonate in high altitude natives and in natives with chronic mountain sickness. *Journal of Clinical Investigation*, **43**, 2303.

Monge, C. C. (1966) Natural acclimatization to high altitudes: clinical conditions. In: *Life at High Altitudes*, p. 46.

Washington: Pan American Organization. Scientific Publication No. 140.

Monge, M. C. (1928) La enfermedad de los Andes, sindromes eritrémicos. *Anales de la Facultad de Medicina de Lima*, **11**, 314.

Monge, M. C. (1943) Chronic mountain sickness. *Physiological Reviews*, **23**, 166.

Monge, M. C. & Monge, C. C. (1966) In: *High Altitude Diseases, Mechanism and Management*. Springfield, Illinois: Charles C. Thomas.

Peñaloza, D. & Sime, F. (1971) Chronic cor pulmonale due to loss of altitude acclimatization (chronic mountain sickness). *American Journal of Medicine*, **50**, 728.

Peñaloza, D., Sime, F. & Ruiz, L. (1971) Cor pulmonale in chronic mountain sickness: Present concept of Monge's disease. In: *High Altitude Physiology: Cardiac and Respiratory Aspects*. Ciba Foundation Symposium. Edited by R. Porter & J. Knight. Edinburgh: Churchill Livingstone.

Reategui-López, L. (1969) Soroche cronico. Observaciones realisadas en el Cuzco en 30 casos. *Revista Peruana de Cardiologia*, **15**, 45.

Recavarren, S. & Arias-Stella, J. (1964) Right ventricular hypertrophy in people born and living at high altitudes. *British Heart Journal*, **26**, 806.

Richter, T., West, J. R. & Fishman, A. P. (1957) The syndrome of alveolar hypoventilation and diminished sensitivity of the respiratory center. *New England Journal of Medicine*, **256**, 1165.

Rodman, T. & Close, H. P. (1959) The primary hypoventilation syndrome. *American Journal of Medicine*, **26**, 808.

Severinghaus, J. W., Bainton, C. R. & Carcelen, A. (1966) Respiratory insensitivity to hypoxia in chronically hypoxic man. *Respiration Physiology*, **1**, 308.

Sime, F., Monge, C. & Whittembury, J. (1975) Age as a cause of chronic mountain sickness (Monge's disease). *International Journal of Biometeorology*, **19**, 93.

Sorbini, C. A., Grassi, V., Solinas, E. & Muiesan, G. (1968) Arterial oxygen tension in relation to age in healthy subjects. *Respiration*, **25**, 3.

Whittembury, J. & Monge, C. C. (1972) High altitude, haematocrit and age. *Nature (London)* **238**, 278.

17

Systemic circulation

As we have seen many of the features of acclimatization are concerned with modifying the transport of oxygen so that the tissues are protected as far as possible from the consequences of the low partial pressure of the gas in the ambient air. We have considered not only the modifications which occur in ventilation and pulmonary diffusion (Chapter 5) and in tissue diffusion (Chapter 7), but also the haematological changes (Chapter 6). The cardiovascular system may also influence the supply of oxygen to the tissues by changes in the cardiac output and in the distribution of blood flow in the body.

Cardiac output

Resting stroke volume and cardiac output start to fall immediately on arrival at high altitude, reach a minimum by the third day and show a secondary fall by about the tenth day according to Hoon et al (1977). These workers estimated stroke volume and cardiac output of 50 healthy lowlanders by the non-invasive technique of electrical impedance plethysmography, and then airlifted them to 3660 m so that serial estimations could be carried out until the tenth day of exposure to high altitude, and for up to five days after return to sea level. At high altitude the mean stroke volume was 63 ml, this representing a reduction of 19.4 per cent of sea level values. At sea level the mean cardiac output was 4.92 l min⁻¹ and on arrival in the mountains this value fell by 19.5 per cent. By the third day of exposure to high altitude the reduction in cardiac output was 24.8 per cent and by the tenth day a maximum decline of 26.2 per cent was found. On return to sea level the cardiac output returned to normal. The reduction in stroke volume is not fully compensated by tachycardia and this is why the overall effect of the chronic hypoxia of high altitude is to reduce cardiac output. Earlier studies had suggested that on acute exposure to high altitude the stroke volume does not alter (Stenberg et al, 1966) and coupled with the characteristic tachycardia this was reported as resulting in an increased cardiac output which did not persist for more than a few days (Klausen, 1966; Vogel and Harris, 1967). However, confirmatory work from the Andes by Sime and his colleagues (1974) also showed a reduction of 10 per cent in stroke index at rest during the first hour of ascent to 2380 m but a further reduction of up to 20 per cent by the fifth day of exposure.

In healthy Andean residents, however, the cardiac output has been reported as normal rather than diminished (Peñaloza et al, 1963). The resting cardiac output of the lowlander resident on a long-term basis in the mountains is also normal. However, return to sea level for only 10 days results in a pronounced augmentation in the stroke volume of such long-term residents (Hartley, 1971). There has been one report (Rotta et al, 1956) that highlanders, like those ascending to high altitude, have a diminished stroke volume and cardiac output.

There is evidence to suggest that exposure to high altitude prevents maximal cardiovascular performance. The pronounced reduction in maximum cardiac output is more apparent in sojourners than in native highlanders (Hartley et al, 1967; Pugh, 1964). There is also a reduction in maximal heart rate which is related to both the altitude and the duration of exposure to it. Thus while the reduction in maximal heart rate is seen inconsistently at 4300 m it occurs regularly at 5800 m (Hartley,

1971). On exercise at high altitude cardiac output increases disproportionately about 20 per cent beyond that caused by the same work load at sea level (Lenfant and Sullivan, 1971).

Mechanisms of the effects of high altitude on cardiac output

During the initial phase of exposure to high altitude the predominant factor which works in the direction of increasing cardiac output is tachycardia, and this appears to be an expression of increased sympathetic activity which is reflected by a rise in the level of plasma and urinary catecholamines (Cunningham et al, 1965) (Chapter 25).

The cause of the reduction in stroke volume and cardiac output is more controversial. A reduced venous return to the heart due to redistribution of blood (see below) is a possibility. More likely, however, is a direct depressing action of hypoxia on the myocardium (Alexander et al, 1967). The contractility of the myocardium of goats exposed to reduced barometric pressure is depressed and contributes significantly to the reduction in stroke volume. It is of interest to note that Tucker and his associates (1976) could demonstrate this depression of myocardial contractility only by removal by chemical blockade of the beta sympathetic drive referred to above. This suggests that the hypoxic myocardial depression of high altitude may be partially or even completely overcome by increased sympathetic stimulation.

Left ventricular function in newcomers to high altitude was studied by non-invasive methods on normal volunteers airlifted or taken by road to 3660 m (Balasubramanian et al, 1978). A significant reduction of all indices of left ventricular function was observed from the second day of induction to high altitude despite increased urinary catecholamine excretion. On return to sea level all the values returned to normal by the third day. Permanent residents of high altitude had normal left ventricular function and temporary residents a moderate depression. These findings confirm that left ventricular dysfunction occurs on exposure to the hypoxia of high altitude.

Redistribution of blood flow

The diminished cardiac output characteristic of prolonged residence at high altitude is advantageous in the sense that the work load of the heart is not increased (Lenfant and Sullivan, 1971). However, this smaller systemic flow of blood is redistributed so that vital organs and muscles are able to function efficiently in the face of chronic hypoxia. Figure 17.1 derived from data of Finch and Lenfant (1972) expresses diagrammatically the fact that the proportion of the total cardiac output received by an organ does not equate with the proportions of the total oxygen being transported. Some tissues such as the skin have modest requirements of oxygen compared to organs like the heart and this is evidenced by the low extraction rate of oxygen also shown in Figure 17.1. At high altitude there is a redistribution of blood away from such areas to increase the oxygen reservoir for the remainder of the body. This decrease in the cutaneous circulation both in indigenous natives and in long-standing sojourners has been noted by Martineaud and associates (1969) (Chapter 23). This change is most pronounced when the temperature of the skin is initially high and the flow increased. As will be seen the fraction of the total blood carried in the skin is small but its redistribution under such conditions of chronic hypoxia is probably of importance.

In the same way there is a decreased renal flow at high altitude both in highlanders (Becker et al, 1957) and in sojourners (Pauli et al, 1968). However, there is a close relation between packed cell volume and the decrease in renal plasma flow (Lozano and Monge, 1965) so in effect there is a normal amount of oxygen being transported to the kidneys, the fall in renal flow being matched by the increased arterial oxygen content. Renal function remains unimpaired with increased filtration rate to perform its rôle in acclimatization at high altitude.

The effect of high altitude on the coronary arterial flow is somewhat surprising when one recalls from Figure 17.1 that the extraction of oxygen from blood in the coronary blood vessels is very considerable so that there would appear to be but little physiological reserve for any decrease in coronary arterial flow. Nevertheless, there now seems little doubt that the coronary arterial flow is decreased at high altitude (Grover et al, 1970). This is certainly in line with the general decrease in cardiac output at high altitude to which we have referred above. This intriguing question of the diminished coronary arterial flow in

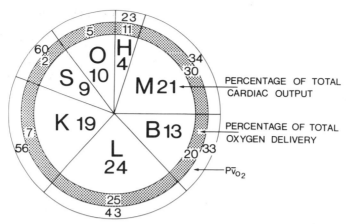

Fig. 17.1 The distribution of blood flow and oxygen to the various tissues of the body. The inner circle represents the total cardiac output and the segments into which it has been divided represent the percentage of the total output received by each organ. The figure indicated in the surrounding ring is the percentage of the total oxygen delivery being received by the corresponding organ. The figure in the outermost ring is the value for $P\bar{v}_{O_2}$ indicating the extraction rate of oxygen from the blood in that organ. Thus the heart, H, receives 4 per cent of the cardiac output and 11 per cent of the available oxygen, while the oxygen saturation of blood in the coronary veins is only 23 per cent indicating a high extraction rate of oxygen. In contrast the skin, S, receives 9 per cent of the cardiac output, but only 2 per cent of the available oxygen, the $P\bar{v}_{O_2}$ being 60 per cent, indicating a low extraction rate of oxygen. H = heart; M = muscle; B = brain; L = liver; K = kidneys; S = skin; O = other tissues. (Based on data from Finch and Lenfant, 1972).

the highlander is described at length in Chapter 18. As we shall see there it seems likely that during prolonged residence at high altitude increased myocardial vascularization counteracts the effects of diminished coronary arterial flow thus avoiding the dangers of myocardial ischaemia and its sequelae.

Blood volume

Another important factor involving the systemic circulation in those living at high altitude is the increase in blood volume which occurs. The total blood volume increases from about 80 ml/kg body weight at sea level to some 100 ml/kg body weight (Fig. 17.2) (Hurtado, 1964). We may in passing note that the total blood volume is some 40 per cent above this in subjects with the clinical syndrome referred to as 'Monge's disease' (Chapter 16) (Hurtado, 1964). The basis for this increase in total blood volume is not in the plasma volume which in fact is lower in the highlander. It lies in the red cell volume which is increased in the highlander and greatly so in Monge's disease (Fig. 17.2). Such changes in plasma and red cell volume take place soon after arrival at high altitude as will be seen from the data of Merino (1950) expressed diagrammatically in Figure 17.3. After only one to three weeks' residence at high

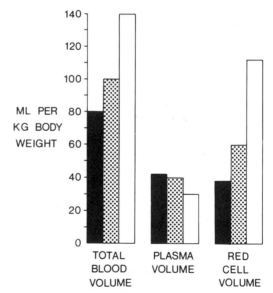

Fig. 17.2 The total blood volume, plasma volume and red cell volume in subjects living at sea level (filled column), at 4540 m (stippled column) and in patients with Monge's disease (open column). (Based on data from Hurtado, 1964.) At high altitude there is an increase in total blood volume and red cell volume. These changes are exaggerated in Monge's disease.

altitude there is already a detectable increase in red cell volume and a decrease in plasma volume (Fig. 17.3). The reversible nature of the changes in

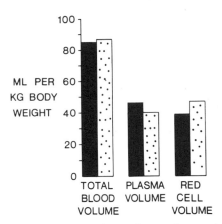

Fig. 17.3 The total blood volume, plasma volume and red cell volume in subjects at sea level (filled column) and after residence at 4540 m for one to three weeks (light stippling). After such a short period the changes illustrated in Fig. 17.2 are already taking place. (Based on data from Merino, 1950).

volume are indicated by data on the red cell and plasma volume from natives of Huancayo (3260 m) who first ascended to Morococha (4540 m) and then descended to Lima (150 m) (Monge and Monge, 1966) (Fig. 17.4). On the ascent there was a definite increase in total blood volume due to increased red cell volume. On subsequent descent to sea level there was a fall in total blood volume and red cell volume which fell progressively over eight weeks (Fig. 17.4). The ratio of plasma volume to red cell

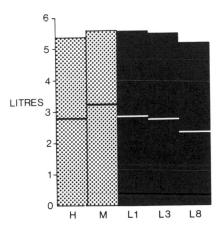

Fig. 17.4 The red cell volume (below the line) and plasma volume (above the line) in natives of Huancayo (3260 m), H, who ascended to Morococha (4540 m), M, and then descended to Lima where studies of the blood volumes are carried out after one week (L1), three weeks (L3) and eight weeks (L8). (Based on data from Monge and Monge, 1966).

volume in camelids indigenous to high altitude is much higher than in the highlander (Fig. 17.5). Thus in man the ratio is 0.6, in the alpaca 1.9, in the llama 1.6, and in the vicuña 1.8. Hence in the highlander the systemic circulation is characterized by hypervolaemia due to an increase in red cell volume.

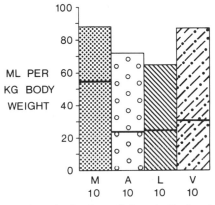

Fig. 17.5 The red cell volume (below the line) and plasma volume (above the line) in 10 men (M), 10 alpacas (A), 10 llamas (L), and 10 vicuñas (V). At high altitude the ratio of plasma to red cell volume is disproportionately high in indigenous camelids compared to highlanders. (Based on data from Reynafarje, 1966).

The effects of carbon dioxide on oxygen transport

So far we have seen that at high altitude the body usually transports oxygen by a lower plasma volume and a higher haemoglobin level. In the first few days the red cell mass does not rise but the haemoglobin level rises by haemoconcentration. Later blood volume rises due to increased red cell mass with a fall in plasma volume. At the same time, as we have just seen, there is a lower stroke volume. The studies of Grover and his colleagues (1976) showed that the addition of carbon dioxide to the inspired air at high altitude alters the response of the heart and systemic circulation. Thus it prevents respiratory alkalosis and negative water balance and prevents contraction of plasma volume and haemoconcentration. As we note in Chapter 23, prevention of hypocapnia prevents systemic vasoconstriction. Hence the breathing of carbon dioxide at high altitude prevents the early increase in haemoglobin concentration and exaggerates the arterial hypoxaemia. Consequently transport of the same quantity of oxygen occurs at a somewhat higher stroke volume and a reduced arterial oxygen content.

Peripheral oedema

On a mountaineering expedition to Mt. Kinabalu (4100 m), in Borneo, all six members of the party developed pronounced peripheral oedema (Sheridan and Sheridan, 1970). The climbers, one woman and five men aged 22 to 30 years, spent sequentially two days at 3500 m, three days at 3800 m, and five days at 4000 m. On the sixth day of the climb all members of the party showed oedema which was most obvious on the face, backs of hands, and ankles. Measurements of ankle circumference showed that oedema increased until the tenth day. Descent to 1550 m was made in less than five hours and during that time there was a dramatic loss of oedema in all members of the party. Frequent halts had to be made because of the diuresis. There was an average decrease in ankle circumference of 5 cm.

Seven years later Hackett and Rennie (1977) observed the same peripheral oedema in 10 hikers at Pheriche (4240 m) on the approach route to Mt. Everest. Seven of the cases occurred in women the mean age being 45 years. In the majority the oedema was facial and especially periorbital (Fig. 17.6). Oedema of the hands was also common and in two instances there was severe oedema of the legs. There was no relationship between the oedema and the menstrual cycle or the taking of oral contraceptives. Two of the worst cases were postmenopausal and not on replacement therapy.

It is of interest that five of the seven women, and all three men, affected by peripheral oedema had loud pulmonary râles. One of the women had anasarca in addition. One of the two women without râles had been treated with diuretics for dyspnoea. Two of the affected subjects had retinal haemorrhages and two had papilloedema. The peripheral oedema was not associated with increased proteinuria.

It is apparent from these reports that peripheral oedema is common in newcomers at high altitude but it is not so clear whether the oedema is related to hypoxia or exercise. Williams and his colleagues (1979) have studied the effect of exercise involved in seven consecutive days of hill-walking on fluid homeostasis at an altitude of only up to 1100 m in the Welsh hills. In five subjects they found a fall in packed cell volume reaching a maximum of 11 per cent by the fifth day and a retention of sodium. After

Fig. 17.6 Systemic oedema of high altitude. This woman developed facial and periorbital oedema whilst on a trekking holiday in the Himalayas to altitudes approaching 4500 m. Outlines of goggles on her face left pits in the oedematous tissues. The facial oedema persisted for the entire four weeks that she was in the Himalayas but disappeared immediately on her descent to Kathmandu (1200 m). We are grateful to Mr W. Butt for this photograph and to Mrs Enid Ellison for allowing us to publish it.

an initial loss of water on the first day there was a modest retention reaching a cumulative maximum on the fifth day. It was calculated that by the end of the hill-walk there was an average increase of 22 per cent in plasma volume, an increase of 17 per cent in interstitial fluid and a decrease of 8 per cent in intracellular fluid volume. These changes were associated in all five subjects with facial oedema. Pitting oedema of the lower leg was present in some subjects and was pronounced in one. These workers consider that on prolonged exercise of this type there is an increase in the capacity of the vascular space, presumably the capillary beds of working muscles and possibly the venous beds, and it is this which underlies the increase in plasma volume. Hence they believe that the increase in extracellular fluid indicated by their study explains the dependent and peri-orbital oedema noted after the strenuous exercise involved in hill-walking or mountain-climbing. Hence at high altitude the effects of exercise and hypoxia on fluid retention and fluid shifts may be additive. Their results give support to the impression that exercise on ascent to mountainous areas is a factor in the genesis of high altitude pulmonary oedema.

Systemic blood pressure

The behaviour of the systemic blood pressure at high altitude was reviewed by Hultgren in 1970. It is now generally accepted that chronic hypoxia has a relaxing effect on smooth muscle so that the effect on the arterial media is to produce vasodilation. In the case of the pulmonary arterial tree this primary vasodilating effect of hypoxia is reversed, perhaps by the influence of an intermediary agent, so that pulmonary vasoconstriction and pulmonary arterial hypertension result as we have seen in Chapter 11. However, in the systemic circulation the primary vasodilating effect occurs without modification, leading to a lowering of systemic blood pressure at high altitude.

Systemic blood pressure in native highlanders

Early observations by Peruvian physicians such as Monge that natives of the Peruvian Andes had lower systemic blood pressures than subjects living at sea level in Peru were subsequently confirmed in 1967 by Marticorena, Severino and Chávez. They found that systemic systolic pressure in each age group was lower than in two groups of normal subjects in the United States studied by Comstock (1957) and Master et al (1950), whose values were similar to those found previously in sea-level residents of Peru. It is apparent from the data of Marticorena et al (1967) that while a rise in systemic systolic blood pressure is evident with age in residents at sea level, this trend is not seen in native Peruvian highlanders. Peñaloza (1971) refers to an investigation by Chávez (1965) at La Oroya (3750 m) who found that the levels of systemic blood pressure in 300 subjects born and living in the area were somewhat lower than those found at sea level.

Tibetans living at 2080 m also have a low systemic systolic blood pressure up to the age of 40 years but compared to Peruvian highlanders living at 3750 m their systolic pressure rises after this age (Fig. 17.7) (Sehgal et al, 1968). There is certainly a suggestion

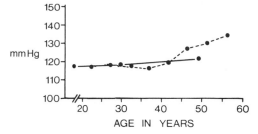

Fig. 17.7   Systemic systolic blood pressure in Quechuas living at 3750 m (continuous line) and Tibetans living at 2080 m (dotted line). Note the rise in blood pressure in the Tibetans at the age of 45 years. (Based on data from Sehgal et al, 1968).

that the high altitude environment in some way protects against elevation of systemic blood pressure in later life. Permanent residence at high altitude appears to be equally effective in maintaining a slightly lower systemic blood pressure in white Caucasians as well as in native Peruvian highlanders. Thus, in the United States, Appleton (1967) studied 2782 high school students living at 1220 m, 1980 m and 2350 m. The mean systolic pressure in the residents at the highest of these three altitudes was 119 mmHg compared to 124 mmHg in those from the lower elevation. Permanent residence at these elevations had, however, little effect on the level of diastolic blood pressure.

During the period 1967 to 1973 surveys of systemic blood pressure were carried out in Peru by Ruiz and Peñaloza (1977) on two communities at sea level (Puente Piedra and Infantas) and on three communities in the range of 4100 m to 4360 m above sea level (Milpo, Colquijirca and Cercapuquio). 4359 subjects were studied at sea level and 3055 at high altitude. The prevalence of systolic and diastolic hypertension in these five communities is shown in Figures 17.8 and 17.9 taking the normal upper limit of systolic blood pressure to be 160 mmHg and the normal upper limit of diastolic pressure to be 95 mmHg. The prevalence of systemic systolic hypertension in males was at least twelve times higher at sea level than at high altitude. The difference was even greater in females. Systemic diastolic hypertension was also commoner at sea level but the differences were less striking than

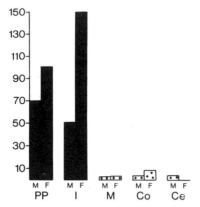

Fig. 17.8 Prevalence of systemic systolic hypertension (>160 mmHg) in two communities at sea level (solid columns) and in three communities at high altitude (stippled columns) in Peru. Prevalence expressed as age-adjusted rate per thousand of population. PP = Puente Piedra, I = Infantas, M = Milpo (4100 m), Co = Colquijirca (4260 m) and Ce = Cercapuquio (4360 m). (Based on data from Ruiz and Peñaloza 1977).

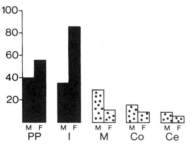

Fig. 17.9 Prevalence of systemic diastolic hypertension (>95 mmHg) in the five communities referred to in Fig. 17.8. (Based on data from Ruiz and Peñaloza, 1977).

for systolic pressure. At sea level systemic hypertension tends to be more frequent in females and systolic hypertension is commoner than diastolic. At high altitude systemic hypertension is more frequent in men and diastolic hypertension is commoner (Ruiz and Peñaloza, 1977). This emphasises the different natural history of systemic hypertension at high altitude.

Effect on systemic blood pressure of acute exposure to high altitude

A fall in systemic blood pressure is also found in sea-level subjects moving to high mountainous areas. It occurs even on acute exposure to high altitude as we have already noted in Chapter 14. Bulstrode (1975) studied the systemic blood

pressure in seven unacclimatized soldiers on acute exposure to 4880 m on Mount Kenya and in seven acclimatized members of a mountain rescue team including four Kikuyu. He found that there was a drop in systolic blood pressure on ascent which was reversed on return to base camp. The drop in systolic blood pressure was more marked in the unacclimatized group, from a mean of 129 to 119 mmHg, than in the acclimatized group from 130 to 125 mmHg. The changes in the diastolic blood pressure were not as large as those occurring in the systolic pressure. In the unacclimatized group the diastolic pressure fell from 84 to 79 mmHg whereas in the acclimatized group it rose from 84 to 88 mmHg.

Effect of prolonged residence at high altitude on the systemic blood pressure of sea-level subjects

After a period of residence of a year at elevation most sea-level subjects show an obvious diminution in systemic blood pressure. Thus Rotta et al (1956) found that both systolic and diastolic pressure were diminished after living for this time at 4500 m. After several years' residence in the mountains not inconsiderable decrements in systemic pressure may be recorded. Marticorena et al (1969) studied a hundred men who had been born at sea level but who lived at 3750 m in the Peruvian Andes for 2 to 15 years. None were Peruvian and the majority came from the United States, Canada and Britain. Blood pressure measurements were obtained from records

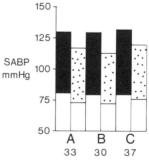

Fig. 17.10 Initial and final systemic blood pressure in 100 sea-level subjects residing at an altitude of 3750 m for periods ranging from 2 to 15 years. (From data in Peñaloza, 1971). A = Residence at high altitude from 2 to 5 years (33 subjects); B = Residence at high altitude from 6 to 9 years (30 subjects); C = Residence at high altitude from 10 to 15 years (37 subjects); Filled column = initial systemic blood pressures; Lightly-stippled column = final systemic blood pressures.

of annual physical examinations. A decrease of 10 mmHg or more in systemic systolic pressure was observed in 56 per cent of the subjects and in diastolic pressure in 46 per cent of them. Furthermore, the longer the residence at high altitude, the greater was the decrease in systolic pressure. Indeed the systolic pressure at the final examination was lower than that of healthy Peruvian natives living at the same altitude. Peñaloza (1971) also refers to the changes in systemic blood pressure found in sea-level males between 25 and 66 years residing at high altitudes at the same town of La Oroya for periods ranging from two to 15 years. These are summarised in Figure 17.10.

The amelioration of systemic hypertension at high altitude

There is some evidence to suggest that systemic hypertension is ameliorated at high altitude. Peñaloza (1971) reported that some patients with the condition who lived at La Oroya (3750 m) for from two to 15 years showed a fall in systemic blood pressure. However, there was no improvement in four patients in whom the diastolic pressure exceeded 95 mmHg. This suggests that, when the basis for the hypertension is vasoconstrictive and muscular, the chronic hypoxia exerts a beneficial vasodilating effect. In cases with more severe organic lesions in the arteries the dilating effect of hypoxia is less effective. There is evidence to suggest that chemodectomas occurring in highlanders may lead to systemic hypertension, perhaps as a result of catecholamine secretion (Saldaña et al, 1973).

Factors involved in the lowering of systemic blood pressure at high altitude

It is widely accepted that in general, hypoxia relaxes vascular smooth muscle. In the pulmonary circulation the vasodilating action is overcome by the more powerful action of an intermediate agent,

perhaps the perivascular mast cell, which induces constriction. However, in the systemic circulation the relaxing effect predominates. Supporting evidence for the operation of an environmental rather than a genetic factor is that after a long residence at high altitude white Caucasian males from sea level show the same lowering of systemic blood pressure as seen in native Quechua highlanders. It seems clear that this factor is hypoxia. The fall in blood pressure in residents at high altitude is greater in the systolic than in the diastolic component. Peñaloza (1971) believes that whereas hypoxia lowers the systolic pressure by a relaxing effect on arterial muscle it may counteract its effect on the diastolic pressure by inducing polycythaemia which raises blood viscosity and peripheral vascular resistance. There appears to be some supporting evidence for the idea of a relation between the levels of polycythaemia and systemic diastolic blood pressure in the fact that in women below the age of forty years at high altitude there is some diminution of systemic diastolic pressure and this may be related to the lower degree of polycythaemia brought about by menstrual loss. In contrast, in cases of Monge's disease (see Chapter 16) there is a severe degree of polycythaemia and diastolic hypertension in the systemic circulation.

The left ventricle at high altitude

Peñaloza (1971) believes that healthy people born and living in an environment of chronic hypoxia at high altitude do not show left ventricular hypertrophy. Enlargement of this chamber is, however, thought by him to occur in Monge's disease (Chapter 16), in which there is severe hypoxaemia and exaggerated polycythaemia. Factors contributing to left ventricular hypertrophy in this syndrome might be hypervolaemia, described above, elevation of the diastolic blood pressure, and myocardial impairment due to a direct effect of severe hypoxia on the myocardial fibres.

REFERENCES

Alexander, J. K., Hartley, H., Modelski, M. & Grover, R. F. (1967) Reduction of stroke volume during exercise in man following ascent to 3100 m altitude. *Journal of Applied Physiology*, **23**, 849.

Appleton, F. (1967) Possible influence of altitude on blood pressure. *Circulation*, **36**, Suppl. 11, 55.
Balasubramanian, V., Mathew, O. P., Tiwari, S. C., Behl, A., Sharma, S. C. & Hoon, R. S. (1978) Alterations in left

ventricular function in normal man on exposure to high altitude (3658 m). *British Heart Journal*, **40**, 276.

Becker, F. L., Schilling, J. A. & Harvey, R. B. (1957) Renal function in man acclimatized to high altitude. *Journal of Applied Physiology*, **10**, 79.

Bulstrode, C. J. K. (1975) A preliminary study into factors predisposing mountaineers to high altitude pulmonary oedema. *Journal of the Royal Naval Medical Service*, **61**, 101.

Chávez, A. (1965) *Presión arterial en altura*. Br. Thesis, Universidad Nacional Mayor de San Marcos, Facultad de Medicina, Lima.

Comstock, G. (1957) An epidemiologic study of blood pressure levels in a biracial community in the Southern United States. *American Journal of Hygiene*, **65**, 271.

Cunningham, W. I., Becker, F. J. & Kreuzer, F. (1965) Catecholamines in plasma and urine at high altitude. *Journal of Applied Physiology*, **20**, 607.

Finch, C. A. & Lenfant, C. (1972) Oxygen transport in man. *New England Journal of Medicine*, **286**, 407.

Grover, R. F., Lufschanowski, R. & Alexander, J. K. (1970) Decreased coronary blood flow in man following ascent to high altitude. In: *Hypoxia, High Altitude and the Heart*, Edited by J. H. K. Vogel Aspen. Colorado. Advances in Cardiology, 5, 32. Basel: S. Karger.

Grover, R. F., Reeves, J. T., Maher, J. T., McCullough, E., Cruz, J. C., Denniston, J. C. & Cymerman, A. (1976) Maintained stroke volume but impaired arterial oxygenation in man at high altitude with supplemental CO_2. *Circulation Research*, **38**, 391.

Hackett, P. & Rennie, D. (1977) Acute mountain sickness. *Lancet*, **i**, 491.

Hartley, H. (1971) Effects of high-altitude environment on the cardiovascular system of man. *Journal of the American Medical Association*, **215**, 241.

Hartley, H., Alexander, J. K., Modelski, M. & Grover, R. F. (1967) Subnormal cardiac output at rest and during exercise in residents at 3100 m altitude. *Journal of Applied Physiology*, **23**, 839.

Hoon, R. S., Balasubramanian, V., Mathew, O. P., Tiwari, S. C., Sharma, S. C. & Chadha, K. S. (1977) Effect of high-altitude exposure for 10 days on stroke volume and cardiac output. *American Journal of Physiology: Respiratory, Environmental and Exercise Physiology*, **42**, 722.

Hultgren, H. N. (1970) Reduction of systemic arterial blood pressure at high altitude. In: *Hypoxia, High Altitude and the Heart*. Edited by J. H. K. Vogel. Aspen. Colorado. Advances in Cardiology, 5, 49. Basel: S. Karger.

Hurtado, A. (1964) Some physiologic and clinical aspects of life at high altitudes. In: *Aging of the Lung*, p. 257. Edited by L. Cander & J. H. Moyer. New York: Grune and Stratton.

Klausen, K. (1966) Cardiac output in man at rest and work during and after acclimatization to 3800 m. *Journal of Applied Physiology*, **21**, 609.

Lenfant, C. & Sullivan, K. (1971) Adaptation to high altitude. *New England Journal of Medicine*, **284**, 1298.

Lozano, R. & Monge, M. C. (1965) Renal function in high altitude natives and in natives with chronic mountain sickness. *Journal of Applied Physiology*, **20**, 1026.

Marticorena, E., Severino, J. & Chávez, A. (1967) Presión arterial sistemica en el nativo de altura. *Arch. Inst. Biol. Andina*, **2**, 18.

Marticorena, E., Ruiz, L., Severino, J., Galvez, J. & Peñaloza, D. (1969) Systemic blood pressure in white men born at sea level: Changes after long residence at high altitudes. *American Journal of Cardiology*, **23**, 364.

Martineaud, J. P., Durand, J., Coudert, J. & Seroussi, S. (1969) La circulation cutanée au cours de l'adaptation a l'altitude. *Pflügers Archiv. (European Journal of Physiology)*, **310**, 264.

Master, A., Dublin, L. & Marks, H. (1950) The normal blood pressure range and its clinical implications. *Journal of the American Medical Association*, **143**, 1464.

Merino, C. F. (1950) Studies on blood formation and destruction in the polycythemia of high altitude. *Blood*, **5**, 1.

Monge, M. C. & Monge, C. C. (1966) In: *High-Altitude Diseases. Mechanism and Management*, p. 62. Springfield, Illinois: Charles C. Thomas.

Pauli, H. G., Truniger, B., Larsen, J. K. & Mulhausen, R. O. (1968) Renal function during prolonged exposure to hypoxia and carbon monoxide. 1. Gomerular filtration and plasma flow. *The Scandinavian Journal of Clinical and Laboratory Investigations*, **22**, Supplement **103**, 55.

Peñaloza, D. (1971) In: *High Altitude Physiology. Cardiac and Respiratory Aspects*, p. 169. Ciba Foundation Symposium. Edited by R. Porter & J. Knight. Edinburgh and London: Churchill Livingstone.

Peñaloza, D., Sime, F., Banchero, N., Gamboa, R., Cruz, J. & Marticorena, E. (1963) Pulmonary hypertension in healthy man born and living at high altitudes. *American Journal of Cardiology*, **11**, 150.

Pugh, L. G. C. E. (1964) Cardiac output in muscular exercise at 5800 m (19 000 ft). *Journal of Applied Physiology*, **19**, 441.

Reynafarje, C. (1966) Physiological patterns: hematological aspects. In: *Life at High Altitudes*, Scientific Publication No. 140. p. 32. Washington: Pan American Health Organization.

Rotta, A., Cánepa, A., Hurtado, A., Velasquez, T. & Chávez, R. (1956) Pulmonary circulation at sea level and at high altitudes. *Journal of Applied Physiology*, **9**, 328.

Ruiz, L. & Peñaloza, D. (1977) Altitude and hypertension. *Mayo Clinic Proceedings*, **52**, 442.

Saldaña, M. J., Salem, L. E. & Travezan, R. (1973) High altitude hypoxia and chemodectomas. *Human Pathology*, **4**, 251.

Sehgal, A., Krishan, I., Malhotra, R. & Gupta, H. (1968) Observations on the blood pressure of Tibetans. *Circulation*, **37**, 36.

Sheridan, J. W. & Sheridan, R. (1970) Tropical high-altitude peripheral oedema. *Lancet*, **i**, 242.

Sime, F. D., Peñaloza, D., Ruiz, L., Gonzales, N., Covarrubias, E. & Postigo, R. (1974) Hypoxemia, pulmonary hypertension, and low cardiac output in newcomers at low altitude. *Journal of Applied Physiology*, **36**, 561.

Stenberg, J., Ekblom, B. & Messin, R. (1966) Hemodynamic response to work at simulated altitude, 4000 m. *Journal of Applied Physiology*, **21**, 1589.

Tucker, C. E., James, W. E., Berry, M. A., Johnstone, C. J. & Grover, R. F. (1976) Depressed myocardial function in the goat at high altitude. *Journal of Applied Physiology*, **41**, 356.

Vogel, J. A. & Harris, C. W. (1967) Cardiopulmonary responses of resting man during early exposure to high altitude. *Journal of Applied Physiology*, **22**, 1124.

Williams, E. S., Ward, M. P., Milledge, J. S., Withey, W. R., Older, M. W. J. and Forsling, M. L. (1979) Effect of the exercise of seven consecutive days hill-walking on fluid homeostasis. *Clinical Science*, **56**, 305.

18

Coronary circulation and electrocardiography

Coronary blood flow at high altitude

After ten days at an altitude of 3100 m the newcomer from sea level experiences a diminution of the coronary blood flow by some 32 per cent (Grover et al, 1976). At the same time there is an increase in extraction of oxygen from coronary arterial blood by some 28 per cent, maintaining delivery of oxygen to the myocardium. Although these authors found a decrease in the oxygen saturation of coronary sinus blood, there was no change in the oxygen tension, implying a decrease in the affinity of haemoglobin for oxygen. Such observations are consistent with the view that coronary blood flow is regulated to maintain constant myocardial tissue oxygen tension. The lack of any fall in PO_2 in coronary sinus blood implies that myocardial hypoxia does not develop.

The coronary blood flow is also diminished in people living permanently at high altitude. Moret (1971) measured coronary flow in two groups living at La Paz (3700 m) and Cerro de Pasco (4330 m) and contrasted his findings with those in subjects living at sea level. The coronary blood flow was diminished by some 30 per cent, the value falling from a mean of 71.7 ml min^{-1}/100 g of left ventricle at sea level to 49.1 ml min^{-1}/100 g of left ventricle at 4330 m. Coronary vascular resistance was also found to be elevated at high altitude. Hence the magnitude of the decrease in coronary flow in highlanders is very similar to that found in newcomers to high altitude by Grover et al (1976). What is more there is, once again, virtually no change in oxygen tension in coronary sinus blood (Moret, 1971).

Coronary blood flow in Monge's disease

The coronary blood flow in four patients with Monge's disease (Chapter 16) was found to be higher than in healthy highlanders (Moret, 1971). The mean coronary flow was found to be 63.1 ml min^{-1}/100 g of left ventricle. This may be an expression of increased left ventricular work in this condition. As a result of the higher haematocrit in this syndrome the supply of oxygen to the myocardium is higher since the desaturation of blood in the coronary sinus is within normal limits.

Incidence of coronary artery disease at high altitude

Since diminished coronary flow in those living at high altitude appears to be balanced by increased extraction of oxygen from coronary arterial blood with resulting maintenance of myocardial tissue oxygen tension it is perhaps not surprising that highlanders show little evidence of myocardial ischaemia. Reliable necropsy data are difficult to come by in such isolated areas as the Andes. However, such as they are, they suggest that both coronary artery disease and myocardial infarction are decidedly uncommon at high altitude. Arias-Stella and Topilsky (1971) refer to a study of Ramos et al (1967) who found not a single case of myocardial infarction or significant coronary artery disease in a consecutive series of 300 necropsies carried out at 4330 m. Epidemiological studies at Milpo (4100 m) have revealed that angina of effort and electrocardiographic evidence of myocardial ischaemia are less common that at sea level (Ruiz et al, 1969). It has to be kept in mind in assessing the low incidence of coronary artery disease at high altitude that the native highlanders are 'low-risk subjects'. They tend to have low levels of cholesterol, total lipids, triglycerides and beta-lipoproteins in their blood (Peñaloza, 1971). Quechua Indians tend to smoke infrequently. We have already noted the rarity at high altitude of such predisposing diseases as systemic hypertension (Chapter 17) and obesity (Chapter 4). Native highlanders tend to be active physically and they are not subjected to the same stresses as the dweller in a modern urban environment in a developed country. Peñaloza (1971) believes that the predominant blood group O of the Quechua Indians (Chapter 4) tends to be associated with lesser susceptibility to ischaemic heart disease.

There is some evidence that there is a progressive decline in mortality from coronary arterial disease in white Caucasians with increasing altitude of residence. Thus Mortimer and his colleagues (1977) carried out epidemiological studies in New Mexico where inhabited areas range from 914 to 2135 m. They compared age-adjusted mortality rates for coronary heart disease for white men and women for the years 1957 to 1970 in five sets of counties, grouped by altitude in 305 m increments. The results showed a serial decline in mortality from the lowest to the highest altitude for males but not for females (Fig. 18.1). Mortimer and his associates

ALTITUDE m

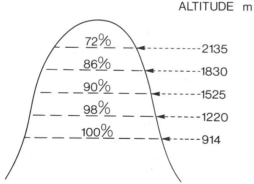

72%	----------2135
86%	----------1830
90%	----------1525
98%	----------1220
100%	----------914

Fig. 18.1 Incidence of coronary heart disease in Caucasian men at various altitudes compared to that at 914 m in New Mexico for the years 1957–1970. (Based on data from Mortimer et al, 1977).

thought it possible that adjustment to living at high altitude by the lowlanders was incomplete so that ordinary daily activities demanded greater exercise than when undertaken at lower altitudes. The increased physical activity was thought to be beneficial in avoiding coronary thrombosis.

Increased vascularization of the myocardium at high altitude

A more intense vascularization of the myocardium of highlanders has been reported by Arias-Stella and Topilsky (1971). At necropsy they injected neoprene latex into the coronary arterial tree in highlanders (4330 m) and in 10 lowlanders of comparable age. The number of secondary branches from both main coronary arteries was counted on the casts obtained. At sea level the mean number was 56.4 but at high altitude it was 79.6. Standard photographs were taken of each specimen at identical distances and magnifications so that the areas occupied by ramifications of the smaller coronary arterial branches could be compared. In the specimens from sea level the range of area occupied by the ramifications was 33 to 52 per cent but in the hearts from high altitude it was 55 to 58 per cent. Further evidence to support the concept that highlanders have a more abundant coronary vasculature than sea-level subjects comes from the work of Carmelino (1970). He has demonstrated by stereoangiography of hearts from subjects at sea level and at Puno (3500 m) that at high altitude the distribution of branches is predominantly from the right coronary artery. His work also reveals a statistically significant greater number of coronary arterial branches of the first order, together with a greater number of intercoronary anastomoses.

The increased vascularization of the myocardium extends down to capillary level. We have already noted in Chapter 7 that one of the features of acclimatization at tissue level is increased capillary density. So it appears to be in the myocardium for Valdivia (1962) found an increased number of blood capillaries per myocardial fibre in guinea pigs kept in decompression chambers compared to those at sea level. A larger capillary area was found in the hearts of puppies born at 6000 m (Becker et al, 1955). It is not certain whether this increased number of coronary blood vessels, both of arterial and capillary dimensions, is perfused at rest but it seems likely that the increased vascular network represents a greater coronary reserve and may explain in part at least the infrequency of myocardial infarction at high altitude.

On the other hand Clark and Smith (1978) found no increase in capillary density in the myocardium of rats subjected to simulated high altitude. They believe that the myocardium contains an optimal capillary density and that hypoxia should not be expected to act as a stimulus for capillary proliferation provided an adequate pressure gradient of oxygen exists between capillaries and the centre of the myofibres. Certainly the increased extraction of oxygen from coronary arterial blood reported by Grover and his colleagues (1976) may balance the diminished coronary blood flow at high altitude and maintain a constant myocardial tissue oxygen tension. They find it difficult to reconcile an increased vascularity of the heart with a diminished coronary blood flow. It is possible that their inability to demonstrate increased vascularization of the myocardium is a reflection of the fact that their experiment lasted only 34 days. It seems to us that this is an inadequate amount of time to allow a significant proliferation of capillaries. It is not to be compared to the age of the native highlanders studied by Arias-Stella and Topilsky (1971). We believe there are dangers in too readily applying the results of very short term experiments employing simulated high altitude to the long-term problems of natural altitude, as we have already said in relation to enlargement of the carotid bodies (Chapter 8).

The electrocardiogram at high altitude

There have been several studies of the changes in the electrocardiogram associated with high altitude and it has to be kept in mind that they have been carried out on unlike groups of subjects under different circumstances. Thus the early studies of Peñaloza and Echevarria (1957) and Peñaloza (1958) were mainly concerned with changes induced by passive transport of subjects from sea level to 4330 m and the further modifications during prolonged residence at those heights. Another group of investigations on the effect of simulated high altitude has been carried out in decompression chambers, particularly at centres of aviation medicine because of the importance of the findings in relation to airmen. Jackson and Davies (1960) carried out a study of the electrocardiogram of the mountaineer at high altitude climbing under the stress of acute hypoxia and load-bearing at an extreme elevation of 5800 m on Ama Dablam in the Himalayas. Milledge (1963) carried out a similar study on Makalu at 7320 m. All these studies are concerned with electrocardiographic changes in lowlanders being acutely exposed to high altitude. Extensive studies on the electrocardiograms of native highlanders in the newborn, in infants, children, adolescents and adults have been made by Peñaloza and his colleagues (1959 and 1961).

Electrocardiograms on acute exposure to high altitude

Electrocardiograms taken on subjects exposed acutely to high altitude confirm the very rapid heart rates which occur on exercise under such conditions (Chapter 14). Jackson (1975) recorded an e.c.g. on a fit 25-year-old man climbing between 5180 m and 5800 m in the Sierra Nevada de Santa Marta and it shows a heart rate of 208 per minute (Fig. 18.2).

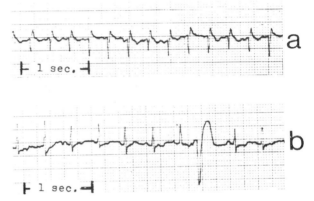

Fig. 18.2 Electrocardiograms recorded on magnetic tape during climbing at altitudes between 5200 m and 5800 m. The upper trace, a, is from a fit young man of 25 years and shows a sinus tachycardia of 208 per minute. The lower trace, b, is from a man of 37 years and shows sinus tachycardia with a ventricular ectopic beat. (After Jackson, 1975).

Changes in rhythm are surprisingly uncommon on acute exposure to high altitude (Jackson, 1975), apart from ventricular extrasystoles, one of which is shown in Figure 18.2. Atrial ectopic beats or rhythms also occur (Jackson, 1975).

Jackson and Davies (1960) studied 12 mountaineers and considered them as comprising a 'European group' and a 'Sherpa group'. The former consisted of six Europeans normally resident at sea level and one Nepalese of Indian descent living at 1330 m. The latter consisted of five Sherpas whose home was at 3660 m. Bearing these two different groups in mind we may now consider the principal electrocardiographic changes that occurred in this party on its ascent to 5800 m. They are summarized diagrammatically in Figure 18.3.

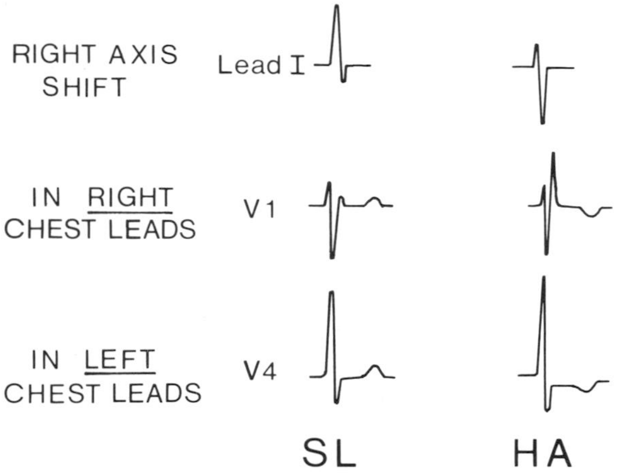

RIGHT AXIS SHIFT   Lead I

IN RIGHT CHEST LEADS   V1

IN LEFT CHEST LEADS   V4

SL   HA

Fig. 18.3 Diagram to illustrate the principal changes which occur in the electrocardiogram of those acutely exposed to high altitude. Complexes as seen at sea level, SL, are contrasted with those which occur at high altitude, HA. At high altitude there is right axis deviation with a prominent S wave in lead I consistent with acute right ventricular overload and right ventricular hypertrophy. There is also T wave inversion in the *right* precordial leads which is also in keeping with the increased load on the right ventricular myocardium. This change and right axis deviation has been reported in European and Sherpa climbers. At high altitude there is lowering of the T wave in the *left* precordial leads possibly indicative of ischaemia of the left myocardium; this has been found in European but not in Sherpa climbers. (After Jackson, 1975).

Right axis deviation

An extreme degree of right axis deviation in the standard limb leads developed and was proportional to the altitude reached. The overall voltage of the QRS complex was greater than at sea level. With increasing altitude S in lead I became more prominent and even dominant. There was a clockwise rotation of the QRS vector in the frontal plane of as much as 50 degrees which is greater than can be produced by deep inspiration in a normal subject. These changes occurred in both European and Sherpa groups. Jackson and Davies (1960) consider the possible explanations for these electrocardiographic changes at high altitude. They think that the clockwise rotation of a vertical heart around its longitudinal axis may be related to the hyperventilation which characterizes residence in the mountains and which is associated with a lower

position of the diaphragm. The normal lead III at high altitude resembles that on full inspiration at sea level. However, the changes are not dissimilar to those occurring with pulmonary embolism or pneumothorax. Since right axis shift may develop irrespective of whether the pneumothorax is on the right or left side it is probably related to acute right ventricular overload. Hence all the changes in the QRS complex referred to so far may reflect the increased right ventricular work necessitated by pulmonary vasoconstriction induced by hypoxia as described in Chapter 11. Milledge (1963) also found right axis deviation and agrees with the interpretation that this is due to right ventricular overload.

Milledge (1963) found that initially on exposure to high altitude the administration of oxygen caused reversal of the right axis deviation. Subsequent inhalations of oxygen did not produce this effect. This is clearly related to the nature of the elevated pulmonary vascular resistance as we describe in Chapters 11 and 28. Initially due entirely to hypoxic pulmonary vasoconstriction it is subsequently based on the less easily reversible factors of polycythaemia and muscularization of the terminal portions of the pulmonary arterial tree (Fig. 28.1).

T wave inversion in the right precordial leads

The European and Sherpa climbers also shared in showing lowering or inversion of the T wave in right chest leads. These changes tended to increase in extent if the stay at high altitude was prolonged for some weeks. The lower position of the heart is not responsible for such changes since deep inspiration causes T in VI to become more upright rather than the reverse (Jackson and Davies, 1960). Almost certainly these changes in the right precordial leads are in keeping with the increased load on the right ventricular myocardium.

T wave changes in the left precordial leads

Significantly these changes occurred in all the European but in none of the Sherpa group. They comprised lowering of the T wave in the left precordial leads. There was some associated depression of the ST segment in the two oldest members of the European party who were 44 and 53 years of age. These changes over the left ventricle have a different significance from those over the right ventricle (Fig. 18.3). They very likely indicate ischaemia of the left myocardium under the hypoxic conditions of life at high altitude. The fact that they occur in lowlanders but not in native highlanders is in keeping with the view that we have advanced earlier in this chapter that the blood supply to the myocardium is better in the highlander. The superior vascularization of the heart muscle described there is the basis for the absence of inversion of the T waves.

Electrocardiographic changes associated with right ventricular hypertrophy

Eventually one would anticipate the electrocardiographic changes of right ventricular hypertrophy with dominant R waves in the right chest leads but this has not been reported in visiting mountaineers (Jackson and Davies, 1960). In contrast electrocardiographic evidence of right ventricular hypertrophy is forthcoming in native highlanders both in childhood and in adult life as we shall see later in this chapter (Peñaloza and his colleagues, 1959, 1961).

The electrocardiographic changes at high altitude and hypokalaemia

Jackson (1968) suggests that the electrocardiographic changes at high altitude are not unlike those seen in hypokalaemia. As we note in Chapter 14 potassium depletion has been regarded as characteristic of acute exposure to the hypoxia of high altitude, being produced by a variety of factors ranging from increased adrenal cortical activity to diminished potassium intake.

Other changes

Jackson and Davies (1960) found no changes in the amplitude of the P waves, and no changes in the P–R, QRS and Q–T intervals in Europeans or Sherpas. These findings were confirmed by Milledge (1963). It is of interest that the electrocardiograms of mountaineers who have climbed as high as 7440 m often show very little evidence of the severe physiological stress they are under (Milledge, 1963; Jackson, 1975).

Electrocardiograms in native highlanders

Extensive studies of the influence of high altitudes on the electrical activity of the heart in native highlanders have been made by Peñaloza and his colleagues in Lima. They carried out electrocardiographic and vectorcardiographic observations in the newborn, in infants, and in children (Peñaloza et al, 1959) and during adolescence and adulthood (Peñaloza et al, 1961).

In newborn, infants and children

The electrocardiograms of 540 normal children were studied by Peñaloza and his colleagues (1959), 350 of them at sea level and 190 at 4540 m. A comparative study was made in five age groups ranging from newborn to 14 years of age. In the newborn the electrical activity of the heart was similar at sea level and at high altitude. In both environments the newborn shows normally a right ventricular preponderance. At sea level this decreases rapidly and is replaced by a physiological left ventricular preponderance. However, at high altitudes the right ventricular preponderance remains throughout infancy and childhood, manifesting itself as an accentuated right ÂQRS deviation and a tall R wave in lead VI. According to Peñaloza et al (1959) another important characteristic of the ventricular activation process is an increased magnitude of the terminal QRS vectors. This finding, and not a special position of the heart, determines both the tall late R wave in lead aVR and the deep or predominant S waves found in the standard limb leads and in the left precordial leads. The positive T waves in the right precordial leads are also related to right ventricular overload.

There is a delay in the normal evolution of the ventricular activation process which occurs at sea level so there is an absence of transitional patterns which are seen frequently at low altitudes between three months and three years of age. However, when highland children are taken down to sea level the ventricular activation process is accelerated and vectors corresponding to left ventricular activation increase in magnitude and transitional patterns are seen. The morbid anatomical basis for these electrocardiographic findings is apparent in Chapter 11, where we describe the persistence from birth of mild pulmonary hypertension and muscularization of the terminal portions of the pulmonary arterial tree. The associated increase in pulmonary vascular resistance leads to persistent right ventricular hypertrophy from birth. This persistence of a heavy right ventricle causes the T wave, which becomes negative within a few hours to days after birth at both sea level and at high altitude, to become positive again in highland infants. At sea level ÂQRS is generally orientated to the left by the age of four months but at high altitude pronounced right ÂQRS deviation remains throughout infancy and childhood.

Peñaloza and his colleagues (1959) believe the characteristic features of the electrocardiogram in normal children at high altitude are reminiscent of those found in pulmonary stenosis with a right ventricular systolic pressure similar to that in the systemic circulation, as occurs in some cases of pure pulmonary stenosis and in cases with an overriding aorta. In both there is pronounced right ÂQRS deviation, a tall R wave in lead VI and positive T waves in the right precordial leads. There are, however, differences and for details of these the reader is referred to the paper by Peñaloza et al (1959).

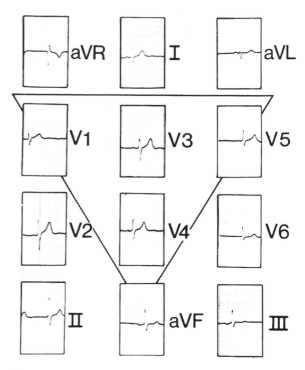

Fig. 18.4 Electrocardiogram in a male Indian resident, 23 years of age, of La Paz, Bolivia (3800 m). It demonstrates that the e.c.g. of native highlanders may show little or no abnormality. It shows a normal axis deviation. The R wave in precordial lead VI is of greater magnitude than the S wave and there is a small secondary 'r' wave in this lead. In a young man of this age these findings could be indicative of early right ventricular hypertrophy.

In adolescence and adult life

Electrocardiograms and vectorcardiograms were studied by Peñaloza et al (1961) in 550 normal subjects, 300 of them at sea level and 250 at 4540 m. Comparative studies were made in three age groups. The electrocardiograms of adolescents and adults who live permanently at high altitude are not similar to those of subjects of comparable age who live at sea level. In the adult highlander right ventricular preponderance is less than in children at the same altitudes but the physiological left ventricular preponderance seen at sea level does not occur even in the older adults. As in the case of the children this electrical right ventricular preponderance is explained by the morbid anatomical events outlined in Chapter 11.

Whereas electrocardiographic and vectorcardiographic patterns are almost stereotyped in infancy and childhood at high altitudes, they are highly variable in adult inhabitants of the same altitude (Fig. 18.4). Five principal QRS patterns are described by Peñaloza et al (1961) as occurring in adult highlanders according to the spatial SÂ QRS orientation. They do not represent different types of ventricular activation. Instead they appear to be varieties of an activation process which has two main characteristics. These are a delay in the pattern of development of the QRS changes that normally occur with ageing, and an increasing magnitude of the terminal QRS vectors. For details of the principal electrocardiographic patterns seen the reader is referred to Peñaloza et al (1961).

Differences in electrocardiographic findings in natives of the Andes and Tibet

Jackson (1975) makes the interesting point that in two Himalayan population surveys at about 3660 m there was a surprising lack of electrocardiographic evidence of right ventricular hypertrophy. The average of the mean frontal QRS axis of 117 adults in Sola Khumbu and Lunana differed by less than 10° from the average of 74 healthy adults from Edinburgh. In the Andes, however, the average axis of 200 adults at 4660 m and at sea level differed by 70° (Peñaloza et al, 1961). A possible explanation for this difference put forward by Jackson is that the Tibetans have lived at these heights for so many thousands of years that they have outbred the reactivity of the pulmonary arteries to hypoxia and that the Andeans may in genetic time be relative newcomers to the environment (Jackson, 1975). We enlarge on this problem of the different biological status of the Sherpa and the Quechua in Chapter 27. We have already contrasted the Andean highlanders to residents of Leadville, Colorado on the same basis in Chapter 11.

REFERENCES

Arias-Stella, J. & Topilsky, M. (1971) Anatomy of the coronary circulation at high altitude. In: *High Altitude Physiology: Cardiac and Respiratory Aspects*, p. 149. Edited by R. Porter & J. Knight. Edinburgh: Churchill Livingstone.

Becker, E. I.., Cooper, R. G. & Hataway, G. D. (1955) Capillary vascularization in puppies born at a simulated altitude of 20,000 feet. *Journal of Applied Physiology*, **8**, 166.

Carmelino, M. (1970) Tesis de Bachiller, Universidad Peruana Cayetano Heredia, Lima.

Clarke, D. R. & Smith, P. (1978) Capillary density and muscle fibre size in the hearts of rats subjected to simulated high altitude. *Cardiovascular Research*, **12**, 578.

Grover, R. F., Lufschanowski, R. & Alexander, J. K. (1976) Alterations in the coronary circulation of man following ascent to 3100 m altitude. *Journal of Applied Physiology*, **41**, 832.

Jackson, F. (1968) The heart at high altitude. *British Heart Journal*, **30**, 291.

Jackson, F. S. (1975) Hypoxia and the heart. In: *Mountain Medicine and Physiology*, p. 99. Proceedings of a Symposium sponsored by the Alpine Club, February 1975. Edited by E. S. Williams, M. P. Ward & C. Clarke.

Jackson, F. & Davies, H. (1960) The electrocardiogram of the mountaineer at high altitude. *British Heart Journal*, **22**, 671.

Milledge, J. S. (1963) Electrocardiographic changes at high altitude. *British Heart Journal*, **25**, 291.

Mortimer, E. A., Monson, R. R. & MacMahon, B. (1977) Reduction in mortality from coronary heart disease in men residing at high altitude. *New England Journal of Medicine*, **296**, 581.

Moret, P. R. (1971) Coronary blood flow and myocardial metabolism in man at high altitude. In: *High Altitude Physiology: Cardiac and Respiratory Aspects*, p. 131. Ciba Foundation Symposium. Edited by R. Porter & J. Knight. Edinburgh: Churchill Livingstone.

Peñaloza, D. (1958) Electrocardiographic changes observed during the first month of residence at high altitude. Report No. 58–90 August, 1958. School of Aviation Medicine. U.S. Air Force, Randolph Base, Texas.

Peñaloza, D. (1971) In: *High Altitude Physiology: Cardiac and Respiratory Aspects*, p. 156. Ciba Foundation Symposium. Edited by R. Porter & J. Knight. Edinburgh: Churchill Livingstone.

Peñaloza, D. & Echevarria, M. (1957) Electrocardiographic observations on ten subjects at sea level and during one year of residence at high altitudes. *American Heart Journal*, **54**, 811.

Peñaloza, D., Gamboa, R., Dyer, J., Echevarria, M. & Marticorena, E. (1959) The influence of high altitudes on the electrical activity of the heart. 1. Electrocardiographic and vectorcardiographic observations in the newborn, infants and children. *American Heart Journal*, **59**, 111.

Peñaloza, D., Gamboa, R., Marticorena, E., Echevarria, M., Dyer, J. & Guitierrez, E. (1961) The influence of high altitudes on the electrical activity of the heart. 2. Electrocardiographic and vectorcardiographic observations in adolescence and adulthood. *American Heart Journal*, **61**, 101.

Ramos, A., Krüger, H., Muro, M. & Arias-Stella, J. (1967) *Boletin de la Oficina Sanitoria Panamericano*, **62**, 496.

Ruiz, L., Figueroa, M., Horna, C. & Peñaloza, D. (1969) Prevalencia de la hipertension arterial y cardiopatia isquémica en los grandes alturas. *Archivos del Instituto de Cardiologia de Mexico*, **39**, 476.

Valdivia, E. (1962) Total capillary bed of the myocardium in chronic hypoxia. *Federation Proceedings: Federation of American Societies for Experimental Biology*, **21**, 221.

19

Myocardial metabolism

Peter Harris

The provision of cellular energy

The processes whereby the many functions of the myocardial cell derive their energy depend on the continuous formation of high-energy phosphate bonds. The major requirement for high-energy phosphate bonds in the heart is the contractile process in which the conversion of energy from chemical bond to the production of a heart beat is mediated through myofibrillar ATPase. All other functions of the cell, however, including the maintenance of electrolyte gradients, depend on a supply of energy in the form of high-energy phosphate bonds.

Such bonds are generated either by the oxidative phosphorylation of ADP to ATP in the mitochondria or at substrate level in the glycolytic pathway. The former process requires oxygen; the latter does not. Both processes depend on the removal of hydrogen from metabolic intermediaries. In the case of oxidative phosphorylation the energy derived from the oxidation of the hydrogen to form water is utilized to form ATP (Fig. 7.8).

In the absence of oxygen, glycolytic phosphorylation depends on the removal of hydrogen by the formation of lactate from pyruvate. The production of ATP by glycolysis is considerably less efficient than that by oxidative phosphorylation. In the absence of oxygen the conversion of one molecule of glucose to lactate results in the formation of two molecules of ATP. In the presence of oxygen the combustion of one molecule of glucose through the glycolytic pathway, the Krebs cycle and the respiratory chain results in the formation of 38 molecules of ATP (Figs. 7.9 and 7.10).

Glycolysis depends on the catabolism of glucose derived from extracellular sources or of glycogen stored within the myocardial cell. The glycolytic chain ends in pyruvate by a series of enzymic reactions which take place in the cytoplasm. In the presence of oxygen, pyruvate is converted by oxidative decarboxylation to acetyl coenzyme A which then enters the Krebs cycle. These processes occur within the mitochondria.

While glycolysis depends on an external supply of glucose or an internal store of glycogen, oxidative catabolism can proceed either from a supply of carbohydrate or from a supply of free fatty acids. In fact the heart normally utilizes free fatty acids preferentially as a source of energy. After entry from the plasma to the myocardial cell the free fatty acids are converted to fatty acyl coenzyme A, transferred into the mitochondria by a transport mechanism which involves the presence of carnitine, and converted to acetyl coenzyme A by the process of β-oxidation.

Thus the oxidation of both pyruvate and of fatty acyl coenzyme A ends with acetyl coenzyme A which enters the Krebs cycle. During these two oxidative processes, hydrogen is removed and passed on to coenzymes or flavoproteins which thereby become converted to a reduced form. Similarly, passage through the Krebs cycle results in the removal of hydrogen which is passed on to coenzymes or flavoprotein. It is the hydrogen formed in these ways that is utilized within the mitochondria to drive the respiratory chain.

The respiratory chain consists of a series of interlinked reactions in which one link becomes oxidized at the expense of the reduction of the next. The direct combination of hydrogen and oxygen to form water would liberate energy in so violent a form as to be biologically unmanageable. Instead, the release of energy is divided into small amounts at each link in the chain. At three such links the energy released is utilized to synthesize a molecule of ATP from ADP and inorganic phosphate. The transfer of ADP and ATP into and out of the mitochondria depends on a specific translocase system which allows the high energy phosphate groupings formed from oxidative phosphorylation to be available for the various extramitochondrial functions of the cell.

Dependence of the myocardium on oxygen

Striated muscle fibres exist in two main forms, white and red. In many animals individual muscles are recognizably pale or dark (Chapter 7). In man all skeletal muscles are a mixture of both white and red fibres. White fibres contain little myoglobin and few mitochondria. Their supply of energy comes predominantly from glycolysis. Functionally they are adapted to rapid, brief activities.

Muscles that are functionally required to maintain tension for postural purposes or a rhythmic contraction as in the heart tend to be red (Chapter 7). The heart may be regarded as the prototype of red muscle. Red muscle contains a high concentration of myoglobin and many mitochondria. Its supply of energy comes predominantly from oxidative phosphorylation, and free fatty acid is utilized in preference to glucose.

The great dependence of the myocardium on a supply of oxygen is evidenced by the low oxygen saturation of the blood in the coronary sinus. In man this is in the region of 25 per cent (Chapter 17, Fig. 17.1) whereas the mixed venous blood is in the region of 70 per cent. This means that some three quarters of the content of oxygen is removed from the blood during passage through the myocardium.

Looked at another way, the PO_2 of the coronary sinus blood is in the region of 15 to 20 mmHg. It follows that the myocardial tissue PO_2 cannot be higher than this. Between the PO_2 in the extracellular fluid and the PO_2 in the vicinity of the intramitochondrial respiratory chain there must also exist a diffusion gradient the magnitude of which is not known. Thus, by comparison with the arterial PO_2, alveolar PO_2 and atmospheric PO_2, the PO_2 in the vicinity of the myocardial respiratory chain is very low. Fortunately, however, the PO_2 below which the supply of oxygen becomes rate-limiting for oxidative phosphorylation in cardiac mitochondria is known to be remarkably low and in the region of 0.5 mmHg (Tenney and Ou, 1969).

The diffusion gradient for oxygen between the blood and the mitochondria will be influenced by the richness of the blood capillary network, which determines the average length of the diffusion pathway (Chapter 7). In addition, the myoglobin in the cytoplasm has an important transport function. The dissociation curve for myoglobin is a rectangular hyperbola, in contrast to the classically sigmoid shape of the curve for haemoglobin. In this way myoglobin retains oxygen in loose combination at a PO_2 at which haemoglobin readily gives it up (Chapter 7). This allows oxygen to be carried in relatively large quantities at a low PO_2 across the cytoplasm. It is only at the very low PO_2 that probably exists at the surface of the mitochondria that myoglobin releases its oxygen.

With the increased mechanical work of the heart which accompanies physical exercise there may be a many-fold increase in the rate of hydrolysis of ATP by the myofibrils. This results in the increased production of ADP. Normally the respiratory chain is abundantly supplied with hydrogen and, if the supply of oxygen is not rate-limiting, the rate of oxidative phosphorylation is determined by the availability of ADP and inorganic phosphate. Thus there exists a beautifully adjusted control mechanism which allows the provision of ATP in relation to its rate of hydrolysis.

The increased activity of the respiratory chain leads to an increased uptake of oxygen. The increased uptake of oxygen by the heart which accompanies physical exercise is accomplished by an augmented coronary blood flow so that the oxygen saturation of the coronary sinus blood does not change (Harris et al, 1964). It seems, therefore, that under normal circumstances the supply of oxygen to the myocardium during exercise is well provided for. This is accomplished without the production of lactate and the lactate:pyruvate ratio in the coronary sinus blood is lower than that in the arterial blood (Harris et al, 1964).

Different forms of tissue hypoxia

The source of hypoxia with which this book is concerned is a reduced PO_2 in the atmosphere. It may, however, help to put the mechanism into perspective by considering other ways in which tissue may be subjected to a chronic shortage of oxygen. Alveolar underventilation due to disease may lead to alveolar hypoxia in the presence of normal atmospheric conditions. A reduced arterial PO_2 may be caused by an inequality of ventilation: perfusion ratios in the lung or by a low pulmonary diffusing capacity. Anatomical right–left shunts from congenital heart disease will also reduce the arterial PO_2 for prolonged periods. The presence of anaemia may impair the oxygen-carrying capacity of the blood. If this is due to a deficiency of iron there may also be a reduction in myoglobin and in the iron-containing cytochromes which are essential constituents of the respiratory chain. Hence the lessons learned from studies at high altitude may have applications over a range of diseases in populations at sea level.

Myocardial response to acute hypoxia (Opie, 1968, 1969)

Energy supply

When, in the isolated heart, the PO_2 of the coronary arterial blood is suddenly and severely reduced, a number of metabolic mechanisms come into play. Since, under these circumstances, the supply of oxygen now limits the function of the respiratory chain, there is a reduction of entry of hydrogen into the other end of the chain and a build up of reduced forms of coenzymes and flavoproteins. This, in turn, limits the oxidative decarboxylation of pyruvate and the β-oxidation of fatty acyl coenzyme A. Because of this the concentrations of pyruvate and fatty acyl coenzyme A increase. The uptake of fatty acid by the myocardium becomes diminished since the pathways involved in fatty acid catabolism have no alternative way of disposing of hydrogen. Glycolysis, on the other hand, can continue by means of the production of lactate from pyruvate which affords a mechanism for the disposal of hydrogen.

Just as the entry of hydrogen into the respiratory chain is limited by severe hypoxia, so is the entry of ADP and the rate of oxidative phosphorylation falls. This leads to an increase in the concentration of ADP and a decrease in the concentration of ATP.

The increase in the concentration of ADP (and AMP) in turn has an important influence on glycolysis.

The enzyme phosphofructokinase which converts fructose-6-phosphate to fructose-1-6-diphosphate is a rate-limiting step in the glycolytic chain. It is normally strongly inhibited by ATP, but ADP and AMP release the inhibition. Thus, when these hydrolysis products rise in acute hypoxia, phosphofructokinase activity is stimulated. In this way glycolysis is not only permitted to continue in the absence of oxygen because of the production of lactate; it is effectively stimulated because of the disinhibition of phosphofructokinase. The effects of acute hypoxia on the uptake of free fatty acid and glucose by the myocardium are, therefore, opposite; the uptake of free fatty acid is diminished while that of glucose is augmented.

In addition to the production of lactate from pyruvate, the production of α-glycerophosphate from dihydroxyacetone phosphate can also act as a mechanism for the removal of hydrogen. With an increased glycolysis the rate of formation of dihydroxyacetone phosphate will be increased so that the concentration of α-glycerophosphate will rise. This, together with the increased availability of fatty acyl coenzyme A during acute hypoxia, will favour the formation of glycerides which may accumulate as lipid droplets in the cell.

Protein synthesis

The effects of acute hypoxia on the isolated, perfused heart seem to be inhibitory to protein synthesis (Jefferson et al, 1971). The rate of incorporation of amino acid into total myocardial protein or myosin is diminished. Transport of amino acids into the cell is unaffected and the inhibition seems likely to be due to a decreased availability of high-energy phosphate groups required for the activation of amino acids or for their translation into protein (Jefferson et al, 1971).

The specific effects of acute hypoxia on the synthesis and turnover of myocardial mitochondria have been studied by Rabinowitz and colleagues (1971) in rats exposed to 4 per cent oxygen for six hours. Their results suggest that there is an increased rate of destruction of mitochondria during hypoxia, followed by a transient increase in the rate of synthesis.

Myocardial acclimatization to chronic hypoxia

The term 'acclimatization' is used to mean the changes that occur in a sea-level resident when subjected to chronic hypoxia. This is in distinction from the term 'adaptation' which will be used to imply genetic selection of a strain suited to hypoxic conditions.

Transport of oxygen to the myocardium

Hurtado (1971) has summarized the processes of acclimatization which help to maintain the tissue PO_2 in the presence of the chronic atmospheric hypoxia of high altitude. Hyperventilation lowers alveolar PCO_2 and increases alveolar PO_2 (Chapter 5). There is a decrease in the alveolar-arterial difference in PO_2 (Chapter 5). Polycythaemia develops, allowing the arterial blood to maintain a high content of oxygen in the presence of a lower than normal PO_2 (Chapter 6). The content of 2,3-diphosphoglycerate increases in the red cells and shifts the dissociation curve for haemoglobin to the right, thus facilitating the release of oxygen to the tissues (Lenfant et al, 1968; Chapter 6).

Indirect influences of chronic hypoxia on the heart

The effects on the heart of chronic hypobaric conditions are not simply the direct effects of oxygen lack on the myocardial cell. There are, in addition, secondary effects on the heart due to actions of hypoxia elsewhere in the body. Chief among these is the development of hypoxic pulmonary hypertension which specifically increases the afterload on the right ventricle and causes right ventricular hypertrophy (Chapter 11). The left ventricle may also be slightly hypertrophied, possibly because of a tendency to an increased systemic arterial pressure and to the presence of polycythaemia (Chapter 16). Thus the specific effects of work hypertrophy have carefully to be disentangled from those that might be attributed to hypoxia alone. In addition, various influences on the heart occur from other tissues or functions. Chronic hypoxia causes anorexia and this has an important influence on cardiac size (Gloster et al, 1972a). Changes in endocrine function occur, involving the sympatho-adrenal system and the thyroid gland (Surks, 1969; Chapter 25). In the field climatic conditions other than the barometric pressure tend to differ at high altitude and exposure to cold may be particularly important (Chapters 22 and 30).

Transient physiological changes also affect the demands on the heart during acclimatization. For a few days the cardiac output during exercise is increased for a given level of oxygen uptake by the body (Vogel et al, 1967), but this has returned to normal or less than normal (Grover and Alexander, 1971) by two or three weeks.

Myocardial blood flow

The studies of Grover and Alexander (1971) on normal men showed a diminution in the myocardial blood flow after 10 days' residence at 3100 m. Such a result is surprising in view of the vasodilatory effects of acute hypoxia in the heart. There was an increase in the myocardial arterio-venous difference in oxygen so that the myocardial uptake of oxygen diminished to a lesser degree. The PO_2 of the coronary sinus blood was approximately 18 mmHg at both altitudes, both at rest and during exercise. There is no indication, therefore, from these studies that the myocardial cells were operating at a low PO_2 even during exercise.

Myocardial vascularization

Miller and Hale (1970) found an increase in the number of capillaries per square millimetre cross-sectional area of myocardium in rats kept for eight weeks in a hypobaric chamber simulating an altitude of 5500 m. The increased capillary density affected both ventricles to approximately the same extent. The production of polycythaemia by injections of cobalt or repeated infusions of blood caused some increase in myocardial capillary density but not to the same extent as that found under hypobaric conditions. The increased number of capillaries would have the effect of reducing the distance for diffusion of oxygen from the blood to the mitochondria and would thus facilitate the transfer of oxygen. Becker et al (1955) reported a similar increase in capillary density in puppies newly born at high altitude of a bitch that had been resident there for two years. There was no distinction between ventricles. The intraluminal volume of the coronary arterial tree almost doubled in young rats exposed two hours a day for 15 days to a simulated altitude of 6600 m (Kerr et al, 1965).

Myoglobin

A number of authors has provided evidence of an increased concentration of myoglobin in the myocardium during acclimatization to chronic hypoxia (Chapter 7). In rats kept in a hypobaric chamber, Poupa and his colleagues (1966) showed an increased concentration of myoglobin in the left ventricle. Anthony et al, (1959) reported an increased concentration of myoglobin in similar experiments but did not distinguish between ventricles. Vaughan and Pace (1956) also observed an increase in the myoglobin concentration in rats transported to an altitude of 3700 m but did not distinguish between right and left ventricles. These authors made the interesting observation that there was a greater increase in the concentration of myoglobin in rats of the same strain that had been born and bred at high altitude.

Mitochondria

Ou and Tenney (1970) have described an increase in the numbers of mitochondria in the myocardium of Hereford cattle reared at sea level (Chapter 7). The sites of ventricular sampling were not specified. The method employed was one of differential centrifugation and counting of mitochondria from hearts stored in dry ice for five days.

Lipid

The accumulation of lipid by the myocardium under chronic hypobaric conditions appears to differ greatly between species. We found, for instance, no effect in the rat (Gloster et al, 1972b) but a distinct accumulation of neutral lipid in the guinea pig (Gloster et al, 1974). It is possible, as discussed earlier, that the accumulation of neutral lipid, presumably mainly triglyceride, is due to an increased availability of α-glycerophosphate consequent on an increased rate of glycolysis. The lack of effect in the rat is consistent with the ultrastructural studies of Bischoff and colleagues (1969) referred to later.

Mitochondrial enzymes

The metabolic response of the myocardium to acute, severe hypoxia, discussed earlier, is determined by changes in substrate and cofactor concentrations and the modification of enzyme activities. Over a longer period it might be anticipated that changes in emphasis of the importance of different metabolic pathways might be determined by changes in enzyme concentrations. The theoretical basis of such an anticipated effect would be similar to the development or induction of specific enzyme activities in other organs. Unfortunately, there is a paucity of data on this aspect and it is still not entirely clear whether such specific changes occur in the myocardium as an acclimatization to chronic hypoxia.

When guinea pigs were maintained at an atmospheric pressure of 400 mmHg for 14 days, there was a slight increase in the activity of succinate dehydrogenase in myocardial homogenates when compared with that from control animals receiving the same diet (Barrie and Harris, 1976). The effect was more obvious in the right ventricle and did not reach a level of significance in the left ventricle. After 28 days the activity of succinate dehydrogenase in the hypobaric rats was slightly and non-significantly lower than that in the corresponding ventricle of the controls.

Other enzymes partly or wholly associated with mitochondrial activities were studied at 28 days. These were malate dehydrogenase, aspartate aminotransferase, mitochondrial glycerol-3-phosphate dehydrogenase and β-hydroxyacyl coenzyme A dehydrogenase. The activities of all these enzymes were slightly lower than in the corresponding ventricle of the controls. Only in the case of mitochondrial glycerol-3-phosphate dehydrogenase activity in the right ventricle, however, did this reach a level of statistical significance.

Vergnes (1973) and Walpurger et al (1970) have described an increase in myocardial succinate dehydrogenase activity in rats during chronic hypoxia. Tenny and Ou (1969) reported an increase in the activity of succinate dehydrogenase and NADH oxidase in cattle reared at high altitude.

Glycolytic enzymes

We have studied the activity of the glycolytic

enzymes in myocardial homogenates of guinea pigs kept at an atmospheric pressure of 400 mmHg for 28 days (Barrie and Harris, 1976). In these studies, the two ventricles were assayed separately and compared with normobaric controls. Two groups of controls were used, one free-fed and the other restricted to the diet chosen by the hypobaric animals. In this way it was possible to distinguish the effects of the anorexia caused by hypobaric conditions. The right ventricle of the hypobaric animals was greatly hypertrophied while the left ventricle was only slightly increased in weight, and yet both ventricles received the same hypoxic blood. The effects of hypertrophy could, therefore, also be distinguished by comparing the ratio of right to left ventricular activities between the hypobaric animals and the restricted diet controls. This left a comparison of left ventricular enzyme activities between the hypobaric animals and their restricted diet controls as the nearest approach one could make to the study of the effects of hypoxia alone. The design of the study has been given in some detail to emphasize the point, discussed later, that it is only in this factorial fashion that the pure effects of hypoxia may be separated. Even so, many other secondary effects of hypobaric conditions on the heart remain unaccounted for.

In this way the effects of anorexia were shown to be a diminution in cardiac weight and in myocardial glycogen phosphorylase and hexokinase activity. The effects attributable to hypertrophy were an increase in the activity of glycogen phosphorylase, phosphoglucomutase, hexokinase, glucosephosphate isomerase, glyceraldehyde-3-phosphate dehydrogenase and phosphoglycerate kinase. These enzymes mediate reactions that would favour flow through the pentose monophosphate shunt, and an increased activity of this pathway might be expected in hypertrophy.

Comparing the left ventricle of the hypobaric animals with that of the restricted diet controls, there was an increase in the activity of hexokinase, aldolase, glyceraldehyde-3-phosphate dehydrogenase and phosphoglycerate kinase. It will be noted that three of these four increased activities could be ascribed to the mild degree of hypertrophy of the hypobaric left ventricle. This leaves very little that can with any certainty be ascribed to the effects of hypoxia alone. Both dietary restriction and

hypertrophy were associated with an increase in the proportion of M-subunits of lactate dehydrogenase. This, however, was not found in the hypobaric left ventricle.

Protective effect of chronic hypoxia

Poupa and his colleagues (Poupa et al, 1966; Poupa, 1972) stimulated strips of the right ventricle of rats during and after the induction of acute hypoxia (see also Chapter 25). They measured the time taken for mechanical contraction to cease during acute hypoxia and the degree of recovery of peak tension subsequently. They found that both these measurements were increased in rats acclimatized at a simulated altitude of 5000 and 7000 m. Sideropenic anaemia also increased the recovery of contractile force after acute hypoxia but had no effect on the time taken for mechanical contraction to cease at the beginning of acute hypoxia. McGrath et al (1969) in similar experiments showed that the developed tension decreased and the resting tension increased with continued stimulation of strips of the right ventricle under hypoxic conditions. These effects were delayed in rats acclimatized to high altitude. The influence of acclimatization was removed by iodoacetate. This substance inhibits glyceraldehyde-3-phosphate dehydrogenase and the influence of acclimatization was, therefore, thought to be due to an increased glycolytic capacity.

Poupa (1972) showed that the severity of myocardial necrosis caused by isoprenaline in rats could be reduced by acclimatization to simulated high altitude. The reduction in severity of the lesions applied to both ventricles.

Myocardial hypertrophy and protein synthesis

The difficulties in distinguishing between the effects of hypertrophy and chronic hypoxia have been considered in some detail with regard to the glycolytic enzymes. Similar difficulties may occur with regard to some of the other factors that have been discussed. For instance, work hypertrophy of the myocardium is also associated with an increased concentration of myoglobin (Adler et al, 1971), and an increase in mitochondrial content occurs early in the development of ventricular hypertrophy (Hatt et al, 1970). Lipid droplets appear (Hatt et al, 1970).

Hypertrophy alone, however, does not seem to be associated with a proliferation of myocardial capillaries (Shipley et al, 1937; Roberts and Wearn, 1941) unless the increased cardiac work is imposed early in life (Shipley et al, 1937).

Nowhere has the distinction between the processes associated with work hypertrophy and the effects of chronic cellular hypoxia been more difficult than in the consideration of protein synthesis in the heart. Hypertrophy itself occurs as a result of an increased rate of protein synthesis. There are adequate physiological reasons to explain the right ventricular hypertrophy that occurs in chronic hypoxia, and the much smaller degree of left ventricular hypertrophy may also have physiological explanations.

Evidence, which has been referred to above, suggests that acute hypoxia causes an inhibition of protein synthesis in the heart. Meerson (1972, 1975) has produced evidence that chronic hypoxia may be associated with an increase in myocardial protein synthesis. In his experiments, rats were kept in a hypobaric chamber at a simulated altitude of 7000 m for six hours each day for 40 days. By the tenth day the weight of the right ventricle had increased by 80 per cent and that of the left ventricle by 28 per cent. The maximal metabolic changes occurred at about 20 days. At this time there was an increased incorporation of radioactive amino acids into myocardial protein, indicating an increased protein synthesis. The right ventricle was affected more than the left and the increase in labelling of mitochondrial protein was greater than that of nuclear protein. The concentration of nuclear DNA did not change in the myocardium and autoradiography showed that the incorporation of ^{3}H-thymidine into nuclei occurred only in connective tissue cells. By contrast the concentration of mitochondrial DNA increased, as did the incorporation of ^{3}H-thymidine into mitochondrial DNA. These changes occurred more in the right ventricle than the left. The concentration of RNA also increased.

Meerson points to the similarity between the effects of work hypertrophy and chronic hypoxia on the protein synthetic mechanisms of the heart and suggests that, in both cases, the stimulus to an increased protein synthesis is a deficiency of high-energy phosphate groups. In the case of

hypertrophy this is due to an increased hydrolysis of ATP; in the case of hypoxia it is due to a decreased oxidative phosphorylation. The cell is able to respond only to the deficit in ATP and cannot distinguish the different courses of events that have given rise to the deficit. Meerson produces evidence from tissues outside the heart in support of his theory. In the heart, however, there are, as discussed above, pathological reasons for the development of myocardial hypertrophy (Chapter 11) and these affect the right ventricle more than the left.

In rats exposed to an atmospheric pressure of 400 mmHg for seven days we have found an increase in the ability of the right ventricular cytoplasm to facilitate the incorporation of amino acid into protein in the presence of a standard ribosomal preparation (Gibson and Harris, 1972). This, however, was clearly attributable to the development of right ventricular hypertrophy. There was no effect on the activity of the enzyme systems that activate amino acids and mediate their incorporation into protein on the ribosomes. In similar studies (Krelhaus et al, 1975) we showed a transient decrease in ornithine decarboxylase activity lasting about four days in both ventricles of rats exposed to 400 mmHg. The design of the study was similar to that on glycolytic enzymes described earlier. Its factorial nature made it possible to eliminate the separate effects of hypertrophy and anorexia so that the reduction in ornithine decarboxylase activity could be ascribed to hypoxia. This enzyme controls the synthesis of polyamines, which appear to play an important rôle in hypertrophy of tissues. Thus the fall in ornithine decarboxylase activity might explain why acute hypoxia diminishes protein synthesis while the transient nature of the fall in activity will allow an increased protein synthesis and hypertrophy ultimately to develop with more prolonged hypoxia.

Ultrastructure

The effects of chronic hypoxia on the structure of the myocardium have been reviewed by Ferrans and Roberts (1972). Studies on rats (Sulkin and Sulkin, 1965) kept for 42 days in a hypoxic environment showed that morphological changes developed when the oxygen concentration was less than 6.5 per cent. The changes consisted of scattered discrete areas of atrophy of muscle and loss of striations. The mitochondria were enlarged and the cristae separate and degenerate, while the matrix was filled with a dense substance. There was dilatation of the sarcoplasmic reticulum and degeneration of myofibrils. The intercalated disc spaces were widened. The focal nature of the changes on light microscopy is noteworthy and similar to the distribution shown by lipid histochemistry (Gloster et al, 1974).

Bischoff and colleagues (1969) kept dogs, rabbits and rats at an altitude of 4300 m for five months. They found swelling and fusion of mitochondria with loss of cristae. The sarcoplasmic reticulum was dilated. An accumulation of lipid droplets and glycogen could be seen. The accumulation of glycogen is in contrast to the depletion caused by acute hypoxia. In addition to the changes in the myocytes, the capillary endothelial cells were swollen. Of the three species, rats were hardly affected.

Some caution needs to be observed in the interpretation of the above findings. The procedure of diffusion fixation used can give rise to artefacts, which include swelling and destruction of mitochondria, destruction of myofibrils, dilatation of sarcoplasmic reticulum and separation of intercalated discs. Although separate specimens were taken from the right and left ventricles, it is not clear to what extent the changes occurred in individual ventricles. Some of the changes are similar to those occurring during hypertrophy or in animals subjected to prolonged daily exercise (Ferrans and Roberts, 1972).

Studies on populations indigenous to high altitude

Such studies have been considered separately from those of acclimatization since it is only by a careful comparison of the indigenous with an acclimatized population that any effects may be attributable to true adaptation.

Transport of oxygen to the myocardium

The differences in the mechanisms of transport of oxygen that are found between populations indigenous to sea level and high altitude are similar to the changes that occur during acclimatization described above.

Myocardial blood flow

Moret (1971, 1972) found an average myocardial blood flow of 72 ml/min/100 g in men indigenous to sea level at Lima. In men indigenous to La Paz at 3800 m the average myocardial blood flow was 55 ml/min/100 g. In men indigenous to Cerro de Pasco at 4300 m it was 49 ml/min/100 g. The findings are similar to those found during acclimatization discussed above.

The oxygen supply to the myocardium was reduced by about one third at high altitude. The average oxygen uptake by the myocardium was 8.7 ml/min/100 g at sea level, 7.1 ml/min/100 g at 3800 m and 6.8 ml/min/100 g at 4300 m. Despite the low oxygen uptake, the external work of the heart was not reduced.

The P_{O_2} of the coronary sinus blood was 20 mmHg at sea level and 17 mmHg at 3800 m and 4300 m, so that, by this criterion, no significant degree of tissue hypoxia was occurring in the high altitude myocardium.

Utilization of substrates

In the studies of Moret (1971, 1972) measurements were made of the arteriovenous differences of glucose, lactate, pyruvate and free fatty acids. The arterial concentration of glucose was slightly lower at high altitude and the myocardial uptake tended to decrease. The arterial concentrations of lactate and pyruvate were increased at high altitude and the uptake of each tended to increase. The lactate:pyruvate ratio of arterial and coronary sinus blood was not substantially changed at high altitude, suggesting that the redox potential of the myocardial cell was the same. The arterial concentration of free fatty acids was similar at high and low altitudes while the myocardial uptake tended to decrease at high altitude. Mensen de Silva and Cazorla (1973) found that the myocardial concentration of lactate and α-glycerophosphate was lower in indigenous high altitude guinea pigs than in sea-level guinea pigs.

Myocardial vascularization

Valdivia (1958) found an increased capillarity of red skeletal muscles in guinea pigs indigenous to high altitude (Chapter 7). The myocardium was not studied. Arias-Stella and Topilsky (1971) compared casts of the coronary arteries of human residents at high and low altitude in Peru and found an increased arterial network in both ventricles at high altitude (Chapter 18).

Myoglobin

An increased concentration of myoglobin was found in the myocardium of dogs living at high altitude when compared with those at sea level (Hurtado et al, 1937). The distribution between the ventricles was not reported. The myoglobin concentration of the hearts of guinea pigs indigenous to high altitude has, however, been found to be less than that of sea-level guinea pigs transported to high altitude (Tappan and Reynafarje, 1957). Moreover, removal of indigenous high altitude animals to sea level did not result in a reduction of cardiac myoglobin. Among a variety of wild rodents (Reynafarje and Morrison, 1962) the myocardial myoglobin concentration of high altitude strains was not greatly different from low altitude strains.

Mitochondria

Kearney (1973), working in Heath's laboratory, was unable to show any quantitative difference in the myocardial mitochondria of rabbits and guinea pigs from Cerro de Pasco (4300 m) and from sea level (Chapter 7). Random electron micrographs were analysed by stereological methods. Mitochondrial volume expressed as a percentage of the cytoplasmic volume was the same at the two altitudes (Fig. 7.4) as was the number of mitochondria (Fig. 7.5), and the surface area of the outer and inner mitochondrial membranes per millilitre of cytoplasm (Figs. 7.6 and 7.7). No difference was observed between the right and left ventricles. Reynafarje (1971–72) compared the respiration of cardiac mitochondria from guinea pigs native to high altitude and sea level. The only difference he was able to show was that in the high altitude group there was a lower apparent Michaelis constant for ADP, which implies that the high altitude organelles were able to respire more at lower concentrations of ADP.

Lipid

We have studied the lipid composition of the myocardium of guinea pigs, rabbits and dogs indigenous to high altitude in the Andes and compared them with the same species at sea level in London (Gloster et al, 1971). The sea-level guinea pigs and rabbits were laboratory strains. High altitude animals of all three species showed an increased myocardial content of total lipid, phospholipid, sphingomyelin and cholesterol. The results are in contrast to the observations on the acclimatization of guinea pigs (Gloster et al, 1974) discussed above, in which there was an increase in neutral lipid but not in phospholipid under chronic hypobaric conditions.

Myocardial enzyme activities

Evidence is conflicting concerning the differences in myocardial enzyme activities that may exist between animals indigenous to high and low altitude. Reynafarje (1961) studied myocardial homogenates of guinea pigs indigenous to high altitude and compared them with a group at sea level. Some of the sea-level animals were laboratory stock. Care was taken to control the diet and the temperature and humidity of the environment. There was an increase in the activity of NADH oxidase, NADPH oxidase and transhydrogenase in the high altitude animals. No distinction was made between right and left ventricles.

We also studied myocardial homogenates from guinea pigs, rabbits and dogs indigenous to high altitude in the Andes and compared them with the same species at approximately sea level in London (Harris et al, 1970). The sea-level guinea pigs and rabbits were laboratory stock. Separate measurements were made from the right and left ventricles. In all three species the high altitude animals showed a higher activity of succinate dehydrogenase but there was no substantial difference in lactate dehydrogenase. No difference was found between the ventricles.

The weakness of the above study lay in the comparison of inbred laboratory animals at sea level with outbred domestic animals at high altitude. In a subsequent investigation (Barrie and Harris, 1976) we have compared outbred guinea pigs and rabbits at the two levels. The enzymes selected were: phosphofructokinase, glyceraldehyde-3-phosphate dehydrogenase, lactate dehydrogenase, α-hydroxybutyrate dehydrogenase, aspartate aminotransferase, β-hydroxyacyl coenzyme A dehydrogenase, succinate dehydrogenase and mitochondrial glycerol-3-phosphate dehydrogenase. There was no difference in the activities of any of these enzymes between the two groups of guinea pigs.

Certain differences, on the other hand, were found between the two groups of rabbits. They were similar for both ventricles but more obvious in the right ventricle. The glycolytic capacity was increased in the high altitude myocardium, as indicated by higher glyceraldehyde-3-phosphate dehydrogenase and phosphofructokinase activities. Lactate dehydrogenase activity was also higher while α-hydroxybutyrate dehydrogenase activity was unchanged. This would indicate an increased proportion of M-subunits of the lactate dehydrogenase tetramer. The activity of succinate dehydrogenase was not increased at high altitude but that of another oxidative mitochondrial enzyme, β-hydroxybutyrate dehydrogenase was. Asparate aminotransferase activity was also increased at high altitude.

The conflicting results just summarized serve to indicate the complexity of factors involved in addition to high altitude. Differences between strains and between ventricles, the effect of diet and of environmental conditions other than high altitude may all play a part.

Adaptation and acclimatization

Having reviewed the changes that occur during acclimatization and the differences that have been found between indigenous high altitude and low altitude populations, we are now in a position to consider the evidence for adaptation. Before doing so it may be helpful to summarize the sort of evidence required to substantiate the existence of adaptation.

In the first place it has to be appreciated that the existence of adaptation by natural selection necessarily implies the emergence of a strain that is distinct from the other strains of the species. If one compares a particular strain of a species with another strain of the same species, one may well find differences that do not necessarily imply the operation of natural selection in response to the environment. In order to support the existence of a

process of genetic adaptation it will be necessary to show that the high altitude strain differs in a specific respect from a number of other strains of the same species or, more strongly, that a similar difference between high and low altitude strains exists in several different species.

Assuming that a difference of this nature is found between high and low altitude strains, it is then necessary to consider whether it is explicable by acclimatization rather than adaptation. The distinction is simple if such a difference in function does not occur during acclimatization, or if the function being studied changes in the opposite direction during acclimatization.

If the function changes in the same direction during acclimatization, the proof of the existence of adaptation will have to rest on the degree of the function or on the lack of alteration when the high altitude strain is transferred to low altitude. In considering the influence of acclimatization it is necessary to bear in mind that acclimatization from birth may have different effects from acclimatization in adult life.

The preceding considerations have assumed that a change in function has occurred during adaptation. The opposite may, in fact, be the case, so that the high altitude adapted strain fails to show the characteristics that develop during the acclimatization of sea-level strains.

Table 19.1 summarizes the changes that have been observed in indigenous high altitude populations and during acclimatization of sea-level strains. In addition are shown the effects of work hypertrophy of the heart. A comparison of the last two columns gives some indication of the degree to which the changes occurring during acclimatization might be ascribed to work hypertrophy. A comparison of the first two columns summarizes the evidence available concerning the existence of adaptation.

As far as the presence of polycythaemia and an increased concentration of 2-3-diphosphoglycerate in the red cells are concerned, there is clear evidence of an acclimatization process (Chapter 27). The evidence is against adaptation since the haematocrit is known to decrease when high altitude natives descend to low altitude. An increase in the density of the myocardial capillary network seems to be due to acclimatization and not hypertrophy. There is no

Table 19.1 Changes observed in indigenous high altitude populations and during acclimatization of sea-level strains

	Indigenous population	Acclimatization	Work hypertrophy
Polycythaemia	+	+	0
2-3-Diphosphoglycerate	+	+	0
Coronary arteries	+	+	+
Capillaries	+	+	0
Blood flow	−	−	
Coronary sinus P_{O_2}	0	0	
Oxygen uptake	−	−	
Myoglobin	+	+	+
Mitochondria	0	+ early	+ early
Lipid	+ phospho	+ neutral (species-dependent)	lipid droplets
Enzymes	species-dependent	changes	similar changes
Protective effect	not known	+	not known
Protein synthesis	not known	+	+
Ultrastructure	? 0	changes	similar changes

+ = increase;   − = decrease;   0 = unchanged

evidence of adaptation in this respect.

The decrease in the myocardial blood flow and the myocardial oxygen uptake is likely to be due to acclimatization and not adaptation. One notes the lack of change in the coronary sinus P_{O_2}. If this is so, why, at least at rest, should oxidative metabolism of the myocardium change?

An increase in the concentration of myoglobin is shown in all three columns in Table 19.1, although it may be noted that there is evidence that the myoglobin concentration of indigenous high altitude strains is less than that which occurs during acclimatization of sea-level strains. There is some doubt whether the effects during acclimatization can be distinguished from those of work hypertrophy.

An increase in mitochondrial mass seems to occur transiently both during acclimatization and hypertrophy. Other structural changes occurring during acclimatization are also similar to those found in work hypertrophy. It seems likely, from our own experience, that the ultrastructure of the hearts of indigenous high altitude species does not differ from that of sea-level animals. There is room for a series of careful studies on this aspect, using perfusion fixation and morphometric techniques.

The accumulation of lipid by the heart during acclimatization is dependent on the species and involves neutral lipid. The differences shown between high and low altitude populations involved mostly phospholipid and are, therefore, not comparable. Such a discrepancy could be evidence towards the existence of adaptation.

The changes in myocardial enzyme activity during acclimatization are not clearly distinguished from those of work hypertrophy. Studies of indigenous populations showed differences at high altitude for the rabbit but not the guinea pig. This could be evidence of adaptation on the part of the guinea pig.

The protective effect of chronic hypoxia has been studied only during acclimatization. The mechanism of an increase of rate of myocardial protein synthesis during acclimatization is (inevitably) indistinguishable from that due to work hypertrophy.

Summary

Obvious changes of acclimatization of a sustained nature affect the bone marrow and the myocardial capillaries. Evidence within the cardiac myocyte is much less clear-cut. There are difficulties in distinguishing the effects of cellular acclimatization to chronic hypoxia from those of work hypertrophy imposed by physiological mechanisms outside the heart. To show the existence of genetic adaptation is a subtle and difficult matter but what evidence is available on this point is fragmentary and weak.

REFERENCES

Adler, C. P., Schlütter, G. & Sandritter, W. (1971) Ultraviolettmikrospecktrophotometrische unter Suchungen am Cytoplasma von Herzmuskelzellen. *Beiträge zur Pathologie* (Stuttgart), **143**, 126.

Anthony, A., Ackerman, R. & Strother, G. K. (1959) Effects of high altitude acclimatization on rat myoglobin. Changes in myoglobin content of skeletal and cardiac muscle. *American Journal of Physiology*, **196**, 512.

Arias-Stella, J. & Topilsky, M. (1971) Anatomy of the coronary circulation at high altitude. In: *High Altitude Physiology: Cardiac and Respiratory Aspects*, p. 149. Edited by R. Porter & J. Knight. Edinburgh: Churchill Livingstone.

Barrie, S. E. & Harris, P. (1976) Effects of chronic hypoxia and dietary restriction on myocardial enzyme activities. *American Journal of Physiology*, **231**, 1308.

Becker, E. L., Cooper, R. G. & Hataway, G. D. (1955) Capillary vascularization in puppies born at a simulated altitude of 20,000 feet. *Journal of Applied Physiology*, **8**, 166.

Bischoff, M. B., Dean, W. D., Bucci, T. J. & Frics, L. A. (1969) Ultrastructural changes in myocardium of animals after five months' residence at 14,110 feet. *Federation Proceedings: Federation of American Societies for Experimental Biology*, **28**, 1268.

Ferrans, V. J. & Roberts, W. C. (1971–72) Myocardial ultrastructure in acute and chronic hypoxia. *Cardiology*, **56**, 114.

Gibson, K. & Harris, P. (1972) Effects of hypobaric oxygenation, hypertrophy and diet on some myocardial cytoplasmic factors concerned with protein synthesis. *Journal of Molecular and Cellular Cardiology*, **4**, 651.

Gloster, J., Hasleton, P. S., Harris, P. & Heath, D. (1974) Effects of chronic hypoxia and diet on the weight and lipid content of viscera in the guinea pig. *Environmental Physiology and Biochemistry*, **4**, 251.

Gloster, J., Heath, D. & Harris, P. (1972a) The influence of diet on the effects of a reduced atmospheric pressure in the rat. *Environmental Physiology and Biochemistry*, **2**, 117.

Gloster, J., Heath, D. & Harris, P. (1972b) Lipid composition of the heart and lungs of rats during exposure to a low atmospheric pressure. *Environmental Physiology and Biochemistry*, **2**, 125.

Gloster, J., Oertel, Y., Heath, D., Arias-Stella, J. & Harris, P. (1971) The lipid composition of the myocardium of animals indigenous to high and low altitude. *Environmental Physiology*, **1**, 77.

Grover, R. F. & Alexander, J. K. (1971–72) Cardiac performance and the coronary circulation of man in chronic hypoxia. *Cardiology*, **56**, 197.

Harris, P., Castillo, Y., Gibson, K., Heath, D. & Arias-Stella, J. (1970) Succinic and lactic dehydrogenase activity in myocardial homogenates from animals at high and low altitude. *Journal of Molecular and Cellular Cardiology*, **1**, 189.

Harris, P., Howel-Jones, J., Bateman, M., Chlouverakis, C. & Gloster, J. (1964) Metabolism of the myocardium at rest and during exercise in patients with rheumatic heart disease. *Clinical Science*, **26**, 145.

Hatt, P. Y., Berjal, G., Moravec, J. & Swynghedauw, B. (1970) Heart failure: An electron microscopic study of the left ventricular papillary muscle in aortic insufficiency in the rabbit. *Journal of Molecular and Cellular Cardiology*, **1**, 235.

Hurtado, A. (1971) The influence of high altitude on physiology. In: *High Altitude Physiology: Cardiac and Respiratory Aspects*, p. 3. Edited By R. Porter & J. Knight. Edinburgh: Churchill Livingstone.

Hurtado, A., Rotta, A., Merino, C. & Pon, J. (1937) Studies of myohemoglobin at high altitudes. *American Journal of the Medical Sciences*, **194**, 708.

Jefferson, L. S., Wolpert, E. B., Giger, K. E. & Morgan, H. E. (1971) Regulation of protein synthesis in heart muscle. *Journal of Biological Chemistry*, **246**, 2171.

Kearney, M. S. (1973) Ultrastructural changes in the heart at high altitude. *Pathologia et Microbiologia*, **39**, 258.

Kerr, A., Jr., Diasio, R. B. & Bommer, W. J. (1965) Effect of altitude (hypoxia) on coronary artery size in the white rat. *American Heart Journal*, **69**, 841.

Krelhaus, W., Gibson, K. & Harris, P. (1975) The effects of hypertrophy, hypobaric conditions and diet on myocardial ornithine decarboxylase activity. *Journal of Molecular and Cellular Cardiology*, **7**, 63.

Lenfant, C., Torrance, J., English, E., Finch, C. A., Reynafarje, C., Ramos, J. & Faura, J. (1968) Effect of altitude on oxygen binding by hemoglobin and on organic phosphate levels. *Journal of Clinical Investigation*, **47**, 2652.

McGrath, J. J., Bullard, R. W. & Komiver, G. K. (1969) Functional adaptation in cardiac and skeletal muscle after exposure to simulated high altitude. *Federation Proceedings: Federation of American Societies for Experimental Biology*, **28**, 1307.

Meerson, F. Z. (1971–72) Role of the synthesis of nucleic acids and proteins in the adaptation of the organism to altitude hypoxia. *Cardiology*, **56**, 173.

Meerson, F. Z. (1975) Role of synthesis of nucleic acids and protein in adaptation to the external environment. *Physiological Reviews*, 55, 70.

Mensen de Silva, E. & Cazorla, A. (1973) Lactate, α-GP and Krebs cycle in sea-level and high altitude native guinea-pigs. *American Journal of Physiology*, **224**, 669.

Miller, A. T. Jr. & Hale, D. M. (1970) Increased vascularity of brain, heart and skeletal muscle of polycythemic rats. *American Journal of Physiology*, **219**, 702.

Moret, P. R. (1971) Coronary blood flow and myocardial metabolism in man at high altitude. In: *High Altitude Physiology: Cardiac and Respiratory Aspects*, p. 131. Edited by R. Porter & J. Knight. Edinburgh: Churchill Livingstone.

Moret, P. R. (1971–72) Myocardial metabolic changes in chronic hypoxia. *Cardiology*, **56**, 161.

Opie, L. H. (1968) Metabolism of the heart in health and disease. Part I. *American Heart Journal*, **76**, 685.

Opie, L. H. (1969) Metabolism of the heart in health and disease. Part II. *American Heart Journal*, **77**, 100.

Ou, L. C. & Tenney, S. M. (1970) Properties of mitochondria from hearts of cattle acclimatized to high altitude. *Respiration Physiology*, **8**, 151.

Poupa, O. (1971–72) Anoxic tolerance of the heart muscle in different types of chronic hypoxia. *Cardiology*, **56**, 188.

Poupa, O., Krofta, K., Procházka, J. & Turek, Z. (1966) Acclimation to simulated high altitude and acute cardiac necrosis. *Federation Proceedings: Federation of American Societies for Experimental Biology*, **25**, 1243.

Poupa, O., Krofta, K., Rakušan, K., Procházka, J., Ráol, J. & Barbashova, Z. I. (1966) Myoglobin content of the heart and resistance of the isolated myocardium to anoxia *in vitro* during adaptation to high altitude hypoxia. *Physiologia Bohemoslovenica (Praha)*, **15**, 450.

Rabinowitz, M., Aschenbrenner, V., Albin, R., Gross, N. J., Zak, R. & Nair, K. G. (1971) Synthesis and turnover of heart mitochondria in normal, hypertrophied and hypoxic rats. In:

Cardiac Hypertrophy, p. 283. Edited by N. R. Alpert. New York & London: Academic Press.

Reynafarje, B. (1971/72) Effect of chronic hypoxia on the kinetics of energy transformation in heart mitochondria. *Cardiology*, **56**, 206.

Reynafarje, B. D. (1961) Pyridine nucleotide oxidases and transhydrogenase in acclimatization to high altitude. *American Journal of Physiology*, **200**, 351.

Reynafarje, B. & Morrison, P. (1962) Myoglobin levels in some tissues from wild Peruvian rodents native to high altitude. *Journal of Biological Chemistry*, **237**, 2861.

Roberts, J. T. & Wearn, J. T. (1941) Quantitative changes in capillary muscle relationship in human hearts during normal growth and hypertrophy. *American Heart Journal*, **21**, 617.

Shipley, R. A., Shipley, L. J. & Wearn, J. T. (1937) The capillary supply in normal and hypertrophied hearts of rabbits. *Journal of Experimental Medicine*, **65**, 29.

Sulkin, N. M. & Sulkin, D. F. (1965) An electron microscopic study of the effects of chronic hypoxia on cardiac muscle, hepatic, and autonomic ganglion cells. *Laboratory Investigation*, **14**, 1523.

Surks, M. I. (1969) In: *Biomedicine Problems of High Terrestrial Elevations*, p. 186. Edited by A. H. Hegnauer, U. S. Army Research Institute of Environmental Medicine.

Tappan, D. V. & Reynafarje, B. D. (1957) Tissue pigment manifestations of adaptation to high altitudes. *American Journal of Physiology*, **190**, 99.

Tenney, S. M. & Ou, L. C. (1969) In: *Biomedicine Problems of High Terrestrial Elevations*, p. 160. Edited by A. H. Hegnauer U.S. Army Research Institute of Environmental Medicine.

Valdivia, E. (1958) Total capillary bed in striated muscle of guinea pigs native to Peruvian mountains. *American Journal of Physiology*, **194**, 585.

Vaughan, B. E. & Pace, N. (1956) Changes in myoglobin content of the high altitude acclimatized rat. *American Journal of Physiology*, **185**, 549.

Vergnes, H. (1973) Thesis. Université Paul-Sabatier, Toulouse.

Vogel, J. A., Hansen, J. E. & Harris, C. W. (1967) Cardiovascular responses in man during exhaustive work at sea level and high altitude. *Journal of Applied Physiology*, **23**, 251.

Walpurger, G., Schlaak, M. & Jipp, P. (1970) Mitochondrialer und extramitochondrialer Myokardstoffwechsel bei der Herzhypertrophie infolge chronischer Sauerstoffmangel-atmung. *Zeitschrift für Kreislaufforschung*, **59**, 643.

20

Alimentary canal

Barodentalgia

During the Second World War it was noted that some air crew personnel complained of dental pain when flying at high altitudes and when entering low-pressure chambers (Harvey, 1943; Joseph et al, 1943). This was attributed by Coons (1943) to nitrogen bubbles in the dental pulp. Some dental authorities (Stones, 1966), however, think it likely that in most cases of this type there is some pathological condition in a vital pulp. This may be secondary to caries or to a filling which has not previously given rise to symptoms. Orban and Ritchey (1945) investigated 250 cases of decompression toothache and studied histological sections of 75 of the teeth involved. They found oedematous pulps in 16, acute pulpitis in 17, chronic pulpitis in 15, non-vital pulps in 7 and normal pulps in 3. Seventeen cases were regarded as unclassifiable. They consider that teeth with normal pulps do not give rise to pain during decompression whether they are intact, carious or filled.

Use of reversible hydrocolloid in dentistry at high altitude

Reversible hydrocolloid impression material is used in dentistry as a flexible gel that will deform in an elastic fashion and produce an accurate impression record (Fine et al, 1977). It is introduced into the mouth as a viscous fluid on special impression trays, and gels in position when cool water is run through the tray. Impressions are removed intact since the flexibility of the gel is sufficient to allow withdrawal over extremely sharp undercuts with no perceptible permanent distortions. It is important for the dentist to bring the reversible hydrocolloid to an adequate liquefaction temperature and at sea level this is accomplished by placing the material in boiling water at 100°C for 15 minutes. However, when the hydrocolloid is placed in boiling water at high altitude, the temperature is lower than 100°C because of the lower barometric pressure and the material becomes grainy and non-homogeneous and appears to cause distortions in impressions. At high altitude the use of propylene glycol in the liquefying bath, either alone or as an additive to water, is effective in attaining the necessary temperature.

Oral haemorrhage

On a visit to the Andes one of us (DRW) developed crops of small haemorrhages in the mucosa of the buccal cavity at altitudes exceeding 4000 m (Fig. 20.1). They were up to 2 mm in diameter. This tendency to bleed at high altitude has already been seen to occur in the finger nails of the native

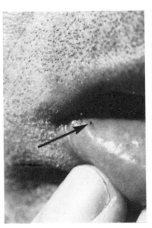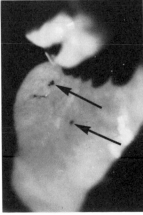

Fig. 20.1 Small haemorrhages (arrows) occurring in one of the authors (DRW) while at Cerro de Pasco (4330 m). (a: left) on the inner aspect of the everted lower lip. (b: right) on the buccal mucosa.

highlander (Chapter 4) and of the subject with Monge's disease (Chapter 16). Retinal haemorrhage at high altitude is described in Chapter 31. As early as 1913 Ravenhill reported that epistaxis occurred in some cases of acute mountain sickness. He described having met an elderly man who at high altitude bled 'not only from the nose, but from every mucous membrane of which he was possessed'. As we shall now see, it seems likely that haemorrhage also occurs readily in the gastric mucosa.

Peptic ulceration at high altitude

Garrido-Klinge and Peña (1959) found that the incidence of peptic ulceration in a group of 17 500 labourers, chiefly miners, born and living in the Andes at altitudes between 3050 m and 4880 m was 0.4 per cent of the insured population or 1.85 per cent of patients admitted to hospital. These figures are similar to those found in other countries according to the medical literature. Vargas (1967) also found that the incidence of peptic ulcer in the native highlander of Peru is no different from that of sea-level residents. In fact Singh and his colleagues (1977) found that the incidence of peptic ulcer in Indian soldiers stationed over a prolonged period at high altitude in the Himalayas is lower than at sea level (Fig. 20.2). Their studies showed that gastric

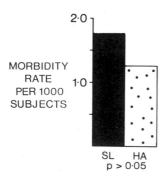

Fig. 20.2 The morbidity rate per thousand subjects from peptic ulcer among 20 000 soldiers stationed between altitudes of 3692 and 5538 m (stippled column marked HA) and 130 700 stationed on the plains (filled columns marked SL). (Based on data from Singh et al, 1977).

acidity diminishes on arrival at high altitude and remains low during the remainder of the stay in mountainous areas. Gastric motility is not altered. Pepsin activity is low for a week after arrival at high altitude but reverts subsequently to sea-level values.

The predominance of gastric ulceration and spontaneous haemorrhage

The incidence of gastric and duodenal ulcers in patients attending the Obrero Hospital in La Oroya at 3750 m in the Peruvian Andes was studied by Garrido-Klinge and Peña (1959). They found that the proportion of gastric ulcers is greater at high altitude than at sea level (Fig. 20.3). Furthermore, the incidence of haemorrhage from these ulcers is considerably greater than at low altitude (Fig. 20.4).

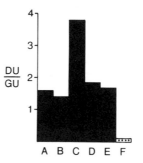

Fig. 20.3 The ratio of duodenal to gastric ulcer ($\frac{DU}{GU}$) at low and high altitude. The data shown in the first five columns are derived from studies at low altitude, that in A, B, C coming from the United States, that in D from Europe, and that in E from Peru. The data in column F comes from high altitude in Peru. It is seen that gastric ulcer is much commoner than duodenal ulcer at high altitude in the Andes. A = Data from Robertson and Hargis (1925); B = Feldman and Weinberg (1951); C = Waskow (1950); D = Knutsen and Selvaag (1947); E = Garrido-Klinge and Peña (1959) Lima 150 m; F = Garrido-Klinge and Peña (1959) La Oroya (3750 m).

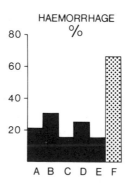

Fig. 20.4 The incidence of haemorrhage in cases of peptic ulcer diagnosed at low and high altitude. The data shown in the first five columns are derived from studies at low altitude, that in A to D coming from the United States and that in E coming from Peru. It is seen that the incidence of haemorrhage as a presenting symptom of peptic ulcer is greater at high altitude. A = Data from Jennison (1938); B = Bockus (1944); C = Wherritt; D = Ivy et al (1950); E = Garrido-Klinge and Peña (1959) Lima (150 m); F = Garrido-Klinge and Peña (1959) La Oroya (3750 m).

Finally, the age of onset of gastric ulceration in the native highlander is usually early, no fewer than 48 per cent of the cases being in the third decade. It is possible that spontaneous haemorrhages in the gastric mucosa, similar to those in the mouth, may account for the high incidence of haemorrhages reported in patients with peptic ulcer at high altitude.

The native populations of mountainous areas have dietary habits that may be of some importance in the aetiology of the increased incidence of gastric ulceration. In the Andes the diet is composed largely of cereals insufficiently cooked on account of the low barometric pressure and many of the population are unable to chew it adequately due to loss of teeth as a result of inadequate dental care from childhood (Garrido-Klinge and Peña, 1959). The Quechua Indians are very fond of potent spices and flavourings on their food and an inferior rough alcohol is consumed by many in considerable quantity. Many are addicted to the chewing of coca leaves from which cocaine is obtained. It will be seen from this imposing list that this native population is chronically exposed to a wide variety of potent gastric irritants so that both dietary and vascular factors could operate in bringing about this increased incidence of gastric ulceration.

Steele (1971) also refers to indigestion being common among the Sherpas and he also relates this to their disposition to drink large quantities of rough, home-brewed spirit, *arak*. Three of these native highlanders accompanying a party climbing Mount Everest had to be brought down as they were suffering from acute peptic ulceration with severe epigastric pain, nausea and vomiting. In view of what has been said above concerning haemorrhage into the gastric mucosa it is of interest to note that one Sherpa had a brisk haemetemesis of about a litre after a heavy drinking bout (Steele, 1971). Vargas (1967) has also reported that at high altitude gastric ulcers are commoner than the duodenal variety but he found the ratio to be much lower at 2.5 : 1 than the ratio reported by Garrido-Klinge and Peña (1959) (20 : 1). He confirmed that the most frequent complication of peptic ulcer at high altitude is haemorrhage which is much commoner than at sea level. Garrido-Klinge and Peña (1959) did not find high levels of acidity in the cases they reported. Increased 'stress' has been considered to be of

importance in the aetiology of duodenal ulcer but it would appear that the hazards and hardships of prolonged residence at high altitude with its exposure to hypoxia, cold and so on do not constitute 'stress' in this sense since the incidence of duodenal ulcer is not increased at high altitude.

Chronic hypoxia and peptic ulceration

Of all the possible factors for the increased incidence of gastric ulceration in high altitude residents referred to above we must give further consideration to that of chronic hypoxia. The initial impression is that in some way hypoxia predisposes to peptic ulceration since many studies have shown that the incidence of peptic ulcer in patients with pulmonary emphysema is high. In Thurlbeck's series (1963) 23 per cent of his patients with severe emphysema had a chronic peptic ulcer. However, the increased incidence of such ulcers in emphysematous subjects is linked to age and sex. Thus Cohen and Jenney (1962) found that in a series of patients dying of pulmonary tuberculosis the incidence of peptic ulcers was the same as in a group of patients with emphysema with the same age and sex distribution. Thurlbeck (1963) found no statistically significant association between emphysema and chronic peptic ulcer. The increased incidence of chronic peptic ulcer in patients with severe pulmonary emphysema is thought to be due to the fact that the great majority of them are men who are older than the random necropsy population.

Blood flow to the stomach on exposure to hypoxia

Although the stomach and duodenum receive their blood supply from the same major arteries, they show different local vascular responses to an acute exposure to hypoxia. Thus when Broadie and his colleagues (1975) exposed 22 mongrel puppies to hypoxia they found that there was no alteration in the blood flow to the duodenum but a striking increase of 88 per cent in the perfusion of the gastric antrum. At the same time there was a reduction of 49 per cent in vascular resistance in the antral area. The blood flow to the remainder of the stomach was not affected.

Splanchnic blood flow and liver metabolism

Studies on Indians of the Bolivian Altiplano living at 4000 m by Capderou and his colleagues (1977) show that splanchnic blood flow and liver metabolism are not modified by the lifelong exposure to hypoxia. They found that splanchnic blood flow fell within normal sea-level values of 1.42 to 1.70 l min^{-1}, the mean value being 1.55 l min^{-1}. Estimates expressed per square metre body surface area to take into account the smaller stature of the highlander were also normal. Blood pressure in the hepatic vein of highlanders (10 mmHg) is similar to that in lowlanders. Splanchnic arteriovenous oxygen difference shows no abnormality. The uptake of indocyanine green, a sensitive and specific test of liver function, was found to be not diminished in highlanders by Capderou et al (1977). This confirmed the previous finding of Berendsohn (1962) that hepatic function is essentially normal in Peruvian highlanders living at 4540 m. Ramsoe and his colleagues (1970) found that hepatic function was normal in newcomers to the mountain environment during the first week at 3500 m.

In sea-level man there is a fall in splanchnic blood flow as part of the response of the circulation to physical exercise to provide a better perfusion and oxygen supply of the exercising muscles. The lactate produced by the exercising muscles is largely removed by increased liver gluconeogenesis. Native highlanders are capable of hard, efficient physical work at high altitude and in carrying this out produce a much lower arterial lactate than sea-level man. This could clearly arise by a lower production of lactate by the muscles or an increased conversion into glycogen in the liver. Capderou et al (1977) tentatively conclude that lactate produced by exercising muscles in highlanders is utilized for liver gluconeogenesis as demonstrated in lowlanders.

Intestinal absorption and hypoxia

Milledge (1972) is of the opinion that loss of body weight is associated with hypoxaemia and he notes that children with cyanotic congenital heart disease are stunted and underweight. Patients with severe emphysema and hypoxaemia also may lose weight. Milledge (1972) studied the absorption of xylose from the small intestine in 16 patients with varying degrees of arterial oxygen desaturation due to either congenital heart anomalies or chronic lung disease. There was a significant correlation between xylose absorption and the systemic arterial saturation. Absorption was decreased in the cases with more severe desaturation. In nine cases hypoxia was relieved by the administration of oxygen or by surgery and repeat testing in such instances showed a statistically significant increase in xylose absorption, the mean increase being 11.7 per cent. It has to be borne in mind that increased metabolism of xylose or increased tubular reabsorption of xylose could account for the same experimental findings. Milledge reported that administration of oxygen for only eight hours increased xylose absorption suggesting that the effect of hypoxia on the intestinal mucosa is likely to be of a biochemical nature rather than in any change in cell population. However, other hypotheses have been advanced to explain the loss of body weight at high altitude. Klain and Hannon (1970) found evidence of a transient abnormality of protein metabolism and suggested that this might contribute to the loss of weight at high altitude. Van Liere et al (1948) showed a decreased motility of the small intestine. Chinn and Hannon (1970) discuss the possibility of disturbances of intermediary metabolism causing loss of weight at high altitude. Pugh (1962) notes that on the Himalayan expedition of 1960 to 1961 there was a tendency towards bulky and greasy stools whenever the fat intake was increased; this suggested to him that there may be a possible disturbance of fat absorption at great altitudes. The basis for the characteristic loss of body weight at high altitude is considered further in the following chapter.

Intestinal volvulus

According to Monge and Monge (1966), Ovando (1962) and Frisancho (1959) state that there is an increased incidence of intestinal volvulus at high altitude.

Haemorrhoids

Piles are known to occur commonly during climbing expeditions at high altitude. Rosedale (1973) describes them as a source of much trouble and disablement on Himalayan expeditions. He notes that all the members of the Everest expedition in that year were equipped with suppositories and given advice on the avoidance of predisposing constipation. Pugh (1962) reported that piles were a common complaint on the Himalayan expedition led by Sir Edmund Hillary in 1960 to 1961. One case of ischio-rectal abscess also occurred. Steele (1971) says that four members of the International Himalayan Expedition in that year had prolapsed piles. He relates this to increased abdominal pressure from overbreathing and carrying heavy loads. Bonington (1971) refers to the advisability of treating early piles before climbing to high altitude.

Diseases of the alimentary system in air travellers

The simulated cabin altitude in modern pressurized aircraft is 1520 m to 2130 m. In accordance with Boyle's law, pressure and volume vary inversely so that as atmospheric pressure is reduced with altitude any gas in the body cavities will expand. Thus at an altitude of 1830 m, simulated in modern aircraft, 100 ml of air will increase in volume to 130 ml (Peffers, 1978). Hence patients who have undergone abdominal surgery with even a mild degree of post operative ileus will experience extreme discomfort if they travel too soon, even after simple operations such as appendicectomy. Though fully ambulant, to avoid complications they should not attempt to travel for about 10 days. Colostomy patients will find that their colostomy becomes over-active due to the expansion of gases and they should be prepared with sufficient dressings for the journey, have a mild bowel sedative and medication to increase the bulk of the stool. Those who have had air introduced to any body cavity for diagnostic or therapeutic purposes, such as pneumoperitoneum, should not travel for at least seven days. The same safeguard should apply to patients who have been subjected to ventriculography. Patients with emphysematous bullae run the risk of the danger of spontaneous pneumothorax. With any induced pneumothorax the lung should be at least three-quarters expanded before travel, to avoid mediastinal embarrassment (Peffers, 1978). Indeed any patient with cysts or collections of air in the body is at some risk when he takes a trip in an aircraft.

REFERENCES

Berendsohn, S. (1962) Hepatic function at high altitudes. *Archives of Internal Medicine* 109, 56.

Bockus, H. L. (1944) *Gastroenterology*. Philadelphia: Saunders.

Bonington, C. (1971) In: *Annapurna South Face*, p. 288. London: Cassell

Bonnington, C. (1973) *Everest South-West Face*. London: Hodder & Stoughton.

Broadie, T., Devedas, M., Rysavy, J., Leonard, A. S. & Delaney, J.P. (1975) Gastroduodenal blood flow in stress and hypoxia: an experimental approach. *Surgical Forum*, 26, 397.

Capderou, A., Polianski, J., Mensch-Dechene, J., Drouet, L., Antezana, G., Zetter, M. & Lockhart, A. (1977) Splanchnic blood flow O$_2$ consumption, removal of lactate, and output of glucose in highlanders. *Journal of Applied Physiology: Respiratory, Environmental and Exercise Physiology*. 43, 204.

Chinn, K. S. K. & Hannon, J. P. (1970) Effects of diet and altitude on the body composition of rats. *Journal of Nutrition*, 100, 732.

Cohen, A. C. & Jenney, F. S. (1962) The frequency of peptic ulcer in patients with chronic pulmonary emphyscma. *American Review of Respiratory Diseases*, 85, 130.

Coons, D. S. (1943) Aeronautical dentistry. *Journal of the Canadian Dental Association*, 9, 320.

Feldman, M. & Weinberg, T. (1951) Healing of peptic ulcer. *American Journal of Digestive Diseases*, 18, 295.

Fine, T. D., Ito, R. J., Christensen, G. J & Cunningham, P. R. (1977) A method for using reversible hydrocolloid at high altitudes. *Journal of Prosthetic Dentistry*, 38, 294.

Frisancho, D. (1959) El colon y sus particularidades en el hombre del altiplano. *Boletin de la Academia Peruana de Cirugia*, 12, 6.

Garrido-Klinge, G. & Peña, L. (1959) The gastroduodenal ulcer in high altitudes (Peruvian Andes). *Gastroenterology*, 37, 390.

Harvey, W. (1943) Tooth temperature with reference to dental pain while flying. *British Dental Journal*, 75, 221.

Ivy, A. C., Grossman, M. I. & Backrach, W. H. (1950) *Peptic Ulcer*. New York: Blakiston.

Jennison, J. (1938) Observations made in a group of employees with duodenal ulcer. *American Journal of the Medical Sciences*, 196, 654.

Joseph, T. V., Gell, C. F., Carr, R. M. & Shelesnyak, M. C. (1943) Toothache and the aviator: study of tooth pain provoked by simulated high altitude runs in low pressure chamber. *United States Naval Medical Bulletin*, 41, 643.

Klain, G. J. & Hannon, J.P. (1970) High altitude and protein metabolism in the rat. *Proceedings of the Society for Experimental Biology and Medicine*, (N.Y.), 134, 1000.

Knutsen, B. & Selvaag, O. (1947) The incidence of the peptic ulcer: An investigation of the population of the town of Dramen. *Acta medica Scandinavica*, 196, 341.

Milledge, J. S. (1972) Arterial oxygen desaturation and intestinal absorption of xylose. *British Medical Journal*, iii, 557.

Monge, M. C. & Monge, C. C. (1966) In: *High-Altitude Diseases. Mechanism and Management*, p. 14. Springfield, Illinois: Charles C. Thomas.

Orban, B. & Ritchey, B. T. (1945) Toothache under conditions simulating high altitude flight. *Journal of the American Dental Association*, 32, 145.

Ovando, R. (1962) Estudio comparativo del volvulo del sigmoide,

Costa y Sierra. *Boletin de la Academia Peruana de Cirugia*, 15, 143.

Peffers, A. S. R. (1978) Carriage of invalids by air. *Journal of the Royal College of Physicians of London*, 12, 136.

Pugh, L. G. C. E. (1962) Physiological and medical aspects of the Himalayan Scientific and Mountaineering Expedition, 1960–61. *British Medical Journal*, ii, 621.

Ramsoe, K., Jarnum, S., Preisig, R., Tauber, J., Tygstrup, N. & Westergaard, H. (1970) Liver function and blood flow at high altitude. *Journal of Applied Physiology*, 28, 725.

Ravenhill, T. H. (1913) Some experiences of mountain sickness in the Andes. *Journal of Tropical Medicine and Hygiene*, 16, 313.

Robertson, H. E. & Hargis, E. H. (1925) Duodenal ulcers: an anatomic study. *Medical Clinics of North America*, 8, 1065.

Rosedale, B. (1973) p. 336 in Bonington, C. (See above.)

Singh, I., Chohan, I. S., Lal, M., Khanna, P. K., Srivastava, M. C., Nanda, R. B., Lambda, J. S. & Malhotra, M. S. (1977). Effects on high altitude stay on the incidence of common diseases in man. *International Journal of Biometeorology*, 21, 93.

Steclc, P. (1971) Medicine on Mount Everest 1971. *Lancet*, ii, 32.

Stones, H. H. (1966) In: *Oral and Dental Disease*, 5th edn. Edited by E. D. Farmer & F. E. Lawton. Edinburgh: Livingstone.

Thurlbeck, W. M. (1963) A Clinico-pathological study of emphysema in an American hospital. *Thorax*, 18, 59.

Van Liere, E. J., Crabtree, W. V., Northrup, D. W. & Stickney, J. G. (1948) Effect of anoxic anoxia on propulsive motility of the small intestine. *Proceedings of the Society for Experimental Biology and Medicine*, (N.Y.) 67, 331.

Vargas, A. C. (1967) Peptic ulcer in the native Peruvian. *Proceedings of the Third World Congress of Gastroenterology*. Tokyo: Nankodo.

Waskow, W. (1950) Incidence of gastric and duodenal ulcer. *Wisconsin Medical Journal*, 49, 683.

Wherritt, B. H. Quoted by Bockus, H. L.

21

Body weight and nutrition; renal function and electrolytes

Most newcomers to high altitude lose weight. We believe this is because the nausea and headache so commonly experienced on ascending into the mountains deaden appetite. There may be additional factors such as diminished intestinal absorption secondary to hypoxia as described by Milledge (1972) which we consider in Chapter 20. Nevertheless, from our personal experience in the Andes and from our experimental studies in the laboratory (Gloster et al, 1972) we conclude that anorexia and hypophagia are the basis for loss of weight at high altitude. The diminished intake of food for short periods of time at high altitude leads initially to loss of stored body fat. More prolonged stay at extreme elevation leads to emaciation with loss of muscle and body nitrogen.

Hypophagia

The anorexia and hypophagia of high altitude have been studied by Hannon et al (1976). They studied the intake of nutrients by eight female students ranging in age from 18 to 23 years over a period of four days at 140 m and over a subsequent period of seven days at 4300 m. Hypophagia was most pronounced during the first three days of exposure to high altitude when there was a transient decrease in the consumption of a wide range of nutrients, including protein, carbohydrate, fat, sodium, calcium, phosphorus, vitamin A, thiamin and niacin. There was a more sustained decrease in the consumption of potassium and ascorbic acid. The degree of anorexia and hypophagia was related to the severity of acute mountain sickness. During the first three days at 4300 m the consumption of calories was reduced by as much as 40 per cent. A reduction of calorie intake of this magnitude is roughly

comparable to that reported for soldiers at the same altitude by Surks et al (1966), Johnson et al (1969), and Whitten et al (1968). At lower altitudes or under conditions of high physical activity anorexia and hypophagia may not be so pronounced. The hypophagia observed in man can be induced experimentally. We subjected six young male guinea pigs to a simulated altitude of 5000 m for 14 days and contrasted their food intake with that of six controls of the same mean body weight (Gloster et al, 1974). The total weight of food consumed by the test animals was only 63 per cent of that eaten by the controls (Fig. 21.1).

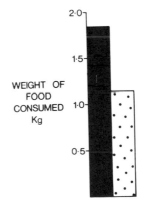

WEIGHT OF FOOD CONSUMED Kg

Fig. 21.1 The total weight of food consumed by six young male guinea pigs over a period of 14 days while they were exposed to a barometric pressure of 400 mmHg, simulating an altitude of 5000 m (stippled column). It was considerably less than that eaten by six control animals at sea-level barometric pressure (filled column). (From data of Gloster et al, 1974).

Distaste for fat

In the studies of Hannon et al (1976) it was noted that on acute exposure to high altitude carbohydrate is chosen in preference to fat in food intake (Fig.

21.2). This same distaste for fat has also been remarked upon by Ward (1975) in climbers and we comment on this further in Chapter 30. Increased amounts of carbohydrate may enhance tolerance to high altitude (King et al, 1945), reduce the severity of acute mountain sickness, and improve physical performance (Consolazio et al, 1969).

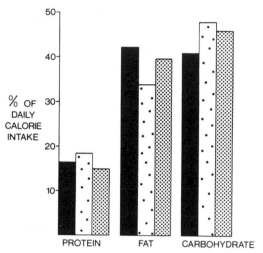

Fig. 21.2 The percentage of the daily calorie intake occupied by protein, fat, and carbohydrate during a four-day period at low altitude (filled columns), during the first three days at 4300 m (lightly-stippled columns), and over the last four days at 4300 m (stippled columns). It will be seen that on initial exposure to high altitude there is a diminished intake of fat, but an increased intake of carbohydrate (From data of Hannon et al, 1976).

Loss of body weight

The diminished intake of food is unable to provide the level of energy required to sustain activity at high altitude and the deficit has to be made up by catabolism of body tissues. As a result loss of body weight occurs. In the women studied by Hannon et al (1976) the loss of weight after seven days at 4300 m was 1.76 per cent of total body weight. In soldiers stopping for 12 days at the same altitude the loss of weight reported was 3.49 per cent (Surks et al, 1966), and 5.0 per cent (Krzywicki et al, 1969). The loss of body weight could be attributable to loss of water, body fat or protein (Krzywicki et al, 1969). We consider the question of hydration below. It seems likely that the initial weight loss is primarily due to loss of stored body fat (Surks et al, 1966) and Hannon et al (1969a) have reported a reduction in skinfold thickness and limb circumference in eight

women exposed to 4300 m for 10 weeks. The initial loss of stored fat is followed by wasting of muscle. Ward (1975) refers to the severe weight loss in climbers at extreme altitudes due to protein breakdown. He comments on the emaciated appearance of the members of the 1953 Everest expedition who ascended to 7920 m. During the Himalayan expedition of 1960 to 1961 all members of the party lost weight at the rate of 0.45 to 1.36 kg a week at 5790 m. By the end of the expedition weight losses ranged from 6.4 to 9.0 kg (Pugh, 1962). Negative nitrogen balance was reported in males acutely exposed to 4300 m for eight days (Surks, 1966) and 28 days (Consolazio et al, 1968). There is also a sustained reduction in ammonia excretion to about half of the value at sea level (Hannon et al, 1976) and this has been held to compensate for the alkalosis produced by hyperventilation (Schonhölzer and Gross, 1948).

Nair and Prakash (1972) claim that it is the simultaneous exposure to cold and hypoxia that induces a significant loss of body weight during the second and third weeks of exposure to high altitude. Indeed they claim that hypoxia alone does not lead to loss of weight. They relate this difference to the increase in basal metabolism which they showed to occur with the combined stimulus of cold and hypoxia (Nair et al, 1971). Their conclusions do not seem to be supported by our animal studies referred to above, which demonstrated loss of weight induced by simulated high altitude alone without the additional factor of cold.

Experimental evidence as to the cause of loss of body weight at high altitude

We have studied this problem in an experiment employing three groups of rats. A test group was exposed to a subatmospheric pressure of 400 mmHg for 14 days (Gloster et al, 1972). Two groups of controls were studied. A free-fed control group was allowed an unlimited amount of food. A food-restricted group was given each day the same amount of food eaten by the test animals in the previous 24 hours. The body weights of the three groups of adult rats over the course of the experiment are shown in Figure 21.3. It will be noted that, as in man, there is a fall in body weight in the hypoxic rats. However, the food-restricted

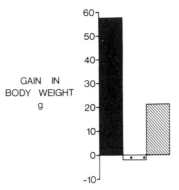

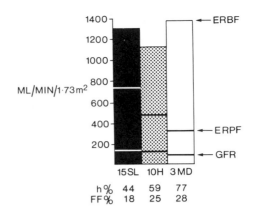

Fig. 21.3 The effect of hypobaric conditions and food intake on whole body weight in rats. The gain in body weight of a group of free-fed rats over a period of 14 days at sea level is indicated by the filled column. When rats were exposed to a barometric pressure of 400 mmHg, simulating an altitude of 5000 m for the same length of time, they actually lost weight (stippled column). However, failure to gain weight normally was also shown by a third group of rats (hatched column) kept at sea-level barometric pressure but given a restricted diet equivalent in amount to that eaten by the hypoxic test rats.

Fig. 21.4 Effective renal blood flow (ERBF), effective renal plasma flow (ERPF), and glomerular filtration rate (GFR), in 15 subjects from sea level, in 10 healthy highlanders, and in 3 subjects with Monge's disease. The studies were carried out at an altitude of 4500 m. For each group the values for haematocrit (h%) and filtration fraction (FF%) are shown at the foot of the column. (Based on data from Monge et al, 1969).

control rats kept at sea-level barometric pressure also failed to gain weight normally. In the hypoxic group there was a gradual recovery in body weight, although the initial starting weight was not regained by the end of the experiment. Such experimental studies suggest that the anorexia and hypophagia of high altitude can account entirely for the loss of body weight.

Renal function

The kidney is able to maintain adequate tubular function at high altitude in the presence of considerable reduction of plasma flow, greatly increased blood viscosity and pronounced hypoxaemia. Thus there does not appear to be any impairment of reabsorption of bicarbonate or sodium, or of secretion of hydrogen ions (Monge et al, 1969). When renal haemodynamics are studied in native highlanders and patients with Monge's disease, they show a reduction of plasma flow and a lesser fall in glomerular filtration rate (Fig. 21:4). Hence there is an elevated filtration fraction in the highlanders which becomes even more pronounced on the development of chronic mountain sickness (Fig. 21.4.). In this way glomerular filtration rate is not reduced to levels that would prove of clinical significance. The study of Monge and his colleagues

(1969) revealed a positive correlation between haematocrit and filtration fraction (Fig. 21.4). They believe that the increased filtration fraction can be explained by increased resistance at the level of the efferent arteriole and that this resistance could be the result of high blood viscosity due to the elevated haematocrit. A fall in creatinine clearance has also been reported at high altitude (Rennie and Joseph, 1970; Rennie et al, 1972; Epstein and Saruta, 1972).

Although renal function copes so well with the altered conditions of life at high altitude, it has become apparent that the kidneys do not escape disturbance altogether. Exposure to high altitude induces proteinuria and brings about glomerular enlargement and we shall consider both of these features now.

Proteinuria

There is an increased excretion of protein in the urine in native highlanders (Fig. 21.5) Rennie and his colleagues (1971b) found that, although this increase barely exceeded physiological levels, it was unequivocal. They studied young male residents of Yauricocha (4640 m), who had lived in the town for at least two years, and found that the proteinuria was associated with normal creatinine clearance rates, thus suggesting that it was independent of glomerular filtration rate. A slightly older group of men from San Cristobel (4710 m) was found to have

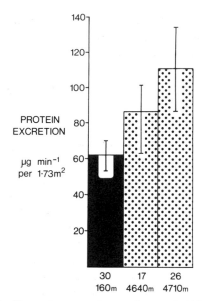

140-

120-

100-

PROTEIN
EXCRETION   80-

μg min⁻¹
per 1·73m²   60-

40-

20-

30   17   26
160m   4640m   4710m

Fig. 21.5 Mean urinary protein excretion rates in 30 lowlanders at 160 m (filled column), in 17 native highlanders at 4640 m (stippled column), and in 26 native highlanders at 4710 m (stippled column). Protein excretion is increased in highlanders. (From data of Rennie et al, 1971b).

a depression of creatinine clearance associated with their proteinuria and this was considered to be related to their higher degree of polycythaemia. In native highlanders it has proved impossible to demonstrate any deficiency in oxygen uptake by the kidney (Rennie et al, 1971a). Hence proteinuria in highlanders is not directly and obviously related to a deprivation of oxygen in the kidney.

Proteinuria is also found in climbers undergoing acclimatization. The protein concentration in early-morning urine specimens collected from 13 climbers undergoing acclimatization was found to correlate significantly with the altitude when adjusted for urine concentration effects (Rennie and Joseph, 1970). This was done by employing the ratio of urine protein concentration to urine osmolality. Since there is no relation of early-morning osmolality and altitude the increased urinary protein excretion cannot be due to urine concentration. It is of interest to note that there is a time lag between acute exposure to high altitude and the onset of proteinuria. This recalls the similar time lag before the onset of acute mountain sickness, referred to in Chapter 14. Rennie and Joseph (1970) postulate on this account that hypoxia exerts a delayed effect on renal glomeruli or tubules or both. These early

observations have been confirmed by Pines (1978). He has reported the occurrence of proteinuria in seven climbers who spent six weeks in the Ruwenzori (5110 m), Kilimanjaro (5960 m), and Mt. Kenya (5200 m) mountain areas of East Africa. The mean protein urine concentration in morning specimens as determined by reagent strips was over 100 mg/dl after climbs during the first 12 days. In subsequent climbs it fell to 15 mg/dl. The highest concentrations (100 to 300 mg/dl) were found in five climbers with peripheral oedema and three of them had obvious acute mountain sickness. The proteinuria was a definite response to height and not to exercise for it was not provoked by the most strenuous and exhausting part of the trip below 3000 m. The validity of Pine's report is borne out by his finding similar proteinuria in eight climbers on Nanda Devi (7820 m) in the Himalayas. Once again the proteinuria occurred during earlier climbs to heights, diminishing with repeated ascents. Of great interest is the fact that the proteinuria was greatest in three climbers who developed acute mountain sickness and in whom 'showers of late inspiratory crackles' were heard. Clearly the findings of Rennie and Joseph (1970) and Pines (1978) implicate the kidney in high altitude acclimatization and illness.

Renal glomeruli in highlanders

Patients with cyanotic congenital heart disease in the same way also develop proteinuria, the severity and extent of which correlates with the length and severity of systemic arterial unsaturation (Rennie et al, 1971b). It was noted that such patients had enlarged renal glomeruli (Meeson and Litton, 1953; Spear, 1960) and this stimulated Naeye (1965) to study the histology of the kidney in 15 children who died in Leadville, Colorado (3100 m), and compare it with that of 80 sea-level children. Using a planimeter, he determined the relative areas of 20 to 40 glomeruli in each child, and the mean glomerular area was thus calculated for each case. He found that at birth children at high altitude have renal glomeruli of normal size but disproportionate enlargement occurs in childhood compared to sea-level subjects. Naeye (1965) determined the 'parenchymal cell density' in individual glomeruli by dividing the number of parenchymal cells by total glomerular area. This proved to be the same in native highlanders as in sea-level subjects. Hence the glomerular enlargement found at high altitude appears to be due to a proliferation of normal glomerular elements. This is precisely what is found in patients with cyanotic congenital heart disease (Naeye, 1965). It is conceivable that the glomerular enlargement in highlanders is in some way associated with the increased blood volume, heightened blood viscosity, and systemic arterial hypoxaemia characteristic of life at high altitude.

Low total serum protein levels

Six of the 49 subjects studied by Rennie et al (1971b) showed low total serum protein levels. In five of these there were raised plasma creatinine levels and low creatinine clearance. Such findings may possibly have a basis in malnutrition in native highlanders.

Body nutrient requirements at high altitude

In general the period of low intake of nutrients in lowlanders during acute exposure to high altitude is so brief as to present no nutritional problem. Evidence as to the need for additional iron to support increased erythropoiesis at high altitude is conflicting (Hornbein, 1962). Female newcomers to altitude, however, have significantly higher haemoglobin levels when given supplemental iron (Hannon et al, 1966). Hannon and his colleagues also think it likely that there may be increased requirements of ascorbic acid, possibly a result of stress-induced enhancement of adrenocortical activity. However, nutrition in native highlanders is commonly inadequate. This is more an expression of the poor socio-economic conditions of isolated mountain communities throughout the world than of any specific effect of high altitude. Picon-Reategui (1978) notes that in the Andes the diet of the highlander is abnormally high in carbohydrate and poor in both fat and protein.

Mineral and vitamin content of the diet of native highlanders

The range of daily intake of some minerals and vitamins in five communities in the Peruvian Andes, as determined by Collazos et al (1954) and Gursky (1969) and as presented by Picon-Reategui (1978) are shown in Table 21.1. Also shown in the same table are the daily need of these nutrients as estimated by the National Academy of Science of the United States (1974). Picon-Reategui (1978) gives some interesting examples of local geographical and social factors that may influence the adequacy of intake of minerals and vitamins. Thus vitamin C intake is intimately related to the availability of fresh potatoes which are consumed in large quantities from April to August, thus ensuring an adequate intake of ascorbic acid, estimated to be 20.5 mg per 100 g of portion of potato. After the seasonal harvest the consumption of the fresh tubers is replaced by the dehydrated form of potatoes, chuño, which containes a very much lower ascorbic acid content of only 1.7 mg per portion. Water boils at a lower temperature at high altitude and this may preserve the content of ascorbic acid in native foods during the process of cooking. This effect of the lower boiling point in mountains may also protect the thiamin content of cooked food.

It will be noted from Table 21.1 that considerable variation in the vitamin A content of the food of native highlanders has been reported by various observers. Not only that but in some instances the intake of vitamin A reported has been totally inadequate. The variation in intake is probably related to the availability of such foods as liver. It is

Table 21.1 Mineral and vitamin contents of the diet in high altitude communities in Peru (Chacan, Vicos, Chillihua, Nuñoa, Sincata) (after Collazos et al, 1954; Gursky, 1969).

	Calcium mg	Phosphorus mg	Iron mg	Vitamin A retinol equivalents	Thiamin mg	Riboflavin mg	Niacin mg	Ascorbic Acid mg
Range of intake	76–870	761–1706	12.1–31.8	1.4–2203.5	1.44–4.16	0.69–1.92	13.06–29.62	10.9–76.9
Daily need*	800	800	10		1.37	1.65	18.1	45

* Based on dietary allowances recommended by National Academy of Sciences, National Research Council, USA (1974).

worthy of note that in the face of these estimates of low vitamin A-intake there have been no reports of clinically detectable hypovitaminosis A in the population and Picon-Reategui (1978) believes that this interesting inconsistency should be investigated.

There is high absorption of calcium from the food ingested by the Quechuas. Picon-Reategui (1978) relates this to the somewhat increased levels of ultraviolet radiation at high altitude which may increase the availability of vitamin D. Calcium absorption may also be aided by the low fat content of the Quechua diet to which we have already alluded above. In spite of the elevated haematocrit of the native highlander and the consequent increased demands for the higher body content of iron, these demands are met over a long period of time. At every altitude there is a balance between blood formation and destruction so that increased daily requirement of iron cannot be detected. It has to be kept in mind that while intake of many minerals and vitamins may be just adequate to meet adult requirements, it may prove insufficient to meet the needs of increased demand such as occur in pregnancy and lactation.

Water

As we have already seen in Chapter 14 mild hypoxia induces polyuria and in those subjects who tolerate the exposure to high altitude well there may be a diuresis that lasts for days. As we shall see below this diuresis is associated with a negative sodium balance. We have already noted previously in Chapter 14, however, that some persons ascending mountains may become oliguric during the first few hours of their exposure to the hypoxic environment. At the same time there is a redistribution of water in the body with the tendency to pulmonary and cerebral oedema and the onset of acute mountain sickness. A positive water balance has been reported during the first two days of exposure to an altitude of 3450 m (Ullman, 1953).

There is controversy as to the state of hydration of the body once this initial stage of acute exposure to high altitude is passed. Many believe that a significant part of the weight loss at high altitude is due to dehydration. Some experimental work has suggested that as much as 95 per cent of the weight loss in rats at high altitude is due to loss of water. Certainly at extreme altitude climbers may become dehydrated due to the increased pulmonary ventilation induced by elevation and exercise, coupled with the low humidity of the ambient air (Pugh, 1962). Consolazio et al (1968) observed a negative water balance in troops after four weeks' exposure to conditions at 4300 m. In other studies Consolazio and his colleagues (1972) and Krzywicki et al (1969) have attributed much of the loss of body weight at high altitude to absolute dehydration with decrease in plasma volume. Consolazio et al (1972) believe that intracellular water *falls* on acute exposure to altitude, whilst extracellular water remains constant. Total body water, therefore, falls according to this view. Bulstrode (1975) lost six kilograms in body weight in rapidly ascending and descending Mount Kenya, and regained it within 24 hours of descent and concluded that this remarkable loss of weight was due to fluid loss.

Such findings are, however, at variance with those of other investigators who report that the decrease in plasma volume referred to above is associated with an *increase* in intracellular water and redistribution of body fluids. The net result of this is that normal body hydration is maintained (Surks et al, 1966; Hannon et al, 1969b; Hannon et al, 1972). Hannon et al (1976) found little evidence of a net loss of body water during a stay of one week at 4300 m of eight young women. Rather there appeared to be renal water conservation, probably in response to respiratory water loss. It should be noted, however, that the altitude employed in this study is considerably less than the extreme altitudes at which climbers suffer undoubted dehydration. Hannon et al (1976) point out that analysis of carcasses of laboratory animals shows little or no change in body water content on exposure to high altitude (Schnakenberg et al, 1971).

Sodium

It would appear that on exposure to high altitude there is a net loss of sodium, the deficit being larger in men than women, probably because of a greater degree of hypophagia (Hannon et al, 1972). In one group of subjects there was an initial positive sodium balance during the first two days of exposure to an altitude of 3500 m but subsequently there was a negative balance (Slater et al, 1969a and b). Earlier studies of Ullman (1953) had also showed a positive sodium balance during the first two days of exposure to 3450 m with a slight negative balance afterwards. Bulstrode (1975) investigated the changes in electrolyte concentration which occurred in ascending Mount Kenya (5200 m) in unacclimatized British soldiers, and acclimatized members of a mountain rescue team who were Kikuyu or Caucasian in origin. He found that in both groups there was a slight rise in plasma sodium and chloride on ascending the mountain corresponding to the early positive sodium balance reported above. Sutton et al (1977) found no change in plasma concentration and no change in urinary sodium excretion in four men acutely exposed for two days to a simulated altitude of 4760 m.

Potassium

There is a tendency for the body to conserve potassium on acute exposure to high altitude, especially during the first three days (Hannon et al, 1972; Janowski et al, 1969). A positive potassium balance over a five-day period of exposure to 3500 m was observed by Slater et al (1969b). There have been reports of a rise in plasma potassium (Epstein and Saruta, 1973; Frayser et al, 1975; Slater et al, 1969b) but Sutton et al (1977) found no change in plasma potassium concentration in four men exposed to a simulated altitude of 4760 m for two days. They like other observers found a decrease in urinary potassium excretion. The loss of sodium and retention of potassium causes the ratio of urinary sodium to potassium to increase at high altitude. We shall see in Chapter 25 that the rise in total body potassium at high altitude also leads to a rise in the sodium/potassium ratio in saliva. As we shall see in that chapter these changes in sodium and potassium excretion are probably related to the fall in aldosterone secretion which occurs during the early stages of exposure (Ayres et al, 1961; Williams, 1966; Janowski et al, 1969; Hannon et al, 1972). The fall in aldosterone secretion is itself probably related to the increased blood volume that occurs on exposure to high altitude (Williams, 1966) and which we describe in Chapter 17. Body stores of phosphorus are conserved like potassium and they have a significant rôle in the maintenance of plasma ionic balance during this period (Hannon et al, 1972).

REFERENCES

Ayres, P. J., Hurter, R. C., Williams, E. S. & Rundo, J. (1961) Aldosterone excretion and potassium retention in subjects living at high altitude. *Nature*, **191**, 78.

Bulstrode, C. J. K (1975) A preliminary study into factors predisposing mountaineers to high altitude pulmonary oedema. *Journal of the Royal Naval Medical Service*, **61**, 101.

Collazos, C., White, H. C., Huenemann, R. L., Reh, E., White, P. L., Castellanos, A., Benites, R., Bravo, Y., Loo, A., Moscoso, I., Cáceres, C. & Dieseldorff, A. (1954). Dietary surveys in Peru III. Chacán and Vicos. Rural communities in the Peruvian Andes. *Journal of the American Dietetic Association*, **30**, 1222.

Consolazio, C. F., Matoush, L. O., Johnson, H. L. & Daws, T. A. (1968) Protein and water balances of young adults during prolonged exposure to high altitudes (4300 meters). *American Journal of Clinical Nutrition*, **21**, 154.

Consolazio, C. F., Johnson, H. L., Krzywicki, H. J. & Daws, T. A. (1972) Metabolic aspects of acute altitude exposure (4300 meters) in adequately nourished humans. *American Journal of Clinical Nutrition*, **25**, 23.

Consolazio, C. F., Matoush, L. O. Johnson, H. L. Krzywicki, H. J., Daws, T. A & Isaac, G. J. (1969) Effects of high carbohydrate diets on performance and clinical symptomatology after rapid ascent to high altitude. *Federation Proceedings: Federation of American Societies for Experimental Biology*, **28**, 937.

Epstein, M. & Saruta, T. (1972) Effect of simulated high altitude on renin-aldosterone and Na homeostasis in normal man. *Journal of Applied Physiology*, **33**, 204.

Epstein, M. & Saruta, T. (1973) Effects of an hypoxic hypobaric environment on renin-aldosterone in normal man. *Journal of Applied Physiology*, **34**, 49.

Frayser, R., Rennie, I. D., Gray, G. W. & Houston, C. S. (1975) Hormonal and electrolyte response to exposure to 17,500 ft. *Journal of Applied Physiology*, **38**, 635.

Gloster, J., Hasleton, P. S., Harris, P. & Heath, D. (1974) Effects of chronic hypoxia and diet on the weight and lipid content of viscera in the guinea-pig. *Environmental Physiology and Biochemistry*, **4**, 251.

Gloster, J., Heath, D. & Harris, P. (1972) The influence of diet on the effects of a reduced atmospheric pressure in the rat. *Environmental Physiology and Biochemistry*, **2**, 117.

Gursky, M. J. (1969) *Dietary survey of three Peruvian highland communities*. Master's thesis in Anthropology, Pennsylvania State University.

Hannon, J. P., Chinn, K. S. K. & Shields, J. L. (1969b) Effects of acute high altitude exposure on body fluids. *Federation Proceedings: Federation of American Societies for Experimental Biology*, **28**, 1178.

Hannon, J. P., Chinn, K. S. K. & Shields, J. L. (1972) Alterations in serum and extracellular electrolytes during high altitude exposure. *Journal of Applied Physiology*, **31**, 266.

Hannon, J. P., Klain, G. J. Sudman, D. M. & Sullivan, F. J. (1976) Nutritional aspects of high-altitude exposure in women. *American Journal of Clinical Nutrition*, **29**, 604.

Hannon, J. P., Shields, J. L. & Harris, C. W. (1966) High altitude acclimatization of women. In: *The Effects of Altitude of Physical Performance*, p.37. Chicago: The Athletic Institute.

Hannon, J. P., Shields, J. L. & Harris, C. W. (1969a) Anthropometric changes associated with high altitude acclimatization in females. *American Journal of Physical Anthropology*, **31**, 77.

Hornbein, T. F. (1962) Evaluation of iron stores as limiting high altitude polycythaemia. *Journal of Applied Physiology*, **17**, 243.

Janowski, A. H., Whitten, B. W., Shields, J. L. & Hannon, J. P. (1969) Electrolyte patterns and regulation in man during acute exposure to high altitude. *Federation Proceedings: Federation of American Societies for Experimental Biology*, **28**, 1185.

Johnson, H. L., Consolazio, C. F., Matoush, L. O. & Krzywicki, H. J. (1969) Nitrogen and mineral metabolism at altitude. *Federation Proceedings: Federation of American Societies for Experimental Biology*, **28**, 1195.

King, C. G., Bickerman, H. A., Bouvet, W., Harrer, C. J., Oyler, J. R. & Seitz, C. P. (1945) Aviation nutrition studies. 1. Effects of pre-flight meals and in-flight meals of varying composition with respect of carbohydrate, protein and fat. *Journal of Aviation Medicine*, **16**, 69.

Krzywicki, H. J., Consolazio, C. F., Matoush, L.O., Johnson, H. L. & Barnhart, R. A. (1969) Body composition changes during exposure to altitude. *Federation Proceedings: Federation of American Societies for Experimental Biology*, **28**, 1190.

Meeson, H. & Litton, M. A. (1953) Morphology of the kidney in morbus caeruleus. *Archives of Pathology*, **56**, 480.

Milledge, J. S. (1972) Arterial oxygen desaturation and intestinal absorption of xylose. *British Medical Journal*, ii, 557.

Monge, C. C., Losano, R., Marchena, C., Whittembury, J. & Torres, C. (1969) Kidney function in the high altitude native. *Federation Proceedings: Federation of American Societies for Experimental Biology*, **28**, 1199.

Naeye, R. L. (1965) Children at high altitude: pulmonary and renal abnormalities. *Circulation Research*, **16**, 33.

Nair, C. S., Malhotra, M. S. & Gopinath, P. M. (1971) Effect of altitude and cold acclimatization on the basal metabolism in man. *Aerospace Medicine*, **42**, 1056.

Nair, C. S. & Prakash, C. (1972) Effect of acclimatization to altitude and cold on body weight. *Indian Journal of Medical Research*, **60**, 712.

NAS/NRC (1974) *Recommended dietary allowances*. A report of the Food and Nutrition Board, National Research Council, Washington D.C. National Academy of Sciences.

Picón-Reátegui, E. (1978) The food and nutrition of high-altitude populations p. 219 in *The Biology of High-Altitude Peoples* Edited by P. M. Baker. Cambridge: Cambridge University Press.

Pines, A. (1978) High-altitude acclimatization and proteinuria in East Africa. *British Journal of Diseases of the Chest*, **72**, 196.

Pugh, L. G. C. E. (1962) Physiological and medical aspects of Himalayan, Scientific and Mountaineering Expedition 1960–61. *British Medical Journal*, ii, 621.

Rennie, I. D. B. & Joseph, B. J. (1970) Urinary protein excretion in climbers at high altitudes. *Lancet*, i, 1247.

Rennies, D., Frayser, R., Gray, G. and Houston, C. (1972) Urine and plasma proteins in men at 5400 m. *Journal of Applied Physiology*, **32**, 369.

Rennie, I. D. B., Lozano, R., Monge, C., Sime, F. & Whittenbury, J. (1971a) Renal oxygenation in male Peruvian natives living permanently at high altitudes. *Journal of Applied Physiology*, **30**, 450.

Rennie, D., Marticorena, E., Monge, C. & Sirotzky, L. (1971b) Urinary protein excretion in high-altitude residents. *Journal of Applied Physiology*, **31**, 257.

Schnakenberg, D. D., Krabill, L. F. & Weiser, P. C. (1971) The anorexic effect of high altitude on weight gain, nitrogen retention and body composition of rats. *Journal of Nutrition*, **101**, 789.

Schonhölzer, G. & Gross, F. (1948) Die Ammoniakausscheidung in Hochgebirge. *Helvetica Physiologica et Pharmacologica Acta*, **6**, 699.

Slater, J. D. H., Tuffley, R. E., Williams, E. S., Beresford, C. H., Sönksen, P. H., Edwards, R. H. T., Ekins, R. P. & McLaughlin, M. (1969a) Control of aldosterone secretion during acclimatization to hypoxia in man. *Clinical Science*, **37**, 327.

Slater, J. D. H., Tuffley, R. E., Williams, E. S., Edwards, R. H. T., Ekins, R. P., Sönksen, P. H., Beresford, C. H., & McLaughlin, M. (1969b) Potassium retention during the respiratory alkalosis of mild hypoxia in man: its relationship to aldosterone secretion and other metabolic changes. *Clinical Science*, **37**, 311.

Spear, G. S. (1960) Glomerular alterations in cyanotic congenital heart disease. *Bulletin of Johns Hopkins Hospital*, **106**, 347.

Surks, M. I. (1966) Metabolism of human albumin in men during acute exposure to high altitude (14 000 feet). *Journal of Clinical Investigation*, **45**, 1442.

Surks, M. I., Chinn, K. S. K. & Matoush, L. O. (1966) Alterations in body composition in man after acute exposure to high altitude. *Journal of Applied Physiology*, **21**, 1741.

Sutton, J. R., Viol, G. W., Gray, G. W., McFadden, M. & Keane, P.M. (1977) Renin, aldosterone, electrolyte, and cortisol responses to hypoxic decompression. *Journal of Applied Physiology*, **43**, 421.

Ullman, E. A. (1953) Renal water and cation excretion at moderate altitude. *Journal of Physiology*, **120**, 58.

Ward, M. (1975) *Mountain Medicine. A Clinical Study of Cold and High Altitude*. London: Crosby Lockwood Staples.

Whitten, B. K., Hannon, J. P., Klain, G. J. & Chinn, K. S. K. (1968) Effect of high altitude (14 000 ft) on nitrogenous components of human serum. *Metabolism*, **17**, 360.

Williams, E. S. (1966) Electrolyte regulation during the adaptation of humans to life at high altitude. *Proceedings of the Royal Society (B)*, **165**, 266.

22

Cold

In considering man's reaction to a low temperature at high altitude a clear distinction must be made between his short-term adjustment to extreme cold in such activities as climbing and his long-term acclimatization to the low temperature innate in permanent residence in a mountain habitat.

Acute exposure to cold

The human body loses heat to its surroundings by several physical processes (Fig. 22.1) By *convection* heat is transferred directly to the surroundings by movement of air. The body also loses heat by

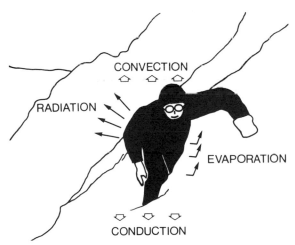

Fig. 22.1 The physical processes by which the body loses heat to its surroundings on acute exposure to cold.

radiation in which electromagnetic energy passes to the environment; direct contact with air is not necessary for this form of heat loss which could take place in a vacuum. As we have already seen in Chapter 2 the body also gains heat by this process

through absorbing solar radiation which is pronounced in the clear skies at high altitude and is exaggerated by reflection from snow. *Conduction* of heat from the body follows direct contact with the surroundings. Finally, *evaporation* of sweat leads to loss of heat in the transfer of energy required to change liquid to a vapour, the so-called latent heat of vaporization. This process is facilitated by the low humidity of high altitude.

Each of these four methods of heat loss carries implications for the implementation of simple precautions in those exposed acutely to the severe cold of high altitude. Loss of heat by convection is restricted by diminishing the area of skin exposed to the surrounding air, by the avoidance of exposure to wind which destroys the still, insulating layer of warm air around the body, and by avoiding the 'bellows effect' of clothing loosely fitting around the neck which may force out the warm insulating air. Absorption of solar radiation is increased by the wearing of dark clothes. Avoidance of direct contact of the body with snow will cut down heat loss due to conduction. So too will keeping dry, for water is a good conductor of heat and wet clothes lose their power for insulation. Furthermore, wet clothes require much heat to evaporate their water and this is drawn from the body. As we have noted in Chapter 2 humidity is low at high altitude so that sweating is easy. The water can recondense on outer clothing and this again requires body heat for its evaporation. For all these reasons wet clothes on acute exposure to cold at high altitude can be dangerous.

Long-term acclimatization to cold

Long-term acclimatization to cold is less important

than might be thought. This is because man largely meets his climatic problems by evading them. At high altitude this is achieved by the wearing of warm clothing which simulates the fur, feathers or hair of animals. Man's immediate response to cold on arriving at high altitude is to conserve heat by reducing the rate of blood flow to the skin (Fig. 22.2) with some erection of hairs leading to increased insulation. Once the skin temperature falls below a thermoneutral zone of between 25°C and 35°C, cold-sensitive receptors in the skin, and probably in the viscera too, send stimuli via the spinal cord to the thermostatic centre in the hypothalamus (Ward, 1975). This initiates stimulation of the motor centre for shivering (Fig. 22.2).

thermoneutrality, human infants and certain newborn and adult mamals can increase heat production without shivering (Brück, 1961). In rats, cessation of shivering does not cause a fall in oxygen consumption suggesting that there must be an alternative form of heat production (Davis, 1974). Rats acclimatized to cold whose muscles are paralysed by curare may actually increase their oxygen consumption in the cold.

Brown fat

The mechanisms involved in non-shivering thermogenesis have yet to be clarified at a cellular or biochemical level but the primary tissue responsible

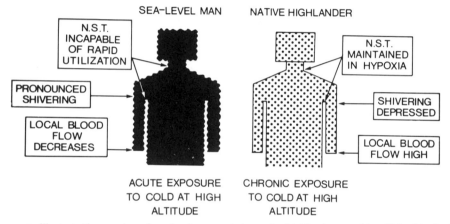

Fig. 22.2 Diagram to illustrate the reaction of sea-level man and the native highlander to cold at high altitude. N. S. T. = Non-shivering thermogenesis.

This coordinated movement of voluntary skeletal muscle under involuntary nervous control may increase heat production threefold (Ward, 1975). In this way cold stimulates heat production and aids survival in the hostile environment. Shivering is the characteristic acute response of the subject who is not acclimatized to cold. On mountains it is typical of the climber or newcomer to high altitude (Fig. 22.2).

Non-shivering thermogenesis

Long-term acclimatization to cold is largely metabolic in nature and brought about by a process called 'non-shivering thermogenesis'. It is well known that, when exposed to temperatures below

for this increase in metabolic rate has been identified as 'brown fat' (Smith and Horowitz, 1969). It is deposited in the abdomen, and in the cervical, interscapular and axillary regions. The rôle of non-shivering thermogenesis in long-term acclimatization to cold in mountainous regions is modified by the fact that the process consumes oxygen and of course the major adverse factor of life at high altitude is hypoxia. Indeed the increase in oxygen consumption normally accompanying exposure to cold is suppressed in some small mammals and their newborn under hypoxic conditions. The decreased metabolic response is thought to be due to depression of non-shivering thermogenesis presumably because there is insufficient oxygen in the ambient air to support it fully. However, it has been shown in the deer mouse (*Peromyscus*) *after acclimatization* to hypoxia heat production under cold exposure returns to normal (Roberts et al, 1969). This is accompanied by hyperplasia of brown fat. Furthermore, as we shall see below, non-shivering thermogenesis appears to operate in infant highlanders.

Acclimatization to cold in man at high altitude

The studies of Davis (1974) at Ladakh (3510 m) confirm that shivering at high altitude is characteristic of the newcomer to the mountains but not of the highlander. He studied five Indians from sea level, five Indian lowlanders who had become acclimatized to high altitude, and five Tibetan refugees native to altitudes of 3960 m to 4880 m. Measurements of rectal temperature, skin temperature, oxygen consumption and shivering rate were taken in response to a standard nude cold exposure of 2°C for 60 minutes. All ten Indian lowlanders complained of intense pain in the extremities and half of them could not complete the hour. The shivering rate was significantly increased in all of them, those who were already acclimatized to mountain conditions faring somewhat better than the sea-level subjects. In contrast the Tibetans had no difficulty in completing the test and they showed a much lower shivering rate. These five subjects were native highlanders and as we shall see in Chapter 27 there is some question as to whether they are adapted rather than acclimatized to mountain conditions. It seems likely that such highlanders employ non-shivering thermogenesis to some extent as part of their long-term acclimatization to cold, although the process seems to be aerobically-dependent and likely to be depressed somewhat by the hypoxia of high altitude as we have mentioned above (Fig. 22.2). Ward (1975) notes that aboriginal bushmen may sleep virtually naked in a cold environment without shivering so it is clear that their long-term acclimatization is largely metabolic and based on non-shivering thermogenesis but such bushmen are not exposed at the same time to hypoxia.

Thermoregulation in highland infants

In an attempt to assess thermal conditions in human infants naturally exposed to cold and hypoxia, Dufour et al (1976) studied skin temperature on three infants in the High Andes. They chose the nape of the neck as an indirect measure of heat production by brown adipose tissue. The Quechua infants were aged respectively, 8, 9 and 15 months and came from Nuñoa (4000 m). A disk thermistor probe was applied to six skin sites, (forehead, nape at the level of the seventh cervical vertebra, chest over manubrium, right lateral thigh and dorsal surfaces of the right hand and foot) every hour from 0700 to 1900 hours. They found that the nape temperatures were warmer than other skin sites, suggesting that in infant highlanders the brown adipose tissue associated with non-shivering thermogenesis is metabolically active despite the reduced oxygen availability at high altitude. Thus in naturally-acclimatized infants at high altitude non-shivering thermogenesis appears to be as important in thermoregulation as it is at sea level. There is also a local acclimatization to cold and we shall consider this now.

Local reaction to cold in highlanders

As we have already noted above, most highlanders avoid generalised cold stress through appropriate clothing and bedding. However, their faces and extremities do not escape exposure to the cold inherent in life on high mountains. Thus in Andean highlanders the rectal temperature is normal but the surface temperature of the extremities is lower (Hanna, 1970). Nevertheless, this lowered peripheral temperature is still higher than that of Caucasians exposed to the same environmental conditions. This is because the naturally acclimatized highlander (Chapter 27) maintains a high level of blood flow to the limbs (Hanna, 1970) (Fig. 22.2). The same is true of Eskimos (Elsner et al, 1960). Thus it would appear that local manifestations of acclimatization to cold are vascular rather than metabolic in nature. The precise physiological mechanisms that allow the microcirculation of the hand to be maintained under conditions which normally lead to vasoconstriction are not fully understood. Jones and his colleagues (1976) studied the effects of these changes in peripheral blood flow in the daily lives of Andean Indians at Nuñoa (4000 m). They investigated thermal responses during acute exposure to cold in washing clothing in a river, or creating a diversion of a stream. Men maintained equal temperatures in hands and feet during the exposure period but the hands of women were slightly warmer than their feet. Women consistently maintained higher temperatures in the extremities than did men (Fig. 22.3).

employ non-shivering thermogenesis which requires oxygen itself and which increases metabolic rate which enhances the severity of the hypoxia. Tissue changes in such rats acclimatized to cold but intolerant to the hypoxia of high altitude include severe depletion of glycogen in the liver, fatty changes in muscle, and lesions in the myocardium.

Fig. 22.3   Two Quechua women washing in an Andean stream at 4200 m. Native highlanders have a high local blood flow in the limbs. It has been reported that women consistently maintain higher temperatures in the extremities than men. The woman on the left is carrying a lamb in her shawl.

Local reaction to cold in lowlanders undergoing acclimatization

It has been suggested that there is a negative cross-acclimatization to cold and hypoxia in newcomers to high altitude. The heat output from the hand was studied in 16 healthy lowlanders between 22 and 28 years of age at Delhi and at 3300 m (Nair et al, 1973). When the subjects were kept warm, there was no change in the heat output for the first three weeks of acclimatization to hypoxia. However, as soon as they were exposed to cold there was a significant *reduction* in heat output.

Influence of cold on acclimatization to hypoxia

Rats already acclimatized to cold are said to be less tolerant to the hypoxia of high altitude than controls (Atland et al, 1973). This is probably because they

Cold and basal metabolic rate during early acclimatization to high altitude

An attempt has been made to distinguish the effects of cold and hypoxia on the basal metabolic rate of newcomers to high altitude (Nair et al, 1971a; Nair et al, 1971b). Twenty healthy subjects between the ages of 22 and 28 years were studied at sea level at room temperature of 25°C to 28°C and were then flown to an altitude of 3350 m where they were restudied. There half of them (Group A, Fig. 22.4) had their basal metabolic rate measured daily while they were kept for three weeks in warm clothing in a building, the temperature of which was maintained in the range of 25°C to 29°C. During this period the stimulus presumed to be operating was hypoxia alone. Subsequently they were exposed to the combined stimuli of cold and hypoxia for three weeks being kept in a tent for six hours a day, clothed in cotton shirts and pants and exposed to a cold temperature range of 6°C to 11°C. The other half (Group B, Fig. 22.4) were subjected to the same two environments in reverse order. During the same

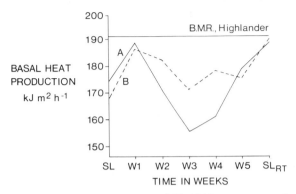

Fig. 22.4 Basal heat production (kJ/m²/h) in two groups of lowlanders studied at sea level (SL) and at 3350 m (for up to six weeks, W1, W2 etc.) A = Half of the 20 subjects subjected to hypoxia and warmth for three weeks and then to hypoxia and cold, both periods of time at 3350 m. B = The other half of the 20 subjects subjected first to hypoxia and cold for three weeks and then to hypoxia and warmth both periods of time at 3350 m. (SL_RT = Return to sea level. On exposure to high altitude there is an increase in BMR for a week which subsequently falls for two weeks only to rise again after a period of three weeks at high altitude. This is related in the text to the effects of adrenal steroids and thyroxine. The level of basal metabolic rate in the highlander in the Andes is also shown. The lowest metabolic rate during sleep can be reckoned to be of the order of 145 kJ/m²/h. The maximum steady state energy output is about ten times greater than this.

investigation the basal metabolic rate of five native highlanders between the ages of 21 and 29 years were studied.

Both groups showed an elevation of basal metabolism by the end of the first week at high altitude to approach the level found in native highlanders (Fig. 22.4). This was ascribed by Nair et al (1971a) to the stress of acute exposure to hypoxia with associated sympathetic activity and stimulation of the adrenal cortex (Chapter 25). Subsequently by the end of the third week of exposure to altitude the group initially exposed to hypoxia alone showed a significant fall in basal metabolic rate. It has been suggested that such a fall is brought about by a diminution in thyroid activity (Chapter 25). Beckwitt et al (1966) and Consolazio et al (1966) have also reported that in newcomers to high altitude there is a decline in heat production after an initial rise.

In the group initially exposed to cold, however, (B in Fig. 22.4), this secondary fall in metabolic rate does not take place to anything like the same extent. This may be because the cold stimulates thyroid activity. Once the cold stimulus was withdrawn, however, there was no fall in basal heat production and these subjects maintained a metabolic rate approaching that of the highlander (Fig. 22.4). Hence simultaneous exposure to hypoxia and cold appears to achieve a faster elevation of basal metabolism and a more rapid acclimatization.

Cold and basal metabolic rate in later stages of acclimatization

As acclimatization progresses there is a lessening of suppression of thyroid activity with an increase in basal metabolic rate (Fig. 22.4). It should be noted that the rise in basal metabolic rate occurs only two weeks after the simultaneous stimulus of cold so either there is a time lag in the response to thyroid hormone, or the rise in basal metabolic rate is not directly attributable to it. Basal heat production is significantly raised after sustained exposure to hypoxia at high altitude (Fig. 22.4), and probably should be regarded as a feature of acclimatization (Gill and Pugh, 1964; Nair et al, 1971a). Other authors have reported no change or a slight increase in basal metabolism at high altitude (Stickney and Van Liere, 1953; Houston and Riley, 1947; Grover, 1963).

Basal metabolism in the highlander

Basal metabolism in the five highlanders studied by Nair et al (1971a) was found to be somewhat elevated compared to that of lowlanders (Fig. 22.4) Studies on miners at 4540 m have supported the concept of elevation of basal metabolism to be part of long-term acclimatization (Picón-Reátegui, 1961).

Diet-induced thermogenesis

It is well known that eating stimulates the metabolic rate. This 'special dynamic action' or 'diet-induced thermogenesis' depends on the type of food eaten. For proteins it is 30 per cent, for fat 5 per cent and for a mixed diet 10 per cent. Stock and Miller (1976) reported a study of the thermic response to a meal in exercising Europeans at London and at Adi Arkai (1500 m) and Debarek (3000 m) in Simien villages in the Ethiopian highlands, and in exercising Ethiopian highlanders in the two villages named. The thermic response to a meal is expressed as the difference in the rate of energy expenditure (Δk J min^{-1}). The usual resting thermic response in resting subjects is 0.7 to 0.8 k J min^{-1} but at sea level after exercise it was 1.8 k J min^{-1}. At high altitude this metabolic response to food is absent and at 3000 m is significantly lower than zero (-0.51 k J min^{-1}). Ethiopian highlanders show that same negative response. High altitude has no significant effect on the thermic response at rest. Blatteis and Lutherer (1976) have also shown that the calorigenic response to cold after eating is reduced on exposure to high altitude in man and small mammals. This effect is not moderated by acclimatization to high altitude but is reversible immediately on descent to sea level.

The studies of Stock and Miller (1976) suggest that the degree of hypoxia at an altitude of 3000 m depresses the metabolic response to food only when combined with the additional stress of exercise. This depressed diet-induced thermogenesis conserves oxygen and may thus represent a feature of acclimatization to diminished oxygen supply. It has the disadvantage that the extra heat normally released after a meal is no longer available for maintaining body temperature in cold environments. As we have already seen, depression of cold-induced non-shivering thermogenesis is also brought about by high altitude.

Exposure to extreme cold

Exposure to extreme cold is a hazard for high altitude climbers. Frostbite is a common outcome of assaults on very high mountains and we describe this condition together with the problems of hypothermia and snow-blindness in Chapters 30 and 33.

REFERENCES

Altland, P. D., Highman, B. & Sellner, R. G. (1973) Tolerance of cold-acclimated and unacclimated rats to hypoxia at 1.7°C. *International Journal of Biometeorology*, **17**, 59.

Beckwitt, H. J., Sarks, M. I. & Chidsey, C. A. (1966) Basal metabolism, thyroid and sympathetic activity in man at high altitude. *Federation Proceedings: Federation of American Societies for Experimental Biology*, **25**, 399.

Blatteis, C. M. & Lutherer, L. O. (1976) Effect of altitude exposure on thermoregulatory response of man to cold. *Journal of Applied Physiology*, **41**, 848.

Brück, K. (1961) Temperature regulation in the new born infant. *Biologia neonatorum*, **3**, 65.

Consolazio, C. F., Matoush, L.O. & Nelson, R. A. (1966) Energy metabolism in maximum and submaximum performance at high altitude. *Federation Proceedings: Federation of American Societies for Experimental Biology*, **25**, 1380.

Davis, T. R. A. (1974) Effects of cold on animals and man. *Progress in Biometeorology*, **I**, Part IA, 215.

Dufour, D. L., Little, M. A & Brooke Thomas, R. (1976) Skin temperature at the nape in infants at high altitude. *American Journal of Physical Anthropology*, **44**, 91.

Elsner, R. W., Nelms, J. D. & Irving, L. (1960) Circulation of heat to the hands of Arctic Indians. *Journal of Applied Physiology*, **15**, 662.

Gill, M. B. & Pugh, L. G. C.E. (1964) Basal metabolism and respiration in man living at 5800 m (19 000 feet). *Journal of Applied Physiology*, **19**, 949.

Grover, R. F. (1963) Basal oxygen uptake of man at high altitude. *Journal of Applied Physiology*, **18**, 909.

Hanna, J. M. (1970) A comparison of laboratory and field studies of cold responses. *American Journal of Physical Anthropology*, **32**, 227.

Houston, C. S. & Riley, R. L. (1947) Respiratory and circulation changes during acclimatization to altitude. *American Journal of Physiology*, **149**, 565.

Jones, R. E., Little, M. A., Thomas, R. B., Hoff, C. J. & Dufour, D. L. (1976) Local cold exposure of Andean Indians during normal and simulated activities. *American Journal of Physical Anthropology*, **44**, 305.

Nair, C. S., Malhotra, M. S. & Gopinath, P. M. (1971a) Effect of altitude and cold acclimatization on the basal metabolism in man. *Aerospace Medicine*, **42**, 1056.

Nair, C. S., Malhotra, M.S., Gopinath, P. M. & Mathew, L. (1971b) Effect of acclimatization to altitude and cold on basal heart rate, blood pressure, respiration and breath-holding in man. *Aerospace Medicine*, **42**, 851.

Nair, C. S., Malhotra, M. S. & Tiwari, O. P. (1973) Heat output from the hand of men during acclimatization to altitude and cold. *International Journal of Biometeorology*, **17**, 95.

Picón-Reátegui, E. (1961) Basal metabolic rate and body composition at high altitude. *Journal of Applied Physiology*, **16**, 431.

Roberts, J. C., Hock, R. J. & Smith, R. Em. (1969) Effects of altitude on brown fat and metabolism of the deer mouse, Peromyscus. *Federation Proceedings: Federation of American Societies for Experimental Biology*, **28**, 1065.

Smith, R. Em. & Horowitz, B. A. (1969) Brown fat and thermogenesis. *Physiological Reviews*, **49**, 330.

Stickney, J. C. & & Van Liere, E. J. (1953) Acclimatization to oxygen tension. *Physiological Reviews*, **33**, 13.

Stock, M. J. & Miller, D. S. (1976) Dietary-induced thermogenesis of high and low altitudes. *Proceedings of the Royal Society of London*, **194**, 57.

Ward, M. (1975) In: *Mountain Medicine. A Clinical Study of Cold and High Altitude*. London: Crosby Lockwood Staples.

23

The skin and nails

The skin is of interest in high altitude studies because on acute exposure to hypoxia it is concerned in the redistribution of blood away from those areas where the extraction of oxygen is low to more vital areas such as the myocardium and voluntary muscles where the oxygen extraction rate is high and a large supply of the gas needed. In addition the skin of those in mountain areas is subjected to other stimuli such as ultraviolet radiation (Chapter 2).

The rôle of the skin in the redistribution of blood during early acclimatization

Changes occur in the cutaneous circulation on initial exposure to high altitude and appear to modify the flow and volume of blood in other organs. The circulation in the right hand, as representative of the cutaneous vascular bed, was studied in 7 Europeans and 32 Bolivian highlanders at low altitude in Paris (50 m) and Santa-Cruz (400 m) respectively and/or at high altitude in La Paz (3750 m) and Chorolque (5200 m) (Durand and Martineaud, 1971). Such studies show that in lowlanders cutaneous flow is reduced immediately on arrival at high altitude and then remains steady at a diminished level; there is a further significant reduction in cutaneous flow on further ascent. Durand and Martineaud (1971) interpret this reduction in cutaneous flow at high altitude as the result of arteriolar constriction but we refer to evidence below that this may be the result of hypocapnia rather than hypoxia.

As well as changes in the resistance vessels of the skin there appear to be alterations in function of the *capacitance vessels comprising capillaries and veins*. Vascular pressure-volume curves obtained by plethysmography show a decrease in compliance at high altitude and this seems to reflect an increase in the tone of the capacitance vessels rather than a large volume of blood in them. Decrease in distensibility occurs immediately on arrival at high altitude but even after a month is greater than in highlanders. The reduction in cutaneous blood flow and volume at high altitude is well established and a reduction in blood flow to the skin in sojourners at high altitude is also suggested by the studies of Elsner et al (1964), Roy et al (1968), Weil et al (1969) and Wood and Roy (1970)

The reduction in cutaneous blood flow is significant only when the skin temperature is above 33°C. Durand and Martineaud (1971) estimate that the fall in cutaneous flow redistributes less than 100 ml/min or some two per cent of the total resting cardiac output. Hence the effect of high altitude on *skin perfusion* should not play any significant physiological rôle except on unusual exposure to heat. Increase in *tone of the capacitance* vessels is the more important physiological entity at high altitude. There is a pronounced *overshoot* of this venomotor reponse in highlanders returning from a lower altitude.

In Chapter 15 we consider the possibility that a central sympathetic discharge may be of importance in the aetiology of high altitude pulmonary oedema, shifting blood from the high-resistance systemic to the low-resistance pulmonary circulation with resultant pulmonary hypertension, lung haemorrhage and malperfusion· and maldistribution of ventilation. The increased tone in the capacitance vessels and the decrease in systemic venous compliance detected by Weil et al (1969) is possibly of significance in the aetiology of high altitude pulmonary oedema. Wood and Roy (1970) found that vascular compliance of the forearm was less in patients with high altitude pulmonary oedema. Cruz et al (1976) confirmed this sustained systemic venoconstriction at high altitude and investigated whether it was related to hypoxia or hypocapnia.

Relation of sustained systemic venoconstriction at high altitude to hypoxia and hypocapnia

Cruz et al (1976) sought to determine whether the major cause of the sustained systemic venoconstriction at high altitude is hypoxia or hypocapnia. Five male subjects were exposed to a simulated altitude of 4000 to 4400 m with supplemental CO_2 (3.77 per cent) in a hypobaric chamber for four days. Similar alveolar oxygen tensions were obtained in four control subjects exposed to 3500 m to 4100 m without CO_2. A water-filled plethysmograph was used to determine forearm flow and venous compliance. Venous compliance at high altitude fell in both groups. Forearm flow and resistance were unaltered at altitude in the group with supplementation with carbon dioxide while flow decreased and resistance increased in the hypocapnic groups after 72 hours exposure. It was concluded that hypoxia is responsible for decreasing venous compliance whereas hypocapnia is responsible for increasing resistance and decreasing flow.

Triple response

When the skin is exposed to ultraviolet radiation, histamine is formed from the amino acid histidine and released from skin cells damaged by ultraviolet radiation, (Ellinger, 1963). Hence with the increased u.v. radiation in the mountains one would expect an increase in tissue histamine and an exaggeration of the triple response but in our experience this is not so. On an expedition to the Andes of Peru the triple response in one of us (DRW) was found to be diminished at high altitude (4330 m). It is conceivable that the diminution of the triple response is in some way connected with the haemodynamic redistribution of the blood away from the skin described above.

Acute effects of ultraviolet radiation

The skin reacts to the increased ultraviolet radiation of high altitude. The effects may be acute, in the case of the lowlander newly arrived in the mountains, or chronic, in the case of the highlander. The acute effect takes the form of sunburn which may produce only a vivid erythema or proceed to vesiculation and crusting. The erythema is not immediate and the time for it to appear is inversely proportional to the intensity of the radiation. After one to three days it fades and is replaced by pigmentation. These changes do not differ qualitatively from sunburn as seen at sea level but they are quantitatively greater. In severe cases there may be considerable oedema of the skin. Natives of high altitude, especially if they are peasants engaged in long hours in the fields exposed to fierce sunlight, develop pigmentation, thickening and furrowing of the skin. This is prominent on the face and on the back of the neck where it has been referred to by dermatologists as *cutis rhomboidalis nuchae*. Such thickening of the keratin layer of the skin may be regarded as a form of acclimatization protecting the skin against ultraviolet radiation.

The studies of Finsen (1899) originally demonstrated that the important component of sunlight which produces effects on the skin is ultraviolet radiation. Subsequent studies (Ellinger, 1963) demonstrated that not all wavelengths are of the same effectiveness in the production of erythema and subsequent pigmentation. There appear to be two areas which produce maximal erythemic responses, one at 380 nm and one at 303 to 297 nm. Ultraviolet of longer wavelengths produces more pigmentation than the shorter wavelengths (Peemöller, 1928). The fair-skinned are more sensitive than the dark and this in itself is likely to be a valuable asset of the mongoloid people living at high altitude, although this is a racial characteristic rather than an evolutionary adaptation.

Prolonged exposure to ultraviolet radiation

It is well known that chronic exposure of skin to sunlight irrespective of the altitude leads to histological changes in both the dermis and epidermis. Degenerative changes occur in the upper third of the dermis appearing as a basophilic degeneration of collagen to form an amorphous granular material in sections stained with haematoxylin and eosin. Such affected areas are found to consist of thick interwoven fibres of what is termed 'elastoid' on applying elastic tissue stains. The nature of this substance and the question as to whether it is derived from collagen or elastic tissue are still unsettled (Lever, 1967).

High altitude dermatopathy

Eguren (1972) believes that the histological changes induced in the skin of the highlander are so characteristic that he terms them 'high altitude dermatopathy'. He studied biopsy specimens from the lobe of the ear from 100 native inhabitants of Condoroma (4850 m) in southern Peru. Most of these subjects were young miners. He also examined biopsy specimens of skin from 50 citizens of Arequipa (2370 m) and, for comparison, from 50 natives of Camaná (20 m). From these studies it would appear that hyperkeratosis is characteristic of the skin of the native highlander. Horny keratin plugs may be found around hair follicles and sweat glands. There is some atrophy of the underlying Malpighian layer and sometimes there is focal dyskeratosis. Pronounced atrophy of this layer is associated with exaggerated hyperkeratosis. Characteristically there is loss of the interpapillary epithelial evaginations so that the basal layer tends to even out horizontally. The skin shows hyperpigmentation with hyperplasia of melanocytes. There is leakage of melanin both into the overlying cells of the Malpighian layer and into the underlying dermis. The phagocytosis of melanin granules in the dermis by a layer of enlarged, overloaded chromatophores in effect increases the thickness of the pigmented layer of the skin. The basal cells commonly show hydropic degeneration with local thickenings and disruption.

Within the dermis there are areas of solar degeneration which we have already described above as occurring at sea level. Eguren (1972) also described dilatation of blood vessels and lymphatics in the dermis. Accumulations of lymphocytes occur around these dilated blood vessels and also around the horny keratin plugs. There is some evidence that these histological changes are partially reversible after several years' residence at sea level.

Factors involved in the pathogenesis of high altitude dermatopathy

There is a close resemblance between Eguren's 'high altitude dermatopathy' and the familiar solar degeneration and keratosis which occur at sea level so there seems little doubt that the histological changes observed in both conditions are due to excessive exposure to ultraviolet radiation. Indeed solar degeneration and keratosis and 'high altitude dermatopathy' may be one and the same thing. Electron microscopic studies by Everett et al (1971) suggest that the lesions due to repeated exposure to ultraviolet radiation start in the epidermis. Early ultrastructural changes include loss of cellular cohesion, increased formation of keratin, increased size of melanosomes and overloading of cells by melanin. Actinic elastosis in the dermis appears to be due to a direct action of ultraviolet radiation on dermal collagen and elastic tissue. Eguren (1972) thinks that the hyperpigmentation and leakage of melanin into the dermis, where it is taken up by melanophores, constitute dermal acclimatization to high altitude. He believes the increased thickness and density of melanin form an increased barrier to ultraviolet radiation. A subsidiary protection may be an increased layer of keratin.

Skin tumours

Prolonged exposure to ultraviolet radiation will lead to changes in the epidermis consisting of hyperkeratosis and parakeratosis which may advance to papillomatosis and after many years to squamous carcinoma. Ultraviolet radiation will produce skin tumours at sea level, if the intensity of radiation is high enough. Thus in Australia where there is a great amount of sunlight and a predominantly white population, skin tumours account for more than half of all diagnosed cancers (Mackie and Mackie, 1963). Furthermore these tumours occur predominantly on parts of the body maximally exposed to sunlight. Mackie and Mackie (1963) suggest that three categories of subject can be recognised dependent on the melanin content of the skin. Type one has little or no melanin, type two has well developed layers of pigment in the basal layer and in type three there is a large amount of melanin in the epidermis. Cancer is most likely to occur in fair-skinned people so the pigmented highlander, whose physical features we describe in Chapter 4 is less likely to develop a malignant tumour of the skin in spite of the increased amount of ultraviolet radiation to which he is exposed (Chapter 2). Hence it is not surprising that Krüger and Arias-Stella (1964) have come to the conclusion that squamous carcinoma of the skin is not unduly common in highlanders. They think an added protection is the Andean custom of wearing wide-brimmed hats from youth which may shield the wearer from ultraviolet and other forms of radiation referred to earlier. They carried out a study of the twenty malignant neoplasms that occurred in Cerro de Pasco during the years 1961 and 1962 when 152 biopsy and surgical specimens and 185 necropsies became available for study. Included were three malignant melanomas of the skin. One occurred on the upper lip of a woman of Yanahuanca (3540 m), the others on the toes in a man from Milpo (3990 m), and in a native of Yanahuanca. A basal cell carcinoma was found on the right lower eyelid of an aged mestizo male from Ninacaca (4110 m). They did not find a single case of squamous cell carcinoma.

Cancer at high altitude

It would seem appropriate at this point to consider the incidence of cancer in general at high altitude. As we have noted in Chapter 8 the benign tumour of the carotid body, the chemodectoma, appears to be commoner in the Andes. This is a special case, however, for this tumour originates in glomic tissue which responds to hypoxia by hyperplasia. Indeed, as we have noted in Chapter 8, some authors regard the 'chemodectoma' of high altitude merely as an unusually exuberant hyperplasia of the carotid body in response to the hypoxic stimulus. Pathologists with personal experience of the practice of morbid anatomy at high altitude tend to remain sceptical about anecdotal evidence on the incidence of various forms of neoplastic disease at high altitude which is supported neither by histopathological studies nor by statistical analysis. As a result of their study of neoplastic disease at Cerro de Pasco referred to above Krüger and Arias-Stella (1964) came to the conclusion that the physical environment of mountainous regions and the genetic constitution and nutritional and social habits of the highlanders do not appear to increase the incidence of tumours at high altitude.

Burton (1975) goes further in believing that there is a statistically significant *negative* correlation of

cancer with altitude for age groups above 60 years. His opinion is based on statistical data derived on the registered incidence of cancer in five continents by the World Health Organization and other bodies. Burton found a significant negative regression of total cancer rates with altitude above 500 m. This rate appeared to decrease on the average by about 100 cases per 100 000 population for each 1000 m of altitude. According to him at very high altitudes the cancer rate may be as low as one third of that seen at sea level.

Clearly caution has to be exercised in interpreting such statistical data because the average age of subjects living at high altitude is lower than that of populations at sea level. Nevertheless, Burton stresses his awareness of this objection and relies entirely on age-specific rates to support his views. When he compared the incidence of cancer in the six regions of highest altitude with those in the six lowest areas, he found that the significant negative correlation occurred only in subject over 60 years of age. The protective effect of high altitude was aparently non-specific for it applied to all classes of tumour. Another possible pitfall in the interpretation of data on tumour registration is that the apparent lower incidence of cancer at high altitude could merely be an expression of inadequate diagnosis of tumours in isolated mountain communities. However, some areas at moderate altitude such as Johannesburg and Alberta practise sophisticated medicine.

Burton (1975) speculates that the postulated effects of high altitude on the incidence of cancer may be related to disturbances of intraceullular pH occurring secondary to the process of acclimatization. He notes that Eagle (1973) has studied the rate of protein sythesis at various levels of pH with a maximal growth rate at a particular pH, characteristic for a particular cell, with a falling off either side of the optimum.

Splinter haemorrhages

Earlier in this book (Chapters 4 and 16) we have noted that splinter haemorrhages occur in the finger nails in healthy highlanders (Fig. 23.1) and in an even more pronounced way in sufferers from Monge's disease (Fig. 23.2). It is of interest that they have also been reported as occurring in a florid

 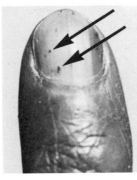

Fig. 23.1 (left) Finger tip of healthy male Quechua indian, aged 23 years, showing a small haemorrhage in the nail (arrow). It is situated at the line of separation of the nail-bed from the underlying nail-plate and this resembles the nail-haemorrhages of healthy lowlanders shown in Fig. 23.3. Such bleeding into the finger-nails of healthy highlanders is probably traumatic in origin.

Fig. 23.2 (right) Finger tip of a male Quechua indian, aged 35 years, with Monge's disease. There are haemorrhages (arrows) in the substance of the nail-plate and they are scattered from the base to the edge of the nail. This distribution appears to be characteristic of extreme hypoxaemia and the resultant greatly elevated haematocrit.

fashion in climbers at an extreme altitude by Rennie (1974).

Splinter haemorrhages beneath the nails constitute a classical clinical sign which was originally associated with a diagnosis of subacute bacterial endocarditis over half a century ago by Horder (1920). Two alternative hypotheses have been advanced to explain their appearance. Some have ascribed the haemorrhages to increased capillary fragility (Lewis, 1942; White, 1947); others believe them to be embolic in origin (Bramwell and King, 1942). By 1956 Wood had come to the conclusion that splinter haemorrhages are not diagnostic of bacterial endocarditis and this was confirmed in subsequent studies by Platts and Greaves (1958) and Gross and Tall (1963). Haemorrhages in the nails occur commonly in both lowlanders and highlanders.

Haemorrhages in the nails of lowlanders

Platts and Greaves (1958) examined the finger-nails of 429 patients and normal subjects and 35 cadavers. They found haemorrhages in many diseases including uncomplicated mitral stenosis, systemic

arterial emboli, rheumatic fever without mitral stenosis, malignant neoplasms, 'pulmonary disease' and in many other conditions. Not only did they find nail-haemorrhages in a wide range of diseases other than subacute bacterial endocarditis but also in several of their healthy acquaintances. We have been able to confirm this in several members of the staff of our laboratory mostly healthy young technicians and secretaries. One of the authors (DH) frequently has nail-haemorrhages and we were able to follow their site of origin and fate over a period of six months (Heath and Williams, 1978). They are characteristically linear in shape and occur at the line of separation of the nail from the underlying nail bed (Fig. 23.3). They form in the

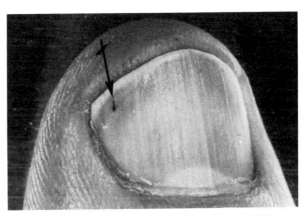

Fig. 23.3 Nail-haemorrhage in one of the authors (DH) at sea level. It is typical of the lowlander in being linear in shape and occurring at the line of separation of the nail-plate from the underlying nail-bed (arrow). This type of haemorrhage appears to be related to minor trauma.

deep part of the nail substance and pass distally with the nail so that eventually they become available for histological examination in nail-parings. As they pass distally they become paler and move to a more superficial level in the nail-plate. In this way they are comparable to the small haemorrhages which occur in the heel of children and which are known as 'black heel'. As a consequence nail-haemorrhages in healthy lowlanders are largely confined to the crescent of nail plate distal to the line of separation from the underlying nail-bed.

Nail parings which contain splinter haemorrhages give the colour reactions of altered blood such as leuco-aniline blue peroxidase. We have studied the histological appearances of nail haemorrhages in methacrylate-embedded sections (Heath and Williams, 1978). They consist of a homogenous mass of altered blood embedded in a layer of squamous cells adherent to the under surface of the cut nail. The haemorrhages seem to arise from capillaries in dermal papillae that could easily be damaged from trauma.

In view of the characteristic line of origin of these haemorrhages at the separation of nail-plate from underlying nail-bed we think it likely that they are traumatic in origin. This impression is strengthened by the fact that nail-haemorrhages are commoner in the right hand in right-handed people and in the first three fingers which are used more extensively than the fourth and fifth (Heath and Williams, 1978). Platts and Greaves (1958) commented on the finding of haemorrhages in the nails of an emphysematous diabetic who was employed in making packing cases. Gross and Tall (1963) also believed the haemorrhages to be traumatic in origin. They examined 267 hospital in-patients for splinter haemorrhages and found them in 19.1 per cent of subjects. They were found to occur more often in those patients whose occupations or activities exposed their hands to frequent trauma. They were also found more often in those patients who had recently been admitted to hospital, their incidence tending to decrease with increasing length of duration in hospital. They suggested that they may be traumatic in origin. Other details which were recorded such as age, sex, temperature, systemic blood-pressure, capillary fragility and haematuria, proved to be irrelevant.

Haemorrhages in the nails of highlanders

Haemorrhages are as common in the nails of highlanders as they are in lowlanders and moreover their distribution is the same, occurring as linear streaks at the line of separation of the nail-plate from the underlying nail-bed (Fig. 23.1). We recently carried out a study of the incidence and distribution of nail-haemorrhages in subjects born and living at Lima (150 m), Cuzco (3400 m), and La Raya (4200 m). The incidence of nail-haemorrhages at these three elevations was 34.9, 43.0, and 57.9 per cent. In view of their situation it seems very likely that nail-haemorrhages in healthy highlanders are traumatic in origin.

In contrast, in subjects with Monge's disease the nail-haemorrhages tend to have a different distribution, occurring throughout the nail, even to its base (Fig. 23.2). We think it likely that they are related to hypoxaemia and a high haematocrit leading to sludging of erythrocytes in small capillaries and associated bleeding. Haemorrhages of the same type and distribution occur in the nails of those patients with cyanotic congenital heart disease who have a high haematocrit but not in those who have recently had a therapeutic venesection. This seems to confirm the association of nail-haemorrhages of this distribution with an elevated haematocrit.

Nail-haemorrhages in lowlanders climbing at extreme altitude

Rennie (1974) reported that on an expedition to Dhaulagiri, Nepal (8170 m), the sixth highest mountain in the world, he developed numerous splinter haemorrhages in his fingernails. They appeared spontaneously during a restful evening at 5880 m. The following morning he had nearly 50 thin red longitudinal streaks under his finger and thumb nails. They occurred near the distal part of the nail bed. In the previous five weeks he had been engaged in manual work transporting loads of 18 to 67 kg from 760 m to 5880 m. He had no purpura or ecchymoses. Eight fellow climbers had no splinter haemorrhages, three had one each, two showed two each, one had three and one four. It seems to us that climbing at high altitude provides an appropriate combination of a rising haemoglobin level and repeated trauma to the hands to produce nail-haemorrhages.

REFERENCES

Bramwell, C. & King, J. T. (1942) In: *Principles and Practice of Cardiology*, p. 386. London: Oxford University Press.
Burton, A. C. (1975) Cancer and altitude. Does intracellular pH regulate cell division? *European Journal of Cancer*, 11, 365.
Cruz, J. C., Grover, R. F., Reeves, J. T., Maher, J. T.,

Cymerman, A. & Denniston, J. C. (1976) Sustained venoconstriction in man supplemented with CO_2 at high altitude. *Journal of Applied Physiology*, 40, 96.
Durand, J. & Martineaud, J. P. (1971) Resistance and capacitance vessels of the skin in permanent and temporary

residents at high altitude. In: *High Altitude Physiology: Cardiac and Respiratory Aspects*, p.159. Ciba Foundation Symposium. Edited by R. Porter & J. Knight. Edinburgh: Churchill Livingstone.

Eagle, H. (1973) The effect of environmental pH on the growth of normal and malignant cells. *Journal of Cell Physiology*, **82**, 1.

Eguren, V. L. (1972) *Morfologia histologica de la piel expuesta en grandes altitudes. Dermatopatio de altura.* Doctoral Thesis. Universidad Nacional Mayor de San Marcos, Lima.

Ellinger, F. P. (1963) Biological effects of ultraviolet radiation. In: *Medical Biometeorology*, p. 338. Edited by S. W. Tromp. Amsterdam: Elsevier Publishing Company.

Elsner, R. W., Bolstad, A. & Forno, E. (1964) In: *The Physiological Effects of High Altitude*, p. 217. Edited by W. H. Weihe. Oxford: Pergamon Press.

Everett, M. A., Nordquist, J. & Olson, R. (1971) Ultrastructure of human epidermis following chronic sun exposure. *British Journal of Dermatology*, **84**, 248.

Finsen, N. R. (1899) *Über die Bedeutung der chemischen Strahlen des Lichtes für die Medizin und Biologie.* Leipzig: F. C. W. Vogel.

Gross, N. J. & Tall, R. (1963) Clinical significance of splinter haemorrhages. *British Medical Journal*, **ii**, 1496.

Heath, D. & Williams, D. R. (1978) Nail haemorrhages. *British Heart Journal*, **40**, 1300.

Horder, T. (1920) Discussion on the clinical significance and course of subacute bacterial endocarditis. *British Medical Journal*, **ii**, 301.

Krüger, H. & Arias-Stella, J. (1964) Malignant tumours in high altitude people. *Cancer*, **17**, 1340.

Lever, W. F. (1967) In: *Histopathology of the Skin*, 4th Edition, pp. 265 & 497. London: Pitman Medical.

Lewis, T. (1942) In: *Diseases of the Heart*, 3rd Edition, p. 89. London: Macmillan.

Mackie, B. S. & Mackie, I. C. (1963) Cancer of the skin. In: *Medical Biometeorology*, p. 481. Edited by S. W. Tromp. Amsterdam: Elsevier Publishing Company.

Peemöller, F. (1928) Die physiologische Bedeutung des Pigments. *Strahlentherapie*, **28**, 168.

Platts, M. M. & Greaves, M. S. (1958) Splinter haemorrhages. *British Medical Journal*, **ii**, 143.

Rennie, D. (1974) Splinter haemorrhages at high altitude. *Journal of the American Medical Association*, **228,** 974.

Roy, S. B., Guleria, J. S., Khanna, P. K., Talwar, J. R., Manchanda, S. C., Pande, J. N., Kaushik, V. S., Subba, P. S. & Wood, J. E. (1968) Intermediate circulatory response to high altitude hypoxia in man. *Nature*, **217**, 1177.

Weil, J. V., Battock, D. J., Grover, R. F. & Chidsey, C. A. (1969) Venoconstriction in man upon ascent to high altitude: studies on potential mechanisms. *Federation Proceedings: Federation of American Societies for Experimental Biology*, **28**, 1160.

White, P. D. (1947) In: *Heart Disease*, 3rd edition, p. 360. New York, Macmillan.

Wood, P. H. (1956) *Diseases of the Heart and Circulation*, 2nd Edition, p. 648. London: Eyre and Spottiswoode.

Wood, J. E. & Roy, S. B. (1970) The relationship of peripheral venomotor responses to high altitude pulmonary oedema in man. *American Journal of the Medical Sciences*, **259**, 56.

24

Infection at high altitude

Bacteria in mountain air

The number of bacteria in ambient air decreases with altitude. They are relatively rare even at elevations as moderate as 500 m but can be found even at such extreme altitudes of 12 000 m as a result of upwinds (Rippel and Baldes, 1952). A study on the Jungfraujoch (3450 m) showed that despite the large number of tourists very few bacteria were present in the air, the species represented being *Bacillus brevis, B. mesentericus, B. subtilis*, and diplococci (Keck and Buchmeiser, 1964). The diminished barometric pressure does not appear to affect the growth of bacteria (Keck and Buchmeiser, 1964) but direct exposure to the increased solar radiation of the mountains severely inhibits the growth of some bacteria. Thus *Staphylococcus aureus* is greatly inhibited but *Escherichia coli* is much more resistant. The inhibitory effect of solar radiation is due to its ultraviolet components (Nusshag, 1954).

Microbial flora at high altitude

The microbial flora in man is uninfluenced by the high altitude environment. The microbial flora of the skin, throat and faeces of 16 subjects at 4300 m has been studied by Weiser and his colleagues (1969). In the throat were *Mycoplasma* species and harmless commensals and on the skin were *Staphylococcus, Candida, Aspergillus*, diphtheroids and α and β haemolytic *Streptococci*. Similar microorganisms are found at sea level. The average total organisms per g of faeces at 4300 m was 63.6 million as contrasted to 58 million at sea level. There are thus no qualitative or quantitative changes in microbial flora at high altitude.

Immune response in animals

The antigen-antibody response mechanism is sensitive to changes in the environment including exposure to high altitude (Dryer, 1966). The immune response of actively immunized animals is increased at high altitude. Haemagglutination titres and levels of circulating antibody demonstrated by quantitative precipitation analysis, after primary and secondary immunization of rabbits with 10 mg bovine serum albumin, proved to be greater at 3230 m than at 1610 m (Trapani, 1966). This author points out that the effect of high altitude is comparable to that of treatment with thyroxine and opposite to the effects of exposure to cold, thyroidectomy or adrenalectomy. The depressant effect of cold is of interest for animals acclimatized to the low temperature of high altitude have more brown fat (Chapter 22) and it is possible that this biochemically active tissue may use up metabolites which are thus no longer available for antibody synthesis (Dryer, 1966).

The immune response has been reported as increased in both indigenous high altitude and acclimatized animals. Thus the immune response of mountain-bred guinea pigs to bovine serum albumin proved to be greater than that of a colony bred at 1610 m (Trapani, 1966). Trapani and Jordan (1962) found that after a single intravenous injection of bovine serum albumin into rabbits which were acclimatized for 30 days at 4310 m precipitins could be detected at 4 to 7 days and at higher level compared to 10 to 14 days in sea-level rabbits. At the end of 10 weeks, when circulating antibody had reached low levels, the acclimatized rabbits were given a second injection of albumin, to determine if their stay at high altitude had any lasting effect on immune response. The time of the anamnestic response proved to be similar to that of the controls but the animals previously exposed to high altitude had higher precipitin levels even though they had been living at the lower altitude for approximately six weeks before challenge.

In the same way Altland et al (1963) found there was a significant increase in the anti-sheep haemolysin titre in polycythaemic male rabbits kept for five weeks at 4880 m 8 to 10 days after their primary immunization with sheep erythrocytes. This increased capacity for producing antibody was lost two months after exposure to high altitude. Splenectomy lowered but did not prevent the significant increase in haemolysin production and this led Altland et al (1963) to postulate that it was related to a hyperplasia of the reticuloendothelial system induced by exposure to high altitude. Antibody decay rates show no change in animals acclimatized to high altitude.

Immune response in animals to viruses and bacteria at high altitude

Animals subjected to simulated high altitude appear to be more resistant to infection by viruses but less resistant to bacteria (Berry and Mitchell, 1953; Berry et al, 1955; Berry and Smythe, 1964). When mice are infected intranasally with pneumotropic PR8 virus suspension at altitudes of 1610 m and 3230 m survival is greater at the higher altitude (Trapani, 1966). Titres of anti-virus antibody are greater in guinea pigs acclimatized to high altitude (Trapani, 1966). The rate of proliferation of influenza virus in the intact hypoxic mouse is low compared with the normally-oxygenated one (Kalter and Tepperman, 1952). The intricate syntheses involved in virus reproduction require oxidative mechanisms closely related to those of the host cell as a source of energy and it seems that a sudden impairment in their efficiency interferes with the ability of virus particles to reproduce.

In contrast at first on exposure to high altitude tubercle bacilli proliferate in the lungs and spleen of guinea pigs more rapidly than in sea-level controls. However, after a month at high altitude the growth of tubercle bacilli falls off rapidly and this may be because the hypoxia of mountainous areas depresses the growth of *Mycobacterium tuberculosis* (Trapani,

1966). High altitude does not diminish the protective effect of BCG.

Immunoglobulins and resistance to viral and bacterial disease in man at high altitude

Increased resistance to viral infection is said to be related to an increase in the synthesis of the immunoglobulin IgG. The graphs included in the paper of Singh et al (1977), which reports a decreased incidence of viral disease in Indian soldiers stationed for long periods in the Himalayas (see below) (Fig. 24.1), certainly suggest higher levels of IgG on arrival in mountainous areas consistent with increased anti-viral activity at high altitude.

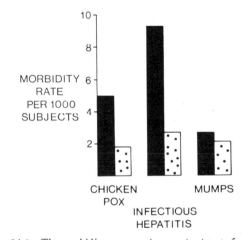

Fig. 24.1 The morbidity rate per thousand subjects from three viral diseases among 20 000 soldiers stationed between altitudes of 3692 and 5538 m (stippled columns) and 130 700 stationed on the plains (filled columns). (Based on data from Singh et al, 1977).

The levels of IgM are not nearly so raised in proportion and since this class of immunoglobulin is generally regarded as more implicated in defence against bacterial invasion this could be held to be consistent with the view that there would be a decrease in resistance to bacteria at high altitude. Trapani (1969) in fact believed there to be a decreased resistance to bacteria and an increased resistance to viruses at high altitude. However, this does not accord with the experiences of Singh and his colleagues (1977) who found, as we shall see below, that bacterial disease was less common in soldiers stationed in mountainous areas (Fig. 24.2).

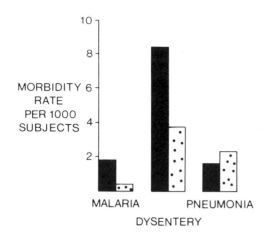

Fig. 24.2 The morbidity rate per thousand subjects from malaria and two bacterial diseases among 20 000 soldiers stationed between altitudes of 3692 and 5538 m (stippled columns) and 130 700 stationed on the plains (filled columns). (Based on data from Singh et al, 1977).

Levels of IgA are raised at high altitude (Chohan et al, 1975). Since this immunoglobulin is dominantly present in secretions such as colostrum, saliva, tears, nasal mucus, tracheobronchial secretions and the lumen of the small intestine, it is likely that increased IgA may assume a protective rôle at mucous surfaces in those in mountainous regions.

In high altitude pulmonary oedema all three immunoglobulins are raised further. According to Singh and Chohan (1972) IgG and IgM may get absorbed on to the surface of platelets altering their mobility and increasing aggregation. IgG may promote the release of platelet factor 3 and ADP, the implications of which are referred to in Chapter 10.

The low incidence of infectious disease in soldiers in the Himalayas

Indian troops were stationed for prolonged periods between altitudes of 3692 and 5538 m after the border incident with China in 1962. This enabled studies of the incidence of common diseases, including bacterial and viral disease, to be made on 20 000 men stationed at high altitude and 130 700 men stationed on the plains (Singh et al, 1977). There was a lower incidence at high altitude of many infections of bacterial, viral and protozoal origin. Only amoebic hepatitis and lobar pneumonia were commoner in the mountainous areas.

Although the incidence of infectious hepatitis is

lower in the soldiers in the Himalayas, the incidence of hepatitis-associated surface antigen is higher among native highlanders than sea-level residents. At high altitude, because of the low environmental temperature, there is no active transmission of malaria so that cases in mountainous areas are either relapse cases or those which have incubatory disease at the time of induction to high altitude. Intravascular coagulation which occurs during the malarial infection probably protects the parasites in various organs and is responsible for severity of disease, resistance to therapy and relapses. The low incidence of relapse at high altitude may be due to increased fibrinolytic activity at high altitude, which prevents intravascular coagulation, thus making the malarial parasites more vulnerable to the increased immune response at high altitude.

Effect of inhibitors or intermediates of tricarboxylic acid cycle

Berry (1956) studied the comparative susceptibility of mice at sea level and at a simulated altitude of 6100 m for three weeks to infection with *S. typhimurium* and *D. pneumoniae* after they had been injected with inhibitors or intermediates of the tricarboxylic acid cycle. The mice at simulated high altitude infected with *S. typhimurium* survived for a significantly shorter period when they were injected with oxaloacetate or citrate. The high altitude mice injected with *D. pneumoniae* survived for shorter periods when injected with malonate. Clearly mice undergoing acclimatization undergo changes in metabolism which alter their resistance to infection.

Daniel Carrion's experiment

There is one bacterial infection so characteristic of the Andes and of such historical interest as to demand more detailed description. It occurs in highlanders of Peru, Colombia and Ecuador and it has become associated for all time with the name of Daniel Alcides Carrion, medical student (Fig. 24.3). As a boy Daniel made frequent trips through the Peruvian mountains with his uncle, Manuel Ungaro, going to school in Lima from his home in Cerro de Pasco (4330 m). During these journeys he saw people with wart-like nodules on the skin and mucous membranes. It was well known that the

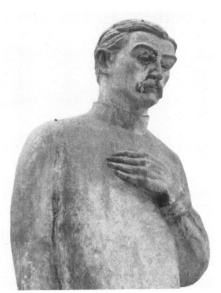

Fig. 24.3 Daniel Alcides Carrion (1857–1885), from a memorial outside the Dos de Mayo Hospital in Lima, Peru. (We are indebted to Miss Lesley Taggart for this picture).

disease was restricted to the steep Peruvian valleys as it had been for centuries (Sutton, 1971). Indeed modern research has revealed a case of this disease in a mummy of the Tiahuanoco culture of Southern Peru, the subject having apparently been sacrificed and some of his organs used in rituals (Allison et al, 1974). When Daniel became a medical student, he became vaguely aware that the cutaneous eruption of this mountain disease was preceded by fever, anaemia and joint pains and he became obsessed with the need for such early diagnosis of the warts so that treatment could be effective (Schultz, 1968). It seemed to him that the most direct way of determining the early signs and symptoms of Peruvian warts was to inject fragments of wart into himself and study the effects. On the morning of August 27th, 1885, he carried out this procedure in the Nuestra Señora de las Mercedes ward of the Dos de Mayo Hospital in Lima (Schultz, 1968). We are very familiar with this hospital with its central tropical garden with religious statues from which radiate the wards. In bed number 5 was a 14-year-old boy, Carmen Paredes, with a wart on his right eyebrow. The medical student tried to inoculate himself with material from this wart but he was not dextrous enough to complete the task on his own so a young doctor, Dr Chavez, completed the

task. Daniel Carrion was delighted with the success of this minor operation and began to keep a diary of any symptoms that might appear.

On September 17th he began to feel unwell and had pain in his left ankle. Two days later he became ill with fever, chattering of the teeth, abdominal cramp, and pains in all the bones and joints of the body (Schultz, 1968). His urine became dark and scanty and he became jaundiced. By September 26th he was too weak to record his symptoms any longer so his fellow medical students took over the task so that the project for his thesis should not be disturbed. At this stage Daniel Carrion felt ill but had the satisfaction, as he thought, that he was at last discovering the symptoms that precede the eruption of Peruvian warts. Soon, however, he became dangerously ill. A systolic murmur became audible over the base of the heart and could be felt over the carotid arteries. The pulse became rapid, he developed abdominal pain and diarrhoea, and vomited frequently. The onset of these serious symptoms led him to recall the events of his boyhood. He remembered that, when the workers had striven to drive a railway through the Andes at La Oroya in 1871, many of them had died from fever and anaemia. This fatal condition was unknown to the medical profession and so, in deference to the route, was designated 'Oroya fever'. It began to dawn on Daniel that as a consequence of his inoculation he was developing not the symptoms of the invasive stage of the verruga but the dreaded Oroya fever. By October 3rd he was still in his lodgings and was visited by a Dr Flores who found that his total red cell count had fallen to just over $1.0 \times 10^{12}/l$. He was sent to hospital for a blood tranfusion and became delirious. On October 5th, 39 days after inoculation and after 18 days of illness he became comatose. Looking at one of his friends, he muttered 'Enrique, c'est fini' and died at 11.30 p.m. He was just 28 years of age (Fig. 24.3).

Daniel Carrion thus demonstrated that verruga peruana and Oroya fever are but phases of one and the same disease. A historic medical advance had been made but Lima was appalled. The experiment was described by a prominent Peruvian physician as a 'horrible act by a naïve young man that disgraces the profession'. The necropsy was a scene of terrible confusion for the body had already begun to decompose. Professor Villar made an eloquent plea in defence of martyrs and initial charges against Dr Chavez were dropped. The remains of the dead medical student were carried on the shoulders of his classmates, past his University to the cemetery (Schultz, 1968).

The small pleomorphic microorganism that causes this severe, usually fatal, febrile anaemia, known as Oroya fever, was first seen by Barton in 1909 in or on red cells. Early in the disease the organism presents in a bacillary form and hence it has come to be designated *Bartonella bacilliformis*. Members of the 1937 Harvard Expedition to Peru were able to confirm that this organism is the cause of both Oroya fever and verruga peruana and the results of their observations on various aspects of the disease by Pinkerton, Weinman and Hertig appeared in five consecutive papers included in the bibliography at the end of this chapter. Thus Pinkerton and Weinman (1937a) were able to grow *B. bacilliformis* in leptospira medium from blood or tissues from patients with Oroya fever or from the cutaneous lesions of verruga peruana and found it behaved as a facultative intracellular parasite when cultivated *in vitro* with growing or surviving guinea pig mesenchymal cells. The different forms of Carrion's disease as seen in man were reproduced in rhesus monkeys by Weinman and Pinkerton (1937a).

The various forms of Carrion's disease

Carrion's disease occurs along the Andean range in Colombia, Ecuador, Peru and Chile. Clinically it presents incubative, invasive, pre-eruptive and eruptive stages (Ricketts, 1949). The incubation period varies from three to 14 weeks. After that time the causative organism, *B. bacilliformis*, parasitises the red blood cells (Fig. 24.4) leading to a rapidly progressive febrile anaemia. The anaemia is macrocytic and frequently hypochromic. There is no spherocytosis, and the saline fragility of the erythrocytes is normal (Ricketts, 1949). Nearly every erythrocyte is involved so that the red cell count is commonly below $1.0 \times 10^{12}/l$. The anaemia does not respond to liver extract.

The pathological lesions in fatal cases of Oroya Fever are those of a severe infectious haemolytic anaemia. The skin and conjunctiva may be icteric.

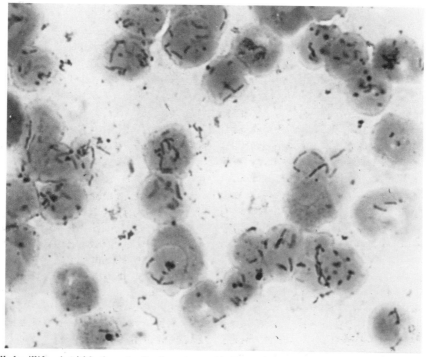

Fig. 24.4 *Bartonella bacilliformis* within the red cells of a patient with Oroya fever. Both bacilliform and coccal forms are seen. Adjacent rods give a criss-cross appearance in places. (Blood film. Giemsa's stain, × 1500) (This figure and Figs. 24.5 and 24.6 kindly provided by Dr D. Weinman).

There is red hyperplasia of the bone marrow. The liver shows fatty infiltration. The reticulo-endothelial cells of the body, in lymph nodes, spleen, bone marrow and hepatic Küpffer cells are crowded with *B. bacilliformis* which form spherical clusters in the cytoplasm (Pinkerton and Weinman, 1937b). They are also to be found in the adrenal (Fig. 24.5) pancreas, thyroid, testicle and kidney (Pinkerton and Weinman, 1937b) and there may be degeneration of renal tubules. More than 90 per cent

extremities of *B. bacilliformis* appear thicker and darker in colour when stained reddish-violet by Giemsa's methods. Up to 90 per cent of erythrocytes may be infected and as many as six organisms may be found in one red cell. Often the organisms are distributed in rows of three or more suggesting prior segmentation. Frequently adjacent rods are at an angle giving a V, Y or criss-cross appearance (Weinman, 1944). This organism is actively motile, possessing flagella which are always unipolar and

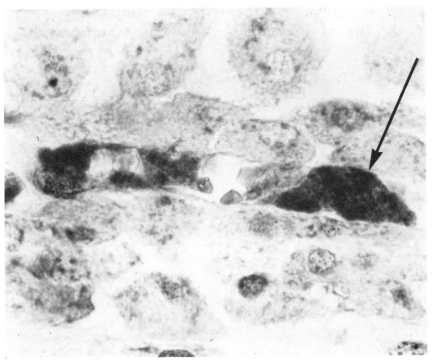

Fig. 24.5 Clusters of *B. bacilliformis* (arrow) within endothelial cells lining small capillaries in the adrenal cortex. (Wolbach's modification of Giemsa's stain after Regaud fixation, × 1200).

of the cells contain organisms (Pinkerton and Weinman, 1937b). The laboratory diagnosis is readily made in the anaemic stage by finding the organism in the red cells in Giemsa-stained blood smears.

   B. bacilliformis is polymorphous in human blood. Usually it is found in red cells as minute Gram-negative rods, 0.3 to 2.5 μm in length and 0.25 to 0.50 μm in width (Weinman, 1944) (Fig. 24.4). However, coccal forms 0.75 μm in diameter, also occur especially when the organism is about to disappear from the blood. At such times they may also show hour-glass and pear-shaped forms. The

which vary in number from one to four and in length from 3 to 10 μm. No spores or capsules are seen. Cultures are obtained in semi-solid leptospira medium (Noguchi and Battistini, 1926) and on solid blood media but not in ordinary broth or agar. The optimum temperature is 25 to 28°C. The organism can survive in blood or cultures for many weeks, months or even years after recovery. During the acute phase of the disease the organisms disappear in a few days from the erythrocytes and they change from the 'bacilliform' to the 'coccoid' form. This tendency to degenerate from sharply-stained bacillary rods to clusters of less discrete, coccoid, granular or amorphous forms was also seen in *Bartonella* grown in mesenchymal cells of the guinea pig (Pinkerton and Weinman, 1937a).

The *pre-eruptive stage* is characterised by pains in joints, muscle, and bones. Inflammatory reactions in various organs, such as parotitis and phlebitis, occur most often in this stage. The bacterial infection may also involve the central or peripheral nervous system and the meninges (Ricketts, 1949). Encephalitis may occur.

Finally *the eruptive stage* occurs several weeks after recovery from the severe febrile anaemia. Cherry-red nodules, referred to as verrugas, appear in the skin. They are 2 to 20 mm in diameter and are located chiefly on the extremities and face but may be situated anywhere on the skin. Such lesions constitute the 'verruga peruana' which was investigated by Carrion. The nodules are soft and deep-red in colour when sectioned. Histologically they resemble a capillary haemangioma, with blood vessels of small calibre, proliferating endothelial cells, and organisms (Fig. 24.6). *B. bacilliformis* is usually found in large numbers when Giemsa's stain is used. Large inclusions may be found in the endothelial cells and may represent clusters of degenerating organisms (Pinkerton and Weinman, 1937b). Patients in the eruptive phase have little in the way of systemic symptoms. Verruga peruana usually confers immunity for a lifetime, although sometimes it is only transitory.

Persons may carry *B. bacilliformis* without any signs or symptoms of Carrion's disease. Weinman and Pinkerton (1937b) showed that while some carriers residing in Callahuanca remembered having had verruga peruana some months before others gave no such history. Most carriers endure living conditions which make them only too accessible to haematophagous arthropods. Hence they constitute an important reservoir of *B. bacilliformis*. The insect vectors for Carrion's disease are two species of sandfly, *Phlebotomus verrucarum* and *P. noguchii* (Hertig, 1937). It is likely that transmission of the disease by *Phlebotomus* is nocturnal. The sandflies leave caves for houses after twilight. Intermediate mammalian hosts have not been discovered. Other representatives of *Bartonella* at high altitude are *B. muris* in rats, *B. canis* in dogs, and *B. tyzzori* in guinea pigs.

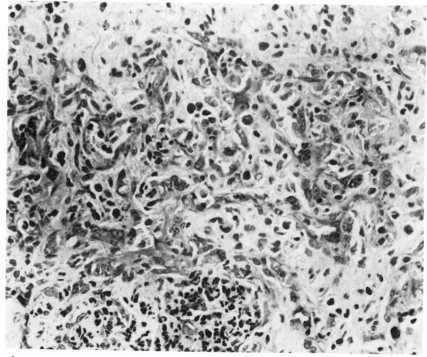

Fig. 24.6 Section of a verruga peruana showing it to consist of small blood capillaries and proliferating endothelial cells, thus resembling a capillary haemangioma. In the lower part of the figure is seen an infiltrate of acute inflammatory cells. (Wolbach's modification of Giemsa's stain after Regaud fixation, × 375).

Electron microscopy of the verruga

The ultrastructure of the verrugas was studied by Recavarren and Lumbreras (1972). They found that while most of the histiocytes between the endothelium-lined capillaries had clear cytoplasm some contained numerous lamellar structures which they thought might represent destroyed organisms. Intact organisms within the cells were not seen but in places cytoplasmic extensions of clear cells around *B. bacilliformis* were seen and were interpreted as suggestive of phagocytosis.

Viral disease at high altitude

The highlands of Bolivia constitute one of the widely separated areas of the world afflicted by 'haemorrhagic fevers'. Although this group of diseases is regarded as infectious and caused by a virus, direct spread from person to person is not a predominant feature (Wilson and Miles, 1974). Bolivian haemorrhagic fever is a representative of this class of condition which occurs in the Andes.

Bolivian haemorrhagic fever

The disease was first recognised in the Department of Beni, Bolivia, in 1959. Four years later a laboratory was established in San Joaquin to study the aetiology, epidemiology and ecology of the disease which was epidemic in that town and proving to have a mortality rate of over 20 per cent. It is caused by the Machupo virus (Figs. 24.7 and 24.8) which has been isolated from human cases during the acute fever (Johnson et al, 1965) and from the wild rodent, *Calomys callosus*, which suffers from a chronic infection excreting virus in the saliva and urine (Johnson et al, 1966). No arthropod vector has been found. It is closely related to Junin virus and other members of the Tacaribe complex of viruses (Fig. 24.9). These viruses have now been assembled into a taxonomic group called arenaviruses (Rowe et al, 1970), which also includes Lassa fever and lymphocytic choriomeningitis viruses (Murphy et al, 1969 and 1970). The name 'arenaviruses' (from *arenosus*, L. sandy) was proposed to reflect the characteristic fine granules

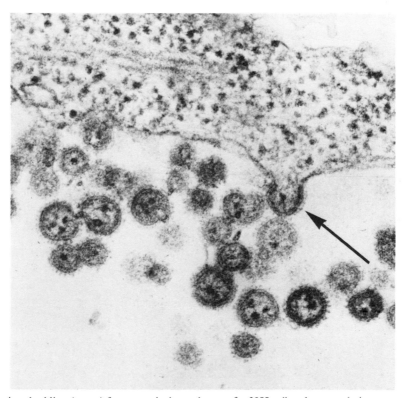

Fig. 24.7 Machupo virus budding (arrow) from marginal membrane of a 33H cell and accumulating extracellularly. (× 114 000). This figure and Fig. 24.8 reproduced from Murphy et al, 1969).

seen in the virion in ultrathin sections. These RNA viruses share a group-specific antigen. The virus

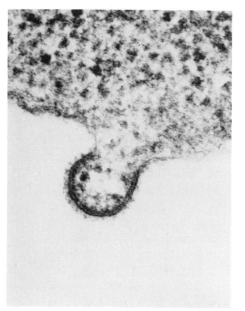

Fig. 24.8   Machupo virus budding from a Raji cell (× 140 000).

particles are round, oval or irregular in shape and range in diameter from 50 to 300 nm, usually between 110 to 130 nm. The particles consist of a dense, well-defined unit membrane envelope with closely spaced projections and an unstructured interior containing a variable number of electron-dense granules. These granules form the most striking unique feature of these viruses; they are 20 to 30 nm in diameter and closely resemble host ribosomes. Virus particles are formed by budding, chiefly from plasma membranes. The pathology of human haemorrhagic fever caused by Machupo virus is generalized lymphadenopathy with haemorrhagic necrosis and haemorrhages and moderate congestion in many organs (Andrewes and Pereira, 1972).

Haemorrhagic exanthem of Bolivia

There has been one report of an epidemic of a haemorrhagic exanthematous disease which occurred in highland Bolivian recruit soldiers during May

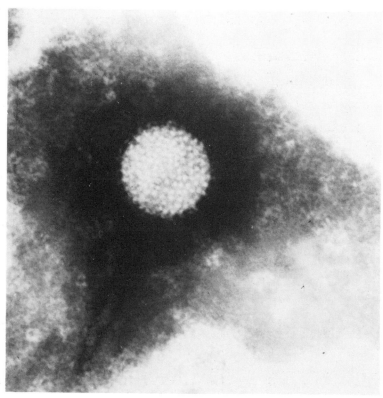

Fig. 24.9   Tacaribe virus particle in negative contrast from ultracentrifuged Vero cell culture supernatant; rather closely spaced surface projections cover its surface (× 330 000). (Reproduced from Murphy et al, 1970).

1967 (Noble et al, 1974). After lifelong residence at high altitude the men had been transferred to a tropical area. This disease was initially considered to be an outbreak of an atypical Bolivian haemorrhagic fever, described above, or even of haemorrhagic smallpox. However, exhaustive clinical and epidemiological studies did not reveal any evidence of man-to-man transmission and serological studies failed to substantiate a viral or other infectious aetiology. Noble et al (1974) came to the conclusion that the syndrome was similar to reported cases of haemorrhagic vesicular lesions following bites from small black flies (*Simuliidae*, locally called 'marigui'). They point out that species of black flies are present on the Altiplano but these bite small mammals not man. Hence the soldiers, native to high altitude, never had the opportunity to develop immunity to the allergens of these tropical insects. The hypersensitivity angiitis may be enhanced by the concurrent vascular fragility and hyperaemia which characterises the highlander and of which we have noted examples throughout this book. In addition to headache and malaise many of those affected showed haemorrhagic manifestations such as bleeding gums, haematuria or gastrointestinal haemorrhage.

REFERENCES

Allison, M. J., Pezzia, A., Gerszten, E. & Mendoza, D. (1974) A case of Carrion's Disease associated with human sacrifice from the Huari Culture of Southern Peru. *American Journal of Physical Anthropology*, **41**, 295.

Altland, P. D., Highman, B. & Smith, F. (1963) Immune response in rabbits exposed to high altitude. *Journal of Infectious Diseases*, **113**, 228.

Andrewes, C. & Pereira, H. G. (1972) Arenavirus. In: *Viruses of Vertebrates*, Third Edition, p. 171. London: Baillière Tindall.

Barton, A. L. (1909) Descripción de elementas endo-globulares hallados en los enfermos de fiebre verrucosa (Articulo preliminar). *Crón. Méd., Lima*, **26**, 7.

Berry, L. J. (1956) Susceptibility to infection as influenced by acclimatization to altitude and Krebs cycle inhibitors and intermediates. *Journal of Infectious Diseases*, **98**, 21.

Berry, L. J. & Mitchell, R. B. (1953) Influence of simulated altitude on resistance—susceptibility to *S. typhimurium* infection to mice. *Texas Reports on Biology and Medicine*, **11**, 379.

Berry, L. J., Mitchell, R. B. & Rubenstein, D. (1955) Effect of acclimatization to altitude on susceptibility of mice to influenza. A virus infection. *Proceedings of the Society for Experimental Biology and Medicine*, **88**, 543.

Berry, L. J. & Smythe, D. S. (1964) Effect of cold on blood clearance of carbon and bacteria of different virulence. *Journal of the Reticuloendothelial Society*, **1**, 405.

Chohan, I. S., Singh, I., Balakrishnan, K. & Talwar, G. P. (1975) Immune response in human subjects at high altitude. *International Journal of Biometeorology*, **8**, 137.

Dryer, R. L. (1966) In Discussion of the paper by Trapani (1966) *Federation Proceedings. Federation of American Societies for Experimental Biology*, **25**, 1260.

Hertig, M. (1937) Carrion's Disease. V Studies on Phlebotomus as the possible vector. *Proceedings of the Society for Experimental Biology and Medicine*, **37**, 598.

Johnson, K. M., Kuns, M. L., Mackenzie, R. B., Webb, P. A. & Yunker, C. E. (1966) Isolation of Machupo virus from wild rodent *Calomys callosus*. *American Journal of Tropical Medicine and Hygiene*, **15**, 103.

Johnson, K. M., Wiebenga, N. H., Mackenzie, R. B., Kuns, M. L., Tauraso, N. M., Shelokov, A., Webb, P. A., Justines, G. & Beye, H. K. (1965) Virus isolations from human cases of hemorrhagic fever in Bolivia. *Proceedings of the Society for Experimental Biology and Medicine*, **118**, 113.

Kalter, S. S. & Tepperman, J. (1952) Influenza virus proliferation in hypoxic mice. *Science*, **115**, 621.

Keck, G. & Buchmeiser, R. (1964) Existence and growth of bacteria at high altitude. In: *The Physiological Effects of High Altitude*, p. 153. Edited by W. H. Weihe. Oxford: Pergamon Press.

Murphy, F. A., Webb, P. A., Johnson, K. M. & Whitfield, S. G. (1969) Morphological comparison of Machupo with lymphocytic choriomeningitis. Virus: Basis for a new taxonomic group. *Journal of Virology*, **4**, 535.

Murphy, F. A., Webb, P. A., Johnson, K. M., Whitfield, S. G. & Chappel, W. A. (1970) Arenoviruses in Vero cells: ultrastructural studies. *Journal of Virology*, **6**, 507.

Noble, J., Valverde, L., Eguia, O. E., Serrate, O. & Antezana, E. (1974) Hemorrhagic exanthem of Bolivia. Studies of an unusual hemorrhagic disease in high altitude dwellers at sea level. *American Journal of Epidemiology*, **99**, 123.

Noguchi, H. & Battistini, T. S. (1926) Etiology of Oroya Fever. 1. Cultivation of Bartonella bacilliformis. *Journal of Experimental Medicine*, **43**, 851.

Nusshag, W. (1954) In: *Hygiene der Haustiere*, p. 86 Leipzig: Verlag S. Hirzel.

Pinkerton, H., & Weinman, D. (1937a) Carrion's Disease. 1. Behaviour of the etiological agent without cells growing or surviving *in vitro*. *Proceedings of the Society for Experimental Biology and Medicine*, **37**, 587.

Pinkerton, H. & Weinman, D. (1937b) Carrion's Disease. II. Comparative morphology of the etiological agent in Oroya fever and *Verruga peruana*. *Proceedings of the Society for Experimental Biology and Medicine*, **37**, 591.

Recavarren, S. & Lumbreras, H. (1972) Pathogenesis of the verruga of Carrion's disease. *American Journal of Pathology*, **66**, 461.

Ricketts, W. E. (1949) Clinical manifestations of Carrion's Disease. *Archives of Internal Medicine*, **84**, 751.

Rippel, A. & Baldes, A. (1952) In: *Grundriss der Mikrobiologie*, p. 281. Berlin: Springer Verlag.

Rowe, W. P., Murphy, F. A., Bergold, G. H., Casals, J., Hotchin, J., Johnson, K. M., Lehmann-Grube, F., Mims, C. A., Traub, E. & Webb, P. A. (1970) Arenoviruses: proposed name for a newly defined virus group. *Journal of Virology*, **5**, 651.

Schultz, M. G. (1968) Daniel Carrion's Experiment. *New England Journal of Medicine*, **278**, 1323.

Singh, I. & Chohan, I. S. (1972) Abnormalities of blood coagulation at high altitude. *International Journal of Biometeorology*, **16**, 283.

Singh, I., Chohan, I. S., Lal, M., Khanna, P. K., Srivastava, M. C., Nanda, R. B., Lamba, J. S. & Malhotra, M. S. (1977) Effects of high altitude stay on the incidence of common diseases in man. *International Journal of Biometeorology*, **21**, 93.

Sutton, J. (1971) Daniel Carrion and Oroya Fever. *Medical Journal of Australia*, **2**, 589.

Trapani, I. L. (1966) Altitude, temperature and the immune response. *Federation Proceedings, Federation of American Societies for Experimental Biology*, **25**, 1254.

Trapani, I. L. (1969) Environment, infection, and immunoglobulin synthesis. *Federation Proceedings, Federation of American Societies for Experimental Biology*, **28**, 1104.

Trapani, I. L. & Jordan, R. T. (1962) Antibody formation in rabbits adapted to high altitude. *Federation Proceedings,*

Federation of American Societies for Experimental Biology, **21**, 25.

Weinman, D. (1944) Infectious anemias due to Bartonella and related red cell parasites. *Transactions of the American Philosophical Society*, **33**, 243.

Weinman, D. & Pinkerton, H. (1937a) Carrion's Disease. III. Experimental production in animals. *Proceedings of the Society for Experimental Biology and Medicine*, **37**, 594.

Weinman, D. & Pinkerton, H. (1937b) Carrion's Disease. IV. Natural sources of Bartonella in the endemic zone. *Proceedings of the Society for Experimental Biology and Medicine*, **37**, 596.

Weiser, O. L., Peoples, N. J., Tull, A. H. & Morse, W. C. (1969) Effect of altitude on the microbiota of man. *Federation Proceedings, Federation of American Societies for Experimental Biology*, **28**, 1107.

Wilson, G. S. & Miles, A. (1974) In: *Topley and Wilson's Principles of Bacteriology, Virology and Immunology*, Sixth Edition. p. 2565. London: Edward Arnold.

25

Endocrines

Adrenal cortex

Ascending to high altitude stimulates the adrenal cortex. The effects are those of a non-specific stress but the precise nature of this is not known. Tromp (1964) believed that low temperature is largely responsible but thought that other factors such as increased intensity of solar or ultraviolet radiation may also be involved. The result of this stress is a secretion of 17-hydroxycorticosteroids (17-OHCS) and catecholamines. Halhuber and Gabl (1964) studied the relation between adrenocortical activity and time of exposure to high altitude. They kept five male students aged 20 to 25 years under observation for four weeks at 2000 m. The activity of the adrenal cortex, as demonstrated by 17-OHCS excretion in the urine, increased sharply during the first week of exposure to high altitude. It decreased during the second week, varied widely during the third and returned to sea-level values during the fourth. Klein (1964) studied adrenocortical activity by determining the level of free 17-OHCS in the plasma and the number of eosinophils in the blood, in men subjected to acute stresses before and after a month's sojourn at 6200 m in the Andes. The stresses employed were oxygen deficiency in a decompression chamber, heat, cold, or physical work on a bicycle ergometer. He found that on exposure to high altitude and before acclimatization there was a short but distinct rise in 17-OHCS associated with a large decrease in eosinophils. The resting level of corticosteroids in the plasma was still elevated by some 60 per cent even five weeks after a month's stay at high altitude. There was, however, a diminished response to the acute stresses. This non-specific adrenocortical activity seems to be independent of other components of acclimatization to high altitude. Mackinnon et al (1963) studied the

effects of rapidly transporting men and women from 1000 to 4330 m. They too found a rapid rise in the excretion of hydroxycorticosteroids which returned to normal by the fifth day. Moncloa et al (1965) also noted a transitory rise in 17-OHCS and in cortisol secretion rate on the sudden exposure of 10 young men to an altitude of 4300 m for two weeks. They found that the level of urinary 17-ketosteroids was not significantly modified. These observed changes in the urinary excretion of 17-OHCS agreed with the previous work of Timiras et al (1957). Sutton et al (1977) found elevation of plasma cortisol levels in four men acutely exposed to a simulated altitude of 4760 m.

Moncloa et al (1965) found that on the stimulation of high altitude the response of the adrenal cortex is not maximal. This could indicate either that at high altitude the adrenal cortex is not subject to intense stimulation or that it is incapable of a greater response. In an attempt to answer this question Moncloa et al (1966) studied the response to ACTH stimulation after dexamethasone inhibition in young adult males acutely exposed to an altitude of 4270 m. Their results indicated that the adrenal cortex can be further stimulated with exogenous ACTH during the first day of exposure to an environment of low oxygen tension. A higher titre of ACTH may be necessary to maintain normal adrenal cortical activity at high altitude (Siri et al, 1969). In a subsequent investigation Moncloa and his colleagues (1968) found that the mean plasma cortisol concentration increases from 9.9 ± 0.6 to 15.5 ± 3.1 μg/dl on the second day of exposure to an altitude of 4330 m of young male sea-level natives. Abnormally high levels of 17-OHCS in the urine occur in climbers on exposure to extreme altitude exceeding 8000 m (Siri et al, 1969).

Adrenal medulla

The activity of the adrenal medulla is also increased during early acclimatization to high altitude. Urinary noradrenaline excretion increased in men taken from sea level to 3800 m and during two weeks' stay noradrenaline excretion doubled (Pace et al, 1964). Klain (1972) found a transitory rise in catecholamine turnover in rats on their exposure to an altitude of 4300 m. Myles (1972) found that catecholamines were increased at simulated altitudes of 5500 m and 7300 m. Adrenal medullary activity in high altitude natives is, however, normal.

There is some evidence to relate the level of catecholamines in the urine with the incidence of acute mountain sickness (Hoon et al, 1976). Forty-seven subjects were exposed to an altitude of 3660 m in their investigation and 29 of them developed acute mountain sickness. The pattern of catecholamine excretion in these two groups was different at high altitude. Those with symptoms showed an immediate rise of 30 per cent above the mean sea-level value of 44.7 μg/24 hours in their urinary catecholamines on exposure. On subsequent days the excretion showed a steady and sustained increase until the tenth day of exposure. The asymptomatic group on the other hand showed but an insignificant increase in catecholamine excretion on arrival at high altitude or subsequently at any stage during their stay in the mountains. On return to sea level the symptomatic group showed a sudden decline in catecholamine excretion and reached initial control values on the fourth day. Hoon and his associates (1976) think it likely that hypersecretion of catecholamines may increase systemic vascular resistance leading to an extensive shift of blood to the pulmonary circulation increasing the likelihood of development of acute mountain sickness and even high altitude pulmonary oedema. A similar investigation was undertaken on 58 subjects who were taken to high altitude slowly (Hoon et al. 1977) Twenty-five lowlanders ascended from 1800 to 3658 m in 50 hours and the remainder in six hours. None of them developed acute mountain sickness and their urinary catecholamine excretion remained normal during 10 days' stay at high altitude.

Aldosterone

The effects of altitude on aldosterone secretion are more specific. Ayres et al (1961) and Williams (1966) found that exposure to high altitude caused a rise in the sodium/potassium ratio in saliva, a rise in total body potassium, and a fall in urine aldosterone secretion. Jung et al (1971) found that altitude caused a fall in blood aldosterone levels in older subjects, but not in younger ones. Sutton et al (1977) reported a decrease in plasma and urinary aldosterone levels in acute exposure to simulated high altitude. Williams related the fall in aldosterone to the increased blood volume that occurs in acclimatization to altitude. This leads to stimulation of the stretch receptors in the right atrium and stretching these receptors is known to depress aldosterone secretion (Anderson et al, 1959). In harmony with these facts is the demonstration of Hartroft et al (1969) that dogs, rabbits, and rats kept at 1520 m to 4260 m for five months show a reduction in the width of the zona glomerulosa, the source of aldosterone. On return to sea level aldosterone levels return to normal (Ayres et al, 1961). As we note in Chapter 17 the sytemic blood pressure is diminished at high altitude and it is conceivable that this is influenced by the lowered secretion of aldosterone, an increase of which is associated with systemic hypertension. Aldosterone appears to regulate sodium and water balance by its action exerted mainly on the distal renal tubule stimulating sodium reabsorption and potassium excretion; hence at high altitude the decreased aldosterone results in potassium retention.

The reduction in aldosterone secretion has been confirmed in climbers. The acid-labile aldosterone excretion in urine over a period of 24 hours was measured by Pines and his colleagues (1977) in five climbers during an ascent of a 7500 m mountain in the Hindu Kush. Representative samples from 24-hour specimens were preserved from all five subjects at 5400 m and the mean value was 2.52 μg/24 hours as contrasted to a normal excretion of 13.4 μg/24 hours at sea level. At 6600 m the mean value for the daily excretion of the hormone had fallen to 1.01 μg. Aldosterone excretion in two climbers with episodes of peripheral oedema at 6600 m fell to 0.49 μg/24 hours.

Spironolactone and acute mountain sickness

Spironolactone is one of a series of steroid derivatives having a lactone ring at position 17. It is devoid of direct effects on electrolytes but antagonises endogenous and exogenous mineralocorticoids by a simple competitive mechanism (Burgen and Mitchell, 1975). There has been one report that spironolactone is useful in prophylaxis against acute mountain sickness (Currie et al, 1976). They found that 13 adults trekking in Nepal in 1974 to altitudes between 4300 m and 5500 m remained free from acute mountain sickness while taking spironolactone as a prophylactic measure. Two years previously five of these adults trekking at similar altitudes, but without treatment, had suffered from acute mountain sickness. The regime used was spironolactone in a dosage of 25 mg three times a day for two days preceding and during the periods spent at altitudes above 3000 m. However, McFarlane (1976) considers the use of spironolactone in the prevention of acute mountain sickness to be theoretically unsound. He points out that renin-aldosterone axis suppression occurs when lowlanders are exposed to high altitude and those who suffer the most severe symptoms of acute mountain sickness show the greatest decrease in aldosterone excretion. Spironolactone, as an anti-aldosterone, might accentuate this process. Currie (1977) subsequently refuted these objections. He postulates that in the first days at high altitude, the initial shift of blood volume centrally overshadows the reduction of plasma volume which occurs consistently, leading to suppression of aldosterone secretion. Later, he believes, the reduced plasma volume, aggravated by dehydration from prolonged hyperventilation, becomes the dominant stimulus, leading to a *rise* in plasma aldosterone levels. Hence antidiuresis occurs in the presence of overloaded pulmonary and cerebral circulations, a dangerous combination. Hence Currie (1977) is led to the view that the theoretical basis for the prophylactic use of spironolactone in acute mountain sickness is that it counteracts the positive aldosterone response to the plasma volume depletion at high altitude. Snell and Cordner (1977) have also reported on the success of spironolactone as a prophylactic against acute mountain sickness during trekking in the Everest-Khumbu area.

Renin-angiotensin system

It has been suggested that the renin-angiotensin system is involved in the development of muscularization of the terminal portion of the pulmonary arterial tree in chronic hypoxia (Berkhov, 1974).

Rats exposed to chronic alveolar hypoxia show increased activation of the renin system (Gould and Goodman, 1970). Mice deprived of oxygen develop increased granulation of the juxtaglomerular apparatus and elevated levels of angiotensin I-converting enzyme in lungs and serum during the second week of exposure (Molteni et al, 1974). After two to three days of hypoxia in both man and rats there is a fall in plasma concentration (Gould and Goodman, 1970; Hogan et al, 1973). Sutton et al (1977) also found no significant change in plasma renin activity on acute exposure to simulated high altitude. However, in rats during the second week of exposure there is an increase over normal levels returning to normal by the end of the third week (Gould and Goodman, 1970). It is of interest to bear in mind that renin is formed in the same organ, and perhaps even the same cell that synthesizes erythropoetin where the stimulus is known to be hypoxia (Eepson and McCarry, 1968; Gould et al, 1968). Thus with renal hypoxia one might anticipate an increased secretion of renin.

The infusion of angiotensin I or II into the pulmonary circulation of the dog caused a threefold increase in the vasoconstrictive response to hypoxia (Alexander et al, 1976). Zakheim et al (1975) found that blockade of angiotensin I conversion by SQ 20881, a synthetic nonapeptide, significantly reduced in Sprague Dawley rats the right ventricular hypertrophy and pulmonary vascular changes of chronic alveolar hypoxia described in Chapter 11. Since this nonapeptide is a specific competitive inhibitor of angiotensin I-converting enzyme, these experimental findings suggest that chronic hypoxia exerts its effects through the agency of the renin-angiotensin system.

Zakheim et al (1975) consider that angiotensin II is necessary in the development of the pulmonary vascular structural changes of chronic alveolar hypoxia. It remains to be determined whether these effects are secondary to increased levels of angiotensin II or a hypoxia-induced increase in the sensitivity of pulmonary arterial smooth muscle to angiotensin II. Angiotensin-converting enzyme in the lung is closely associated with the endothelial cells of the pulmonary vasculature. If alveolar air is able to control the conversion of angiotensin I in the pulmonary vascular bed adjacent to it, the amount of perfusion of the area could be regulated by the amount of angiotensin II produced (Alexander et al, 1976).

The anterior pituitary

Nelson and Cons (1975) studied the growth and level of activity of the pituitary and thyroid glands in rats born and raised at a simulated altitude of 3800 m. From birth until 40 days of age the test rats were significantly lighter than controls. The most striking impairment in growth was in females, and at the age of 40 days they showed significantly lighter pituitary glands, ovaries and uteri. The pituitary contained less growth hormone (GH) (Nelson et al, 1968) but more follicle-stimulating hormone (FSH) and luteinizing hormone (LH) than in sea-level controls. Thyrotropic stimulating hormone (TSH) was found in equal concentration in the pituitaries of high altitude and sea-level rats. The plasma levels of GH, FSH, LH and TSH were the same in test and control animals. The uptake of ^{131}I by the thyroid and the level of plasma protein-bound ^{131}I were significantly reduced in the rats born and spending their infancy in hypoxic conditions. Thus it would appear that the continuous exposure of developing female rats to environmental hypoxia impairs the function of the anterior pituitary, retards body growth, and delays reproductive maturation. The reduction of the weight of the endocrine organs has been confirmed by workers over a considerable period (Moore and Price, 1948; Timiras and Woolley, 1966). Delayed brain maturation has been reported by Timiras and Woolley (1966).

The posterior pituitary

Diuresis develops in persistent but mild hypoxia and appears to be due to diminution of circulating antidiuretic hormone (Silvette, 1943). This probably results from inhibitory impulses originating from the receptors in the left atrium when it is distended by the increased blood volume resulting from acclimatization to high altitude (Chapter 17) (Henry and Pearce, 1956; Henry et al, 1956). However, as we note in Chapter 14, the onset of acute mountain sickness appears to be associated with oliguria. This seems to be linked with a sudden discharge of antidiuretic hormone (Brun et al, 1945; Noble and Taylor, 1953), as suggested in the link-diagram shown as Figure 14.6.

Thyroid

The thyroid gland is one of the main regulators of oxygen consumption and hence one would expect its function at high altitude to be modified. Most studies have been carried out on animals and in conditions of acute hypoxia. Much less is known about thyroid function in man exposed to chronic hypoxia at high altitude. There are two complicating factors in the study of the effect of high altitude on thyroid function under natural conditions. The first is that mountain ranges are often deficient of iodine as we discuss below. The second is that a cold environment affects thyroid function both in man (Williams, 1974) and laboratory animals. These variables are much easier to control in experimental animals than in man.

In experimental animals, reports on the immediate effects of low pressure on thyroid function are conflicting. Thus Verzár et al (1952) found a temporary fall in thyroid ^{131}I uptake in rats exposed to a pressure of 250 to 380 mmHg. Mulvey and Macaione (1969) on the other hand found no change in the ^{131}I uptake in rats exposed to a simulated altitude of 4570 m for 17 to 40 hours. Finally, Nelson and Anthony (1966) reported an increase in thyroid ^{131}I uptake lasting up to 60 hours, in rats exposed to a simulated altitude of 5490 m.

In contrast, the experiments of more than three days' duration uniformly show a reduction in thyroid function, which occurs in both animals born and bred at high altitudes such as free-living gophers (Tryon et al, 1968) and experimental rats, and in animals developing in a normal environment and then exposed to low pressures (Gorden et al, 1943; Surks, 1966; Martin et al, 1971; Galton, 1972). The iodine retention in the whole animal and the thyroid concentrating power are decreased. The mechanism of this depression is not clear.

Surks (1969) claimed that TSH injections restored the thyroid function to normal in intact hypoxic rats and thereby deduced that the thyroid changes were secondary to diminished TSH stimulation by the pituitary. In direct contrast, Martin et al (1971) examined the thyroid histology of rats kept at a simulated altitude of 7560 m and thought the changes suggested increased TSH stimulation. They explained the lowered thyroid function as the result of hypoxia directly inhibiting thyroxine synthesis in the thyroid. Galton (1972) assessed the turnover-rate of thyroxine in the peripheral tissues of acclimatized rats and found it to be reduced. He concluded that high altitude reduced requirement for thyroxine in the tissues and thus indirectly brought about a concomitant reduction in hormone synthesis.

Possible significance of the hypothyroid state in acclimatization

The depression of thyroid function in animals at high altitude is thus well established, even though the mechanism of this is still debatable. Martin et al (1972) studied its possible significance in acclimatization. They tested 'myocardial resistance' by the method of McGrath and Bullard (1968). Strips of right ventricular muscle were suspended in de-oxygenated Krebs-Ringer solution and the percentage of normal contractile strength remaining was measured after 200 anoxic contractions. Rats were made hyperthyroid by the injection of thyroxine or hypothyroid by injection of propyl-thiourocil. Either altitude, or drug-induced

hypothyroidism increased the percentage of contractile strength which remained, the so-called 'myocardial anoxic resistance', but exogenous thyroxine abolished such raised resistance. Thus increased myocardial anoxic resistance at high altitudes, which may be a feature of acclimatization, appears to be related to the hypothyroid state. Thyroxine inhibits enzymes which catalyse anaerobic metabolism and thus increases cardiac dependence on aerobic processes. The hypothyroid state has the opposite effect. Cardiac muscle glycogen stores are augmented in hypothyroidism and appear to be important in surviving hypoxic stress. The hypothyroid state appears to have no influence on inducing polycythaemia.

The histology of the thyroid in experimental animals at high altitude

Working in our laboratory Gradwell (1978) made a histological study of the thyroid glands of rats exposed to a simulated altitude of 5500 m for four weeks. He found that compared to sea-level controls the thyroid glands of the test animals showed a statistically significant increase in the amount of colloid and a statistically significant decrease in the amount of follicular epithelium (Figs. 25.1 to 25.3). On the other hand, there was no significant change in the mean diameter of the follicles of the thyroid (Fig. 25.1). These histological changes were considered to be consistent with reduced thyroid activity induced by exposure to the hypoxia of high altitude. Earlier workers reported similar observations and conclusions. Thus Gordon et al (1943) also

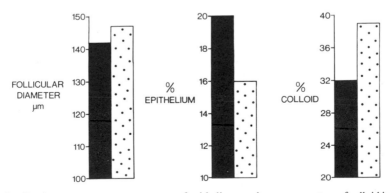

Fig. 25.1 Mean follicular diameter, mean percentage content of epithelium, and mean percentage of colloid in rats at sea level (filled column) and after being subjected to a simulated altitude of 5500 m for four weeks (stippled column). (Based on data from Gradwell, 1978).

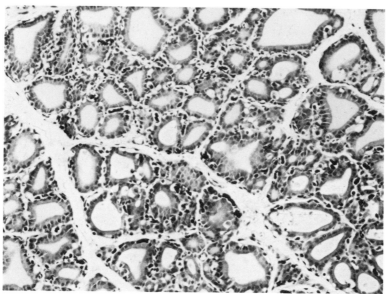

Fig. 25.2   Section of thyroid of control rat at sea level. The small follicles are lined by cuboidal epithelium (Haematoxylin and eosin, × 150).

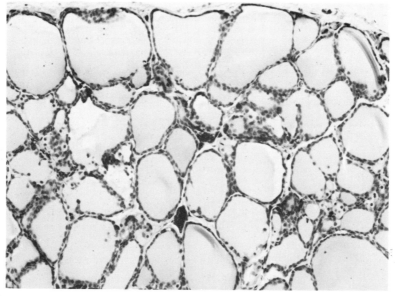

Fig. 25.3   Section of thyroid from a rat exposed to a simulated altitude of 5500 m for four weeks. The follicles are enlarged and lined by flattened epithelium (H. and E., × 150).

found that after an exposure of two to three weeks to a simulated altitude of 7620 m to 8230 m adult rats exhibited mild thyroid hypoplasia wherein the follicles contained slightly more colloid and the epithelium appeared somewhat lower. Surks (1966) studied adult rats exposed for one to 37 days to an altitude of 4300 m and noted large colloid-filled follicles in the thyroid glands and in this case the epithelium was almost squamous in appearance. On the other hand Nelson (1971) reported that in rats born and developing at 3800 m the general appearance of the thyroid gland remained relatively unchanged, the percentage composition of colloid being the same as in sea-level controls. The follicular epithelium was reduced in height but was not flattened as is typical in hypophysectomized animals.

The human thyroid at high altitude

In man the effects of high altitude on the thyroid have been less studied. Dragan and Pop (1969) found a rise in the red cell tri-iodothyronine uptake (Hamolsky test) in 13 sportsmen at 1950 m. This reached a peak at the tenth day but had returned nearly to normal by the end of the third week. They believe that during effort at a medium elevation the thyroid shows increased activity. Moncloa et al. (1966) transported ten young men from sea level to a height of 4300 m and studied thyroid function for 14 days. The basal metabolic rate was unchanged. The 24 hour thyroidal ^{131}I uptake rose significantly (from 34.4 per cent to 51.4 per cent).

In summary the long-term exposure of experimental animals to high altitude causes a lowering of thyroid function, though the mechanism is not clear. This depressed thyroid function mediates the increased anoxic resistance of the heart muscle, but the other features of acclimatization to high altitude do not depend on altered thyroid activity. The effects of high altitude on human thyroid function have been less studied and the results are not clearly defined.

Colloid goitre in mountainous areas

Enlargement of the thyroid gland due to colloidal adenomatous hyperplasia is the result of increased levels of thyroid-stimulating hormone (TSH), brought about by some block in the output of thyroid hormone (Robbins, 1967). Three distinct mechanisms lead to the same histopathological changes. These are a deficiency of iodine, ingestion of goitrogenic foods or compounds that block thyroglobulin synthesis, and genetic defects involving enzymes vital to the elaboration of thyroid hormones. A deficiency of iodine is characteristic of the mountain regions of the world. The lack of exogenous iodine leads to the formation of inadequate amounts of thyroid hormone and excessive TSH stimulation from the pituitary. Endemic goitre (Fig. 25.4) is common where iodine intake falls below 100 µg per day. There is not a

Fig. 25.4   Goitre at high altitude. (a: left) In a villager at Askole (3050 m) in the Karakorum range of Kashmir. This village is of historical interest in that it was the first inhabited spot south of the mountain range reached by Sir Francis Younghusband in 1887 on his walk of 3000 miles from Peking to India via the Gobi desert and the Karakorums. He was disenchanted with Askole, finding the village dirty and the villagers inhospitable. (b: right) In a porter at Skardu (2340 m) situated south of Askole.

universal distribution of the disease throughout the population at risk and this may be because the other two factors referred to above operate unevenly in the community. Thus some members of the group at risk may ingest goitrogenic substances such as turnips or cabbage, which contain antithyroid compounds. Others may have hereditary defects in the synthesis or transport of thyroid hormone. Ward (1975) refers to this patchy distribution of goitre in the population of the mountainous areas of Northern Greece and he points out that Malamos et al (1965) have suggested that the goitres may be genetically determined. The classical studies of McCarrison (1906, 1908) suggested that pollution of drinking water with human and animal excreta brought about the development of goitres in iodine-deficient areas. Work in the same valley of the Karakorum mountains over sixty years later (Chapman et al, 1972) did not substantiate any correlation between bacterial contamination of water supply and goitre incidence. The combination of a diet in the mountains which is rich in calcium and poor in iodine may be important (Hellwig, 1934). For many years goitre has been known as 'Derbyshire neck' indicating the high incidence of goitre in a limestone area.

Although mountainous areas are known to predispose to endemic goitre through lack of iodine intake as we describe above, it is of interest that the effects of high altitude per se appear to counteract the effects of iodine deficiency. Thus both Moncloa (1966) and Fierro-Benitez et al (1969) point out that there is a lower prevalence of goitre at altitudes exceeding 3000 m in Peru. Ward (1975) thinks that the lower incidence of colloid goitre at high altitude is related to a lower peripheral utilisation of thyroxine and the general depression of thyroid function such as we describe above.

Problems in the treatment of goitre at high altitude

Ward (1970) found that the incidence of goitre in the general population at high altitude at Dhankuta in Nepal was no less than 32 per cent. Clearly the treatment of such a common and disfiguring disease is an important problem. In the native highlanders of the Himalayas enormous goitres may compress the recurrent laryngeal nerves and cause gross cosmetic disfigurement. Surgical treatment under such conditions is very difficult for the goitres are often so vascular as to produce severe haemorrhage from dilated veins and arteries, and their size leads to grossly distorted anatomy making operation difficult. In isolated mountainous areas such as Nepal the necessary excellent theatre facilities and blood transfusion services are hard to come by.

Radio-iodine is easy to administer but the strict precautions in its use would be difficult to enforce in an uneducated population. Traditional treatment by thyroid extract offers a better proposition. Ward reports that in Dhankuta 18 patients were treated with thyroid extract in a dose of 60 mg daily, increased by 60 mg weekly until a top dose of 300 mg daily was attained and sustained for three to six months. In the first six weeks of such treatment there were startling reductions in the size of the goitre. Such treatment must of course be followed up either by an adequate intake of iodine to prevent recurrence, or by surgical excision.

Endemic goitre has been known to occur in the mountainous Huon peninsula of eastern New Guinea for many years. A study in 1964 indicated that the inhabitants of the area suffered from a severe iodine deficiency and trials were undertaken to determine whether the injection of iodized oil (iodized oil fluid injection B.P.) would correct that deficiency. Studies were also made to determine what effect injections given in the previous years had had. The results showed that, whereas the urinary

iodine and ^{131}I uptake in New Guineans who had received iodized oil in 1957 were similar in untreated persons yet a single injection of 4 ml iodized oil, containing 2.15 g iodine, appeared substantially to correct iodine deficiency for four to five years (Buttfield and Hetzel, 1967). There was also a significant regression of goitre in all but one of a group of 61 persons with easily visible goitres, within three months of their receiving an injection. No case of thyrotoxicosis or iodism was seen in more than 2000 subjects who were given injections. The injection of iodized oil is particularly recommended for the correction of iodine deficiency in children and in women of child-bearing age whenever the efficacy of other measures is uncertain (Buttfield and Hetzel, 1967). The method is relatively inexpensive and well suited to mass prophylaxis concerning people with a low standard of living. Iodized salt should be made freely available to such mountain populations but it is obvious that this is a far less reliable method of administration of iodine than by injection.

Insulin

The blood sugar rises on acute exposure to high altitude but decreases on chronic exposure. Thus Williams (1975) reported that after an effortless ascent to 3500 m by airlift five men showed an increased concentration of glucose in the blood to values between 115 and 135 mg/dl. Forbes (1936) had earlier reported that fasting blood sugars decreased as men went from sea level to 3660 m and then increased at higher altitudes. In the soldiers studied by Singh et al (1977) the blood sugar was similarly found to be raised at two weeks after arrival at high altitude. However, this rise persisted for only ten months but by two years the blood glucose level was significantly lower than the initial values at sea level. Blume and Pace (1967) also noted this secondary fall in blood glucose levels but in their subjects it occurred after only three weeks exposure at an altitude of 3800 m. Native highlanders have also been said to have reduced blood sugars (Picón-Reátegui et al, 1970).

Both oral and intravenous glucose tolerance tests have shown that glucose utilisation is increased in native highlanders. Studies employing the infusion of 14-C glucose have revealed that this is also the case in recent arrivals at high altitude (Johnson et al, 1974). Associated with the reduced levels of blood sugar is a low glycogen content of the liver (Blume and Pace, 1967; Johnson et al, 1974). This reduction in the liver glycogen store is probably one of the reasons why glucose tolerance tests do not produce as large an elevation in blood glucose levels at high altitude as at sea level. Glycogen synthesis would be stimulated by reduced stores and this would rapidly remove the excess glucose from the blood. In laboratory animals chronically exposed to high altitude there is an increase in the rate of glycogenesis in the liver and also in the muscles and myocardium. The sensitivity to endogenous insulin appears to be increased at high altitude.

Williams (1975) found that in five men passively transported to 3500 m the average blood insulin concentration rose. To give an example, in one subject the mean insulin concentration in micro-units per ml of blood serum rose from 19.00 over four days before ascent, to 21.25 over four days at 3500 m, and then fell to 14.40 over five days on return to sea level. The diet was strictly controlled

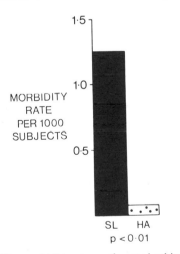

MORBIDITY RATE PER 1000 SUBJECTS

1·5
1·0
0·5

SL   HA
p < 0·01

Fig. 25.5 The morbidity rate per thousand subjects from diabetes mellitus among 20 000 soldiers stationed between altitudes of 3692 and 5538 m (stippled columns) and 130 700 stationed on the plains (filled columns). (Based on data from Singh et al, 1977).

during the whole period of the study, the degree of physical activity was uniform, and the blood samples were all taken after overnight fasting at the same time each day.

The incidence of diabetes mellitus was found to be lower in soldiers stationed at high altitude in the Himalayas than in men stationed on the plains of India (Singh et al, 1977) (Fig. 25.5). These authors postulate at least six mechanisms which may be involved in keeping the incidence of diabetes mellitus low at high altitude. These are a delay in absorption of sugar, increased glycogen synthesis, decreased glycogenolysis, increased sensitivity to endogenous insulin, decreased sensitivity to endogenous ACTH, and increased rate of utilisation of glucose.

The endocrine glands in Monge's disease

Arias-Stella et al (1973) reported that at necropsy on a woman of 48 years there was a nodular goitre with hyperplasia of the adrenal cortex. (Fig. 16.10).

Testosterone

As we shall see in the following chapter at high altitude there are changes in the structure of the testis and in the composition of semen. There are also disturbances of the endocrine testis. In the first place on exposure to high altitude there is an early fall in the urinary excretion of testosterone which is most pronounced by the third day after ascent but which then rapidly returns to normal. Thus in one study by Guerra-Garcia (1971) the 24 hour urinary testosterone excretion was measured in 10 normal men between the age of 19 and 25 years who were exposed to an altitude of 4250 m. The mean value was 99.8 ± 14.3 μg/24 hours and by the third day there was a significant fall to 39.1 ± 7.1 μg. By the seventh day of exposure the mean daily excretion rate of testosterone had recovered to 144 ± 23 μg. High altitude natives have a similar excretion of testosterone in 24 hours amounting to 96.5 ± 10.16 μg.

Although there is a diminished excretion of testosterone on acute exposure to the hypoxia of high altitude, the androgen-producing Leydig cells maintain their normal response to an adequate stimulus (Guerra-Garcia, 1971). The response to human chorionic gonadotrophin in terms of urinary testosterone excretion was normal in four men from sea level during their first days of residence at high altitude. It seems likely that the fall in excretion of urinary testosterone is due to a diminished plasma level of luteinizing hormone. A fall in plasma LH was found by Sobrevilla and Midgley (1968) to occur in men exposed for 28 days to the hypoxic environment of Cerro de Pasco (4330 m).

Urinary gonadotrophins

The fall in urinary testosterone levels is more likely to follow modifications in the FSH/LH ratio rather than any general increase in gonadotrophins. Thus in one study at 4250 m the total urinary gonadotrophin levels in eight native highlanders were similar to those of lowlanders (Donayre, 1966).

REFERENCES

Alexander, J. M., Nyby, M. D. & Jasberg, K. A. (1976) Effect of angiotensin on hypoxic pulmonary vasoconstriction in isolated dog lung. *Journal of Applied Physiology*, **41**, 84.

Anderson, C. H., McCally, M. & Farrell, G. L. (1959) The effects of atrial stretch on aldosterone secretion. *Endocrinology*, **64**, 202.

Arias-Stella, J., Krüger, H. & Recavarren, S. (1973) On the pathology of chronic mountain sickness. *Pathologia et Microbiologia*, **39**, 283.

Ayres, P. J., Hunter, R. C., Williams, E. S. & Rundo, J. (1961) Aldosterone excretion and potassium retention in subjects living at high altitude. *Nature*, **191**, 78.

Berkhov, S. (1974) Hypoxic pulmonary vasoconstriction in the rat. The necessary role of angiotensin II. *Circulation Research*, **35**, 256.

Blume, F. D. & Pace, N. (1967) Effect of translocation to 3800 m altitude on glycolysis in mice. *Journal of Applied Physiology*, **23**, 75.

Brun, C., Knudsen, E. O. E. & Raaschou, F. (1945) On cause of post-syncopal oliguria. *Acta Medica Scandinavica*, **122**, 486.

Burgen, A. S. V. & Mitchell, J. F. (1975) In: *Gaddum's Pharmacology* (Seventh Edition), p. 173. London: Oxford University Press.

Buttfield, I. H. & Hetzel, B. S. (1967) Endemic Goitre in Eastern New Guinea. With special reference to the use of iodized oil in prophylaxis and treatment. *Bulletin of World Health Organization*, **36**, 243.

Chapman, J. A., Grant, I. S., Taylor, G., Mahmud, K., Sardar-Ul-Mulk, & Shahid, M. A. (1972) Endemic goitre in the Gilgit Agency, West Pakistan. *Philosophical Transactions of the Royal Society, Series B*, **263**, 459.

Currie, T. T. (1977) Spironolactone and acute mountain sickness. *Medical Journal of Australia*, **1**, 419.

Currie, T. T., Carter, P. H., Champion, W. L., Fong, G., Francis, J. K., McDonald, I. H., Newing, R. K., Nunn, I. N., Sisson, R. N., Sussex, M. & Zacharin, R. F. (1976) Spironolactone and acute mountain sickness *Medical Journal of Australia*, **2**, 168.

Donayre, J. (1966) Population growth and fertility at high altitude. In: *Life at High Altitudes*, p. 74. Washington: Pan American Health Organization, Scientific Publication No. 140.

Dragan, I. & Pop, T. (1969) Researches concerning the thyroid function during the effort at medium altitude. (Test in vitro with I^{131}). *Journal of Sports Medicine and Physical Fitness*, **9**, 162.

Eepson, J. H. & McCarry, E. E. (1968) Polycythemia and increased erythropoietin production in a patient with hypertrophy of the juxtaglomerular apparatus. *Blood*, **32**, 370.

Fierro-Benitez, R., Wilson, P., DeGroot, L. J. & Ramirez, I. (1969) Edemic goiter and edemic cretinism in the Andean region. *New England Journal of Medicine*, **280**, 296.

Forbes, W. H. (1936) Blood, sugar and glucose tolerance at high altitudes. *American Journal of Physiology*, **116**, 309.

Galton, V. A. (1972) Some effects of altitude on thyroid function. *Endocrinology*, **91**, 1393.

Gould, A. B. & Goodman, S. A. (1970) The effect of hypoxia on the renin-angiotensinogen system. *Labaratory Investigation*, **22**, 443.

Gould, A. B., Keighley, G. & Lowy, P. H. (1968) On the presence of a renin-like activity in erythropoietin preparation. Laboratory Investigation, **18**, 2.

Gordon, A. S., Tornetta, F. J., d'Angelo, S. A. & Charipper, H. A. (1943) Effects of low atmospheric presssures on activity of the thyroid, reproductive system and anterior lobe of the pituitary in the rat. *Endocrinology*, **33**, 366.

Gradwell, E. (1978) Histological changes in the thyroid gland in rats on acclimatization to simulated high altitude. *Journal of Pathology*, **125**, 33.

Guerra-Garcia, R. (1971) Testosterone metabolism in men exposed to high altitude. *Acta Endocrinologica Panama*, **2**, 55.

Halhuber, M. J. & Gabl, F. (1964) 17-OHSC excretion and blood eosinophils at an altitude of 2000 m. In: *The Physiological Effects of High Altitude*, p. 131. Edited by W. H. Weihe. Oxford: Pergamon Press.

Hartroft, P. M., Bischoff, M. B. & Bucci, T. J. (1969) Effects of chronic exposure to high altitude on the juxta-glomerular complex and adrenal cortex of dogs, rabbits, and rats. *Federation Proceedings, Federation of American Societies for Experimental Biology*, **28**, 1234.

Hellwig, C. A. (1934) Experimental colloid goiter. *Endocrinology*, **18**, 197.

Henry, J. P., Gauer, O. H. & Reeves, J. L. (1956) Evidence of atrial location of receptors influencing urine flow. *Circulation Research*, **4**, 85.

Henry, J. P. & Pearce, J. W. (1956) Possible role of cardiac atrial stretch receptors in induction of changes in urine flow. *Journal of Physiology*, **131**, 572.

Higan, R. P. III, Kotchen, T. A., Boyd, A. E. III, & Hartley, H. L. (1973) Effect of altitude on renin-aldosterone system and metabolism of water and electrolytes. *Journal of Applied Physiology*, **35**, 385.

Hoon, R. S., Sharma, S. C., Balasubramanian, V., Chadha, K. S. & Mathew, O.P. (1976) Urinary catecholamine excretion on acute induction to high altitude (3658 m) *Journal of Applied Physiology*, **41**, 631.

Hoon, R. S., Sharma, S. C., Balasubramanian, V. & Chadha, K. S. (1977) Urinary catecholamine excretion in induction to high altitude (3658 m) by air and road. *American Journal of Physiology: Respiratory, Environmental and Exercise Physiology*, **42**, 728.

Johnson, H. L., Consolazio, C. F., Burk, R. F. & Daws, T. A. (1974) Glucose-¹⁴C-UL metabolism in man after abrupt altitude exposure (4300 m) *Aerospace Medicine*, **45**, 849.

Jung, R. C., Dill, D. B., Horton, R. & Horvath, S. M. (1971) Effect of age on plasma aldosterone levels and hemo-concentration at altitude. *Journal of Applied Physiology*, **31**, 593.

Klain, G. L. (1972) Acute high altitude stress and enzymal activities in the rat adrenal medulla. *Endocrinology*, **91**, 1447.

Klein, K. (1964) Contribution to discussion after paper by Halhuber and Gabl (above) In: *The Physiological Effects of High Altitude*, p. 136. Edited by W. H. Weihe. Oxford: Pergamon Press.

Malamos, B., Miras, C., Kostamis, P., Mantzos, J., Kralios, A. C., Rigopoulos, G., Zerefos, N. & Koutras, D. S. (1965) In: *Current Topics in Thyroid Research* Edited by C. Cassano & M. Andreoli. New York: Academic Press.

McCarrison, R. (1906) Observations on endemic goitre in the Chitral and Gilgit Valleys. *Lancet*, **i**, 1110.

McCarrison, R. (1908) Observations of endemic cretinism in the Chitral and Gilgit Valleys. *Lancet*, **ii**, 1275.

McFarlane, A. C. (1976) Spironolactone and acute mountain sickness. *Medical Journal of Australia*, **2**, 923.

McGrath, J. J. & Bullard, R. W. (1968) Altered myocardial performance in response to anoxia after high altitude exposure. *Journal of Applied Physiology*, **25**, 761.

Mackisson, P. C. B., Monk-Jones, M. E. & Fotherby, K. (1963) A study of various indices of adrenocortical activity during 23 days at high altitude. *Journal of Endocrinology*, **26**, 555.

Martin, L. G., Westenberger, G. E. & Bullard, R. W. (1971) Thyroidal changes in the rat during acclimatization to simulated high altitude. *American Journal of Physiology*, **221**, 1057.

Martin, L. G., Westenberger, G. E., Hippensteele, J. E. & Bullard, R. W. (1972) Thyroidal influence on myocardial changes induced by simulated high altitudes. *American Journal of Physiology*, **222**, 1599.

Molteni, A., Zakheim, R. M., Mullis, K. & Mattioli, L. (1974) Effects of chronic hypoxia on lung and serum angiotensin-I-converting enzyme activity. *Proceedings of the Society for Experimental Biology and Medicine*, **147**, 263.

Moncloa, F. (1966) Physiological Patterns: Endocrine factors. In: *Life at High Altitudes*, p. 36. Washington: Pan American Health Organization, Scientific Publication, No. 140.

Moncloa, F., Beteta, L., Velazco, I. & Goñez, C. (1966) ACTH stimulation and dexamethasone inhibition in newcomers to high altitude. *Proceedings of the Society for Experimental Biology and Medicine*, **122**, 1029.

Moncloa, F., Donayre, J., Sobrevilla, L. A., & Guerra-Garcia, R. (1965) Endocrine studies at high altitude: II. Adrenal cortical function in sea level natives exposed to high altitudes (4300 meters) for two weeks. *Journal of Clinical Endocrinology and Metabolism*, **25**, 1640.

Moncloa, F., Guerra-Garcia, R., Subauste, C., Sobrevilla, L. A. & Donayre, J. (1966) Endocrine studies at high altitude. I. Thyroid function in sea level natives exposed for two weeks to an altitude of 4300 meters. *Journal of Clinical Endocrinology and Metabolism*, **26**, 1237.

Moncloa, F., Velasco, I. & Beteta, L. (1968) Plasma cortisol concentration and disappearance rate of 4-₁₄C-cortisol in newcomers to high altitude. *Journal of Clinical Endocrinology and Metabolism*, **28**, 379.

Moore, C. R. & Price, D. (1948) A study at high altitude of

reproduction, growth, sexual maturity, and organ weights. *Journal of Experimental Zoology*, **108**, 171.

Mulvey, P. F. & Macaione, J. M. R. (1969) Thyroidal dysfunction during simulated altitude conditions. *Federation Proceedings, Federation of American Societies for Experimental Biology*, **28**, 1243.

Myles, W. S. (1972) The excretion of 17-hydroxycorticosteroids by rats during exposure to altitude. *International Journal of Biometeorology*, **16**, 367.

Nelson, B. D. & Anthony, A. (1966) Thyroxine biosynthesis and thyroidal uptake of I¹³¹ in rats at the onset of hypoxia exposure. *Proceedings of the Society for Experimental Biology and Medicine*, **121**, 1256.

Nelson, M. L & Cons, J. M. (1975) Pituitary hormones and growth retardation in rats raised at simulated high altitude (3800 m). *Environmental Physiology and Biochemistry*, **5**, 273.

Nelson, M. L., Srebnik, H. H. & Timiras, P. S. (1968) Reduction in pituitary growth hormone and thyroid-stimulating hormone at high altitude. *Excerpta Medica Foundation. International Congress Series*, **157**, 196.

Nelson, M. L. (1971) Thyroid function in immature rats developing at high altitudes. *Environmental Physiology*, **1**, 96.

Noble, R. L. & Taylor, N. B. G. (1953) Antidiuretic substances in human urine after haemorrhage, fainting, dehydration and acceleration. *Journal of Physiology*, **122**, 220.

Pace, N., Griswold, R. L. & Grunbaum, B. W. (1964) Increase in urinary norepinephrine excretion during 14 days sojourn at 3800 meters elevation. *Federation Proceedings, Federation of American Societies for Experimental Biology*, **23**, 521.

Picón-Reátegui, E., Buskirk, E. R. & Baker, P. T. (1970) Blood glucose in high altitude natives and during acclimatization to altitude. *Journal of Applied Physiology*, **29**, 560.

Pines, A., Slater, J. D. H. & Jowett, T. P. (1977) The kidney and aldosterone in acclimatization at altitude. *British Journal of Diseases of the Chest*, **71**, 203.

Robbins, S. L. (1967) In: *Pathology* (Third edition) p. 1202. Philadelphia: W. B. Saunders.

Silvette, H. (1943) Some effects of low barometric pressure on kidney function in the white rat. *American Journal of Physiology*, **140**, 374.

Singh, I., Chohan, I. S., Lal, M., Khanna, P. K., Srivastava, M. C., Nanda, R. B., Lamba, J. S. & Malhotra, M. S. (1977) Effects of high altitude stay on the incidence of common diseases in Man. *International Journal of Biometeorology*, **21**, 93.

Siri, W. E., Cleveland, A. S. & Blanche, P. (1969) Adrenal gland activity in Mount Everest climbers. *Federation Proceedings, Federation of American Societies for Experimental Biology*, **28**, 1251.

Snell, J. A. & Cordner, E. P. (1977) Spironolactone and acute mountain sickness. *Medical Journal of Australia*, **1**, 828.

Sobrevilla, L. A. & Midgley, A. R. (1968) In: *Gonadotropins*, p. 367. Edited by E. Rosemberg. Los Altos, California: Geron - X.

Surks, M. I. (1966) Effect of hypoxia and high altitude on thyroidal iodine metabolism in the rat. *Endocrinology*, **78**, 307.

Surks, M. I. (1969) Effect of thyrotropin on thyroidal metabolism during hypoxia. *American Journal of Physiology*, **216**, 436.

Sutton, J. R., Viol, G. W., Gray, G. W., McFadden, M. & Keane, P. M. (1977) Renin, aldosterone, electrolyte, and cortisol responses to hypoxic decompression. *Journal of Applied Physiology*, **43**, 421.

Timiras, P. S. & Woolley, D. E. (1966) Functional and morphologic development of brain and other organs of rats at high altitude. *Federation Proceedings, Federation of American Societies for Experimental Biology*, **25**, 1312.

Timiras, P. S., Pace, N. G. & Hwang, C. A. (1957) Plasma and urine-7: hydroxycorticosteroid and urine 17-Ketosteroid levels in man during acclimatization to high altitude. *Federation Proceedings, Federation of American Societies for Experimental Biology*, **16**, 340.

Tromp, S. W. (1964) Contribution to discussion after paper of Halhuber and Gabl (above). In: *The Physiological Effects of High Altitude*, p. 188. Edited by W. H. Weihe. Oxford: Pergamon Press.

Tyron, C. A., Kodric, W. R. & Cunningham, H. M. (1968) Measurement of relative thyroid activity in free-ranging rodents along an altitudinal transect. *Nature*, **218**, 278.

Verzár, F., Sailer, E. & Vidovic, V. (1952) Changes in thyroid activity at low atmospheric pressure and at high altitudes, as tested with I¹³¹. *Journal of Endocrinology*, **8**, 308.

Ward, J. P. (1970) The medical treatment of large group III

goitres with thyroid extract. *British Journal of Surgery*, **57**, 587.

Ward, M. (1975) *Mountain Medicine. A Clinical Study of Cold and High Altitude*. London: Crosby Lockwood Staples.

Williams, E. S. (1966) Electrolyte regulation during the adaptation of humans to life at high altitudes. *Proceedings of the Royal Society (B)*, **165**, 266.

Williams, E. S. (1975) Mountaineering and the endocrine system. In: *Mountain Medicine and Physiology*, p. 38. Edited by C. Clarke, M. Ward & E. Williams. London: Alpine Club.

Williams, R. H. (1974) In: *Textbook of Endocrinology*, 5th Edition, p. 126. Philadelphia: W. B. Saunders.

Zakheim, R. M., Mattioli, L., Molteni, A., Mullis, K. B. & Bartley, J. (1975) Prevention of pulmonary vascular changes of chronic alveolar hypoxia by inhibition of angiotensin I-converting enzyme in the rat. *Laboratory Investigation*, **33**, 57.

26

Fertility and pregnancy

In the course of history some peoples have settled and developed at high altitude. The most celebrated of these were the Incas who established an empire which extended from the northern borders of present-day Ecuador to central Chile, a distance of some 3000 miles, and comprised a population of six million people (Marett, 1969). Their capital city, Cuzco (3400 m), still exists as a thriving town, and the remains of their military and religious buildings are to be found at such sites as Majchu Picchu (2440 m) (Fig. 3.3), and Sacsayhuaman (3580 m) (Fig. 3.8). No doubt the reasons for the development of this empire at high altitude during mediaeval times (1438–1533) was in part religious, in part historical and in part military. The inhospitable nature of the high altitude environment must in itself have given a measure of security against lowland enemies not prepared to pit themselves against the combination of mountain warriors and the effects of mountain air. Indeed from accounts of the battle at Junin (4400 m) in 1824 it would appear that some of the Spanish troops rapidly transported to that elevation from the coast frothed at the mouth, very possibly due to high altitude pulmonary oedema. These early settlers in the Andes discovered precious metals and the mining industry still maintains large active communities in such towns as Cerro de Pasco (4330 m).

The drift from the mountains

However, the military advantages of life in the mountains have now passed into history and the commercial attractions of mining are for the comparatively few. In this century there has been a flight from the unpleasant and potentially harmful environment of high altitude, although demographers have paid surprisingly little attention to this factor. Returns from the census in Peru in 1940 and 1961 (Fig. 26.1) show that in the earlier year two thirds of the population lived above 1750 m but 21 years later only just over half of the population lived above this altitude. If this trend continues, as it is

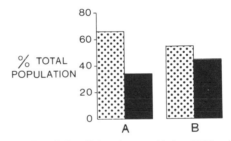

Fig. 26.1 Populations living above and below 1750 m in Peru in 1940 (A) and 1961 (B). Populations living above 1750 m indicated by stippled columns and those below that altitude by filled columns. (After Donayre, 1966).

likely to do, the next census will show that the majority of Peruvians live as lowlanders. Such a steady emigration to the coast without a compensatory immigration from the lowlands will clearly diminish the mountain population. However, the evidence shows that the situation at high altitude is further aggravated by a low birthrate. This is a result of sociological factors and infertility.

Infertility at high altitude

Stycos (1963) maintains that the birthrate is low in any predominantly Indian area because cohabitation between the sexes occurs later than in mestizos. His observations also suggest that sexual relationships are less permanent among the Indians. Another sociological effect which reduces the chances of conception, according to the studies of Heer (1964) in Bolivia and Ecuador, is the sustained hard daily work carried out by the women. Data from a population census in these three countries show that there is a strong negative correlation between birthrate and altitude (James, 1966). There is quaint historical evidence to support this for according to Monge and Monge (1966) the birth of the first Spaniard did not take place until 53 years after the founding of Potosi (4070 m) and was indeed called 'the miracle of Saint Nicholas of Tolentino' according to Father Calancha in 1639. Such a correlation between high altitude and infertility suggests that physiological as well as sociological factors operate at great elevations. Clearly infertility at high altitude could be due to changes in the male or female and we shall consider this problem now.

Changes in the male

Changes in the semen of animals at high altitude

The problem as to which sex is responsible for diminished fertility at high altitude has been of economic importance in Peru. Many ewes were slaughtered on the tacit assumption that they were infertile. However, Monge and San Martin (1942, 1944, 1945) suggested that the responsibility for the infertility was with the males. At sea level one or two rams were used to serve a hundred ewes but at high altitude five or seven males were required. Mature spermatozoa are not affected by the hypoxia of high altitude (Monge and Monge, 1966), and the practical implication of this is that normal healthy rams can be kept at sea level so that acclimatized ewes at high altitude can be inseminated artifically with semen sent from sea-level locations. When the rams themselves are taken to high altitude, however, their semen shows abnormalities. Monge and San Martin (1942) studied the semen of two rams transported to an altitude of 3260 m and found that after 50 days exposure to these conditions there was azoospermia and severe oligospermia with an increase of abnormal forms and low motility. Only one of the animals showed reversibility of these changes after five months of living at high altitude. Of 20 rams kept for 30 days at 3260 m three showed azoospermia and four oligospermia.

Changes in the semen of men at high altitude

Donayre (1966) reports on the effect of exposure to high altitude on the seminal characteristics of 10 men. The semen of these subjects was studied at sea level and after 8 and 13 days exposure to an altitude of 4330 m. His studies with Guerra-Garcia, Moncloa and Sobrevilla showed a pronounced decrease in the sperm count, dropping from a mean of 216.2×10^6/ml to 98.2×10^6/ml (Fig. 26.2a). There was an increase in abnormal forms from zero to 39.3 per cent (Fig. 26.2b), and at the same time there was a decrease in motile forms (Fig. 26.2b). The levels of fructose increased (Fig. 26.2c), probably due to lack of utilization by the non-motile spermatozoa (Donayre, 1966). At the same time the citric acid level in the semen was found to decrease and this was associated with elevation of the pH of the seminal fluid to slight alkalinity. The changes described above reveal that acute exposure to the hypoxia of high altitude brings about mild and reversible damage to the seminiferous epithelium. Since the seminiferous cycle of man averages about 72 days the long term effects of hypoxia on the seminiferous epithelium will require further study.

Subsequent studies were carried out by Donayre (1968a and b) on the semen of nine young healthy lowlanders who were taken to 4270 m for four weeks after a five week control period at sea level. Sperm count, expressed as total number of sperm, dropped gradually from a control level of 7.42×10^8 to the 27th day of the exposure period and continued declining 15 days after descent to sea level. Motile cells decreased from control values of 85.8 per cent to 53.4 per cent at the end of the experimental period, although the percentage of live cells was normal. Return to sea level restored motility after 15 days. The incidence of abnormal forms increased from a control level of 15.5 per cent to 31.6 per cent. This incidence of abnormal forms continued after return to the coast. Initially alterations of the neck and middle piece occurred with the formation of tapering forms. Later structural alterations occurred in the head. Hence it appears that high altitude induces morphological changes in the spermatozoa while the continuing fall in the sperm count suggests gradual damage at an early stage of the spermatogenic cycle. The rapidity with which these changes occur suggests that there

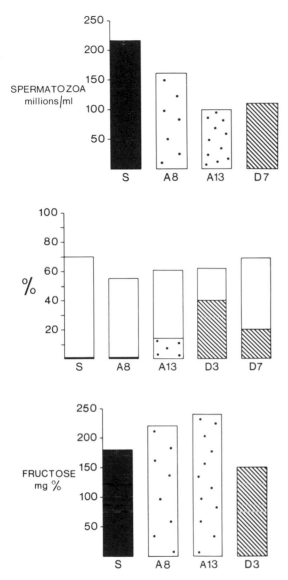

Fig. 26.2 (a) The number of spermatozoa in millions per ml in the semen of 10 men at sea level (S), after 8 (A8) and 13 (A13) days exposure to an altitude of 4330 m and 7 days after return to sea level (D7). Acute exposure to high altitude produces a pronounced decrease in the number of sperms but this effect is reversible after return to sea level. (After Donayre, 1966).

(b) The percentage of motile (open columns) and abnormal forms (hatched and stippled columns) in spermatozoa in semen of 10 men at sea level (S), after 8 (A8) and 13 (A13) days exposure to an altitude of 4330 m and 3 (D3) and 7 (D7) days after descent to sea level. On acute exposure to high altitude there is a reversible fall in the number of motile forms and a reversible rise in the number of abnormal forms. (After Donayre, 1966).

(c) The fructose content of semen, in mg%, of 10 men at sea level (S), after 8 (A8) and 13 (A13) days exposure to an altitude of 4330 m and 3 days (D3) after descent to sea level. On acute exposure to high altitude the fructose content of the semen rises but falls on return to sea level. (After Donayre, 1966).

is an effect on epididymal sperm. The decrease in motility seems to be related to complex metabolic alterations rather than to the changes in the fructose level shown in Fig. 26.2c. The alterations in morphology of the sperms brought about by chronic hypoxia are similar to those induced by viral infections or allergic reactions (MacLeod, 1965) suggesting that the testis has a limited variability of response. It is clear from these studies that hypoxia has profound effects on spermatogenesis.

Histological changes in the testes

The testicles of male rats transported to 4500 m were found to be devoid of germinal epithelium with Sertoli cells replacing spermatogonia (Mori-Chavez, 1936; Monge and Mori-Chavez, 1942). Leydig cells were increased noticeably in number. Similar lesions were also found in rabbits. Guerra-Garcia (1959) showed that the exposure of guinea pigs born and bred at sea level to an altitude of 4330 m for two weeks produces profound alterations in the epithelium of the seminiferous tubules with a marked decrease of all cellular types.

Working with simulated altitudes of 7620 m Altland (1949) found severe destruction of the germinal epithelium starting on the third day of exposure leading to considerable damage at the fourteenth day. The alterations showed no spontaneous improvement during the experiment. However, recovery took place four weeks after the end of the experiment. Simulated altitudes of 5790 m led to pronounced degeneration of the germinal cells of guinea pigs after 60 days (Shettles, 1947). The occurrence of acellular areas and the sloughing of cells in the seminiferous tubules of rats exposed to intermittent low pressure equivalent to elevations of 6550 m has been reported by Altland and Highman (1968). According to Monge and Mori-Chavez (1942) cats and rabbits show various degrees of destruction of the germinal epithelium six months and 15 days respectively after natural exposure to an altitude of 4510 m.

Cold, hypoxia and spermatogenesis

As we have seen in Chapter 2 the factors of cold and hypoxia both operate at high altitude and Donayre (1968a) has attempted to distinguish between their effects on spermatogenesis by exposing rats to high altitude in two groups, one in a room without temperature control (6° to 18°C) and the other in heated rooms (23°C ± 1°C). He found that in the rats exposed to cold, testicular changes occurred after four weeks. They comprised cellular disorganization, sloughing of epithelium, absence of spermatids and the appearance of multinucleated cells. With the passage of time more and more tubules were involved so that up to a quarter of the germinal epithelium was affected. Changes occurred earlier in the epididymis within one to two weeks and mature sperms began to be replaced by an accumulation of immature forms. When the temperature was controlled, spermatogenesis was carried to completion up to the sixth week. By the seventh week changes similar to the 'cold' group began to appear. Again the epididymis showed earlier changes with immature forms being present from the third week. Donayre (1968a) believes two separate processes are induced by high altitude. One is the arrest of spermatogenesis with morphological alterations. The other is interference with migration of sperm to the epididymis which are normally occupied by mature sperm released from the testis but which at high altitude contain abundant immature forms.

Changes in the female

The female reproductive cycle

In non-primate mammalian species there is a period of sexual receptivity at a particular phase of the reproductive cycle. This period coincides with the time of maximal oestrogen secretion prior to ovulation and is known as oestrus or 'heat'. Any changes in the oestrous cycle of animals or the human menstrual cycle are of interest in considering fertility at high altitude since these cycles represent the first stage in reproduction.

Oestrous cycle in the rat

The oestrous cycle of the rat is very sensitive to changes in the environment and the two principal stimuli inherent in exposure to high altitude which affect the cycle are hypoxia and cold. Donayre (1969) has attempted to differentiate between the effects of hypoxia and cold. He studied 120 female rats at sea level for 30 days and found that they had an oestrous cycle of four to five days with an incidence of oestrous days of 35 per cent. His definition of oestrus was the detection of a vaginal smear containing characteristic cornified epithelial cells. The rats were then transported to 4330 m where they were divided into three groups exposed to the high altitude environment for respectively 30, 60 and 90 days. Each group was further subdivided into two. Half were exposed to a temperate environment of 23°C while the remainder were kept in an area where the temperature varied from 9 to 16°C. Both groups showed a fall in the frequency of oestrus on initial exposure to high altitude. In the 'cold group' the incidence of oestrous days fell to 13 per cent but in the 'warm group' it fell to only 27 per cent. However, after this initial fall, the incidence of oestrus rose in both groups to exceed that of sea-level controls. In the warm group the incidence of oestrous days rose to 66 per cent and even in the cold group it rose to 49 per cent. After 50 days the warm rats continued to exhibit sustained oestrus whereas the cold animals showed a fall to values not significantly different to those found at sea level.

On return to sea level the warm animals exposed to high altitude for 30 and 60 days returned to a normal cycle within 10 days. However, the group exposed to high altitude for 90 days turned from sustained oestrus to anoestrus for the 30 days in which they were followed at sea level. The rats exposed to cold in addition to hypoxia showed a more erratic response on return to sea level. The 30-day group showed an increase in oestrus, the 60-day group a fall followed by an increase, while the 90-day group maintained the normal sea-level cycle length achieved at high altitude.

Hence rats exposed to cold in addition to hypoxia showed a lesser incidence of oestrous days at all times than the warm groups exposed to hypoxia alone. After 50 days the warm rats showed a pronounced increase in the incidence of oestrous days while the cold animals had returned to sea-level values. Thus it seems that cold superimposes an anoestrous effect on the pronounced oestrus produced by hypoxia alone. After 50 days the opposing effects of hypoxia and cold produce a situation where the incidence of oestrus is similar to that seen at sea level. Therefore, it seems that these changes in the oestrous cycle cannot wholly account for the decrease in fertility reported at high altitude.

Delayed menarche

Menarche is delayed in school-age girls living at high altitude compared to sea-level residents according to Moncloa, quoted by Donayre (1966). At Cerro de Pasco (4330 m) only 38 per cent of girls aged 13 years had presented their menarche compared to 73 per cent in a comparable age group at sea level.

Menstrual disturbances

Donayre (1966) reports an increased incidence of menstrual disturbances at high altitude. This includes dysmenorrhoea, increased menstrual flow and irregular periods. The majority of the subjects questioned were taking contraceptive pills.

Umbilical arterial oxygen tension at high altitude

It was established many years ago that even at sea level the umbilical arterial oxygen saturation in the fetus is low (Huggett, 1927; Eastman, 1930; Barcroft, 1933). This is an expression of the resistance offered to the diffusion of oxygen by the tissue barrier of the placenta and to its intrinsic oxygen utilization. Indeed in Chapter 11 we have already pointed out that its level has been reported as barely exceeding 20 mmHg which corresponds to an atmospheric oxygen tension of about 60 mmHg which would be found at an elevation of 7500 m. Hence even at sea level the fetus is hypoxaemic and lives under physiological conditions which in some respects resemble those experienced by the native highlander. This being so, one might anticipate that during pregnancy at high altitude, where oxygen tension in the systemic arteries and hence maternal placental capillaries is low, even more pronounced hypoxaemia might be suffered by the fetus, perhaps endangering its very survival.

This interesting problem has been studied by Metcalfe et al (1962) who investigated 21 pregnant ewes carrying fetuses ranging from 52 to 135 days in gestational age at Morococha (4540 m) where they had been bred and pastured. They determined the oxygen and carbon dioxide contents and tensions in the maternal arterial, uterine venous, umbilical arterial and umbilical venous bloods.

First, their studies confirmed that at high altitude there is a distinct reduction in the oxygen tension in the uterine and maternal placental capillaries. Thus while oxygen tension in the maternal uterine capillaries at sea level is 63 mmHg (Barcroft et al, 1940; Kaiser and Cummings, 1957) at an altitude of 4540 m it is 41.3 mmHg (Metcalfe et al, 1962). The effective carbon dioxide tension in uterine capillaries is lowered by 11 mmHg from a range of 35 to 36 mmHg to that of 24 to 25 mmHg.

However, the remarkable fact is that the studies of Metcalfe et al (1962) demonstrated that despite the lowered oxygen tension in the maternal uterine and placental capillaries, the oxygen tension in the umbilical vessels is similar to that reported for fetuses carried by ewes at sea level. Thus the oxygen tension in the umbilical vein at sea level is in the range of 21.3 to 27.5 mmHg with an average value of 24.9 mmHg (Kaiser and Cummings, 1957). At 4540 m its range is 14.2 to 28.1 mmHg with an average value as high as 22.0 mmHg, virtually the same level as that seen at sea level (Metcalfe et al, 1962). The oxygen tension in the umbilical arteries at sea level is in the range of 11.6 to 15.4 mmHg with an average value of 13.4 mmHg, (Kaiser and Cummings, 1957). At 4540 m its range is 10.4 to 14.8 mmHg with an average value of 12.5 mmHg (Metcalfe et al, 1962). Once again this value is not much different from that found at sea level. These results demonstrate that despite the lowered transplacental gradient in oxygen tension the rate at which oxygen reaches the fetal blood per kilogram of tissue supplied is the same at high altitude as at sea level. It seems likely that at high altitude the placenta is modified in some way to decrease the resistance to the diffusion of oxygen across the placental barrier, or to increase the surface area across which gas exchange takes place. We may now consider the modifications in the structure of the placenta which are known to occur at high altitude.

The placenta and the weight of newborns at high altitude

At high altitude newborns weigh less and placentas weigh more. Krüger and Arias-Stella (1970) studied 118 pregnancies at Lima (150 m) and 84 pregnancies at Rio Pallanga (4600 m). They found that fetuses at this altitude had a mean weight some 16 per cent below those from Lima (Fig. 26.3). The difference was more pronounced in female newborns and in multiparous pregnancies. In contrast the average weight of the placenta was 12 per cent higher in the high altitude cases (Fig. 26.3).

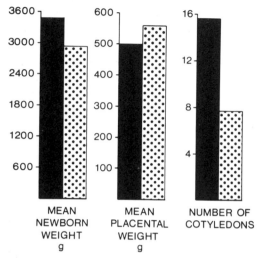

| | MEAN NEWBORN WEIGHT g | MEAN PLACENTAL WEIGHT g | NUMBER OF COTYLEDONS |

Fig. 26.3 Mean newborn weight, mean placental weight, and number of placental cotyledons in 118 pregnancies at sea level and in 84 pregnancies at 4600 m. Data from sea level indicated by filled columns and from high altitude by stippled columns. (After Krüger and Arias-Stella, 1970).

In this case there was no significant difference in relation to the sex of the newborn infant. However, the placental weight in the primiparous at high altitude was 23 per cent higher whereas in multiparous pregnancies it was only 9 per cent higher. There is evidence that the same thing occurs in animals. Prystowsky (1960), in observations of the placenta in sheep living at altitudes between 4570 m and 4880 m, reported that the placentas appeared to be much larger and the fetuses slightly smaller, for a given gestational age, than those of sea-level sheep. Krüger and Arias-Stella (1970) define a 'placental coefficient' as the weight of the placenta related to the weight of the newborn infant. In cases from Lima this coefficient was 0.144 but in those from Rio Pallanga it was 0.192, a highly significant difference. The upper limit of normal for this coefficient at sea level is given as 0.180 by Little (1960). Placentas at sea level have twice as many cotyledons as those from high altitude (Fig. 26.3). Furthermore the sulci between the cotyledons of the high altitude placentas are poorly delineated.

The smaller birth weight at high altitude could be attributed to the lesser heights and weights of the mothers in the Andes and also to the lower nutritional status of the mountain dweller. However, one would expect such factors to result in a lower placental weight whereas in fact as we have seen the placenta weighs more at high altitude. The increased weight of the placenta is not readily explicable since there is no concomitant increase in placental volume. Krüger and Arias-Stella (1970) believe that some of the histological structures of the placenta are more densely distributed per unit of placental volume at high altitudes for according to Monroe Rodriguez (1966) at high altitude the area of intervillous space is smaller than at sea level.

The low birth weight at high altitude has also been studied by Lichty et al (1957) in relation to the problem of prematurity in Colorado. Defining prematurity as a birth weight of less than 2500 g, they found its incidence to be 30.8 per cent for high altitude areas of the state (more than 3050 m), compared with an average of 10.4 per cent for the state as a whole. However, the neonatal mortality was not proportionally high. This prompted them to suggest that the lower limit of birth weight of 2500 g was not applicable to high altitude communities for distinguishing between premature and full term births. In other words babies weighing less than 2500 g at birth are not premature, just lighter. McCullough et al (1977) confirmed that in Colorado there is a progressive decrease in birth weight with increasing altitude. They found that low birth weight was not an adverse factor among *full-term* infants born at higher altitudes. However, the mortality rate among premature babies at altitudes between 2740 and 3100 m was almost double that found at altitudes between 1140 and 2130 m.

As we have already said the greater size of the placenta, probably associated with a greater density of its histological components, may be associated with oxygen uptake for the fetus at high altitude. Thus Howard et al (1957) reported that the oxygen arterial blood saturation and haematocrit value of the newborn at an altitude of 3050 m do not differ from those found in the newborn at sea level. This is an interesting example of adaptation of the placenta as an organ of nutrition and respiration during fetal life. Another factor that has been considered to favour oxygenation of the fetus at high altitude is hyperventilation of the mother during pregnancy. Hellegers et al (1961) have demonstrated with a group of pregnant women living at 4270 m that pregnancy at high altitude is accompanied by a hyperventilatory effect even higher than the hyperventilation already present in women living at high altitudes.

Fetal haemoglobin

Fetal haemoglobin differs from the adult variety in that the polypeptide chains usually termed beta are so different chemically that they are designated gamma chains. It has a greater affinity for oxygen than the adult form and its oxygen-haemoglobin dissociation curve is displaced to the left. Thus at a PO_2 of 20 mmHg fetal haemoglobin is 70 per cent saturated as compared with only 20 per cent saturation for adult haemoglobin. It has been suggested (Barker, 1964) that the alterations to the placenta at high altitude described above may be effective enough to prevent hypoxia of fetal tissues becoming severe enough to induce the formation of more HbF in the human and ovine fetus. However, according to Barker (1964) fetal and neonatal rats, mice, rabbits and puppies exposed for up to 20 hours a day to altitudes which do not inhibit growth, increase their HbF production. Such species as the rat do not increase placental weight at high altitude. Thus it would appear that while some species utilise the placental mechanism for adjustment to the hypoxia of high altitude other species may utilise the alternative HbF response (Barker, 1964).

Development of the chick embryo at high altitude

As we have just seen, in sheep the interposition of maternal systems in the form of the placenta between the fetus and the external environment provides substantial, if not complete, protection from environmental hypoxia encountered at high altitude. This appears to be true of mammalian species in general for Moore and Price (1948) found no abnormality of gestation in rats, mice and hamsters at an elevation of 4340 m, although mortality rates in the newborn were excessive. Developing avian embryos lack any similar protection and decreased hatchability of chicken and turkey eggs at elevations above 1200 m has been

recognised since the last century. The phenomenon was reviewed by Taylor in 1949. The chick embryo is especially valuable for developmental studies at high altitude being available in abundance in standard strains, permitting statistical treatment of results. Also the embryogeny of the chick is well known so that standard data are available for comparison. Sets of fertile eggs were incubated at sea level and at 3100 m and 3800 m by Smith et al (1969). They found that hypoxia leads to an increase in mortality rate throughout incubation but especially at an unusual time, namely during the second week of incubation. Usually embryo mortality occurs either during the first week of incubation, when it is due to difficulties in organogenesis, or during the third week due to 'positioning difficulties', reorientation of the embryo in preparation for the actual hatching process. Hypoxic death to the embryo during mid-incubation appears to be metabolic in nature and resembles that due to moderate hypothermia. It is likely that this similarity in the developing chick embryo is due to its poikilothermy. In contrast, the effects of hypothermia and hypoxia are dissimilar in the homeothermic adult.

Embryonic growth in the chick is slowed by hypoxia but principally in the early and late stages of incubation before the tenth day and after the sixteenth day of incubation (Smith et al, 1969). Hypothermia slows growth in the early phase but moderately stimulates late growth. The 'differentiation rate', that is the chronology of attaining specific developmental stages, is also repressed by hypoxia but in a manner different from the growth effect. Differentiation is mainly affected between the tenth and thirteenth days, whereas growth is most involved after the sixteenth day, when half of embryonic growth normally takes place. The repression of embryonic growth is not uniformly shared by the organs and the most pronounced effect is on the brain (Smith et al, 1969).

Polycythaemia and cardiac hypertrophy in chicks hatched and raised at high altitude

Burton and Smith (1969a) studied the rates of development of polycythaemia and cardiac hypertrophy in the embryo and neonatal chick to determine how chicks hatched and raised at high

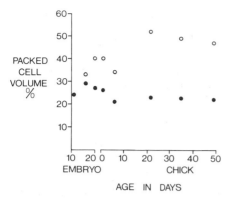

Fig. 26.4 The packed cell volume in chicks hatched and raised at sea level (closed circles) and at 3800 m (open circles). The packed cell volume is higher in the high altitude chicks.

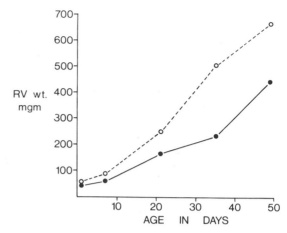

Fig. 26.5 Right ventricular weight in chicks hatched and raised at sea level (closed circles) and at 3800 m (open circles). The weight of the right ventricle is greater in the high altitude chicks.

altitude acclimatize to a hypoxic environment at a critical age for survival (Figs. 26.4 and 26.5). They found that such chicks prepare for their hypoxic environment by an increased haematocrit at hatch, which is a change requiring several days residence at high altitude to achieve (Figs. 26.4 and 26.5).

Burton and Smith (1967) found that the percentage packed cell volume in 30 hens at sea level was 29.2 per cent (± 0.58). In 38 hens after living for one year at White Mountain Research Station (3810 m) it rose to 41.5 per cent (± 0.90). In 19 hens who lived at this altitude for one year and then returned to sea level for an additional year the packed cell volume fell again to 27.5 per cent (± 1.14). The haematocrits of some of these birds at sea level and at 3810 m were chronically altered with the use of sex hormones. Androgens increased the birds' haematocrits and tolerances to acute hypoxia at both elevations, whereas oestrogens had the opposite effects (Burton et al, 1969b).

REFERENCES

Altland, P. D. (1949) Effect of discontinuous exposure to 25,000 feet simulated altitude on growth and reproduction of the albino rat. *Journal of Experimental Zoology*, **110**, 1.

Altland, P. D. & Highman, B. (1968) Sex organ changes and breeding performance of male rats exposed to altitude: Effect of exercise and training. *Journal of Reproduction and Fertility*, **15**, 215.

Barcroft, J. (1933) The conditions of foetal respiration. *Lancet*, ii, 1021.

Barcroft, J., Kennedy, J. A. & Mason, M. F. (1940) Oxygen in the blood of the umbilical vessels of the sheep. *Journal of Physiology*, **97**, 347.

Barker, J. N. (1964) In: *The Physiological Effects of High Altitude*, p. 125. Edited by W. H. Weihe. Oxford: Pergamon Press.

Burton, R. R. & Smith, A. H. (1967) Effect of polycythemia and chronic hypoxia on heart mass in the chicken. *Journal of Applied Physiology*, **22**, 782.

Burton, R. R. & Smith, A. H. (1969a) Induction of cardiac hypertrophy and polycythemia in the developing chick at high altitude. *Federation Proceedings, Federation of American Societies for Experimental Biology*, **28**, 1170.

Burton, R. R., Smith, A. H., Carlisle, J. C. & Sluka, S. J. (1969b) Role of hematocrit, heart mass, and high-altitude exposure in acute hypoxia tolerance. *Journal of Applied Physiology*, **27**, 49.

Calancha, A. de la. (1639) *Crónica Moralizada de la Orden de San Augustin*. Barcelona Vol. 1.

Donayre, J. (1966) Population growth and fertility at high altitude. In: *Life at High Altitudes*, p. 74. Washington: Pan American Health Organization. Scientific Publication, No. 140.

Donayre, J. (1968a) Effect of high altitude on spermatogenesis. *Excerpta Medica International Congress Series No. 184. Proceedings of the Third International Congress of Endocrinology*. Mexico, June 1968.

Donayre, J. (1968b) Endocrine studies at high altitude. IV Seminal changes in men exposed to altitude. *Journal of Reproduction and Fertility*, **16**, 55.

Donayre, J. (1969) The oestrous cycle of rats at high altitude. *Journal of Reproduction and Fertility*, **18**, 29.

Eastman, N. J., (1930) Foetal blood studies: (1) The oxygen relationships of umbilical cord blood at birth. *Bulletin of the Johns Hopkins Hospital*, **47**, 221.

Guerra-Garcia, R. (1959) *Hipófisis, adrenales y testiculo de cobayos a nivel del mar y en la altitud*. Br. Thesis. Universidad Nacional Mayor de San Marcos, Facultad de Medicina.

Heer, D. M. (1964) Fertility differences between Indian and Spanish speaking parts of Andean countries. *Population Studies*, **18**, 71.

Hellegers, A., Metcalfe, J., Huckabee, W., Prystowsky, H.,

Meschia, G. & Barron, D. H. (1961) Alveolar P_{CO_2} and P_{O_2} in pregnant and non-pregnant women at high altitude. *American Journal of Obstetrics and Gynecology*, **82**, 241.

Howard, R. C., Lichty, J. A. & Bruns, P. D. (1957) Studies of babies born at high altitude. (1) Arterial oxygen saturation and haematocrit values at birth. *Diseases of Children*, **93**, 670.

Huggett, A. St. G. (1927) Foetal blood-gas tensions and gas transfusion throughout the placenta of the goat. *Journal of Physiology*, **62**, 373.

James, W. H. (1966) The effect of altitude on fertility in Andean countries. *Population Studies*, **20**, 97.

Kaiser, I. H. & Cummings, J. B. (1957) Hydrogen ion and hemoglobin concentration, carbon dioxide and oxygen content of blood of the pregnant ewe and fetal lamb. *Journal of Applied Physiology*, **10**, 484.

Krüger, H. & Arias-Stella, J. (1970) The placenta and the newborn infant at high altitudes. *American Journal of Obstetrics and Gynecology*, **106**, 586.

Lichty, J. A., Ting, R. Y., Bruns, P. D. & Dyar, E. (1957) Studies of babies born at high altitude. 1. Relation of altitude to birth weight. *Diseases of Children*, **93**, 666.

Little, W. A. (1960) The significance of placental/fetal weight ratios. *American Journal of Obstetrics and Gynecology*, **79**, 134.

MacLeod, J. (1965) Human seminal cytology following the administration of certain anti-spermatogenic compounds. In: *Agents Affecting Fertility*, p. 93. Edited by C. R. Austin & J. S. Perry. Boston, Mass: Little, Brown.

Marett, R. (1969) In: *Peru*, p. 39. London: Ernest Benn.

McCullough, R. E., Reeves, J. T. & Liljegren, R. L. (1977) Fetal growth retardation and increased infant mortality at high altitude. *Archives of Environmental Health*, **32**, 36.

Metcalfe, J., Meschia, G., Hellegers, A., Prystowsky, H., Huckabee, W. & Barron, D. H. (1962) Observations on the placental exchange of the respiratory gases in pregnant ewes at high altitude. *Quarterly Journal of Experimental Physiology*, **47**, 74.

Monge, M. C. & Monge, C. C. (1966) In: *High Altitude Diseases.*

Mechanism and Management. Springfield, Illinois: Charles C. Thomas.

Monge, M. C. & Mori-Chavez, P. (1942) Fisiologia de la reproducción en la altura. *Anales de la Facultad de Medicina de Lima*, **25**, 34.

Monge, M. C. & San Martin, M. (1942) Nota sobre la azoospermia de corneros recien llegados a la altura. *Anales de la Facultad de Medicina de Lima*, **25**, 58.

Monge, M. C. & San Martin, M. (1944) Fisiologia de la reproducción en el altiplano. *Ann. III a Convención Agronómica Lima.*

Monge, M. C. & San Martin, M. (1945) Aclimatación avina en los altiplanos andinos. Infertilidad reversible debida a la acción del viaje maritimo de Magallanes al Callae durante el varano. *Anales de la Facultad de Medicina de Lima*, **28**, 1.

Monroe Rodriguez, A. G. (1966) Br. Thesis. Universidad Nacional de San Augustin de Arequipa, Facultad de Medicina.

Moore, C. R. & Prince, D. (1948) A study at high altitude of reproduction, growth, sexual maturity, and organ weights. *Journal of Experimental Zoology*, **108**, 171.

Mori-Chavez, P. (1936) Manifestacione pulmonares del conejo del llama trarispontado a la altura. *Anales de la Facultad de Medicina de Lima*, **19**, 137.

Prystowsky, H. (1960) In: *The Placenta and Fetal Membranes*, p. 151. Edited by C. A. Villee. Baltimore: Williams and Wilkins.

Shettles, L. B. (1947) Effects of low oxygen tension on fertility in adult male guinea pigs. *Federation Proceedings: Federation of American Societies for Experimental Biology*, **6**, 200.

Smith, A. H., Burton, R. R. & Besch, E. L. (1969) Development of the chick embryo at high altitude. *Federation Proceedings, Federation of American Societies for Experimental Biology*, **28**, 1092.

Stycos, J. C. (1963) Culture and differential fertility in Peru. *Population Studies*, **16**, 257.

Taylor, L. W. (1949) In: *Fertility and Hatchability of Chicken and Turkey Eggs*. New York: Wiley.

27

Adaptation and acclimatization

Not all men and animals living at high altitude are of equal biological status. In this chapter we consider the various groups that can be recognised and the difference in their responses to the hypoxia of high altitude.

'High altitude man', fact or fiction?

The Quechua Indian who is born at high altitude is able to survive in this adverse environment leading an active, healthy life (Figs. 27.1a to c). The life span of the majority of highlanders is not shortened by exposure to the chronic hypoxia inherent in living at these great heights. More likely causes of death in such mountainous areas as the Andes are endemic infectious diseases such as tuberculosis and infectious hepatitis. So effectively in fact does the highlander cope with the chronic hypoxia of his mountain habitat that there has grown up the concept of a 'high altitude man' inherently more capable of surviving and living a normal active life at high altitude.

The clash of opinions is well exemplified in the discussions which took place at a meeting of the Pan American Health Organization Advisory Committee on Medical Research in Washington in 1966. The background to these discussions was the classical view of Monge (1948) that the indigenous

Fig. 27.1 (a) A native of Cuzco (3400 m) whose *natural acclimatization* enables him to lead an active life at high altitude. (b) A Quechua woman of Tarma (3000 m) carrying a load to market. Children are usually carried on the back in a similar manner. Such physical tasks at great elevation are made possible by *natural acclimatization*. (c) A native highlander of Cuzco (3400 m) whose *natural acclimatization* enables him to carry loads so large as almost to engulf him.

races became suited to high altitudes through an age-long process and that 'the Man of the Andes possesses biological characteristics distinct from those of sea-level man'. This view has perhaps not unnaturally been championed by the Peruvian school. Thus Hurtado (1966) (Fig. 27.2) states that 'between a high-altitude native and a man originally from sea level who has been at high altitudes for some time, there are some definite differences', or again 'We observed that no matter how long a man from sea level stays at a high altitude, his efficiency and his tolerance for maximal work were a great deal lower'. This view is supported by some Europeans

Fig. 27.2 Professor Alberto Hurtado, the Father of High Altitude Medicine, talking to one of the authors (D.R.W.) in Lima in August 1979.

like von Muralt (1966) who says 'In the experience of every single European who has worked with Sherpas in the Himalayas, under any given conditions the Sherpa is superior to the European member of the party in physical performance except at the end . . . In Switzerland too I think this fact can be corroborated. We have a group that we call Mountain Guides, the grandsons of the famous mountain guides who took Whymper and others up in the pioneer times of Alpine discovery. My personal experience with these men is that, under extreme conditions, their performance and resistance not only to hypoxia but also to cold and other difficulties are superior to what any sportsman can exhibit'. Hurtado (1964) found that the Quechua had a very high tolerance to maximal exercise on a treadmill at high altitude. He thought a high work capacity is perhaps the best index of natural acclimatization of 'the Man of the Andes'.

However, in contrast Chiodi (1966) doubts this traditional concept and says 'there are few data on permanent biological features that would differentiate an Andean man from his sea level peer'. He believes that many of the biological characteristics of 'the Man of the Andes' are features acquired through a given subject's long residence at high altitudes. We are inclined to favour Chiodi's view and believe that the Quechua Indian is able to cope with his adverse environment so effectively in the main through the process of acclimatization.

Natural acclimatization and partial adaptation in the highlander

Acclimatization is a reversible, non-inheritable change in the anatomy or physiology of an organism which enables it to survive in an alien environment. We consider the multiplicity of these physiological and anatomical changes throughout this book and there is no need to repeat them here. They are both fully developed and effective in the Quechua and together constitute what Hurtado (1966) terms 'natural acclimatization'. Before accepting all the characteristics of the native highlander as features of natural acclimatization we must recognise that the problem is complicated by the fact that the Quechua Indian has certain anatomical attributes which may have arisen as a result of natural selection and to be of physiological advantage at high altitude. The outstanding example is the rounded, fuller chest, described by some authorities as 'barrel-shaped' (Chapter 4) which is said to increase the internal surface area of the lung and hence the blood-gas interface for respiratory exchange. Although the postulated increased internal surface area has never been satisfactorily demonstrated (Chapter 4), it could be argued that the configuration of the chest represents a *partial adaptation* to high altitude. As we shall note later, adaptation, in contrast to acclimatization, is the development of anatomical and biochemical features which are inheritable and of genetic basis, enabling the animal to explore the environment of high altitude to its best advantage.

Acquired acclimatization

The lowlander after residence for a prolonged period at high altitude also shows acclimatization *qualitatively* identical to that which occurs in the

Quechua but quantitatively different. This less effective *acquired acclimatization* is gained slowly and painfully. Velásquez (1964) studied at low altitude and at 4540 m the physical activity of 10 healthy young men who had been born and domiciled permanently at sea level. They were subjected to exercise on a treadmill. Within a month of ascent there was a profound decrease in endurance, falling to a mere 10 per cent of sea-level values in one instance. At the end of a year's residence at high altitude his subjects were still able to do only half as much work as at Lima and individual differences in the acquired acclimatization achieved were great. After this prolonged residence at high altitude considerable improvement was achieved but a return to sea-level capacity was still impossible. The studies of Velásquez (1964) do not indicate whether a sea-level man can eventually equal the physical capabilities of the indigenous high altitude native. Although acquired acclimatization is less efficient than the natural variety, it appears that the components of the two forms are the same.

Accommodation

The third biologically distinct class of subject at high altitude is the recently arrived lowlander who is acutely exposed to hypoxia. He reacts by such physiological changes as hyperventilation and tachycardia as we have seen in the earlier chapters. Such symptoms do not constitute acclimatization and may cause the subject considerable distress. These early reactions have been termed 'accommodation' by von Muralt (1966). They should not be confused with the symptoms of acute mountain sickness which are described in Chapter 14.

Loss of acclimatization

While it is tempting to accept that all the structural changes found in the acclimatized are beneficial, it is salutary to bear in mind that some of them are merely the response of tissues to hypoxia irrespective of whether or not they are advantageous in the high altitude environment. Thus muscularization of the terminal portions of the pulmonary arterial tree leads to pulmonary hypertension and this has been considered by some to lead to a better perfusion of the apices of the lung increasing the blood-gas interface. However, when the identical hypoxic stimulus is applied to rats exposed to simulated high altitude in a hypobaric chamber, if it is severe and prolonged enough, it will lead to death with right ventricular hypertrophy and associated oedema. This would appear to be a surprising outcome for something considered to be a feature of acclimatization. It seems more reasonable to suggest that muscularization of the pulmonary arterial tree is the inevitable effect of hypoxia on this structure irrespective of whether the effect is beneficial or not.

Overaccentuation of some of these features may in fact lead to loss of acclimatization. Thus the cardiovascular form of loss of acclimatization is seen in calves in high mountain ranges in Utah where it is manifested as 'Brisket disease' (Chapter 13). Its basis is persistent accentuated constriction of the naturally muscular bovine pulmonary vasculature. The respiratory form of loss of acclimatization is exemplified by 'Monge's disease' (Chapter 16) which is characterized by exaggerated arterial oxygen unsaturation, heightened haemoglobin levels and so on. These two diseases are examples of the fourth biological group at high altitude characterized by *loss of acclimatization*.

Adaptation in mammals indigenous to high altitude

Some species of animals such as the llama, alpaca, vicuña and yak are indigenous to high altitude (Fig. 27.3). They have lived for thousands of years in their mountain home and thrive in an environment which is unpleasant or dangerous to other species. It is of singular interest to recognise that their physiological characteristics differ significantly from those of acclimatization of the human body to the same environmental circumstances (Table 27.1). Although we are primarily concerned with man in this book, we may briefly consider some of these animals which illustrate the physiological characteristics of a species showing *adaptation* rather than *acclimatization*. For numerous years over countless generations these animals of the great mountains have through a process of natural selection become fully *adapted* to the hypoxia of high altitude. Here we use the term *adaptation* in a precise Darwinian sense rather than in a general sense. It is the

Fig. 27..3 Fully adapted alpacas (*Lama pacos*) grazing at La Raya (4200 m).

development of biochemical, physiological and anatomical features which are heritable and of genetic basis enabling the species to explore the environment of high altitude to its best advantage. Hence in the great mountains one finds living together, both successfully, acclimatized man and adapted indigenous high altitude animals. Clearly both acclimatization and adaptation can overcome satisfactorily the adversities of chronic deprivation

of oxygen but the physiological mechanisms appear to be distinct. Throughout this book we consider the components of acclimatization in man. Here we may dwell on the physiological components of adaptation in indigenous high altitude species. We shall first select the llama (*Lama glama*) for this camelid which is so characteristic of the High Andes of South America has been extensively studied by Banchero et al (1971).

Oxygen transport in the llama

These authors studied the response of three male llamas, 5 to 14 months old, born at sea level, first at 260 m and then again after 5 and 10 weeks at 3420 m. They found that the llama does not hyperventilate like man on exposure to the hypoxia of high altitude (Table 27.1). This is in agreement with the data of Brooks and Tenney (1968) who found that llamas have a low hypoxic threshold so that the hypoxic ventilatory response occurs only when PA_{O_2} drops below 60 mmHg and Pa_{O_2} falls below 40 mmHg. It would seem that for any given altitude and thus any given partial pressure of oxygen in inspired air, the llama operates at a lower Pa_{O_2} than man but at levels of systemic arterial

Table 27.1 Points of contrast between *acclimatization* in the Quechua Indian and *adaptation* in the llama at high altitude

Feature	Quechua	Llama
Major area where chronic oxygen deprivation is overcome	Diminution in magnitude of 'oxygen cascade' (Chapter 5).	Increased extraction of oxygen from tissues (low $P\bar{v}_{O_2}$) (Chapter 7)
Hyperventilation	Pronounced	Minimal or absent
Haematocrit	60 per cent	30 to 40 per cent
2,3-DPG in red cell	Raised level	Weak reaction between haemoglobin molecule and intraerythrocytic phosphates.
Shift of oxygen-haemoglobin dissociation curve	To right (P_{50} 26.8 mmHg)	To left (P_{50} 23.7 mmHg)
Carotid bodies	Enlarged	Not enlarged
Terminal portion of pulmonary arterial tree	Muscularized	Not muscularized
Right ventricle	Hypertrophied	Not hypertrophied

oxygen saturation which equal or exceed those found in man (Banchero et al, 1971). Hence with a Pa_{O_2} of 51 mmHg the llama does not show a sustained ventilatory response because at this level of Pa_{O_2} it maintains a high systemic arterial oxygen saturation of some 92 per cent. These features could be explained by a higher affinity of llama haemoglobin for oxygen and the shape of its oxygen-haemoglobin dissociation curve. When the llama ascends to altitude there is no decrease in the affinity of its haemoglobin for oxygen (P_{50} of 22.7 mmHg at sea level and P_{50} of 23.7 mmHg at 3420 m (Banchero et al, 1971). There is no rightward shift in the oxygen-haemoglobin dissociation curve which remains to the left of that of man, with a P_{50} of 23.7 mmHg compared to 27 mmHg in man.

In Chapter 6 we make reference to 2,3 diphosphoglycerate in the erythrocyte and its rôle in acclimatization to hypoxia. We note there that in the metabolism of glucose to pyruvate in the red blood cell 2,3 DPG is formed and has a tendency to be bound by reduced haemoglobin reducing affinity for oxygen and producing a shift of the oxygen-haemoglobin dissociation curve to the *right*. Increases of about a fifth in the concentration of 2,3 DPG occur in the red cells of man and domesticated guinea pigs at high altitude (Eaton et al, 1969; Torrance et al, 1970; Baumann et al, 1971). However, as Bullard (1972) points out, in animals indigenous to high altitude such as the llama, vicuña, alpaca and the yellow-bellied marmot there is a shift of the oxygen-haemoglobin dissociation curve to the *left*, suggesting that successful adaptation to high altitude in these animals does not involve an increase in concentration or action of 2,3 DPG (Table 27.1). As we note above the most successful residents at high altitude are the hoofed members of the order *Artiodactyla* including the yak, alpaca, vicuña, llama and guanaco and as we describe below there is evidence to suggest that in these species there is a weak reaction between the haemoglobin molecule and intraerythrocytic phosphates which in the guanaco at least have been shown to be present in normal quantities.

Another very characteristic feature of *acclimatization* to high altitude is an increase in blood haemoglobin concentration as we describe in Chapter 6. In that chapter, however, we also point out that llamas and vicuñas born and bred at high altitudes do not show high haemoglobin or haematocrit levels (Table 27.1). Banchero et al (1971) found that llamas born at sea level show a very modest rise in haemoglobin level on residence at 3420 m for 10 weeks and they presume that the mildness of the increase is due to the high systemic arterial saturation that this animal maintains in spite of a low Pa_{O_2}.

So far as adaptation of the llama is concerned the crux of the matter is that, *although the haemoglobin concentration is lower, the oxygen extraction in resting conditions is, as in man, about 40 to 45 ml/l* (Banchero et al, 1971). Since the oxygen content of systemic arterial blood is lower because of the comparatively low haemoglobin concentration the more efficient oxygen extraction implies diminished oxygen content and saturation of systemic venous blood as demonstrated by Banchero et al (1971). Hence in the llama the enhanced affinity of the haemoglobin for oxygen does not diminish the unloading of oxygen to the tissues. The factors concerned in tissue diffusion are described in Chapter 7 but we do not know of any investigations in this area specifically carried out on the tissues of the llama. One possible factor may be that the erythrocytes of the llama are small and numerous (Chapter 6) and have a characteristically ellipsoidal shape (Fig. 6.9) which may offer an increased surface area for oxygen diffusion. In this connection one has to bear in mind that the llama, like the other high altitude animals, is a camelid and the smallness of the red cells may be more a characteristic of the family than a specific altitude adaptation (Bullard, 1972). The reader will recall that in Chapter 4 we faced the same difficulty in assessing the functional significance of certain physical features of the Quechua Indian which may be more an expression of their ethnic background than their adaptation to life at high altitude.

Hence in summary (Table 27.1) it would seem that the physiological basis for adaptation in the llama is the very efficient utilization of oxygen by the tissues and the ability to operate at a $P\bar{v}_{O_2}$ lower than most other species even at sea level (Banchero et al, 1971) (Fig. 27.4). Mechanisms for reducing the oxygen cascade such as hyperventilation (Chapter 5) and increasing the haemoglobin concentration

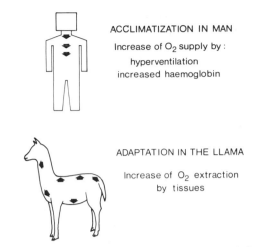

ACCLIMATIZATION IN MAN

Increase of O_2 supply by :
hyperventilation
increased haemoglobin

ADAPTATION IN THE LLAMA

Increase of O_2 extraction
by tissues

Fig. 27.4 Diagram to illustrate the physiological basis for acclimatization in man and adaptation in the llama.

(Chapter 6) which are so important in acclimatization in man, appear to play a very minor rôle in the adaptation of the llama to high altitude (Fig. 27.4).

Other features of adaptation in the llama

The llama shows other adaptations to high altitude which are not concerned with oxygen transport. They are microanatomical and appear to aid survival in an adverse hypoxic environment. Thus the pulmonary arteries of the llama remain thin-walled like those of sea-level man (Heath et al, 1974) and the terminal portion of its pulmonary arterial tree

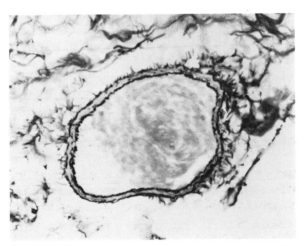

Fig. 27.5 Transverse section of a muscular pulmonary artery from an adult llama showing the thin media of circularly-orientated smooth muscle bounded by internal and external elastic laminae. (Elastic Van Gieson × 375).

does not show the muscularization which characterizes the human and bovine lung at high altitude (Chapters 11 and 13) (Fig. 27.5). The medial thickness of the pulmonary arteries in the llama is less than 4 per cent (Heath et al, 1969). As a result the llama does not develop hypertrophy of its right ventricle, the ratio of left to right ventricular weight being in the range of 3.0 to 3.5 (Heath et al, 1974), which is what one would expect in a normal lowlander. As a result of these structural features which are apparently inappropriate for such a hypoxic environment there is no tendency for this species to develop brisket disease as in calves which are low altitude animals and have a thick-walled, muscular pulmonary vasculature.

Oxygen transport and adaptation to high altitude in the alpaca

We may now consider the features of adaptation to high altitude in a second species, the alpaca (*Lama pacos*). Sillau and his colleagues (1976) studied oxygen transport in this animal at 3300 m and then after a sojourn of three months at sea level. In contrast to the studies of the llama by Banchero et al (1971) referred to above they found a slight diminution of Pa_{CO_2} at high altitude indicating a slight degree of hyperventilation. Just as in the case of the llama the haemoglobin of the alpaca has a high affinity for oxygen, the oxygen-haemoglobin dissociation curve lying to the left of that for man. According to Reynafarje et al (1975) the particular shape of the haemoglobin dissociation curve in high altitude camelids such as the alpaca facilitates oxygen uptake in the lung but permits the release of oxygen in the tissues with almost the same facility as in man. P_{50} was 17.8 mmHg at 3300 m and 19.7 mmHg at sea level, lower values than those found for the llama by Banchero et al (1971). The high affinity for oxygen of alpaca blood maintains a high systemic arterial oxygen saturation of 92 per cent at 3300 m and 95.3 per cent at sea level (Table 27.2). Just like its fellow high altitude camelids, the alpaca has a low haemoglobin level (11.8 g/dl) and a low haematocrit found by Sillau et al (1976) to be 27 per cent, an even lower figure than that given by Reynafarje (1966) (see Chapter 6) (Table 27.2). According to Reynafarje et al (1975) the adult alpaca retains a high percentage of HbF in its blood

Table 27.2   Some physiological data obtained from five alpacas at 3300 m (after Sillau et al, 1976)

Subject	Mean value
Body temperature (°C)	38.5
Pa_{O_2} (mmHg)	52.9
Sa_{O_2} (%)	92.0
Pa_{CO_2} (mmHg)	30.7
$P\bar{v}_{O_2}$ (mmHg)	26.2
$S\bar{v}_{O_2}$ (%)	62.8
Hb (g/dl)	11.8
Haematocrit (%)	27.0
pH a	7.465
pH $\bar{v}$	7.421

(Chapter 9). Thus like the llama, the alpaca appears to be a species that employs more efficient extraction of the smaller quantities of oxygen carried in its blood. It operates a lower $P\bar{v}_{O_2}$ than other mammals, especially at high altitude where its average value is only 26 mmHg (Table 27.2).

In discussing the llama we pointed out possible advantages for diffusion of oxygen to the tissues from the very small ellipsoidal erythrocytes. Earlier work suggests other aspects of the physiology of the alpaca that aids oxygen diffusion in the tissues of this species. The muscle of the alpaca is said to have high myoglobin concentrations the significance of which is discussed in Chapters 7 and 19. The diameter of some of the muscle fibres of the alpaca is as low as 38 μm (Sillau et al, 1976) and we have already seen in Chapter 7 how this aids tissue oxygen diffusion, when combined with increased capillary density. Lactic dehydrogenase activity is said to be increased in alpaca tissues six times above that in human beings (Reynafarje et al, 1975). Alpacas do not hyperventilate appreciably or possess particular cardio-respiratory mechanisms that would aid acclimatization. Like the other high altitude camelids they appear to be *adapted*. We may now give some consideration to the cause of the leftward shift of the oxygen-haemoglobin dissociation curve

which seems to characterize *adaptation* as contrasted to *acclimatization*.

The cause of the high affinity for oxygen of the haemoglobin of geese and camelids indigenous to high altitude

The high oxygen affinity of haemoglobin and the resultant shift to the left of the oxygen-haemoglobin dissociation curve which appear to be advantageous in adjusting to very high elevations, and which are so characteristic of animals showing adaptation rather than acclimatization to this environment, could in theory arise in three distinct ways (Petschow et al, 1977). First, the haemoglobin of the species concerned might have intrinsically a high affinity for oxygen. This appears to be the case in certain sheep and we shall describe this situation in the next section. Second, the concentration of the intraerythrocytic organic phosphates such as 2,3 diphosphoglycerate (2,3 DPG), adenosine triphosphate (ATP), or inositol pentaphosphate (IPP) which decrease the affinity of haemoglobin for oxygen may be low. This occurs in fetal as contrasted to adult pigs. Third, the interaction of intraerythrocytic organic phosphates and haemoglobin might be reduced.

The third explanation appears to be the appropriate one for geese and camelids adapted to high altitude. The partial pressure of oxygen at 50 per cent saturation of haemoglobin (P_{50}), and the concentration of various intraerythrocytic phosphate compounds were measured by Petschow and his colleagues (1977), in one species of goose from high altitude and two species from sea level. The high altitude species chosen for study was the bar-headed goose (*Anser indicus*) which migrates across the Himalayas from India to Tibet at an altitude of 10 000 m where the ambient P_{O_2} is some 50 mmHg (Swan, 1970). A species from the Andes in which similar results might be anticipated is the Bolivian goose, *Chloëphaga melanoptera* (Fig. 27.6). The sea-level species studied were the Canada goose (*Branta canadensis canadensis*) and the greylag goose (*Anser anser*). It was confirmed that there is a shift to the left of the oxygen-haemoglobin dissociation curve of the high altitude species, the P_{50} for the bar-headed goose being 29.7 mmHg while that for the Canada goose was 42.0 mmHg and that for the

Fig. 27.6 *Chlöephaga melanoptera*.

greylag goose was 39.5 mmHg. However, it also became apparent that this increased oxygen affinity of haemoglobin of the high altitude bird was not due to its having a lower level within its erythrocytes of the organic phosphates listed above. Indeed while the concentration of such phosphates in the bar-headed goose was 7.2 μmol/ml RBC that in the greylag goose was 7.9 μmol/ml RBC and that in the Canada goose was 9.1 μmol/ml RBC. The levels of the individual organic phosphates (IPP, ATP, ADP, and 2,3-DPG) were found to be very similar in high altitude and sea-level species. Rather it became clear that the cause of the increased oxygen affinity of the haemoglobin of the Himalayan goose rested on the fact that it reacted more weakly with the organic phosphates than did the haemoglobin of the sea-level species.

The cause of the pronounced affinity of haemoglobin of camelids indigenous to high altitude was found to be of the same basis. Thus Petschow et al (1977) studied the haemoglobin of the guanaco (*Lama guanicoe*) and compared it with that of man. Once again the leftward shift of the dissociation curve was demonstrated, the P_{50} of the guanaco being 22.5 mmHg in contrast with the value of 26.8 mmHg found in man. In the guanaco 2,3-DPG concentration is lowered by only some 20 per cent compared with the levels in man. Once again the high affinity of guanaco blood for oxygen depends more on the weak interaction of the haemoglobin molecule and the organic phosphates.

The affinity for oxygen in the haemoglobin of sheep at high altitude

In the case of sheep at high altitude the high affinity of blood for oxygen appears to be rather an expression of the intrinsic characteristics of the type of haemoglobin present. It is now established that the blood of healthy, adult sheep contains two types of haemoglobin, A and B. These types are inherited as Mendelian traits and are present in about equal amounts in the blood in heterozygotes (Battaglia et al, 1969). In anaemic animals with HbA a third type of haemoglobin, C, appears in large quantities in the blood (Van Vliet and Huisman, 1964). This type releases oxygen to the tissues more readily than HbA (Huisman and Kitchens, 1968). Hence in the sheep we have the unusual situation of one species carrying haemoglobins with widely different haemoglobin oxygen dissociation curves with their differing influences on oxygen transport in the blood and oxygen release to the tissues (Fig. 27.7). Battaglia et

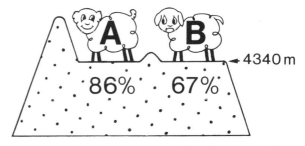

Fig. 27.7 At an altitude of 4340 m sheep homozygote for haemoglobin A show a systemic arterial oxygen saturation of 86%. Sheep homozygote for haemoglobin B show a saturation of 67%. (From data of Battaglia et al, 1969).

al (1969) found that at high altitude (4340 m) sheep carrying HbA had a mean systemic arterial saturation (86.4 per cent), significantly higher than HbB carriers (67.5 per cent) (Fig. 27.7). The carriers of a mixture of HbA and HbB had an intermediate saturation (77.0 per cent). P_{50} in sheep homozygote for HbA is 31 mmHg compared to that of 27 mmHg for sea-level man whereas in sheep homozygote for HbB, P_{50} is closer to 40 mmHg. Haemoglobin B appears to be advantageous to sheep at low altitude for here the P_{O_2} is sufficient to achieve 90 per cent arterial oxygen saturation while at the same time this type of haemoglobin will ensure a plentiful release of oxygen to the tissues. On the other hand at high altitude HbA affords an advantageous uptake of oxygen from the lung, although of course release of oxygen to the tissues is hindered (Fig. 27.7). Even so, as we have already noted, the P_{50} of sheep haemoglobin A is similar to that of blood from man at sea level.

Haematological features of the yak

The yak (*Bos grunniens*) belongs to the same genus as the cow, which as we have already seen in Chapter 13 is susceptible to brisket disease, but it is adapted, rather than acclimatized, to survive for long periods at altitudes of up to 6000 m in Tibet and the Himalayas. Its blood has certain characteristics of that other species adapted to high altitude in the Andes, the llama (*Lama glama*) (Table 27.3). Thus its red cells are small and this might affect blood viscosity or increased surface area for gas diffusion to the tissues but similar indices are found in the Jersey cow and, therefore, it seems unlikely that this

Table 27.3 Haematological observations on the yak (after Adams et al, 1975)

Subject	Value
White cell count	8 to 10 $\times$ 10^9/l^{-1}
% Neutrophils	35
% small lymphocytes	55
% large lymphocytes	6
% monocytes	4
% eosinophils	1 to 2
% basophils	rare
Platelet count	700 $\times$ 10^9/l^{-1}
Hb level (at 4000 m)	13 g/dl
Mean corpuscular volume	43 μm^3
Mean corpuscular haemoglobin	16.2 pg
Ratio of slow to fast Hb	38 : 62
2,3-DPG	0.04μM/gHb
(Man 2,3-DPG)	(10.30 μM/gHb)
P_{50}	26 mmHg

represents a specific form of altitude adaptation. Its level of haemoglobin is not raised (Table 27.3) as in acclimatized man. Levels of 2,3-DPG are extremely low in the yak (Table 27.3) in contrast to those found in man, this once more being an expression of adaptation rather than acclimatization. The P_{50} is similar to that of man (Table 27.3). There is an unusual steepness to the upper part of the oxygen-haemoglobin dissociation curve and this has been regarded as aiding survival at high altitude. As one would anticipate yak erythrocytes show no tendency to sickling (Adams et al, 1975) but it is of considerable interest to note that one of the two forms of yak haemoglobin to be referred to below (Hb slow) is prone to methaemoglobin formation which we have noted in llamas and highlanders in Chapter 9.

Although the yak is an animal adapted to high altitude, it is of singular interest to note that when this species is rendered anaemic by bleeding at high altitude, there is no evidence for the production of any new haemoglobin to aid oxygen transport to the tissues and yield to the cells there. Nevertheless, *Bos grunniens* has two haemoglobins, designated 'fast' and 'slow' (Adams et al, 1975). *Hb slow* is unique to the yak in the genus *Bos* but it is not yet known if it is of any importance in adaptation to high altitude perhaps in its affinity for oxygen. The two yak haemoglobins share a common globin chain. Five yaks studied by Adams et al (1975) had identical ratios of the two haemoglobins (Table 27.3).

Ward (1972) examined a yak killed for food during the ascent of Everest in 1953. On casual observation the right ventricle appeared almost as thick as the left. He attributed this right ventricular hypertrophy to increased resistance to the flow of blood through the lungs at high altitude.

The frogs of Lake Titicaca

Even animals much lower in the evolutionary scale may show physiological features which have enabled them to adapt to life at high altitude. *Telmatobius culeus* is an aquatic frog of Lake Titicaca situated at an altitude of 3810 m in the Andes on the border of Peru and Bolivia. It has become adjusted to the low partial pressure of oxygen in the water and to the coldness of its aquatic environment by means of a variety of developments in its morphology, physiology and behaviour. Thus the surface area of the skin is increased by pronounced numerous large folds which hang from the dorsum, sides and hind legs and these are highly vascularized by subepidermal plexi with cutaneous capillaries which penetrate to the outer layers of the skin. The buccal cavity is also highly vascularized.

The volume of the erythrocytes (394 μm^3) is the smallest reported for any amphibian (Hutchinson et al, 1976), and the erythrocyte count (729 × $10^3/mm^3$) is the highest value known for an anuran (Hutchinson et al, 1976). The oxygen capacity (11.7 vol per cent), haemoglobin (8.1 g/dl), haemoglobin concentration (0.281 pg/μm^3) and haematocrit (27.9 per cent) are all elevated in comparison with most amphibians. The oxygen-haemoglobin dissociation curve is sigmoid and the Bohr factor is small. The P_{50} (15.6 mmHg) is the lowest recorded in an anuran. After removal to a lower altitude at Oklahoma and sojourn there for one to six months, specimens of *Telmatobius culeus* showed a decrease in erythrocyte count (557 × $10^3/mm^3$) and the haemoglobin and haematocrit levels fell (Hutchinson et al, 1976).

The metabolic rate is the lowest reported for a frog and indeed among amphibians only the giant salamanders have lower values. In the natural state the frogs have never been seen to surface for air and they swim from near the surface to the maximum depth of the lake of 280 m. Under these circumstances, and, if prevented from surfacing in hypoxic waters in the laboratory, the frogs ventilate the skin by 'bobbing behaviour'. If the frogs do surface for any reason, they ventilate the small lungs and the metabolic rate increases.

It is clear that this collection of morphological, physiological and behavioural adjustments allow *Telmatobius culeus* to survive in the waters of the high altitude lake where an infinite supply of oxygen exists at low pressure. The specialised skin serves as a gill allowing the animal to obtain sufficient oxygen to meet its metabolic requirements. The high oxygen affinity of the haemoglobin enhances the uptake of what oxygen is available and the high haematocrit and erythrocyte count facilitate the transport of oxygen. The very small erythrocytes very likely aid the supply of oxygen to the tissues. Behavioural adaptations such as the increased contact between skin and water during 'bobbing

behaviour' and the possibility of pulmonary gas exchange are available to supplement oxygen supply, if needed. We have personally studied the behaviour of another high altitude amphibian, *Batrachophrynus macrostomus* which is to be found in lakes around Rancas in Central Peru (Fig. 27.8).

Fig. 27.8 *Batrachophrynus macrostomus*, a high altitude amphibian which we found in lakes near Rancas (4720 m) in Central Peru. Through the kindness of Miss A. G. C. Grandison of the British Museum we were able to establish that the toe web in the specimens we collected is less full than in typical examples of *B. macrostomus*. It is also asymmetrical leaving one or two free digits.

Lack of high altitude adaptation in Sceloporine lizards

Not all lower forms of life living at high altitude appear to be adapted to the environmental conditions. Studies on Sceloporine lizards show that the type and duration of activity fail to show any altitude effect or adaptation to high altitude. The duration of maximal activity and the intensity of anaerobic metabolism were measured in lowland *Sceloporus occidentalis* at low (60 m) and high (3090 m) elevations and compared to those of resident *S. occidentalis* and *S. graciosus* at high altitude (Bennett and Ruben, 1975). Lowland *S. occidentalis* have just as much endurance and qualitatively identical activity at both altitudes even though oxygen pressure at the latter altitude is only two-thirds that at the former. They have as much stamina as high altitude residents of the same species and of *S. graciosus*, a species generally associated with high altitude. The fact that behaviour of all these groups is essentially identical suggests that even if aerobic adaptations are present in high altitude residents, they are not of significance in prolonging activity or increasing stamina. Probably since Sceloporine lizards are highly dependent upon anaerobic metabolism during activity, they are largely independent of external oxygen sources.

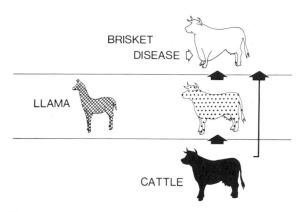

Fig. 27.9 (a) Diagrammatic representation of adaptation and acclimatization to high altitude in animals. Filled figures represent life at sea level, stippled: acclimatization, cross hatching: full adaptation, and open figures: represent loss of acclimatization. The upper and middle compartments represent life at high altitude, the lower compartment, life at sea level. The llama is fully adapted. High altitude cattle are acclimatized but some calves in Utah show the cardiovascular form of lack of acclimatization termed 'Brisket disease'.

Species at high altitude which have to acclimatize

Not all species of animal living on mountains are indigenous to the environment and adapted to high altitude. These have to *acclimatize* to the conditions just like man. Thus both *Apodemus*, (the Russian wood mouse), and *Peromyscus*, (the deer mouse of White Mountain) may show an increased haemoglobin level in response to the hypoxic environment. Such representatives of these species on high mountains are not members of isolated highland populations but form an extension of a common stock largely native to the surrounding lowlands. Altitude or origin and haematocrit level do not correlate (Morrison et al, 1963a and b) as we have seen in this chapter and in Chapter 6.

Application of these principles to a mountain community

We may apply the principles that we have outlined in this chapter to an actual community. As an example of which we have personal experience we may take the small mining town of Cerro de Pasco at an altitude of 4330 m in the Peruvian Andes. Most of the inhabitants are Quechua Indians who were born and lived all their lives in the area (Figs. 4.2a, 4.2c, 4.4a and 4.4b). They illustrate *natural acclimatization* with perhaps *partial adaptation* in the form of a

voluminous chest which allows them to undertake hard physical labour at high altitude (Figs. 27.1b and c). Also employed in the area are miners of Caucasian origin, born on the coast or in the United States who have worked at Cerro for many years. They show *acquired acclimatization* which may be of a high order. Travellers from the coast, either freshly arrived at Cerro or in transit through the Andes to the Amazon are to be seen in the market place where some of them in the throes of *soroche* will be entering the early stages of *accommodation*. In the surrounding altiplano will be herds of llamas and alpacas (Fig. 27.3). Due to *adaptation* they are capable of prolonged and heavy physical labour as beasts of burden. Scattered throughout the population are a few sufferers from the condition of Monge's disease (Fig. 16.2) which represents a respiratory form of loss of *acclimatization*. No to be found in Peru but in Utah, cattle may show a *cardiopulmonary form of loss of acclimatization* in the form of brisket disease (Fig. 13.1). These constitute the five groups of different biological status at high altitude.

These views on adaptation and acclimatization at high altitude are summarised diagrammatically in Figures 27.9a and 27.9b. In these Figures we include the fetus, at sea level or at high altitude, as an acclimatized subject. As we have discussed in

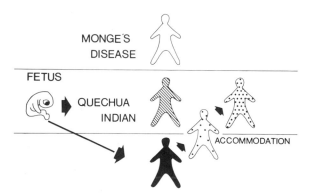

Fig. 27.9 (b) Diagrammatic representation of adaptation and acclimatization to high altitude in man. The key is as in Fig. 27.9a, with the addition that single hatching represents partial adaptation. The Quechua Indian shows natural acclimatization and partial adaptation. The lowlander ascending to high altitude passes through a stage of accommodation to acquired acclimatization. A few Quechuas show the respiratory form of lack of acclimatization termed 'Monge's disease'. The fetus may be considered to be acclimatized to the hypoxaemia of intrauterine life simulating high altitude.

Chapter 11 even at sea level the human fetus is exposed to a degree of hypoxaemia that corresponds to an altitude of 7500 m. At birth the baby at high altitude needs to acclimatize to what is in effect a hyperoxic environment compared to conditions *in utero*. At sea level the sudden acclimatization required to an atmosphere rich in oxygen is even greater.

The Bohr effect in Sherpas (Table 27.4)

Earlier in this book (Chapter 6) we refer to the work of Morpurgo et al (1972) which showed an increased Bohr effect in Quechua Indians which they regarded as a feature of acclimatization. Subsequent studies by this group showed that this was not apparent in Sherpas in Nepal (Table 27.4). In a further paper (Morpurgo et al, 1976) this group employed their data to develop the hypothesis that Quechuas and

Table 27.5 Factors related to oxygen transport by blood in Caucasians, Quechuas and Sherpas (after Morpurgo et al, 1976)

Group	Low, high or extreme altitude	Hb	Oxygen affinity	2,3-DPG
Caucasians	L	O	O	O
	H	+	−	+
Quechuas	L	O	O	O
	H	+	−	+
Sherpas	L	−	−	O
	H	O	+	O
	E	+	...	+

O = Normal sea-level value
+ = Above normal value
− = Below normal value

Table 27.4 The Bohr effect in Europeans and Sherpas (after Morpurgo et al, 1976)

Altitude	P_{50} pH 7.4		P_{50} pH 6.8		Bohr effect	
	low	high (4950 m)	low	high	low	high
Europeans	22.23	21.67	28.69	28.92	6.46	7.25
Sherpas	21.62	21.25	27.38	27.42	5.77	6.17

Sherpas have come to terms with the high altitude environment in two different ways.

Haematological characteristics of the Sherpa

Morpurgo et al (1976) carried out their research in two small Nepalese villages in the Solo-Khumbu region (Kunde at 3800 m and Kumjung at 3900 m). Also studied were Sherpas living permanently at 1200 m. The Sherpas at high altitude have normal levels of haematocrit, most values falling in the range of 41 to 50 per cent, with haemoglobin levels in the range of 14.5 to 17.5 g/dl and a red cell count of 4.0 to 5.0 × 10^{12}/l. These results may be contrasted with those which we have quoted for the acclimatized Quechua (Chapter 6). The levels of 2,3-diphosphoglycerate are not raised in highland Sherpas compared to sea-level subjects (Table 27.5). Sherpas living at Katmandu (1200 m) have lower haemoglobin levels than do Caucasians living at sea

level and Sherpas living at high altitude (Morpurgo et al, 1972) but on ascending to altitudes in excess of 4000 m the Sherpas react to hypoxia by an increase in erythrocyte number, haemoglobin content and 2,3-DPG (Morpurgo et al, 1972) (Table 27.5).

Sherpas in the Himalayas differ from Quechuas in the Andes in showing a pronounced shift of the oxygen-haemoglobin dissociation curve to the *left*, indicating an increased affinity of haemoglobin for oxygen. Mean P_{50} values in Sherpas living permanently at high altitudes was 22.6 mmHg contrasted to 27.0 mmHg in sea-level Caucasians, and 36.7 mmHg in Sherpas living permanently at low altitudes. It is not known how in the Sherpa the oxygen may be discharged efficiently from the better oxygenated blood to the tissues. In contrast, Sherpas living at a lower altitude in Katmandu have an oxygen-haemoglobin dissociation curve strongly shifted to the *right*. The mechanism of the shift to the left is not clear. The presence of abnormal haemoglobins is to be excluded since the electrophoretic patterns of European and Sherpa haemoglobin are indistinguishable (Morpurgo et al, 1972). Morpurgo et al (1976) found that whereas in red cell extracts levels of lactate, total glutathione, ATP and ADP were greatly increased both in Sherpas living at high altitude and acclimatized Caucasians, the 2,3-DPG concentration in Sherpas living at high altitude (4983 n mol/ml red cells) is similar to that of Caucasians living at sea level (4295 n mol). The level of 2,3-DPG rises sharply in acclimatized Caucasians (7170 n mol).

The Sherpa and the Quechua. Are some highlanders acclimatized and others adapted?

The investigations of Morpurgo et al (1976) raise interesting questions. One is inclined to regard all peoples residing permanently above an altitude of 3000 m as 'native highlanders'. However, while this grouping together of such ethnic groups as Quechuas and Sherpas takes recognition of the fact that both live permanently above a certain altitude it does not take into account the all-important fact that such different peoples may have been in such residence for very different lengths of time. As we have developed our arguments throughout this book there is no doubt that the Quechua of the Andes is *acclimatized*. He presents a shift to the right of his oxygen-haemoglobin dissociation curve with a sharp decrease in the affinity of his haemoglobin for oxygen. This is in striking contrast to the situation in the Sherpa as described by Morpurgo and his colleagues (1976). They present a highlander with his oxygen-haemoglobin dissociation curve shifted to the left, his haemoglobin showing a marked affinity for oxygen. Morpurgo and his colleagues relate this to the length of time the two peoples have been domiciled in the mountains. According to them man migrated to America through the Bering Strait probably only 35 000 years ago, the colonization of the Peruvian Andes having taken place only over the past 14 000 years. In contrast it seems likely that man has inhabited Central Asia including Tibet for close on half a million years. In the conventions which we have developed in this chapter it would appear that the Amerindians are *acclimatized* while the Sherpas share certain features with the llama and experimental animals referred to in Chapter 6 who have a 'shift to the left' of the oxygen-haemoglobin dissociation curve and increased affinity of haemoglobin for oxygen and are *adapted* to high altitude. It would appear that the Sherpas, having lived in their mountain home for so much longer, are at a more advanced biological state in coping with their adverse environmental conditions.

Acclimatization from the point of view of the highlander

It is customary to accept sea-level values as 'normal' but it is an interesting exercise to view acclimatization from the other angle. Rahn (1966) expressed this point of view elegantly at a meeting of the Pan American Health Organization in Washington and we quote him here: 'What would the textbook of normal physiology look like if it were written by a fetus? What would it look like if it were written by the Incas, who had their empire and their major city (Figs. 3.3 and 3.8) at an altitude of ten thousand feet and regarded sea level areas as places to which they banished the undesirable citizens? It is easy to list what such a textbook would say about sea-level man. He would obviously be described as anaemic; he would have a relatively high blood volume; he would be hypoventilating and exhibit a hypercarbia . . . The people of Cuzco would worry about the pulmonary hypotension of sea-level man'. We do not accept the view that the physiological features of man at high altitude are normal even though they are exhibited by numerous mountain dwellers throughout the world. We think they represent automatic physiological and anatomical responses to an abnormal lack of oxygen and the higher the altitude of residence the more pronounced the changes are and the greater the likelihood that they will merge into unequivocal disease. Thus as we have seen in Chapter 6 there is a relation between altitude, haemoglobin level and tendency to develop mountain sickness.

REFERENCES

Adams, W. H., Graves, I. L. & Pyakural, S. (1975) Hematologic observations on the yak. *Proceedings of the Society for Experimental Biology and Medicine*, **148**, 701.

Banchero, N., Grover, R. F. & Will, J. A. (1971) Oxygen transport in the llama (*Lama glama*). *Respiration Physiology*, **13**, 102.

Battaglia, F. C., Behrman, R. E., De Lannoy, C. W., Hathaway, W., Makowski, E. L., Meschia, G., Seeds, A. E. & Schruefer, J. J. P. (1969) Exposure to high altitude of sheep with different haemoglobins. *Quarterly Journal of Experimental Physiology*, **54**, 423.

Baumann, R., Bauer, C. & Bartels, H. (1971) Influence of chronic and acute hypoxia on oxygen affinity and red cell 2,3-diphosphoglycerate of rats and guinea pigs. *Respiration Physiology*, **11**, 135.

Bennett, A. F. & Ruben, J. (1975) High altitude adaptation and anaerobiosis in Sceloporine lizards. *Comparative Biochemistry and Physiology*, **50A**, 105.

Brooks, J. G. & Tenney, S. M. (1968) Ventilatory response of llama to hypoxia at sea level and high altitude. *Respiration Physiology*, **5**, 269.

Bullard, R. W. (1972) Vertebrates at altitudes. In: *Physiological Adaptations. Desert and Mountain*, p. 209. Edited by M. K. Yousef, S. M. Horvath & R. W. Bullard. New York: Academic Press.

Chiodi, H. (1966) In: *Life at High Altitudes*, p. 67. Pan American Health Organization, Scientific Publication No. 140. Washington.

Eaton, J. W., Brewer, G. J. & Grover, R. F. (1969) Role of red cell 2,3-diphosphoglycerate in the adaptation of man to altitude. *Journal of Laboratory and Clinical Medicine*, **73**, 603.

Heath, D., Castillo, Y., Arias-Stella, J. & Harris, P. (1969) The small pulmonary arteries of the llama and other domestic animals native to high altitude. *Cardiovascular Research*, **3**, 75.

Heath, D., Smith, P., Williams, D., Harris, P., Arias-Stella, J. & Krüger, K. (1974) The heart and pulmonary vasculature of the llama (*Lama glama*). *Thorax*, **29**, 463.

Huisman, T. H. J. & Kitchens, J. (1968) Oxygen equilibria studies of the hemoglobins from normal and anemic sheep and goats. *American Journal of Physiology*, **215**, 140.

Hurtado, A. (1966) In: *Life at High Altitudes*, p. 68. Pan American Health Organization, Scientific Publication No. 140, Washington.

Hurtado, A. (1964) In: *The Physiological Effects of High Altitude*, pp. 2 and 344. Edited by W. H. Weihe. Oxford: Pergamon Press.

Hutchinson, V. H., Haines, H. B. & Engbretson, G. (1976) Aquatic life at high altitude: respiratory adaptations in the Lake Titicaca frog, *Telmatobius culeus*. *Respiration Physiology*, **27**, 115.

Monge, C. (1948) In: *Acclimatization in the Andes* Baltimore: John Hopkins Press.

Morpurgo, G., Arese, P., Bosia, A., Pescarmona, G. P., Luzzana, M., Modiano, G. & Krishna Ranjit, S. (1976) Sherpas living permanently at high altitude: A new pattern of adaptation. *Proceedings of the National Academy of Sciences of the United States of America*, **73**, 747.

Morpurgo, G., Battaglia, P., Carter, N. D., Modiano, G. & Passi, S. (1972) The Bohr effect and the red cell 2,3-DPG and Hb content in Sherpas and Europeans at low and at high altitude. *Experientia*, **28**, 1280.

Morrison, P. R., Kerst, K. & Rosenmann, M. (1963a) Haematocrit and haemoglobin levels in some Chilean rodents from high and low altitude. *International Journal of Biometeorology*, **7**, 45.

Morrison, P. R., Kerst, K., Reynafarje, C. & Ramos, J. (1963b) Haematocrit and haemoglobin levels in some Peruvian rodents from high and low altitude. *International Journal of Biometeorology*, **7**, 51.

von Muralt, A. (1966) In: *Life at High Altitudes*, pp. 53 and 69. Pan American Health Organization, Scientific Publication No. 140, Washington.

Petschow, D., Würdinger, I., Baumann, R., Duhm, J., Braunitzer, G. & Bauer, C. (1977) Causes of high blood O_2 affinity of animals living at high altitude. *Journal of Applied Physiology*, **42**, 139.

Rahn, H. (1966) In: *Life at High Altitudes*, p. 83. Pan American Health Organization, Scientific Publication No. 140, Washington.

Reynafarje, C., Faura, J., Villavicencio, D., Curaca, A., Reynafarje, B., Oyola, L., Contreras, L., Vallenas, E. & Faura, A. (1975) Oxygen transport of hemoglobin in high altitude animals (*Camelidae*). *Journal of Applied Physiology*, **38**, 806.

Sillau, A. H., Cueva, S., Valenzucla, A., & Candela, E. (1976) O_2 transport in the Alpaca (*Lama pacos*) at sea level and at 3300 m. *Respiration Physiology*, **27**, 147.

Swan, L. W. (1970) Goose of the Himalayas. *Natural History*, **79**, 68.

Torrance, J. D., Lenfant, C., Cruz, J. & Marticorena, E. (1970) Oxygen transport mechanism in residents at high altitude. *Respiration Physiology*, **11**, 1.

Van Vliet, G. & Huisman, T. H. J. (1964) Changes in the haemoglobin types of sheep as a response to anaemia. *Biochemical Journal*, **93**, 401.

Velásquez, T. (1964) Response to physical activity during adaptation to altitude. In: *The Physiological Effects of High Altitude*, p. 289. Edited by W. H. Weihe. Oxford: Pergamon Press.

Ward, M. (1972) In: *In This Short Span*, p. 119. London: Victor Gollancz.

28

The descent to sea level

As we have progressed through the chapters of this book it has become apparent that virtually every system of the body in those living at high altitude undergoes physiological disturbance. Sometimes these changes in function are associated with modifications in structure as well. As examples of this we may recall the microanatomical changes which involve the carotid bodies (Chapter 8), the small pulmonary arteries (Chapter 11), and the pulmonary trunk (Chapter 12). One of the most characteristic features of the great majority of these derangements of form and function is their *reversibility* on descending to live at sea level. An important exception appears to be the impaired sensitivity of the carotid bodies to hypoxic stimuli after residence at high altitude during infancy (Chapter 8). This, however, appears to be the exception that proves the rule. Throughout this volume we have referred to this reversibility in general terms referring, for example, to the rapid clearance of high altitude pulmonary oedema (Chapter 15) and to the general improvement in the clinical condition of patients with Monge's disease on descent to sea level (Chapter 16). In this chapter we wish to consider in greater detail the effect on some haematological and cardiopulmonary parameters in native highlanders of residence at sea level for a prolonged period. In doing this a second principle of *overshoot* appears. Not only do the physiological abnormalities reverse to sea-level values but they may progress beyond normality in the opposite direction. Thus as we shall see the fall in the abnormally high levels of haemoglobin may end in anaemia.

The most complete study of the long-term results of descent to sea level has been made by Sime et al (1971). They studied 11 native highlanders of Cerro de Pasco (4330 m), aged 18 to 23 years. These young men were taken to Morococha (4540 m) where the investigations listed in Table 28.1 were carried out. They were then taken to Lima (150 m) and the investigations were repeated after they had been living at sea level for two years. Where appropriate the investigations were carried out at rest and after exercise.

Anaemia

After two years of continuous residence at sea level high altitude natives show the expected drop in haemoglobin level and haematocrit. Overshoot commonly occurs so that highlanders may be anaemic after this period (Table 28.1). Such anaemia may be a factor in the slight increase in cardiac output which characterizes the descent to sea level. As we note in Chapter 17 at high altitude there is characteristically a hypervolaemia which has a basis in the increased red cell mass (Hurtado et al, 1945; Merino, 1950). On descent to sea level there is a pronounced fall in haematocrit (Chapter 6, Table 28.1) with a proportionate increase in plasma volume maintaining the total blood volume. After some four months the total blood volume falls since, as we have pointed out above, red cell mass falls to subnormal values leading to anaemia. This gradually improves over the following months. These changes in plasma volume associated with fall in haemoglobin level may have some importance in predisposing the highlander at sea level to the onset of pulmonary oedema on his return to the mountains (Chapter 15).

Ventilation

After residence at sea level the characteristic hyperventilation of those living at high altitude disap-

Table 28.1 Changes in haematological and cardiopulmonary parameters in high altitude natives after residence at sea level for two years (after Sime et al, 1971)

	4540 m (Morococha)		150 m (Lima)	
	At rest	On exercise*	At rest	On exercise*
Haemoglobin (g/dl)	18.5±0.55		13.5±0.24	
Haematocrit (%)	55.4±1.6		41.9±0.8	
Mean oxygen uptake (ml min⁻¹ m² BSA)	158±4.7	802±28.8	161±3.7	866±24.0
Ventilation (l min⁻¹ m² BSA)	5.24±0.36	24.74±1.07	4.80±0.25	19.88±0.75
Arterial oxygen saturation (%)	78.48±1.25	69.20±1.28	97.31±0.68	94.33±0.57
Heart rate (beats/min)	77 ±3.9	144±6.9	59 ±2.1	114±3.8
Cardiac Index (l min⁻¹ m² BSA)	3.83±0.21	7.57±0.29	4.32±0.20	8.79+0.32
Stroke Index (ml beat m² BSA)	50.2±2.7		74.2±3.6	
Pulmonary arterial pressure (mmHg)	24±1.6	54±3.7	12±0.6	25±1.1
Pulmonary resistance (dyn s cm⁻⁵)	334±27.4		145±10.7	

* exercise = 300 kg m min⁻¹ m² BSA

pears. As would be anticipated from our observations in Chapter 5 the change in ventilation at sea level is more apparent on exercise. Differences in oxygen uptake in individuals at high altitude and sea level are not significant (Table 28.1) and increments from rest to exercise are very similar for the same work load in two environments. The arterial oxygen saturation rises significantly at sea level.

Bradycardia

The heart rate falls on descent from high altitude and after two years' residence at the coast highlanders may show bradycardia. The slowing of the heart occurs very shortly after arrival at low altitudes. Thus Hartley and his associates (1967) describe a slight fall in heart rate in residents of Leadville, Colorado (3100 m) after only 10 days' residence at sea level. The same rapid slowing of heart rate has been detected during electrocardiographic studies on children and adults taken down

to sea level from the Andes. An acute reduction in heart rate at high altitude following the administration of oxygen has been reported by Hultgren et al (1965).

Cardiac output

The decrease in heart rate just referred to is associated with an increased stroke volume and an increase in cardiac output (Table 28.1). Thus the normal or slightly diminished cardiac output of the Andean native referred to in Chapter 17, described by several authors including Peñaloza et al (1963) and Sime et al (1963) and confirmed by Sime et al (1971), is increased on descent to sea level.

Reversal of pulmonary hypertension

After two years' residence at sea level the levels of pulmonary arterial pressure and vascular resistance are similar to those found in normal sea-level residents. Inhalation of oxygen alone will not produce such a lowering as Grover et al (1966) discovered in their investigation at Leadville, Colorado (3100 m). The reason for this is not far to seek. As we have seen in Chapter 11 the development of pulmonary hypertension in long-standing residents at high altitude is associated with that of hypoxic hypertensive pulmonary vascular disease characterized by muscularization of the terminal portion of the pulmonary arterial tree. Thus time is required for the regression of this muscularization to occur. When a highlander descends to sea level, his pulmonary arterioles will still have muscular walls. Hence it is not surprising that Hartley et al (1967) did not find a significant drop in pulmonary arterial pressure after only 10 days' residence at low altitude. Peñaloza and Sime (1971) found that in three cases of Monge's disease the fall in pulmonary arterial pressure was related to the time spent at sea level and was not normal after 60 days. In one case Grover et al (1966) noted a fall of pulmonary arterial mean pressure of only 17 mmHg after 11 months' residence at sea level. It seems likely that the reduction in pulmonary arterial pressure and resistance on descent to sea level is achieved in three stages (Fig. 28.1) First, there is a relaxation of pulmonary vasoconstriction formerly maintained by the chronic hypoxia of the high altitude environment. Second, there is a progressive fall in polycythaemia. Third and finally, there is a regression of the muscularization of the terminal portions of the pulmonary arterial tree. Considerable rises in pulmonary arterial pressure are to be anticipated while the pulmonary arteries retain their muscular nature.

'Acquired acclimatization to sea level'

Velasquez (1966) described changes of the type reported by Sime et al (1971) as 'acquired acclimatization to sea level'. He refers to a report that Bolivian athletes were unable to repeat at sea level their

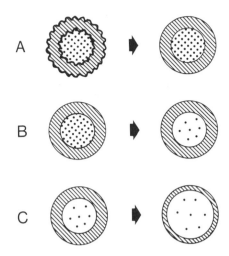

Fig. 28.1 Diagrammatic representation of the immediate and long-term reversibility of pulmonary hypertension in highlanders after descent to, and prolonged residence at, sea level.
A. Immediate reversibility of pulmonary hypertension. Relaxation of pulmonary vasoconstriction which was maintained by the chronic hypoxia of the high altitude environment. There is loss of crenation of elastic laminae which delineate the media of smooth muscle (indicated by hatching). B. Progressive fall in polycythaemia (indicated by fewer dots). C. Long-term reversibility of pulmonary hypertension. Regression of the muscularization of the terminal portions of the pulmonary arterial tree.

previous performances which had usually been at La Paz (3800 m). This is a view to which we do not subscribe. As we state elsewhere in this book we regard the various functional features of acclimatization to high altitudes as normal physiological responses to abnormal environmental stimuli. On descent to sea level these stimuli are removed. There seems to be no convincing evidence that descent to the sea-level environment carries a risk for the highlander. The only hazard appears to be that the increase in plasma volume in the highlander which occurs on his moving to sea level may predispose to the development of high altitude pulmonary oedoema when he returns to the mountains.

REFERENCES

Grover, R. F., Vogel, J. H. K., Voigt, G. C. & Blount, S. G. Jnr. (1966) Reversal of high altitude pulmonary hypertension. *American Journal of Cardiology*, **18**, 928.

Hartley, L. H., Alexander, J. K., Modelski, M. & Grover, R. F. (1967) Subnormal cardiac output at rest and during exercise in residents at 3100 m altitude. *Journal of Applied Physiology*, **23**, 839.

Hultgren, H. N., Kelly, J. & Miller, H. (1965) Effect of oxygen upon pulmonary circulation in acclimatized man at high altitude. *Journal of Applied Physiology*, **20**, 239.

Hurtado, A., Merino, C. & Delgado, E. (1945) Influence of anoxemia on the hemopoietic activity. *Archives of Internal Medicine*, 75, 284.

Merino, C. F. (1950) Studies on blood formation and destruction in the polycythaemia of high altitude. *Blood*, 5, 1.

Peñaloza, D. & Sime, F. (1971) Chronic cor pulmonale due to loss of altitude acclimatization (chronic mountain sickness). *American Journal of Medicine*, 50, 728.

Peñaloza, D., Sime, F., Banchero, N., Gamboa, R., Cruz, J. & Marticorena, E. (1963) Pulmonary hypertension in healthy men born and living at high altitudes. *American Journal of Cardiology*, 11, 150.

Sime, F., Banchero, N., Peñaloza, D., Gamboa, R., Cruz, J. & Marticorena, E. (1963) Pulmonary hypertension in children born and living at high altitudes. *American Journal of Cardiology*, 11, 143.

Sime, F., Peñaloza, D. & Ruiz, L. (1971) Bradycardia, increased cardiac output, and reversal of pulmonary hypertension in altitude natives living at sea level. *British Heart Journal*, 33, 647.

Velasquez, T. (1966) Acquired acclimatization to sea level. In: *Life at High Altitude*, p. 58 Washington: Pan American Health Organization. Scientific Publication No. 140.

29

Athletic performance at moderate altitude

In Chapter 2 we define 'high altitude' for the purposes of this book as an elevation exceeding 3000 m because at this height the environmental hypoxia leads to discernible effects at rest in the majority of subjects. However, under certain circumstances a much lower elevation may prove to be important. Thus, when the international sports meeting is held at only moderate altitude, the performance of the athletes is still influenced by the hypoxic conditions and by any previous acclimatization that they have effected. Hence their capacity to compete successfully depends not only on their individual fitness and attainment but also on their ability to cope with the environment. Such conditions applied to Mexico City at an altitude of 2380 m which was chosen as the venue for the Olympic Games in 1968. There was much discussion before these Games as to whether the participants would be adversely affected in their performance by such an altitude and this discussion continued long after the Games had been completed. The purpose of this chapter is to examine, from one event in the 19th Olympiad and from physiological studies carried out both previously and subsequently, the possibility that moderately high altitude may adversely affect sea-level athletes.

Background to the 19th Olympiad

In October 1963 the International Olympic Committee chose Mexico City as the venue for the 19th Olympiad. Although many of the effects of high altitude on the human body were known at that time, it was not clear whether or not the comparatively moderate elevation of Mexico City would be deleterious to athletes coming from sea

level. According to Coote (1968), a bulletin of the International Olympic Committee stated that 'Mexico refuted the arguments concerning the difficulty of athletes adapting themselves to a high altitude'. It was considered that 48 hours would be sufficient for adequate acclimatization.

However, this initial willingness to accept the factor of increased altitude as being of little consequence gave way to concern on the part of representatives of sea-level countries planning to participate in the Games. In June 1965 the British Olympic Association, on the advice of its medical committee, decided to send a research team to Mexico to study the time necessary to acclimatize and the type of training schedules the athletes would require. Its report concluded that in events requiring endurance, such as track events exceeding 1500 m, four weeks would be the minimal acceptable time for acclimatization. It also concluded that athletes other than those in such endurance events would benefit from acclimatization and that there was no evidence of any risk of permanent injury. Concern over the possible effects of high altitude led to the appearance of a letter in *The Times* in April 1966 signed by 26 British Olympic medallists. In this letter they refer to 'The Mexicans who chose deliberately to ignore the problems of altitude when they put forward their candidature'. The signatories pointed out that natives or long-term residents at high altitude had a great advantage over those ascending from sea level. Bannister, himself an international athlete, said 'The conclusion I draw is that some risk does exist and, however small, it represents a powerful argument for never holding distance events at altitude. I would like to see this written into a rule book for sports administrators' (Bannister, 1966).

Athletic performance at Mexico City

The practical effect of high altitude on athletic performance can be seen by studying athletes who were capable of world record performances at sea level. In the final of the 10 000 m race Ron Clarke, who was the world record holder in this event, finished sixth and in a state of collapse, although he was not reported as suffering from any other illness (Fig. 29.1). The pace had been unremarkable and

Gammoudi of Tunisia, fourth Juan Martinez of Mexico and fifth Nikolay Sviridov of U.S.S.R., who was thought to have been domiciled at Alma-Ata (1000 m) or Leninakan (1500 m). The time of the winner Temu, 29 min 27.4 s, was almost two minutes slower than Clarke's world record.

We note above that the winner of the 10 000 m event at Mexico City was a Kenyan. It has since come to be recognised that Kenyans have

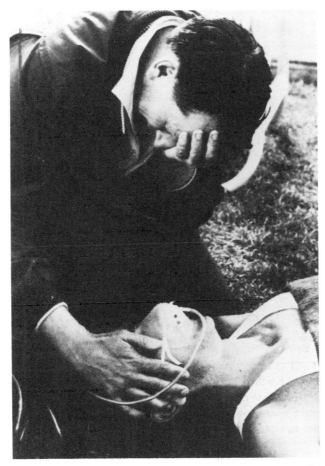

Fig. 29.1 Dr. Corrigan, the Medical Officer to the Australian Olympic team in 1968, weeps as he administers oxygen to Ron Clarke who collapsed as he crossed the finishing line of Men's 10 000 m event at the 1968 Olympic Games held in Mexico City (2380 m). (Popperfoto).

yet Clarke could not accelerate when necessary. The first five places were filled by athletes either native to high altitude or domiciled there for a prolonged period. First was Naftali Temu of Kenya, second Mamo Wolde of Ethiopia, third Mohamed

outstanding ability in middle-distance events. The British Broadcasting Corporation Television Service was so intrigued by this that it sent a Unit to Africa to investigate the matter. Its findings were incorporated into a television programme, *Kenya*

Runner (BBC, 1975). It was found that the successful middle-distance runners originated from two areas of the plateau near Lake Victoria at an altitude of 1500 m to 2000 m. The tribes concerned are the Kalenjin and the Kipsigis (sometimes called Kisii). The natives of these tribes have a physical advantage in that their femora are longer than those of Caucasians. However, what is probably more important is that from early childhood they run long distances to school at this moderate altitude. Those interested in athletics find their way into the Army or the Prison Service during which they are encouraged to undertake resistance training at an altitude of 1520 m. As we have already noted in Chapter 4 the development of these Kenyan children during daily exercise in the mildly hypoxic conditions of life on the plateau may well influence the magnitude of vital capacity achieved.

Lung function and habitual activity in childhood as exemplified by the highlanders of New Guinea

The concept that we have just advanced, that exercise at moderate altitude, in children, improves lung function and thereby athletic performance for distance events, is supported by the investigations of Cotes et al (1973). They studied lung function tests at sea level and at moderate altitude in New Guinea. The sea-level studies were carried out on coastal dwellers from Kaul village on Kar Kar Island approximately 10 miles off the mainland of New Guinea. The highlanders who were 17 to 30 years of age, came from Lufa situated at an altitude of 2000 m, and the studies were carried out in a laboratory situated at 1700 m in Goroka. They studied the lung volume and ventilatory capacity of their young adult male and female subjects but in addition they carried out an investigation of the transfer factor which estimates the diffusing capacity of the lung. The results were carefully standardized for age, height, and in the case of transfer factor, the haemoglobin concentration.

They found that lung function in the coastal dwellers in New Guinea resembled that of people of Indian and West African descent. Thus the inspiratory capacity (Fig. 29.2) and the expiratory reserve volume (Fig. 29.2) were smaller than for comparable Europeans (Fig. 29.3). However, the highlander of this area of of New Guinea was found

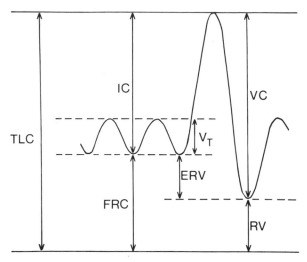

Fig. 29.2 Definitions of the various functional lung compartments referred to in the text. TLC = Total lung capacity; FRC = Functional residual capacity; IC = Inspiratory capacity; ERV = Expiratory reserve volume; V_T = Tidal volume; VC = Vital capacity; RV = Residual volume.

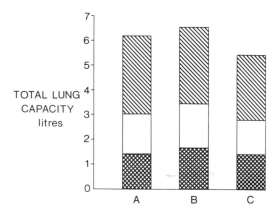

Fig. 29.3 Total lung capacity and its subdivisions for young men (height 1.7 m) in A: Europeans; B: Highlanders in New Guinea; C: Coastal people in New Guinea. The cross hatched area = residual volume. Open area = expiratory reserve volume. Hatched area = inspiratory capacity. (Based on data from Cotes et al, 1973).

to have a larger total lung capacity (Fig. 29.2) similar to that of Europeans and mainly due to a larger inspiratory capacity (Fig. 29.3). The transfer factor, measuring diffusing capacity, was greater in the highlander than in either the native sea-level residents or in Europeans (Fig. 29.4).

As we have already pointed out in Chapter 4 it has been known for many years that the lung volume of highlanders is increased compared with those at sea

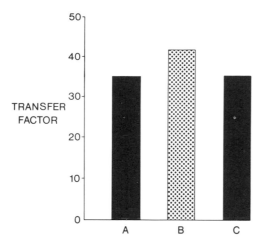

Fig. 29.4 Transfer factor for young men (height 1.7 m) in A: Europeans; B: Highlanders in New Guinea; C: Coastal people in New Guinea. (Based on data from Cotes et al. 1973).

level (De Graff et al, 1970; Cotes and Ward, 1966). What is remarkable is that the increased total lung capacity reported by Cotes and his colleagues (1973) occurred in highlanders living at the moderate altitude of only 2000 m and the greater lung size noted by Woolcock et al (1972) was in subjects living at a mere 1200 m. This raises the interesting possiblity that the unusual lung function of the highlanders of New Guinea depends on an added factor superimposed upon that of moderate altitude. This seems to be the high level of physical activity characteristic of the highlanders of the area (Durnin et al, 1972) and it probably exerts its greatest effect on lung size during infancy, childhood and adolescence when these organs are developing (Cotes et al, 1973). In this respect the physiological responses to exercise in childhood resemble those of the athlete in western countries (Sinnett and Solomon, 1968; Cotes et al, 1972). Hence from New Guinea we have evidence that a population with a lung size comparable to African Negroes, Malayans, Chinese and Indians may increase this to a volume more appropriate to Europeans because of habitual exercise undertaken at moderate altitude during childhood. The improvement in lung function is shown in both sexes. Should these highland populations of underdeveloped countries assume in the future a western style of living, the advantage of increased lung volume and transfer may well be lost (Cotes et al, 1973). The significantly increased

pulmonary diffusing capacity of these highlanders, shown by Cotes et al (1973), may be due to attenuation and increase in area of the alveolar capillary membrane, consequent upon the increase in total lung capacity. Such physiological investigations add weight to the view that highland populations of this type would be placed in an unusually advantageous position with regard to long distance events in international athletics meetings.

Predicted and actual performances at 19th Olympiad

Predictions for athletic performance prior to the Olympiad had been based on comparisons drawn from the results of the 1955 Pan American Games also held at Mexico City, with the results of previous and subsequent Pan American competitions held at sea level (Jokl and Jokl, 1968; Faulkner, 1967). This comparison had shown a small gain in running events of 100 to 400 m (because of a decrease in wind resistance) and a slowing of times for more long-distance endurance events. This slowing amounted to 6 to 7 per cent in the 5000 and 10 000 m events and rose to 17 to 22 per cent in the Marathon (42 000 m). In some instances it was possible to compare the performance of individuals both at sea level and at moderate altitude. While the apparent effect of such altitude varied there was an immediate loss of performance of 6 to 8 per cent. When high altitude natives competed at sea level, the gain in performance was 2 to 4 per cent.

The results from Mexico City were in fact better than those predicted (Faulkner, 1971; Craig, 1969). Twenty-nine per cent of competitors broke world records and the average winning margin was 0.9 per cent below the world record. This compares with the margin of 2.9 per cent below the world record in previous Olympiads held at sea level (Shephard, 1973). To assist the interpretation of the results and to separate short and long term events, Craig (1969) expressed the results as a percentage of the world record plotted against their duration. When a linear regression was applied to both track and swimming events, it showed that, although world records were exceeded in short events, there was a steady decline in events lasting longer than one minute. This fall ranged from 3 per cent at four minutes to 8 per cent at one hour.

DISCUSSION

Limitations of existing knowledge

Information on the effects of hypoxia at altitudes of less than 3000 m is somewhat sparse, for most of the studies on the cardiopulmonary effects of hypoxia are based on much higher altitudes as we have seen earlier in this book. Consequently, it is important to differentiate between the studies carried out on high mountains and those concerned with the more subtle problems of altitudes below 3000 m. In Mexico City the effects of moderate altitude were normally slight and may be confused with the symptoms of anxiety before the contest. Sime et al (1974) have carried out a study at Arequipa in Southern Peru, the altitude of which is similar to Mexico City. They studied eight male amateur soccer players between the ages of 16 and 21 years. These men were subjected to right heart catheterization and studies by bicycle-ergometer at moderate and submaximal exercise (300 and 600 kg m min^{-1} m^2 respectively) initially at sea level and subsequently at Arequipa (2370 m), four to six hours after arrival and, in some instances, after five days sojourn at this altitude.

Ventilation and oxygen uptake

Sime and his colleagues found that minute ventilation did not change at 2370 m at rest but it increased by some 12 per cent on submaximal exercise, compared to sea level. Following ascent systemic arterial oxygen saturation fell 4 per cent at rest and 4 to 7 per cent on exertion. Oxygen uptake decreased 10 per cent at rest and 7 per cent during exercise. This confirmed Pugh's (1967) earlier results in a series of investigations carried out prior to the Olympics at Mexico. The major difference between the two groups was that while Sime's group was composed of amateur soccer players, Pugh's group consisted of six international middle-distance runners. Although they were still technically amateurs, it is reasonable to assume that their intensity and level of training was much higher than in the case of the soccer players. Pugh (1967) also used a bicycle-ergometer and the work load was adjusted so that the subjects could just undertake five minutes work at sea level. The data for maximum oxygen uptake was obtained from one minute gas samples obtained at the third to fourth,

and fourth to fifth minute and usually these samples agreed to within 50 ml indicating a steady intake of oxygen. The initial tests undertaken at Mexico City were on the second day of exposure to the altitude. They showed a mean reduction of 14.6 per cent of oxygen intake for all individuals compared with the mean values at sea level. This reduced with exposure to altitude until by the fourth test undertaken on the 27th day of exposure the reduction was 9.6 per cent. Although all six athletes were of international standing, the individual variation was considerable. Three improved progressively, one showed little improvement and in two cases the results were so varied that little use could be made of them. This wide range was reflected in the ventilatory response. At sea level the minute volume varied from 122 to 196 l/min with a mean of 148 l/min. In Mexico City ventilation increased by between 3 and 22 l/min to a mean value of 160 l/min.

Maximum oxygen consumption in sea-level subjects falls with decreasing barometric pressure and even at this moderate elevation this general principle still holds. The decrease in oxygen uptake occurs in spite of a slight increase in pulmonary ventilation. It is clear, therefore, that the limiting factor in oxygen consumption is not ventilation. It is more likely to be a combination of the lowering of arterial oxygen saturation and the depression of cardiac output referred to below which interfere with the diffusion of oxygen at tissue level (Sime et al, 1974).

Heart rate and cardiac output

The heart rate at an altitude like that of Mexico City does not increase significantly in its response to exercise from that at sea level. Sime et al (1974) found that the heart rate did not change either at rest or during moderate exercise but decreased significantly on submaximal exercise both immediately after ascent and after some days sojourn at high altitude. Pugh (1967) also found no significant differences in heart rate. The mean values for individuals at sea level after exercise varied from 159 to 185 beats per minute with a mean of 168 per minute and at 2270 m after comparable exercise the individual mean range was 147 to 184 beats per minute with a mean of 166 beats per minute. Sime et al (1974) found that the cardiac index was reduced by 10 per cent at rest at 2370 m on the first day but fell by 20 per cent at rest and by 15 per cent on exercise by the fifth day. Since heart rate at moderately high altitude does not differ significantly in its responses from that at sea level it seems more likely that a *diminution in stroke volume* is far more likely to account for the *diminution in cardiac output.* This could arise from a defective contractile myocardial force resulting from a lower coronary arterial tension or flow. However, Sime et al (1974) were unable to find any correlation between the degree of lowering of cardiac output or stroke volume and the altitude or degree of work performed. An alternative view is that there may be a lowering of venous return from a diminished circulating blood volume due to redistribution of blood flow occurring early on exposure to altitude (Chapter 17).

Blood pressures

In Sime's group the pulmonary arterial mean pressure increased 18 per cent at rest and 30 per cent on submaximal exercise above sea-level values. On ascent to 2370 m there was an increase of total pulmonary resistance of 29 per cent and this rose to 42 per cent after five days sojourn at high altitude. Pugh (1967) did not subject his subjects to this type of investigation but measured the systemic blood pressure by auscultation. He obtained values of 200 to 230 mmHg in submaximal exercise at sea level and at altitude. During the period of maximal exercise the systolic pressure was measured by means of a cuff and transducer and showed pressures of up to 300 mmHg.

Newcomers to high altitude are exposed to a degree of hypoxaemia which is different to that found in the long-term residents at the same altitude due to the lower systemic arterial oxygen saturation. Since the extent of hypoxaemia is greater in newcomers to the mountains it might be anticipated that they would experience higher levels of pulmonary arterial pressure secondary to exaggerated hypoxic pulmonary vasoconstriction. According to Sime et al (1974) the reverse is true. It seems likely that the higher degree of pulmonary hypertension in long-standing high altitude residents both at rest and during exercise are related to the muscularization of pulmonary arterioles which we describe in Chapter 11.

In short-duration athletic events the speed at which oxygen debt increases is the main physiological determinant of performance, the event having finished before the athlete reaches the maximum tolerable oxygen debt (Shephard, 1973). In events of a longer duration 50 per cent of the energy expenditure can come from the build up of a large oxygen debt.

Exercise tolerance

At 2370 m the soccer players studied by Sime et al (1974) were exhausted on completing a work intensity of 600 kg m/min/m^2 indicating that submaximal exercise at sea level becomes nearly maximal at high altitudes. Pugh (1967) found that the oxygen debt mechanism was unchanged at Mexico City and that in his opinion a reduction in performance in endurance events arises from a reduction in oxygen uptake. The reduction in work performance must clearly be related to the reduction in oxygen intake but the diminution in cardiac output and arterial oxygen saturation with an increase in pulmonary arterial pressure must play an important rôle in the overall mechanism. The decreased exercise capacity of sea-level subjects at high altitude has also been reported by Grover and Reeves (1966) at a medium altitude of 3100 m and Pugh et al (1964) at an extreme altitude of 5790 m. Certainly all the results of the physiological studies of Pugh (1967) and Sime et al (1974) indicate that moderately high altitude has a deleterious effect on work capacity and suggest that at the altitude of Mexico City submaximal exercise entails a handicap for sea-level athletes.

Sime et al (1974) would classify such an elevation as 'low altitude', considering that the effects of environmental hypoxia below an elevation of 3050 m are mild or indiscernible at rest. We think this is an unfortunate terminology since this is a significant degree of elevation which is capable of handicapping sea-level athletes as their own studies referred to here show. They regard 'medium altitude' as lying in the range of 3050 m to 4270 m and 'great altitude' as an elevation exceeding 4270 m.

Exercise and the highlander

Work by Hurtado (1964) had demonstrated the superiority of the native highlander in exercise under the hypoxic conditions inherent in life at high altitude. In considering the activity for exercise, athletic or otherwise, at high altitude one has to remember that the components are in part those of acclimatization and in part those of habitual exercise in childhood. Hurtado (1964) studied 11 non-athletes at 150 m and 12 native highlanders at 4540 m. They were asked to run on a treadmill until exhausted. The highlanders tolerated the exercise twice as long and showed a greater degree of hyperventilation. They showed smaller increases in their pulse rate and systemic systolic blood pressure. Their blood showed smaller rises in the levels of lactic and pyruvic acid, and of glucose. The oxygen debt was prominently reduced in the native highlander.

Athletes and acute mountain sickness

Acute mountain sickness can be of considerable importance to an international athlete for its many manifestations (Chapter 14) may severely handicap him in any competition. He is particularly vulnerable as he must adhere to his training programme of vigorous exercise despite feelings of headache, sickness, dizziness or any of the other symptoms he may encounter. Fortunately as the symptoms tend to resolve themselves within 48 hours conservative treatment with a temporary lightening of training is generally sufficient.

Altitude training

A joint meeting to discuss 'altitude training' was held between the British Association of Sport and Medicine and the British Olympic Association at the Royal Society of Medicine in London in November 1973. The proceedings were subsequently published in the *British Journal of Sports Medicine* in April 1974 (Bannister, 1974a and b; Brotherhood, 1974; Johnston and Turner, 1974; Keul and Cerny, 1974; Lloyd, 1974; Owen, 1974; Shephard, 1974; Travers and Watson, 1974; Watts, 1974). The optimistic spirit with which this meeting opened was exemplified by the opening remarks by Bannister, at that time Chairman of the Sports Council. He presented the anecdotal evidence which had led many athletes and members of the public to believe that there is some advantage to be gained from training at high altitude. He said 'I remember some early results from a group of American world record holders, including Jim Ryun, who went to altitude for 14 days. After coming down Jim Ryun set up a world record for the mile of 3 min. 51.3 sec., a record in which I have a little personal interest. Five out of six of the other athletes also achieved best performances. They then went to altitude for another 14 days and after coming down Ryun lowered the 1500 metre world record. After a third spell at altitude, five out of six athletes again produced best performances. That was the moment at which it seemed to me that there was something rather special going on'.

However, by the time the meeting concluded in a panel discussion, again chaired by Bannister, it had become only too clear from the papers that had been read, that many of the participants were totally confused as to whether altitude training bestowed any benefits at all, and even so, whether such training was ethically desirable. Many expressed the view that training athletes together in a training-camp atmosphere might in itself be expected to bring advantages excluding altogether the factor of altitude. The truth of the matter is that there have been very few scientifically-controlled comparisons of the efficacy of training at low and high altitude. Athletes training at altitude in the past have not been studied adequately. As Bannister said at the meeting, the objective of athletes is to run well, not to be harnessed to the devices of physiologists (Bannister, 1974b).

One could take the view that the polycythaemia induced by training or residence at high altitude might provide increased oxygen-carrying ability. However, as we have pointed out in Chapter 28 there is a decrease in red cell production and an increase in red cell destruction as soon as the subject descends to sea level or lower altitudes. Furthermore at sea level, erythropoietin becomes undetectable and there is a progressive decrease in the activity of the bone marrow. The consequence of this is that the haematological gains of several years may disappear in a matter of weeks. In addition the disruption in the training programme and the less vigorous training while at altitude may cause a lessening of the subject's overall fitness. For these above reasons athletic training at high altitude cannot be recommended as a panacea for improving physical fitness. Undoubtedly the performance of athletes who are native highlanders suggests that there is some advantage from *growing up as a child* at high altitude as we describe earlier in this chapter but the advantages of short periods at altitude training camps are much more open to dispute.

It seems to us that little improvement in lung function can be expected if the adult fully developed lung is exposed to the hypoxia of high altitude for a period of a few short weeks. Improvement in lung function of the type that could enhance performance in middle or long-distance running at sports meetings must depend on the exposure for many years of the developing lung of the child and adolescent accustomed to the hypoxia and habitual activity of the native highlander at moderate altitude. The advantages derived by the children of native highlanders from their high altitude habitat cannot hope to be achieved in a temporary visit by an adult European. Adams and his colleagues (1975) also share this view. They found from study of 12 trained athletes that hard endurance training at an altitude of 2300 m had no advantage over equivalently severe training at sea level either on the maximum oxygen uptake or on the performance time for a two mile race in already well conditioned middle-distance runners. The athletes were divided into two groups matched for age, for time achieved in a run of two miles, and for maximal oxygen uptake. The first group trained for three weeks at sea level, running 19.3 km per day at 75 per cent of the maximal oxygen uptake at sea level. The second group trained for an equivalent distance at the same intensity at a barometric pressure of ambient air of 586 mmHg. The groups then exchanged sites and followed a training programme of similar intensity. Periodic tests were then carried out on a treadmill approaching exhaustion and reaching maximal oxygen uptake, and in competitive time trials over a distance of two miles. The initial times for the two mile race for the athletes trained at high altitude were 7.2 per cent slower than the sea-level groups. A second trial held at 2300 m showed improved performance in both groups but post-altitude performance was no better than that of sea-level controls. Clearly the training at high altitude had proved to be a failure.

REFERENCES

Adams, W. C., Bernauer, E. M., Dill, D. B & Bomar, J. B. Jr. (1975) Effects of equivalent sea level and altitude training on Vo₂ max and running performance. *Journal of Applied Physiology*, 39, 262.

Bannister, R. (1966) Athletics at altitude. *New Scientist*, 30, 228.

Bannister, R. (1974a) Chairman's opening remarks *British Journal of Sports Medicine*, 8, 3.

Bannister, R. (1974b) Panel discussion *British Journal of Sports Medicine*, 8, 56.

BBC (1975) Television Programme: *Kenya Runner*. December 14th.

Brotherhood, J. R. (1974) Human acclimatization to altitude *British Journal of Sports Medicine*, 8, 5.

Coote, J. (1968) *Olympic Report 1968* London: Robert Hale.

Cotes, J. E., Davies, C. T. M., Patrick, J. M., Reed, J. W. & Saunders, M. J. (1972) Cardio-respiratory response to submaximal exercise; comparison of young adults in New Guinea and U.K. *Ergonomics*, 15, 484.

Cotes, J. E., Saunders, M. J., Adam, J. E. R., Anderson, H. R. & Hall, A. M. (1973) Lung function in coastal and highland New Guineans – comparison with Europeans. *Thorax*, 28, 320.

Cotes, J. E. & Ward, M. P. (1966) Ventilatory capacity in normal Bhutanese. *Journal of Physiology*, 186, 88P–89P.

Craig, A. B. (1969) Olympics 1968: a post mortem. *Medicine and Science in Sports*, I, 177.

DeGraff, A. C., Grover, R. F., Johnson, R. L., Hammond, J. W. & Miller, J. M. (1970) Diffusing capacity of the lung in Caucasians native to 3,100 m. *Journal of Applied Physiology*, 29, 71.

Durnin, J. V. G. A., Ferro-Luzzi, A. & Norgan, N. G. (1972) An investigation of a nutritional engima—studies on coastal and highland populations in New Guinea. *Human Biology in Oceania*, 1, 318.

Faulkner, J. A. (1967) Training for maximum performance at altitude. In: *The Effects of Altitude and Athletic Performance*, p.88 Edited by R. Goddard Chicago: Athletic Institute.

Faulkner, J. A. (1971) Maximum exercise at medium altitude. In: *Frontier of Fitness*, p. 360 Edited by R. J. Shephard. Springfield Illinois: Charles C. Thomas.

Grover, R. F. & Reeves, J. T. (1966) Exercise performance of athletes at sea level and 3100 m meters altitude. *Medicina Thoracalis*, 23, 129.

Hurtado, A. (1964) In: *Aging of the Lung*, p. 270. Edited by L. Cander & J. H. Moyer. New York: Grune and Stratton.

Johnston, T. F. K. & Turner, D. M. (1974) Altitude training and physiological conditions from the practical point of view of the runner. *British Journal of Sports Medicine*, 8, 52.

Jokl, E. & Jokl, P. (1968) The effect of altitude on athletic performance. In: *Exercise and Altitude*, p. 28. Baltimore: University Park Press.

Keul, J. & Cerny, F. C. (1974) The influence of altitude training on muscle metabolism and performance in man. *British Journal of Sports Medicine*, 8, 18.

Lloyd, B. B. (1974) Chairman's introductory remarks. *British Journal of Sports Medicine*, 8, 37.

Owen, J. R. (1974) A preliminary evaluation of altitude training *British Journal of Sports Medicine*, **8**, 9.

Pugh, L. G. C. E. (1967) Athletes at altitude. *Journal of Physiology*, **192**, 619.

Pugh, L. G. C. E., Gill, M. B., Lahiri, S., Milledge, J. S., Ward, M. P., & West, J. B. (1964) Muscular exercise at great altitudes. *Journal of Applied Physiology*, **19**, 431.

Shephard, R. J. (1973) The athlete at high altitude. *Canadian Medical Asosication Journal*, **109**, 207.

Shephard, R. J. (1974) Altitude training camps *British Journal of Sports Medicine*, **8**, 38.

Sime, F., Peñaloza, D., Ruiz, L., Gonzales, N., Covarrubias, E. & Postigo, R. (1974) Hypoxaemia, pulmonary hypertension and low cardiac output in newcomers to low altitude. *Journal of Applied Physiology*, **36**, 561.

Sinnett, P. F. & Solomon, A. (1968) Physical fitness in a New Guinea highland population. *Papua New Guinea Medical Journal*, **2**, 56.

The Times (1966) Letter, p. 9. April 16th.

Travers, P. R. & Watson, R. (1974) Results of altitude training in British track and field athletes, 1972. *British Journal of Sports Medicine*, **8**, 46.

Watts, D. (1974) Altitude—A coach's conclusions *British Journal of Sports Medicine*, **8**, 30.

Woolcock, A. J., Colman, M. H. & Blackburn, C. R. B. (1972) Factors affecting normal values for ventilatory lung function. *American Review of Respiratory Disease*, **106**, 692.

30

Exposure to extreme altitudes

The features of acclimatization described in the earlier chapters enable man to carry on a comparatively normal social and economic life at high altitude. Thus Cerro de Pasco, the centre of the Peruvian mining industry, is situated at 4330 m (Fig. 30.1). However, there is a critical altitude which man can live permanently. Natives miners live there at 5330 m but they climb every day to their work at 5790 m (Fig. 30.1). They refused to occupy a camp built for them at the higher elevation on account of difficulty in sleeping. Just as in Chapter 2 we gave an arbitrary definition of 'high altitude'

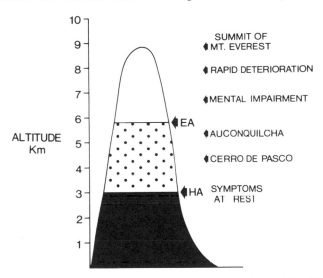

Fig. 30.1 Definition of high altitude (HA) and extreme altitude (EA) on the criteria considered in the text.

above which successful, permanent acclimatization cannot take place and this limit appears to be somewhere around an elevation of 5500 m (Fig. 30.1).

(3000 m) so here we define 'extreme altitude' as one exceeding 5800 m (Fig. 30.1).

'High altitude deterioration' in climbers

'Extreme altitude'

Pugh (1962) from his experience in the Himalayas, is of the opinion that 5790 m is too high for what he calls 'complete adjustment'. The mining community of Auconquilcha in the Andes is of significance in determining the highest altitude at

This is not to say that man cannot survive at elevations considerably greater than this for shorter periods. High altitude climbers deliberately expose themselves to extreme altitudes for short periods but they usually have the advantage of a period of acclimatization while they climb the foothills to place them into position for the assault on the mountain in question. Aviators do not seek to expose themselves to such biological stress but have to accept that in the course of their work they may be suddenly exposed to extreme altitudes without any such protective acclimatization.

Mountaineering at great altitudes is undertaken by a small number of highly motivated people (Chapter 33) who accept exposure to very low temperatures, high winds, low intakes of food and fluid, and severe hypoxia. High altitude climbers expose themselves to the hazards of altitude up to the ultimate summit of 8850 m. At such extreme elevations the changes they experience are pathological rather than physiological. Indeed above an altitude of 5800 m there is a progressive worsening in their mental and physical condition which has been termed 'high altitude deterioration'.

Early signs of deterioration become apparent at 5800 m (Ward, 1975), which is the same altitude at which Peruvian miners refuse to live permanently as we have already noted. Ward points out that during the Makalu expedition of 1960–61 a stay of 90 to 100 days at this elevation led to early deterioration despite the fact that the party enjoyed good living conditions with adequate food and fluid intake with protection from the environment. In general terms high altitude climbers at this elevation show what is in effect an exaggeration of signs and symptoms which may be encountered at much lower elevations. Thus the anorexia which overtakes the majority of people ascending mountains becomes more severe. In fact the higher the altitude and the longer the stay, the greater is the hypophagia with resulting loss of weight. The physiological basis for the weight-loss is considered in Chapter 21. The appetite for sweet foods is said to increase and some climbers may develop craving for certain foods such as tinned salmon or pineapple cubes (Ward, 1975). Dyspnoea, headache, nausea and vomiting, which as we have seen in Chapter 14 are often encountered at much lower altitudes, become more pronounced. There is no fall in the capacity for work and Sherpas may carry loads up to 27 kg (60 lb) all day without undue signs of fatigue. However, already at this altitude there are subtle signs of a falling off of mental capacity. There is a statistically significant change in the efficiency of card-sorting (Gill et al, 1964) of the type described in Chapter 32.

Above 5800 m heavy loads can still be carried and

extreme exertion is possible but such muscular activity is commonly followed by pronounced exhaustion. The normal working day tends to be shortened to six hours. Anorexia and loss of weight may become acute. From 6000 to 6700 m there is a wide variation in performance and feeling of well being from day to day. Muscular fatigue becomes very pronounced and beneficial results in allaying this by the ingestion of easily assimilated carbohydrate such as sugar become apparent.

It is between altitudes of 6700 m and 7900 m that the appearance of significant mental impairment takes place (Ward, 1975). Although routine tasks can still be carried out without too much trouble, any activity requiring initiative takes much longer. At the same time weakness and fatigue become striking and cases of total exhaustion are not uncommon. Weight loss, nausea and dyspnoea become even more exaggerated.

Finally at altitudes of between 7900 and 8850 m one passes from any semblance of physiological acclimatization to high altitude to a state of rapid deterioration, in which mental aberrations occur and survival itself becomes the central issue. Mental depression is common. The ability to perform routine mechanical tasks in an orderly way takes much longer. The insight into one's behaviour is lost. Climbers may embark on foolhardy procedures not having insight into the risks involved. Hallucinations may occur.

The dramatic and disastrous effects that the hallucinations induced by hypoxia may have are illustrated by a description of an ill-fated climb of Aconcagua (6920 m) in Argentina, the highest mountain in the western hemisphere (Shults and Swan, 1975). Eight climbers between 25 and 52 years of age, one a woman, climbed the mountain too rapidly after ascending as much as 610 m in a single day. At 5370 m one developed pulmonary oedema and another cerebral oedema; the doctor accompanying the party remained with them. The remaining five continued the climb but one soon became grossly disorientated. The last four camped at 6400 m and what transpired above this altitude was open to doubt for only two of the climbers survived and their recall of events was fragmentary. When the two survivors arrived back at base camp, they reported having seen on the summit highway equipment, dead mules, skiers and trees. They recalled the presence of voices of an Argentinian mountain patrol that was in fact never there. Such bizarre effects of deprivation of oxygen on the brain have commended themselves to the popular press. One such press report included an interview with Alfredo Magnani, a guide on Aconcagua, who referred to the cerebral symptoms in the following terms: 'It is like dreaming on your feet. I saw a horse dancing once, many years ago' (Lindley, 1975).

At these great heights the combination of coldness and severe hypoxia may lead to loss of consciousness. Oxygen inhalation increases endurance and enables climbers to take an interest in their surroundings. Men have, however, scaled Everest (8850 m) without oxygen (Chapter 32). This extraordinary capacity may be related to the fact that barometric pressure in mountains is higher than would be expected from the international altimeter calibration used in aviation (Pugh, 1962). The atmospheric pressure on the summit of Mount Everest is 250 mmHg which is equivalent to 8380 m on an altimeter scale instead of the true altitude of 8850 m. At these extreme altitudes one has left the realms of the physiology of acclimatization for those of heroic endurance. Prolonged stay at extreme elevation leads to a decrease in intellectual activity for a period after return to sea level. There is often a temporary disturbance of memory but this passes off and permanent mental damage has not been reported (Ward, 1975).

Exercise at extreme altitude

The vigorous exercise necessitated in the climbing of very high mountains cannot be sustained above the altitude at which the maximum uptake of oxygen by the body is less than that required by the contracting muscles. As we have already pointed out in Chapter 5, the quantity of oxygen consumed by the tissues of the body *at rest* each minute is 220 ml to 260 ml (STPD) (Harris and Heath, 1962). This is easily supplied and we have already noted that the respiratory rate at rest at high altitude is not elevated except in those subjects developing severe acute mountain sickness (Chapter 14). Even at extreme altitude the bodily requirements of oxygen can be met *at rest*. Thus Ward (1975) notes that on the first ascent of Mount Everest (8850 m) Sir Edmund Hillary removed his oxygen mask at the summit for

about ten minutes before symptoms occurred. In the Foreword to the first edition of this book Sir Cyril Clarke pointed out that man can endure a night at 8600 m without additional oxygen.

However, the oxygen requirements of the tissues may increase tenfold *on exercise* so that some 2.2 to 2.6 l min⁻¹ is required (Chapter 5). Difficulties now arise because oxygen requirements for a given level of exercise remain the same at high altitude as they are at sea level. At the same time, however, the maximum oxygen uptake of the body per minute falls with increasing altitude. At sea level this maximum uptake is 3.5 to 4.0 l min⁻¹ but at 5800 m, corresponding to our definition of 'extreme altitude', this capacity for uptake of oxygen falls to 2.0 to 2.5 l min⁻¹ (Pugh et al, 1964). This uptake is only just sufficient to provide the quantity of oxygen necessary for severe exercise referred to above. Ward (1975) gives the somewhat lower value of 1.7 l min⁻¹ as the necessary oxygen intake of a man weighing 70 kg during normal climbing activity. This means that as man progresses up into extreme altitude breathing and exercise become progressively more difficult. Ward (1975) reckons that up to about 6100 m the acclimatized climber from sea level can proceed at a normal pace appropriate for Alpine climbing (Fig. 30.2). By the time, however,

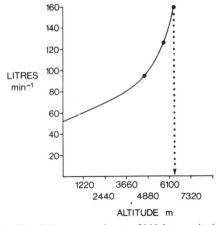

Fig. 30.2   Ventilation at a work rate of 900 kg m min⁻¹, which is approximately the preferred work rate of Alpine climbing. At an altitude of about 6100 m it becomes impossible to climb continuously without oxygen. (Data from Milledge, 1975).

that man reaches an altitude of 7460 m his maximum oxygen uptake is only in the range of 1.3 to 1.5 l min⁻¹ (Pugh et al, 1964). This will no longer sustain necessary muscular activity.

In Figure 30.2 are shown the levels of ventilation with increasing altitude which accompany a work rate of 900 kg m min⁻¹ which may be taken as appropriate for normal Alpine climbing (Milledge, 1975). It is clear from this diagram that at a certain critical range of altitude ventilation becomes inadequate to sustain continuous muscular exercise so that climbing becomes progressively intermittent with the oxygen debt being repaid during periods of rest. The critical altitude at which this takes place is given as 6100 m by Ward (1975) and 5790 m by Milledge (1975). Breathlessness at rest is noted at 5180 m (Milledge, 1975) and eventually at extreme altitudes exceeding 8000 m several breaths may have to be taken for each step. Individuals have scaled Mount Everest without oxygen. Climbing with oxygen becomes worthwhile at 7010 m. Below this altitude oxygen reduces ventilation and makes the climber feel more comfortable but the extra weight burden makes it hardly worthwhile (Milledge, 1975).

As we have noted in Chapter 27, Hurtado (1964) believes the capacity for sustained work and exercise at high altitude is the prime factor which distinguishes the native highlander from the acclimatized lowlander. Certainly, Sherpa porters, like the Quechuas of Hurtado, can still carry loads exceeding half their body weight above 6400 m (Ward, 1975). Even the Sherpas, however, tend to work intermittently.

Lactic acid production

Acute exposure to the hypoxia of high altitude does not diminish the formation of lactic acid during severe exercise so that blood lactate levels are increased. On the other hand it has been known for some years that *chronic* exposure to hypoxia brings about a reduction in the amount of lactate formed (Edwards, 1936). Indeed the greater the altitude the smaller the increase in blood lactate after exhausting exercise. Furthermore, in native highlanders the accumulation of blood lactate during exercise is lower than in sea-level subjects (Fig. 30.3) and its rate of disappearance is faster. Hurtado (1971) reported that, compared to healthy residents at sea level, highlanders living at 4540 m showed lower blood lactate levels and oxygen debt after severe exercise (Fig. 30.3). This finding is suggestive of a

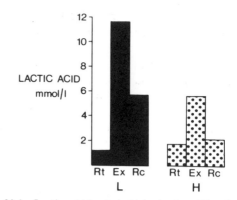

Fig. 30.3   Lactic acid levels in 11 lowlanders (L) and 12 highlanders (H) (at 4540 m) at rest (Rt), on exercise (Ex), and on recovery (Rc). Exercise was specified as running to exhaustion on a treadmill at a speed of 135.3 m/min and gradient of 18.9 per cent. (After Hurtado, 1964).

more aerobic than anaerobic source of energy in the mountain dweller. It could be held to represent a mechanism of acclimatization since the buffer base at high altitude is reduced as we have already pointed out in Chapter 5, (Fig. 5.2) and hence diminution of lactate level would embarrass less such a diminished level of bicarbonate.

Exposure to extreme cold

In Chapter 22 we considered changes in the skin and in body metabolism that occurred in response to the degree of cold met at altitudes between say 3000 m and 4500 m at which long-term residence is possible. Here we may briefly consider the deleterious effects on the body brought about by exposure to the severe cold of extreme altitude. These are pathological conditions most unlikely to be met in long-term residents at high altitude. They will be found far more commonly in high altitude climbers, polar explorers and air crew under abnormal conditions in whom there may be sudden exposure to extreme cold. The effect may be general leading to hypothermia which is defined as a lowering of body temperature below 35°C. The effects may be local. Under dry freezing conditions frostbite may ensue. In a wet and cold, but not necessarily freezing situation, immersion hand or foot may result.

Frostbite

In this condition the tissues freeze with the

formation of intra- or extra-cellular ice crystals. Depending upon the degree of cold and the duration of exposure to it, a shell of skin and underlying tissue of variable depth is frozen. Plasma escapes from underlying blood vessels to form blisters. Sludging of erythrocytes occurs within the blood vessels and this together with associated constriction of the arteries leads to a diminution in the blood supply to the skin and subcutaneous tissues. According to Ward (1975) arteriovenous shunts come into operation so that the supercooled blood from the affected area is unable to enter the general circulation. The net result of this is an increase in the chance of survival but the frozen shell of tissue dies. This turns black and hard and encases the finger or toe damaged by cold. It has been thought to resemble the shell of a tortoise and is sometimes termed a 'carapace'. This peels away in ensuing months to leave tender underlying new epithelium. Connective tissue, tendons and bone are more resistant than skin to cold so that the frostbitten tissues may be moved over still viable tendons. The incidence of frostbite depends on ambient air temperature, length of exposure and 'wind chill' (Chapter 2). A lesser degree of this condition is 'frostnip' when supercooling of the skin leads to blanching, numbness and tingling of the extremities. This process is reversible and can be cured by the simple expedient of exercise on the mountain side (Steele, 1976).

Hypothermia

This process is said to begin when the body temperature falls below 35°C (95°F). It becomes lethal when the temperature of the vital organs falls to about 25°C. Hypothermic subjects may become delirious and confused. As well as impairment of cerebration there are important effects on the heart. When haemoglobin is cooled, it releases oxygen less easily. When these effects are added to diminished coronary blood flow (Chapter 18) the supply of oxygen to the cardiac muscle is barely sufficient for its needs even at rest. The myocardial ischaemia leads to alteration in cardiac rhythm. There is a slowing of heart rate until the cardiac pace-maker fails; ectopic beats or ventricular fibrillation may supervene. Death may result from cardiac arrest (Ward, 1975).

Snow blindness

This condition is caused by damage to the cornea by excessive ultra-violet radiation reflected from the snow. The keratitis may be associated with blistering of the cornea and induces intensive photophobia. Because of the intense pain the affected subject can hardly bear to open his eyes to the light and so is, in effect, blind—hence the term 'snow blindness'. There is intense congestion of the conjunctiva. The condition may be prevented by goggles or dark glasses. In an emergency, horizontal slits may be cut into a piece of cardboard which can be tied around the head with string. The pain and spasm may be treated by anaesthetic drops (Steele, 1976).

Haematological values

Climbers exposed to extreme altitude (5800 m) show macrocytosis and reticulocytosis (Pugh, 1962). The haemoglobin levels attained by high altitude climbers may be considerable when compared with those of native highlanders such as we quote in Chapter 6. Thus Pugh states that the mean of 51 observations on 40 subjects from five expeditions was 20.5 g/dl. After 40 to 50 days mean values of haemoglobin reach a steady level and are then independent of time and altitude. It would seem that at extreme altitude haemoglobin concentration stabilises early and is regulated by changes in plasma volume.

Low humidity and dehydration

In Chapter 14 we consider the pathogenesis of acute mountain sickness and in particular the view that it is the result of oliguria associated with redistribution of blood within the body leading to pulmonary and cerebral oedema. At extreme altitude on the contrary dehydration rather than over-hydration may become an important problem. The air is very dry and respiration is both deep and rapid so that there is an appreciable loss of water through ventilation. The low humidity also allows very free sweating and at 5800 m the water turnover rate is 3.9 l/day compared with 2.9 l/day at sea level (Pugh, 1962). Men engaged in physical activity for seven hours a day at this altitude require 5 litres of water a day. In other words survival at extreme altitude demands a large water requirement. Bonington (1971) refers to the unpopularity amongst high altitude climbers of frusemide administered for the prevention of treatment of acute mountain sickness or pulmonary oedema because it exaggerates the general dehydration already established. Indeed above 7000 m such enhanced dehydration could play a rôle in initiating thrombosis and we shall consider this problem now.

Thrombotic episodes at extreme altitudes

In Chapter 16 we noted that the significant difference in the histology of the pulmonary vasculature in Monge's disease compared to that of the healthy Quechua Indian was superimposed thrombosis no doubt related to the exaggeration of an already elevated haematocrit. The increased viscosity associated with the raised haematocrit predisposes to more serious thrombotic episodes in larger blood vessels. Ward (1975) gives an account of these occurring in high altitude climbers. They include thrombophlebitis in the calf followed by pulmonary thromboembolism, pulmonary thrombosis and infarction, and hemiplegia following cerebral thrombosis. The episodes reported by Ward occurred in comparatively young men between the ages of 23 and 41 years as would be anticipated in followers of this athletic pursuit. With one exception which took place at 4270 m all the thrombotic incidents took place at great heights between 5790 and 7930 m. They were not confined to European climbers; three cases of hemiplegia occurred in Sherpas. At such extreme altitudes above 7000 m pronounced dehydration due to increased respiration takes place and this factor also predisposes to thrombosis. Enforced inactivity in tents on stormbound mountainsides may also play a rôle and Ward (1975) recommends that adequate hydration and exercise should be ensured to prevent thrombosis. He points out that anticoagulants should not be administered until adequate laboratory control is available. Once a person has suffered one serious thrombotic episode at high altitude it would be foolhardy to risk exposure to the same environmental hazards again. Ward recommends that susceptible subjects should not ascend again above 4270 m. Genton et al (1970) refer to the case

of a young climber developing thrombophlebitis and pulmonary thromboembolism during an attempted ascent of Mount Godwin Austen (K2) (8610 m) in 1953. Pugh (1962) reports that Sir Edmund Hillary became ill at 5790 m on Makalu during the Himalayan Expedition of 1960–61. He developed aphasia and right-sided facial palsy, preceded by headache. It seemed likely that he had developed a mild cerebral thrombosis. He was advised not to ascend again beyond 4570 m.

Aviation and high altitude

Modern commercial aircraft are pressurized to avoid the effects of high altitude and an emergency oxygen supply is available should pressurization fail. In unpressurized aeroplanes the crew will develop signs and symptoms of hypoxia if certain critical altitudes are exceeded. The oxygen saturation of arterial blood will fall below the acceptable level of 85 per cent at 3660 m when the aviator is breathing air and at 12 192 m when he is breathing 100 per cent oxygen (Robinson, 1973). With increasing altitude water vapour and carbon dioxide occupy more and more of the volume of the lungs until at 15 240 m they are filled entirely with these gases and even 100 per cent oxygen at this altitude cannot sustain life (Robinson, 1973). Table 30.1 outlines the symptoms experienced by aviators with progressive increase in altitude. We may give brief examples of situations in which aviators have been acutely exposed to extreme altitudes.

Acute exposure to extreme altitude

Early balloonists ascended rapidly to great heights without appreciating the risks to which they were exposing themselves. Robinson (1973) gives an account of such an ascent by Glaisher from Wolverhampton on July 17th, 1862. The balloon was inflated by an unusually light gas mixture specially prepared at the Stafford Road Gasworks and it shot up to 7990 m. Glaisher experienced palpitation, cyanosis and a feeling of sea-sickness. Undeterred, on September 5th of the same year he ascended to 12 180 m and lost the power of his limbs and then became unconscious but he survived. On November 4th, 1927 Gray ascended from Scott Field, Illinois, and rose to 12 940 m. Later that

Table 30.1 Symptoms of hypoxia recorded in aviators with increasing altitude. (After Robinson, 1973)

Range of altitudes (m) at which symptoms occur, when breathing:		Symptoms
Air	100% O_2	
0 to 3050	10 360 to 11 890	Impaired night vision. (See Chapter 31.)
3050 to 4570	11 890 to 12 960	Hyperventilation, Tachycardia, increased cardiac output.
4570 to 6100	12 960 to 13 660	Fatigue, lassitude, headache, sleepiness. Euphoria resembling alcoholic intoxication. Impaired vision. Faulty judgement. Poor memory. Delayed reaction time. Impaired muscle co-ordination. Fine muscle movements impossible.
6100 to 7010	13 660 to 13 870	Loss of consciousness. Failure of respiratory centre. Convulsions. Death.

afternoon his balloon returned to earth with its occupant dead.

Such episodes have not been lost entirely to the era of modern aircraft for, although the pilot is able to climb to great altitudes in safety, occasionally he has to bale out. Robinson (1973) points out that with an open parachute a man would die during the ten minutes or so that it would take to descend from 10 670 to 6100 m. It was apparent that with a free fall and closed parachute the drop could be made in one minute. However, early studies showed that even during such a short period of free fall a test subject became cyanotic and began to lose consciousness after only 30 seconds forgetting to open the valve of an emergency oxygen cylinder. After a further five seconds he became unconscious and moved convulsively. Subsequently one subject survived a parachute jump from an altitude of 12 192 m using an open parachute but having a supply of oxygen from a small cylinder carried in the leg pocket of his flying suit. From these two examples it will be appreciated that very short exposure to an extreme altitude can be lethal.

The symptoms of immediate exposure to the hypoxia of high altitude, as would be experienced by passengers when an aircraft suddenly loses cabin pressure or by experimental subjects in a decompression chamber, have been summarised by Milledge (1975). Acute exposure to an altitude as low as 1220 m gives rise to a purplish tinge to the vision and night vision is measurably affected (see Chapter 31). Sudden subjection to elevations of 1830 m and above causes slight breathlessness on exertion and such dyspnoea is more pronounced the higher the altitude suddenly faced. Sudden exposure to 4880 m brings about tingling of the lips and fingers, feelings of unreality and dizziness. The higher functions of the brain are affected with disturbances of association, memory and fine movement. Difficulty is experienced in recognising the orientation of stylised figures presented to subjects in a decompression chamber (Purves and Ponte, 1977) and writing becomes slurred. Once the initial exposure is to an altitude of 6100 m or above unconsciousness occurs with increasing speed so that only two minutes useful consciousness can be expected after immediate exposure to 8840 m (Fig. 30.4). It is of interest to note that Quechuas normally living at a height of 4270 m can be subjected to a simulated altitude of 9750 m in a decompression chamber without loss of consciousness (Milledge, 1975).

In recent times there has been a remarkable survival of an 18-year-old aerial stowaway in the confined space of the landing gear cell, without pressurization, of a DC-8 which flew for nine hours at an altitude of 8840 m from Havana to Madrid (Pajares and Merayo, 1970). During a period of eight hours the temperature fell to −6°C. On arrival he was in a state of unconsciousness and showed the

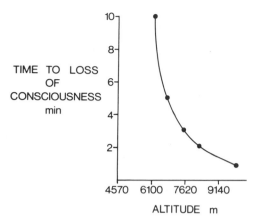

Fig. 30.4   The time in minutes for which consciousness is maintained on sudden exposure to altitudes exceeding 6100 m. (after Milledge, 1975).

features listed in Figure 30.5. Pajares and Merayo (1970) believe that in this extraordinary case the progressive hypothermia in some way protected against the effects of the severe hypoxia.

leads to only a 3 or 4 per cent desaturation of systemic arterial blood which is of no importance to subjects free of heart or lung disease. However, in the presence of diseases which in themselves lead to the anaemic, stagnant, or histotoxic forms of hypoxia, even this mild degree of arterial desaturation may become significant. Thus the physiological and clinical implications of air travel must be considered in patients with obstructive airways disease, restrictive lung disease, anaemia or any degree of failure of the circulation. Each case must be considered on its individual merits. Thus patients with bronchitis, bronchial asthma, or pulmonary emphysema may be allowed to travel by air provided they have a reasonable exercise tolerance at ground level with no gross cardiac decompensation (Peffers, 1978). The myocardium damaged by ischaemia or frank infarction will react poorly to the reduced oxygen tension of high altitude. In patients with severe angina a supply of oxygen should be available so that it can be supplied

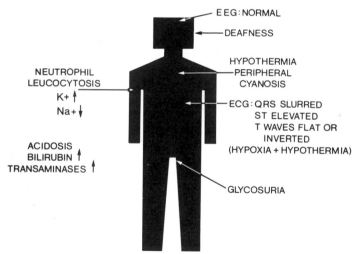

Fig. 30.5   Diagram to illustrate the changes induced in an 18-year-old youth exposed for nine hours to an altitude of 8840 m and a temperature of −6°C.

Air travel and patients with heart disease

As we have noted above, all modern aircraft have pressurized cabins so that irrespective of the height at which they fly the pressure in the cabin rarely falls below that appropriate to an altitude of only 1520 m to 1830 m that is about the altitude of Zermatt in the Swiss Alps. The small reduction of PA_{O_2} brought about by simulation of such relatively low altitudes

from a mask from the moment of take-off (Jackson, 1968). A patient who has had a myocardial infarct in the recent past should have a reasonable functional reserve before being allowed to undertake a long journey by air (Peffers, 1978). Such cases should have successfully passed through convalescence to the stage where they are ambulant and able to walk a distance of a hundred yards at a reasonable pace

(Peffers, 1978). They should be able to climb up to twelve stairs without symptoms. The length of time taken to achieve this degree of recovery obviously varies from one individual to another, and with the severity of the initial attack, but is normally in the region of three to six weeks for uncomplicated cases. Those in severe, uncompensated heart failure should not attempt air travel until the condition is well under control. Nowadays there are not infrequently requests for patients with congenital heart disease to fly to centres of cardiac surgery in various parts of the world. Subjecting such children to simulated high altitude in a pressurized aircraft is usually acceptable, provided additional oxygen for prolonged administration is available (Peffers, 1978). It has to be kept in mind that many patients with cardiopulmonary disease find the strain of flying less than that of a long journey involving tiring changes from ships to trains and cars (Jackson, 1968). A note of caution is necessary for such sick travellers who plan flights to such airports as La Paz in Bolivia where passengers step directly from a simulated altitude of 1520 m to 1830 m in the pressurized aircraft to the natural altitude at the airport of 4110 m. Furthermore in some areas of the world such as the Andes, some tourist flights may have to be taken in unpressurized aircraft.

We have already pointed out the dangers of the simulated high altitude inherent in travel in aircraft to subjects with sickle-cell anaemia or haemoglobin C disease in Chapter 9, and to patients with air-containing cysts or air introduced into the body for diagnostic or therapeutic purposes in Chapter 20.

Principles of rescue at great altitudes

The principles of rescue at high altitude are based on an understanding of the features of pathophysiology of acclimatization outlined in this book. These principles may seem only too obvious to the reader but perhaps for this very reason neglect of them may prove hazardous or fatal. The first is what Wilson (1973) terms elegantly 'the cardinal principle of retreat'. In other words the best management of a patient suffering from one or other of the illnesses attributable to the altitude is to remove him as soon as possible to a lower altitude. Wilson (1973) describes the case of a 21-year-old climber on Mount McKinley in 1971 who became ill at 4400 m and was taken to a nearby but higher shelter at 5200 m rather than take the longer journey down to base camp. He died. The second principle is to have adequate oxygen equipment, to know how to use it, and to ensure that the cylinders contain an adequate supply of oxygen. The authors have had personal experience of not ascertaining this. Third it is dangerous to transport rescue volunteers abruptly from sea level to any elevation above 2400 m. On a rescue operation on Mount McKinley in 1960 two of the rescuers developed acute high altitude pulmonary oedema (Wilson, 1973). Groups of rescuers should be at least three in number.

The ultrastructure of the lung in acute decompression

As we have already seen in Chapter 15 there appear to be ultrastructural features associated with high altitude pulmonary oedema, consisting of the extrusion of oedema vesicles into the pulmonary capillaries (Fig. 15.4). When mammalian lungs are suddenly subjected to the greatly diminished atmospheric pressure of extreme altitudes a very different set of ultrastructural changes ensues. We have studied these events in the lungs of rats exposed suddenly to a barometric pressure equivalent to that of the summit of Mount Everest (Mooi et al, 1978). Under these circumstances there is swelling and destruction of the pneumocytes of granular and membranous types with exposure of the denuded fused basement membrane of the alveolar-capillary wall (Fig. 30.6). At the same time there is what almost amounts to a suction effect on the pulmonary capillaries so that erythrocytes are sucked from them into the interstitial tissues of the lung (Fig. 30.7). It is of interest that the administration of heavy doses of frusemide to such rats prevents these deleterious effects on the pneumocytes but induces a cystic change in the endothelial cells of the pulmonary capillaries (Fig. 30.8).

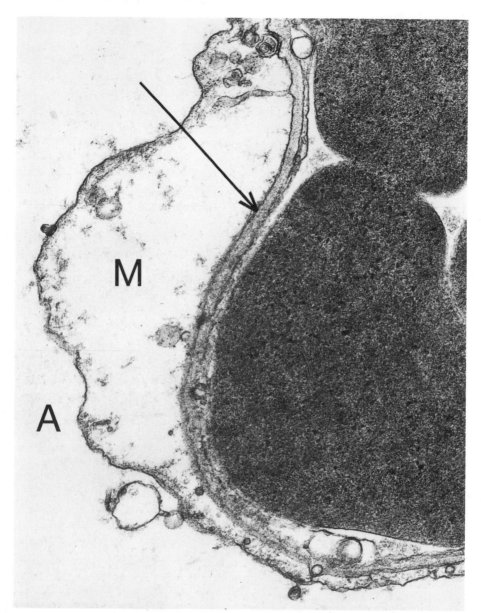

Fig. 30.6   Electron micrograph of alveolar wall from the lung of a Wistar albino rat exposed to acute decompression of 265 mmHg for one hour. This pressure simulates an altitude of approximately 8000 m which is roughly equivalent to the summit of Mount Everest. There is swelling and vacuolation of a membranous pneumocyte, M, lying on the fused basement membrane of the alveolar-capillary wall (arrow). Alveolar space A. ($\times$ 33 750).

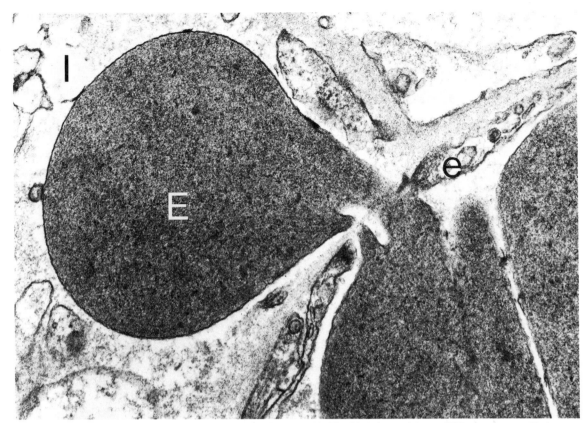

Fig. 30.7   Electron micrograph of lung from the same rat illustrated in Fig. 30.6. An erythrocyte (E) in the process of being extruded from a capillary into the interstitial space (I) of an alveolar wall. The erythrocyte passes through a rupture in the capillary endothelium (e) which otherwise is normal. The interstitium contains oedema fluid (× 45 000).

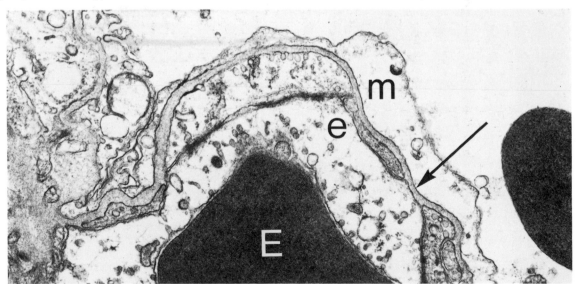

Fig. 30.8   Electron micrograph of lung of rat subjected to the same degree of acute decompression but pre-treated with frusemide. An alveolar capillary contains an erythrocyte (E). The capillary endothelium (e) is swollen and shows a loss of cytoplasmic density. The overlying membranous pneumocyte (m) is also pale and degenerate. The pallor of the cells renders the intervening basal lamina more conspicuous (arrow) but it is nevertheless normal (× 22 500).

REFERENCES

Bonington, C. (1971) In: *Annapurna South Face*, p. 296. London: Cassell.

Edwards, H. T. (1936) Lactic acid at rest and work at high altitude. *American Journal of Physiology*, **116**, 367.

Genton, E., Ross, A. M., Takeda, Y. A. & Vogel, J. H. K. (1970) Alterations in blood coagulation at high altitude. In: *Hypoxia, High Altitude and the Heart*, Edited by J. H. K. Vogel. Aspen. Colorado. Advances in Cardiology, 5, 32. Basel: S. Karger.

Gill, M. B., Poulton, E. C., Carpenter, A., Woodhead, M. M. & Gregory, M. H. P. (1964) Falling efficiency at sorting cards during acclimatization at 19,000 ft. *Nature*, **203**, 436.

Harris, P. & Heath, D. (1962) *The Human Pulmonary Circulation. Its Form and Function in Health and Disease*. 1st Edition. Edinburgh: Livingstone.

Hurtado, A. (1964) In: *The Physiological Effects of High Altitude*. Edited by W. H. Weihe. Oxford: Pergamon Press.

Hurtado, A. (1971) The influence of high altitude on physiology. In: *High Altitude Physiology: Cardiac and Respiratory Aspects*, p. 3. Edited by R. Porter & J. Knight. Edinburgh: Churchill Livingstone.

Jackson, F. (1968) The heart at high altitude. *British Heart Journal*, **30**, 291.

Lindley, R. (1975) Secrets of killer mountain. *Sunday Times*, June 22, 1975.

Milledge, J. S. (1975) Physiological effects of hypoxia. In: *Mountain Medicine and Physiology*, p. 73. Edited by C. Clarke, M. Ward & E. Williams. London: Alpine Club.

Mooi, W., Smith, P. & Heath, D. (1978) The ultrastructural effects of acute decompression on the lung of rats: the influence of frusemide. *Journal of Pathology*, **126**, 189.

Pajares, J. & Merayo, F. (1970) Unique clinical case both of hypoxia and hypothermia studied in an 18 year old aerial stowaway on a flight from Havana to Madrid. *Aerospace Medicine*, **41**, 1416.

Peffers, A. S. R. (1978) Carriage of invalids by air. *Journal of the Royal College of Physicians of London*, **12**, 136.

Pugh, L. G. C. E. (1962) Physiological and medical aspects of the Himalayan scientific and mountaineering expedition 1960–61. *British Medical Journal*, **ii**, 621.

Pugh, L. G. C. E., Gill, M. B., Lahiri, S., Milledge, J. S., Ward, M. P. & West, J. B. (1964) Muscular exercise at great altitudes. *Journal of Applied Physiology*, **19**, 431.

Purves, M. & Ponte, J. (1977) Film: *Hypoxia: and the Role of Peripheral Chemoreceptors*. Directed by: Ponting, D. Distributed by: John Wiley & Sons, London.

Robinson, D. H. (1973) *The Dangerous Sky. A History of Aviation Medicine*. Henley-on-Thames: G. T. Foulis.

Shults, W. T. & Swan, K. C. (1975) High altitude retinopathy in mountain climbers. *Archives of Ophthalmology*, **93**, 404.

Steele, P. (1976) *Medical Care for Mountain Climbers*. London: William Heinemann.

Ward, M. (1975) *Mountain Medicine. A Clinical Study of Cold and High Altitude*. London: Crosby Lockwood Staples.

Wilson, R. (1973) Acute high-altitude illness in mountaineers and problems of rescue. *Annals of Internal Medicine*, **78**, 421.

31

Retina and special senses

The retinal circulation is of considerable interest in high altitude studies because it presents for direct ophthalmoscopic examination a sample of systemic arteries and veins under the stimulus of chronic hypoxia.

The retinal circulation on exposure to hypoxia

Pronounced changes occur in the retinal circulation in most subjects on acute exposure to high altitude. After only two hours at 5330 m retinal arteries and veins increase in diameter by about a fifth (Fig. 31.1) (Frayser et al, 1971). Under these conditions

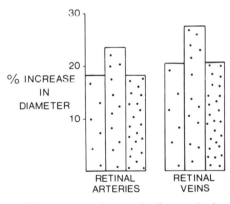

Fig. 31.1 The percentage increase in diameter in the retinal arteries and veins in nine males between 20 and 40 years of age (mean age 25.9 years) after two hours at 5330 m (very lightly stippled columns), and after five days at the same altitude (lightly stippled columns) and in nine males between 21 and 40 years of age (mean age 27.3 years) who had been at 5330 m for five to seven weeks (medium stippled columns). The data in this figure and figs. 31.2 to 31.4 are from Frayser et al (1971).

retinal blood flow increases by some 90 per cent (Fig. 31.2), the retinal circulation time decreases (Fig. 31.3) and the retinal vascular blood volume

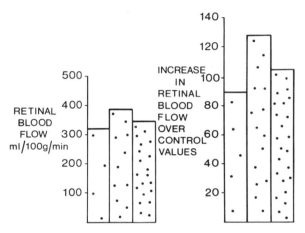

Fig. 31.2 The retinal blood flow expressed in ml/100 g/min and as a percentage increase over control values in the same subjects as in Fig. 31.1. The identification of the columns is as in that figure.

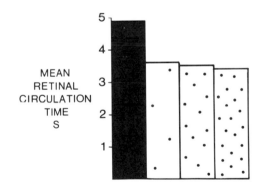

Fig. 31.3 The mean retinal circulation time in seconds in the same subjects as in Fig. 31.1. The identification of the columns is as in that figure with the addition that the filled column are values obtained on nine subjects at 790 m.

increases (Fig. 31.4). These trends increase over the following five days at this altitude and then revert back to the levels found initially on exposure to high altitude (Frayser et al, 1971). Duguet et al (1947)

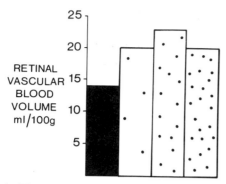

Fig. 31.4 The retinal vascular blood volume (ml/100 g) in the same subjects as in Fig. 31.1. The identification of the columns is as in Fig. 31.3.

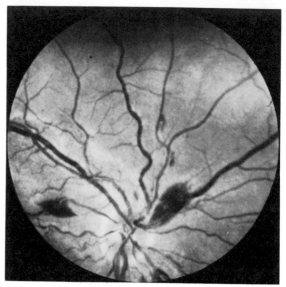

Fig. 31.5 Retinal photograph showing two haemorrhages which occurred in a climber ascending above 5000 m.

had reported earlier that dilatation of the retinal vessels became apparent at 1830 m and maximal at 5490 m. In contradistinction to the later studies of Frayser et al (1971) they found no further increase in vessel size after 15 minutes at this altitude.

Hypocapnia without hypoxia significantly reduces retinal blood flow (Hickam and Frayser, 1966). Thus at an elevation of 5330 m restoration of the systemic arterial oxygen saturation from 70 to 95 per cent at the prevailing Pa_{CO_2} of 24 mmHg will decrease retinal flow by about half to approximately sea-level values (Frayser et al, 1974). There is a significant decrease in retinal flow after nine days at high altitude as compared to five days suggesting that the tone of retinal vessels may become reset at the prevailing CO_2 tension and that hypoxia becomes the predominant controlling factor in regulating retinal flow at high altitude. There is an analogy here to the resetting of the medullary CO_2 receptor to lower partial pressures of carbon dioxide as described in Chapter 5.

Retinal haemorrhages

Retinal haemorrhages (Fig. 31.5) may occur on ascending to above 4000 m. The first report of their occurrence appears to have been made in July 1968 when they were found in two persons working at 5330 m on Mount Logan in Yukon territory. One of them developed papilloedema and became semi-comatose. The following year the Arctic Institute undertook a pilot study of the retinal circulation at high altitude. Twenty-five subjects were studied in groups exposed to different conditions of ascent

(Frayser et al, 1970). Eight mountaineers climbed to the laboratory at 5330 m and stopped there for seven weeks. Seven scientists and 10 volunteers were flown directly to the area and remained 2 to 10 days. Some of them were given acetazolamide before ascent or frusemide for one to three days after arrival. None of the subjects going to high altitude had retinal disease. Three of the eight mountaineers who ascended slowly, developed retinal haemorrhages as did 6 of 17 subjects who ascended rapidly. Hence there was no correlation between the occurrence of retinal haemorrhages and the speed of ascent. Similarly there was no relation to the severity of headache or other symptoms of acute mountain sickness. Neither acetazolamide nor frusemide prevented the retinal haemorrhages. All but one of the affected subjects were unaware of the presence of bleeding. However, in the ninth subject the haemorrhage occurred at the macula and resulted in a scotoma. This man developed the haemorrhage while he was sitting quietly after having been at altitude for nine and a half days. On examination he had papilloedema with prominently tortuous retinal arteries and veins. High as the incidence of retinal haemorrhage (36 per cent) was in this group changes in the blood vessels of the retina of the type described above were even higher. Indeed pronounced hyperaemia around the optic disc, tortuosity and increased diameter of veins and arteries appeared in all subjects within a few days of arrival at high altitude and persisted throughout the whole length of the stay. Frayser et al (1970) injected fluorescein and were unable to detect leakage from the vessels. There was a substantial increase in retinal blood flow with a fall in retinal mean circulation time from 4.9 s at 610 m to 3.4 s at 5330 m. Hence the retinal vessels were maximally dilated in the presence of increased retinal blood flow.

Schumacher and Petajan (1975) studied 39 subjects who spent up to 24 days above 4330 m on Mount McKinley. No fewer than 14 of the group developed retinal haemorrhage which was commoner in those with high altitude headache. There was an association with rapid ascent above 4200 m. 'Quick dashes' to the summit of Mount McKinley from 3000 m gave rise to retinal haemorrhage in six of nine climbers. Neither of the authors of this book developed retinal haemorrhages on rapid ascent to 4330 m in Peru. Retinal haemorrhages are unlikely to occur at rest below 5300 m.

Retinal haemorrhage is now such a well-documented complication of ascent to high altitude that no fewer than three papers on the subject appeared in a single issue of *Archives of Ophthalmology* (Rennie and Morrissey, 1975; Wiedman, 1975; Shults and Swan, 1975). Rennie and Morrissey (1975) described retinal changes which occurred in the members of the American Expedition to Dhaulagiri, Nepal (8170 m), the sixth highest mountain in the world. They studied 15 subjects, retinal photographs being taken at sea level and at 5880 m, after each climber had descended from his highest point. Five Nepali Sherpas were also studied. Vascular engorgement with tortuosity was observed, with an increase of 24 per cent in retinal arterial diameter and an increase of 23 per cent in diameter of retinal veins. Retinal haemorrhages were seen in a third of the American climbers but in none of the Sherpas. Shults and Swan (1975) reported the occurrence of retinal haemorrhages in four of six surviving members of a climbing expedition on Aconcagua (6920 m) in Argentina, the highest mountain in the Western hemisphere. Two were left with permanently disturbed vision with paracentral scotomas. Clarke and Duff (1976) reported the development of retinal haemorrhages in four out of six Britons of the successful 1975 British Everest Expedition who were newcomers to altitudes over 6000 m, in two of 14 Britons who had previously visited these altitudes, and in two of the 75 Sherpas with the Expedition.

Form

Retinal haemorrhages are usually located throughout the fundus without involving the macula. The haemorrhages may be diffuse, punctate, confluent or flame-shaped (Fig. 31.5). Clarke and Duff (1976) described several flame haemorrhages near the left optic disc of a 28-year-old Briton on the successful ascent of the South West face of Everest in 1975. Duff himself developed flame haemorrhages near the right optic disc and gross retinal venous tortuosity. Retinal haemorrhages at high altitude may extend into the vitreous (Wiedman, 1975). Macular haemorrhages are infrequent and may be solitary or occur in conjunction with diffuse haemorrhage into the posterior pole of the fundus.

Relation to intraocular pressure

Rennie and Morrissey (1975) believe that hypoxic vasodilatation makes retinal vessels more vulnerable to sudden rises in intraocular pressure. Frayser et al (1970) point out that intraocular pressure falls with exercise, and they think it possible that sudden increase in systemic blood pressure at high altitude may be transmitted to an embarrassed retinal circulation and result in haemorrhage. In fact there is no evidence that significant alterations in intraocular pressure occur at high altitude. Clarke and Duff (1976) measured the intraocular pressure of climbers ascending Mount Everest with a Perkins tonometer. Readings had to be confined to the early part of the expedition because the investigation was unpopular with the climbers. There was no noticeable fall in intraocular pressure with increasing altitude. During ascent to 6000 m intraocular pressure was within normal limits.

Cause

Clarke and Duff (1976) believe that retinal haemorrhages are related to the increased retinal blood flow and dilatation of retinal vessels. They believe that such bleeding is not necessarily a warning sign of impending cerebral oedema. Nevertheless they recognise that retinal haemorrhage may be associated with bilateral papilloedema. Other authorities believe it likely that a rapid increase in intracranial pressure explains the ocular changes. Sudden intracranial hypertension leads to effusion of cerebrospinal fluid into the optic nerve sheath resulting in compression of the central retinal vein and dilatation of the nerve sheath, swelling of which reduces the venous drainage of the eye by compressing the retino-choroidal anastomosis, producing retinal venous hypertension and haemorrhage (Muller and Deck, 1974; *British Medical Journal*, 1975). Shults and Swan (1975) believe that the extreme physical exertion and concomitant Valsalva manoeuvre required in mountain climbers may have a rôle in the development of retinal haemorrhage.

Course

The majority of high altitude retinal haemorrhages absorb spontaneously without loss of visual acuity. Descent from high altitude is the immediate treatment. Both Everest climbers with flame haemorrhages reported by Clarke and Duff (1976) had normal fundi at 1500 m after leaving the mountain. The use of frusemide or acetazolamide for prevention or therapy is inconclusive (Wiedman, 1975). Subjects in whom retinal haemorrhages have once occurred should be advised against returning above 3050 m so as to avoid recurrencies.

Light sensitivity and visual acuity

At the same time as these haemodynamic disturbances are overtaking the retinal circulation there is impairment of both sensitivity to light and visual acuity. *Light sensitivity* is impaired at elevations as low as 1220 m to 1520 m according to McFarland (1972) who states that at 4880 m ability to see may be reduced to half of sea-level performance. In other words twice as much light is required for perception of a given stimulus under these conditions. Halperin and his associates (1959) showed that on exposure to altitudes of between 2130 m and 5030 m for only three to four hours

diminished *visual acuity* developed but this was reversed within a few minutes of inhaling 100 per cent oxygen. Visual acuity is known to be impaired at altitudes of the order of 4880 m (McFarland and Evans, 1939). The effects of high altitude on visual acuity depend on the intensity of illumination. When this is high, there is little or no impairment of vision until an altitude of some 5490 m is exceeded. By contrast under low illumination visual acuity diminishes at an elevation as low as 2440 m (McFarland and Halperin, 1940), and it is reduced to only half of the sea-level value at 5180 m. During the Battle of Britain night blindness became a problem in fighter pilots. It was almost complete at an altitude of 3660 m but was found to be totally reversible, not by large amounts of vitamin A but by oxygen inhalation (Robinson, 1973). It is clear that the retinal rods, which are responsible for night vision, are very sensitive to hypoxia.

Colour vision

Colour discrimination is said to be increased between 2000 m to 3000 m but decreased from 5000 m to 7000 m (Frantsen and Iusfin, 1958). There has been one report that at 5490 m there is a slight deterioration in the recognition of blue and green (Schmidt and Bingel, 1953).

The middle ear and rapid ascent to high altitude

The middle ear is an air-containing cavity surrounded on all sides by bone (Fig. 31.6). It is lined by mucous membrane and contains the three

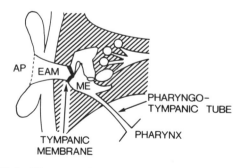

Fig. 31.6 Diagram to show the connections of the pharyngo-tympanic tube which allows equalisation of the pressure in the middle ear, ME, with the atmospheric pressure, AP, in the external auditory meatus, EAM.

auditory ossicles. If this cavity were completely closed, a sudden change in barometric pressure such as occurs on rapid ascent to high altitude would lead to a different pressure on the outside of the tympanic membrane from that on the inside. This dangerous situation is, however, avoided because there is a communication between the middle ear and pharynx (Fig. 31.6). This is the pharyngotympanic (Eustachian) tube. This canal opens at the posterior extremity of the inferior turbinate bone but it is not permanently open. It is shut by a valve which is relaxed during the act of swallowing and while it is open it permits the pressure in the middle ear to become the same as that of the ambient air. Variation in pressure on either side of the ear drum is thus prevented. Nevertheless an uncomfortable feeling of pressure in the ears associated with temporary deafness is not uncommonly associated with rapid ascent to altitude as occurs at take-off in aircraft or in cable cars travelling to the summits of volcanoes or Swiss peaks. Avoidance of this unpleasant sensation is achieved by repeated swallowing and many airlines offer their passengers sweets to chew to equalise pressure on either side of the tympanic membrane. A cold in the head will prevent proper equalisation of pressure due to blockage of the Eustachian tube by exudate or by pressure from without by inflammatory exudate or swollen lymphoid tissue. Such upper respiratory infections may lead to a persistence of aural symptoms for several days after descent from high altitude.

Auditory sensitivity

In contrast auditory sensitivity appears to be resistant to hypoxia. There is no difference between initial threshold values at sea level for both air and bone-conducted hearing and those found after an exposure to a simulated altitude of 4570 m for 30 minutes (Curry and Boys, 1956). Klein et al (1961) found that with bone and air conduction exposure to hypoxia resulted in a diminution in auditory sensitivity at the lower frequencies and a slight improvement at 4096 Hz. A subsequent study by Klein (1961) under similar hypoxic conditions confirmed a significant loss of bone-conducted thresholds at frequencies lower than 1024 Hz but enhancement at 4096 Hz.

Inner ear function

It has been suggested that in normal subjects a reversible alteration of both the cochlear and the vestibular functions occurs at high altitude. Singh and his colleagues (1976) studied 54 healthy Indian troops between the ages of 17 to 30 years at low altitude (210 m) and at an elevation of 3500 m after passive transport there by air. A pure tone audiogram showed loss of both low and high tone on arrival at high altitude. This was reversible and returned to normal after a stay of four days at high altitude. The threshold of hearing under the conditions of the test carried out by Singh et al (1976) was consistently better in native highlanders than in lowlanders but this was possibly attributable to the absence of background noise in the mountains. Some of the soldiers showed signs of vestibular upset on exposure to high altitude in the form of swaying (Fig. 31.7), and spontaneous and

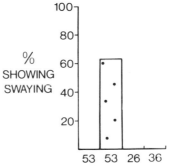

Fig. 31.7 The incidence of swaying when walking at high altitude on a straight line on an inclined floor with eyes closed: 53 lowlanders at 210 m; 53 lowlanders soon after arrival at 3500 m (very lightly stippled column); 26 of these lowlanders six months after acclimatization to high altitude; 36 native highlanders. (Based on data from Singh et al, 1976).

positional nystagmus (Fig. 31.8). The same cochlear and vestibular disturbances were apparent in sufferers from acute mountain sickness and high altitude pulmonary oedema. Singh and his co-workers (1976) point out that abnormalities of the cerebral blood flow run parallel with abnormal findings on audiological and vestibular testing both returning to normal after four to five days sojourn at high altitude. This suggests to them that disturbances of inner ear function are the result of the altered haemodynamics induced by acute exposure to high altitude.

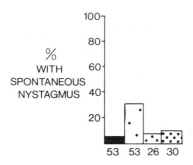

Fig. 31.8   The incidence of spontaneous nystagmus at high altitude. The columns refer to the same subjects as in Fig. 31.7 except that in this instance 30 native highlanders were studied. (Based on data from Singh et al, 1976).

Vestibular disturbance in astronauts

It is of interest to note that the altered haemodynamics of weightlessness in astronauts are also associated with vestibular disturbances which are unlike those of motion sickness (BBC, 1978). The nature of these disturbances is as yet obscure but they improve with practice of living under conditions of zero G. (BBC, 1978).

Gustatory sensitivity

A study by Maga and Lorenz (1972) was designed to evaluate what effect altitude has on a person's ability to detect the four basic tastes, salt (tested by sodium chloride), sour (citric acid), bitter (caffeine) and sweet (sucrose). Increasing molar concentrations of compounds representative of the four tastes were presented to six women and their threshold value for each taste determined. To simulate altitude, a chamber equipped with vacuum and pressure controls and controlled temperature and humidity was used. The altitudes studied were sea level and 1520 m and 3050 m above sea level. When the four tastes were considered together as a unit, a significant difference resulted between sea level and 1520 m since the lower the altitude the more sensitive was the composite taste response. However, no significant difference resulted in going from 1520 m to 3050 m. The study confirms the previous finding of Grandjean (1955) that the thresholds of gustatory sensitivity to glucose, salt, tartaric acid and quinine were diminished during exposure to a real altitude of 3450 m on the Jungfraujoch. These thresholds returned to normal on descent to low altitude. Finkelstein and Pippett (1958) found that exposure to a simulated altitude of 7620 m for two hours with or without administration of 100 per cent oxygen had no effect on taste perception. This is an altitude well in excess of that at which Maga and Lorenz found changes in gustatory sensitivity.

REFERENCES

British Broadcasting Corporation (1978) BBC2 TV Programme 'Zero G'.

British Medical Journal (1975) High altitude retinal haemorrhage. *British Medical Journal*, iii, 663.

Clarke, C. & Duff, J. (1976) Mountain sickness, retinal haemorrhages, and acclimatisation on Mount Everest in 1975. *British Medical Journal*, ii, 495.

Curry, E. T. & Boys, F. (1956) Effects of oxygen on hearing sensitivity of simulated altitudes. *Eye, Ear, Nose and Throat Monthly*, 35, 239.

Duguet, J., Dupont, P. & Baillant, J. P. (1947) The effects of anoxia on retinal vessels and retinal arterial pressure. *Aviation Medicine*, 18, 516.

Finkelstein, B. & Pippett, R. G. (1958) Effect of altitude upon primary taste perception. *Journal of Aviation Medicine*, 29, 386.

Frantsen, B. S. & Iusfin, A. I. (1958) On the alteration of colour sensation under conditions of hypoxia. (Ob izmeneniiakh tsvetaoshchashchenila v usloviiakh gipoksil). *Fiziologicheskii Zhurnal SSSR Imeni J. M. Sechenova*, 44, 519.

Frayser, R., Houston, C. S., Bryan, A. C., Rennie, D. I. & Gray, G. (1970) Retinal haemorrhage at high altitude. *New England Journal of Medicine*, 282, 1183.

Frayser, R., Houston, C. S., Gray, G. W., Bryan, A. C. & Rennie, I. D. (1971) The response of the retinal circulation to altitude. *Archives of Internal Medicine*, 127, 708.

Frayser, R., Gray, G. W. & Houston, C. S. (1974) Control of the retinal circulation at altitude. *Journal of Applied Physiology*, 37, 302.

Grandjean, E. (1955) The effect of altitude on various nervous functions. *Proceedings of the Royal Society*, 143, (B), 12.

Hickam, J. B. & Frayser, R. (1966) Studies of the retinal circulation in man. *Circulation*, 33, 302.

Halperin, M. H., McFarland, R. A., Niven, J. I. & Roughton, F. J. W. (1959) The time course of the effects of carbon monoxide on visual thresholds. *Journal of Physiology*, 146, 583.

Klein, S. J. (1961) Effects of reduced oxygen intake on bone conducted hearing thresholds in a noisy environment. *Perceptual and Motor Skills*, 13, 43.

Klein, S. J., Mendelson, E. S. & Gallagher, T. J. (1961) The effects of reduced oxygen intake on auditory threshold shifts in a quiet environment. *Journal of Comparative and Physiological Psychology*, 54, 401.

Maga, J. A. & Lorenz, K. (1972) Effect of altitude on taste thresholds. *Perceptual and Motor Skills*, 34, 667.

McFarland, R. A. (1972) Psychophysiological implications of life at altitude and including the role of oxygen in the process of aging. In: *Physiological Adaptations. Desert and Mountain*, p. 157. Edited by M. K. Yousef, S. M. Horvath & R. W. Bullard. New York: Academic Press.

McFarland, R. A. & Evans, J. N. (1939) Alterations in dark adaptation under reduced oxygen tensions. *American Journal of Physiology*, **127**, 37.

McFarland, R. A. & Halperin, M. H. (1940) The relation between foveal visual acuity and illumination under reduced oxygen tension. *Journal of General Physiology*, **23**, 613.

Muller, P. J. & Deck, J. H. N. (1973) Intraocular and optic nerve sheath haemorrhage in cases of sudden intracranial hypertension. *Journal of Neurosurgery*, **41**, 160.

Rennie, D. & Morrissey, J. (1975) Retinal changes in Himalayan climbers. *Archives of Ophthalmology*, **93**, 395.

Robinson, D. H. (1973) *The Dangerous Sky. A History of Aviation Medicine*. Henley-on-Thames: G. T. Foulis and Co.

Schmidt, I. & Bingel, A. G. A. (1953) Effect of oxygen deficiency and various other factors on colour saturation thresholds. U.S.A.F. School of Aviation Medicine Medical Reports. Project No. 21–31–022.

Schumacher, G. A. & Petajan, J. H. (1975) High altitude stress and retinal haemorrhage. *Archives of Environmental Health*, **30**, 217.

Shults, W. T. & Swan, K. C. (1975) High altitude retinopathy in mountain climbers. *Archives of Ophthalmology*, **93**, 404.

Singh, D., Kochhar, R. C. & Kacker, S. K. (1976) Effects of high altitude on inner ear functions. *Journal of Laryngology and Otology*, **90**, 1113.

Wiedman, M. (1975) High altitude retinal haemorrhage. *Archives of Ophthalmology*, **93**, 401.

32

Mental reaction and the psyche

Electroencephalographic changes in hypoxia

The electrical activity of the brain is influenced by exposure to hypoxia. Rebuck and his colleagues (1976) found similar electroencephalographic changes in 12 of their 15 studies on six healthy men in the age range of 25 to 42 years. Irregular slow waves appeared in all leads, being of greatest amplitude (50 μV) in the fronto-temporal regions. The 5Hz activity appeared first in the frontal region and then usually became generalized, a point at which the study was completed and oxygen given. The slow wave activity was not produced by eye movements. Any posterior alpha activity of 30 to 40 μV decreased with the onset of hypoxic changes. Variance analysis of the e.e.g. by Ernsting (1963) has indicated that an increase in alpha activity (8 to 16 Hz) may be the first change induced by hypoxia but Rebuck et al (1976) did not detect this on simple visual inspection of the e.e.g. record before the onset of the obvious slow-wave activity.

Variations in Pa_{CO_2} are without effect on the characteristics of the e.e.g. of hypoxia but do influence the level of systemic arterial saturation at which electroencephalographic changes occur. The higher the level of Pa_{CO_2}, the lower the level of Sa_{O_2} at which the hypoxic e.e.g. changes were first detected. Variations in P_{CO_2} probably mediate changes in cerebral oxygenation through alterations in cerebral blood flow. It is now generally agreed that cerebral blood flow is increased by hypoxia or hypercapnia and is decreased by hypocapnia or hyperoxia (Sokoloff, 1959). Such effects may represent a homeostatic mechanism for the removal of carbon dioxide produced. Clearly studies of acute progressive hypoxia are more safely performed with hypercapnia. If eucapnia is present, Sa_{O_2} may be decreased to 75 per cent with little risk of cerebral hypoxia as manifested by the appearance of slow waves in the e.e.g.

The relation of sensory and mental impairment at high altitude

Impairment of sensory functions seems to occur at a much lower altitude than slowing of mental reaction. Thus McFarland (1972) found that visual processes were reduced to unacceptable levels of impairment at an altitude of 3050 m which only just comes within our definition of 'high altitude' (Chapter 2). However, he found that mental tests including pattern-perception and decision-making could be carried out successfully up to altitudes of 5490 m which approach what we would regard as 'extreme altitude' (Chapter 30). Above this elevation there may be a serious impairment of higher cerebral functions.

Memory

When memory is tested by a paired-word association test, or by memory for pattern and position as measured by immediate or delayed recall, it is found to be increasingly impaired with increasing altitude (McFarland, 1938; Malmo and Finan, 1944). Memory is said to be noticeably impaired over 3660 m (McFarland, 1972).

Tests of mental reaction at high altitude

There is a considerable literature on testing the performance of higher cerebral function at high altitude but it has come under penetrating criticism from some authorities like Tune (1964) who finds it less convincing and developed than other areas of physiological research at high altitude. There seems to be a need to relate the results of somewhat artificial experiments to a consideration of whether these imply a significant deterioration of mental performance in daily activities at high altitude.

Bearing these reservations in mind, the results of studies suggest that decision-making tends to be impaired at high altitude. Most of the early studies were concerned with 'choice-reaction time'. Thus McFarland and Dill as early as 1938 found that, when subjects were required to discriminate among five different coloured lights, their reaction times were a positive function of altitude up to 6100 m. In more recent years the tests have become ever more complex. Thus Cahoon (1972) studied the reactions of eight volunteer soldiers after exposure to a simulated altitude of 4570 m for 3, 20, 24 and 45 hours. They were asked to perform four tasks. One was a simple psychomotor task of sorting 96 blank cards into two bins alternately. One was a complex psychomotor task of sorting 96 blank cards into 16 bins sequentially. A simple cognitive task involved sorting the 96 cards into two bins according to whether the central figure was red or green. Finally a complex cognitive task involved sorting 96 cards into 16 bins according to colour, shape and size of central figure and presence or absence of a black dot. The results showed that cognitive tasks showed a greater decrement at 4570 m in speed and accuracy than psychomotor. Complex decision-making tasks were more affected than simple tasks. Speed was sacrificed to maintain accuracy.

Another investigation which suggests that mental performance and decision-making are impaired at high altitude is that of Sharma and his colleagues (1975) who studied the sequential changes in psychomotor performance during prolonged stay at high altitude. Their study was carried out at an altitude of 4000 m in the western Himalayas, the aim being to record sequential changes in eye-hand coordination in terms of speed and accuracy over a period of two years. The subjects were 25 healthy soldiers between 21 and 30 years of age who were natives of the plains of southern India. They were required to move a stylus between a groove 0.5 cm cut to form a multicornered star. Errors were sounded automatically. The time to trace the design was recorded by stopwatch. The test was applied

after 1, 10, 13, 18 and 24 months' residence at high altitude. Psychomotor efficiency was found to be adversely affected both in terms of speed and accuracy during early stages of residence at high altitude. After living in the mountains for 10 months there was a gradual return towards the original level of psychomotor efficiency.

The interpretation of the initial impairment of mental performance as an effect of high altitude and its subsequent recovery as an expression of acclimatization is open to question. The early fall off in mental agility in lowlanders exposed to high altitude could equally well be attributed to the considerable psychological stress they might experience with, for example, the onset of symptoms of acute mountain sickness rather than to a direct effect of hypoxia on higher cerebral function involved in decision-making. The improvement in performance after a few months at high altitude could similarly be easily explained by the growing familiarity of the subjects with the tests rather than by any process of acclimatization.

The effect of motivation

Further studies of Cahoon (1973) are enlightening in that they demonstrate that motivation and training can effectively compensate for the stress imposed by a high altitude atmosphere for monitoring tasks of short duration. Cahoon (1973) sought to determine the effects of high altitude on the performance of a simulated radio-communication task. The subjects studied had to monitor and respond to tapes of simulated radio traffic at four different altitudes namely sea level, 3960 m, 4570 m and 5180 m. Two groups were tested in this manner. The first consisted of nine volunteers. Above an altitude of 3960 m they detected significantly fewer messages, the percentage falling from nearly 80 per cent at 3960 m to only 40 per cent at 5180 m. Furthermore, reaction time increased from 1.3 seconds at 3960 m to 1.8 seconds at 5180 m. Apparently the environmental conditions had led to a deterioration in performance. The second group consisted of highly motivated trained soldiers. Four had received radio training and four had not. The difference in the responses of this group was significant and dramatic. They performed better in all forms of monitoring. At 3960 m they achieved a

95 per cent detection rate of messages and it still exceeded 90 per cent at 5180 m. Their reaction time actually fell from 1.00 seconds at 3960 m to 0.9 seconds at 5180 m. There was a striking difference in attitude as well as performance and the results were not significantly different in radio-trained and non-radio-trained subjects indicating that motivation is a more important factor than training in maintaining performance at high altitude. Such findings do not deny that in extended operations where motivation may drop, say through lack of sleep, it is reasonable to expect that the stress of high altitude will affect performance as in the first group of volunteers described above.

The interaction of drugs and hypoxia at high altitude

Alcohol and hypoxia. An early comment on the influence of alcohol on the development of acute mountain sickness came from Ravenhill (1913). He felt that alcohol accelerated attacks and increased the severity of the symptoms. He stated that 'if a man takes much alcohol on his journey to the heights he nearly always suffers more than the abstemious man. Habitual alcoholics, however, are not necessarily attacked more severely, provided they do not indulge too freely on the journey. Total abstainers, on the other hand, are not by any means exempt'.

In an effort to determine the effects of alcohol upon the time of useful consciousness of a subject when he is suddenly subjected to hypoxic conditions, Nettles and Olson (1965) studied 10 normal subjects at a simulated altitude of 7620 m. After their oxygen masks were removed, they performed various tasks requiring mental and physical coordination until definite hypoxic symptoms were manifested. Later the same subjects were submitted to identical conditions after the administration of 0.5 ml of 100 per cent alcohol per pound of body weight. With blood alcohol levels of between only 22 to 49 mg/100 ml there was a reduction of 38 per cent in the time of useful consciousness under these hypoxic conditions. At the present time, high altitude flight is possible in pressurized aircraft which carry emergency oxygen supplies for the unlikely event of a rapid decompression. An elevated blood alcohol level in passengers thus suddenly exposed to hypoxia might delay their carrying out the simple procedure necessary to adjust to the emergency system and thus might have serious consequences. The interrelation of the effects of alcohol and altitude on man during rest and work has also been studied by Mazess et al (1968).

Marijuana and hypoxia. It is now widely accepted that marijuana or its principal active ingredient delta-9-tetrahydrocannabinol (Δ^9-TCH) impairs the performance of complex behavioural tasks in animals and man (U.S. Department of Health, 1974). Lewis et al (1976) have studied the effects of marijuana on two female adolescent baboons to see if the impairment produced by the drug is potentiated by the hypoxia of simulated high altitude. The animals were trained to perform a delayed matching-to-sample task on red, white and green lamps. Both baboons were subjected to various combinations of oral doses of marijuana, ranging from 0.25 to 2.0 mg per kg body weight and exposure to simulated altitudes of 390 m to 3660 m, before performing the matching tests. The study showed that while accuracy of matching performance was unaffected by the combination of hypoxia and marijuana, the level of work output was greatly reduced. This was due to deterioration in the speed of response.

Psychological effects in the newcomer to high altitude

Psychological stress is not uncommon in the newcomer freshly undergoing acclimatization at altitudes as low as 4500 m. In a person already aware of the reputation of acute mountain sickness and of the dangers of high altitude pulmonary oedema, the instability of the cardiovascular system during early days at high altitude may lead to considerable apprehension.

Mental effects and hallucinations of extreme altitude

Altitudes exceeding 5490 m represent the limit for acceptable performance for mental tests requiring decision-making (McFarland, 1972). At such elevations hypoxia exerts an increasingly severe effect on higher cerebral functions. Attention fluctuates more easily and mental block is common. Calculations are unreliable, judgement becomes faulty and emotional responses are unpredictable. As we have seen in Chapter 30 there is evidence of a falling off in mental capacity in climbers at 5800 m. Above 6700 m there is an onset of significant mental impairment with serious lapses of judgement. At such great altitudes one has left the realms of the physiology of acclimatization for those of heroic endurance and deliberate exposure to extreme conditions.

Hallucinations may occur in climbers at extreme altitude. We have already made reference to these in Chapter 30 but we may consider here two of the most characteristic phenomena. The first is the *phantom companion*. Smythe (1934) gave a most graphic account of this curious sensation and we quote here his experience during the attempted ascent of Everest in 1933. 'All the time that I was climbing alone I had a strong feeling that I was accompanied by a second person. This feeling was so strong that it completely eliminated all loneliness I might otherwise have felt. It even seemed that I was tied to my 'companion' by a rope, and that if I slipped 'he' would hold me. I remember constantly glancing back over my shoulder, and once, when, after reaching my highest point, I stopped to try to eat some mint cake, I carefully divided it and turned round with one half in my hand. It was almost a shock to find no one to whom to give it. It seemed to me that this 'presence' was a strong, helpful and friendly one, and it was not until Camp VI was sighted that the link connecting me, as it seemed at the time to the beyond, was snapped and, although Shipton and the camp were but a few yards away, I felt suddenly alone'. Two members of the British Everest team of 1975 dug a snow-hole on the mountain and spent the night at 8800 m without oxygen. Both reported the curious sensation that they had been accompanied by a third person (Clarke, 1976). The phantom companion at extreme altitudes is probably fabricated in the mind to bring some psychological support in a very insecure situation.

Frank hallucinations may occur at extreme altitudes and that reported by Smythe (1934) during the same ascent has become one that is classically quoted. He reports at an altitude of 8290 m: 'I was still some 200 feet above Camp VI and a considerable distance horizontally from it when, chancing to glance in the direction of the north ridge, I saw two curious-looking objects floating in the sky. They strongly resembled kite-balloons in shape, but one possessed what appeared to be squat under-developed wings, and the other a protuberance suggestive of a beak. They hovered motionless but seemed slowly to pulsate, a pulsation incidentally much slower than my own heart-beats, which is of interest supposing that it was an optical illusion. The two objects were very dark in colour and were silhouetted sharply against the sky, or possibly a background of cloud. So interested was I that I stopped to observe them. My brain appeared to be working normally, and I deliberately put myself through a series of tests. First of all I glanced away. The objects did not follow my vision, but they were still there when I looked back again. Then I looked away again, and this time identified by name a number of peaks, valleys and glaciers by way of a mental test. But when I looked back again, the objects still confronted me'. One is aware that large mountain birds such as lammergeyers frequent the extreme altitudes of the Himalayas (Chapter 3) but it seems most likely that these bird-like creatures observed by Smythe (1934) were hallucinatons. As we note in Chapter 30 dancing horses have been observed high on Aconcagua in Argentina.

Treacherous euphoria

Habeler (1979) gives a graphic account of the mental effects induced by the hypoxia of extreme altitude. He and Messner climbed the main summit of Mount Everest (8850 m) without the use of a supply of oxygen on May 8, 1978. At the summit they found that their attentiveness and concentration declined dramatically. The capacity for clear logical thinking was lost. Slowly they became overwhelmed by a dangerous sense of euphoria. 'I felt somehow light and relaxed, and believed that nothing could happen to me. Undoubtedly, many of the men who have disappeared for ever in the summit region of Everest had also fallen victim to this treacherous euphoria'. After this fleeting sense of triumph Habeler felt exhausted and suddenly afraid of death or severe damage to the brain. This led him to descend as rapidly as possible by sliding down the mountain.

This remarkable exploit certainly demonstrated

that two very fit mountaineers can survive without artificial oxygen at 8850 m but there seems a strong likelihood that during these heroic acts brain damage occurs. We have already seen in several places in this book that hypoxia induces haemorrhages into such disparate sites as the retina, gastric mucosa and nails. It is probable that in severe hypoxia similar haemorrhages occur in the brain. It is of interest that after his heroic climb Habeler suffered nightmares and lapses of memory which persisted to the time of writing his book.

The psyche of the native highlander

So far as we have been able to judge, the effects of hypoxia of high altitude *per se* on the psyche of the native highlander of Peru are minimal. The Indians born and living all their lives at altitudes of up to 4500 m work on the land or in mines, raise families, hold markets, dance and play football. They live normal lives in high mountains; it is their home. The introverted withdrawn character of the Quechuas is an expression of race and ethnic origin rather than hypoxic stress. It is not surprising that Sharma and Malhotra (1976) found that Gurkha soldiers, who originate from Nepal living up to the level of the Alpine zone at 2500 m, showed better psychological tolerance of a stay of ten months at 4000 m than Madrasi and Rajput subjects. So far as the events of daily life are concerned the native highlander is well adjusted to his mountain environment, accepting the hard realities of life at high altitude as commonplace. We shall find that this is not true, however, when we consider the rôle of the mountains in his religious beliefs (Chapter 34).

REFERENCES

Cahoon, R. L. (1972) Simple decision-making at high altitude. *Ergonomics*, 15, 157.

Cahoon, R. L. (1973) Monitoring army radio-communications networks at high altitude. *Perceptual and Motor Skills*, 37, 471.

Clarke, C. (1976) On surviving a bivouac at high altitude. *British Medical Journal*, i, 92.

Ernsting, J. (1963) Some effects of brief profound anoxia upon the central nervous system. In: *Selective Vulnerability of the Brain in Hypoxaemia*, p. 41 Edited by W. H. McMenemy & J. P. Schade. Oxford: Blackwell Scientific Publications.

Habeler, P. (1979) In: *Everest Impossible Victory*, pp 179 and 180. London: Arlington Books.

Lewis, M. K., Ferrado, D. P., Mertens, H. W. & Steen J. A. (1976) Interaction between marihuana and altitude on a complex behavioural task in baboons. *Aviation, Space and Environmental Medicine*, 47, 121.

Malmo, R. B & Finan, J. L. (1944) A comparative study of eight tests in the decompression chamber. *American Journal of Psychology*, 57, 389.

Mazess, R. B., Picòn-Rèategui, E., Thomas, R. D. & Little M. A. (1968) Effects of alcohol and altitude on man during rest and work. *Aerospace Medicine*, 39, 403.

McFarland, R. A. (1938) *The Effects of Oxygen Deprivation (High Altitude) on the Human Organism*. Report 13. Washington, D.C. Department of Commerce, Bureau of Air Commerce, Safety and Planning Division.

McFarland, R. A. (1972) Psychophysiological implications of life at high altitude and including the role of oxygen in the process of aging. In: *Physiological Adaptations. Desert and Mountain*, p. 157. Edited by M. K. Yousef, S. M. Horvath & R. W. Bullard. New York: Academic Press.

McFarland, R. A. & Dill, D. B. (1938) A comparative study of reduced oxygen pressure on man during acclimatization. *Journal of Aviation Medicine*, 9, 18.

Nettles, J. L. & Olson, R. N. (1965) Effects of alcohol on hypoxia. *Journal of the American Medical Association*, 195, 1193.

Ravenhill, T. H. (1913) Some experiences of mountain sickness in the Andes. *Journal of Tropical Medicine, and Hygiene*, 16, 313.

Rebuck, A. S., Davis, C., Longmire, D., Upton, A. R. M. & Powles, A. C. P. (1976) Arterial oxygenation and carbon dioxide tensions in the production of hypoxic electro-encephalographic changes in man *Clinical Science and Molecular Medicine*, 50, 301.

Secretary, Department of Health, Education and Welfare (1974) Marihuana and health. DHEW Report No. 74 - 50 U.S. Government Printing Office, Washington D.C.

Sharma, V. M. & Malhotra, M. S. (1976) Ethnic variations in psychological performance under altitude stress. *Aviation, Space and Environmental Medicine*, 47, 248.

Sharma, V. M., Malhotra, M. S. & Baskaran, A. S. (1975) Variations in psychomotor efficiency during prolonged stay at high altitude. *Ergonomics*, 18, 511.

Smythe, F. S. (1934) The second assault. In: *Everest 1933* (J. Ruttledge, 1934), p. 164. London: Hodder and Stoughton.

Sokoloff, L. (1959) The action of drugs on the cerebral circulation. *Pharmacological Reviews*, 11, 2.

Tune, F. S. (1964) Psychological effects of hypoxia. Review of certain literature from the period 1950 to 1963. *Perceptual and Motor Skills*, 19, 551.

33

The climbers

A love of mountains and a desire to scale them is not innate in man. Europeans during the Middle Ages feared that the mountain peaks shrouded in mist were frequented by monsters such as dragons (Chapter 34). This terror of mountains persisted for centuries and exceptional indeed were such events as the ascent of Mont Aiguille (2100 m) by de Beaupré in the South of France in July 1492, a few weeks before Columbus set sail for the New World. A change in attitude began in the eighteenth century under the influence of people like Albrecht von Haller, a noted Swiss physiologist who was one of the first to identify the carotid bodies, whose rôle in acclimatization to high altitude we discuss in Chapter 8. With the ascent of Mont Blanc (4810 m) by Jacques Balmat and Dr Michel Paccard in August 1786 for the prize offered by de Saussure the era of climbing began. Once the taste for this activity had been whetted, there was a slow but inexorable movement to climb ever higher peaks where the climber had to contend not only with the dangers of the mountain but also with those of high altitude. Elsewhere in this book we note that man largely overcomes his environmental difficulties by avoiding them. Not so the high altitude climber who deliberately exposes himself both to physical injury and to extreme conditions of cold and hypoxia that will not support life indefinitely and which may prove fatal.

The personality of the high altitude climber

Some interesting insights into the personality of British and American high altitude climbers were gained from the analysis of questionnaires completed by them (Ward, 1975). As might be expected they were assessed to be intelligent, self-sufficient and resourceful. Less expected they were also found to be withdrawn and detached. They tended to be aggressive, self-centred and highly competitive. A Polish study (Ryn, 1971) came to similar conclusions. Nicolson (1975) also accepts this analysis of the personality of many high altitude climbers and believes that for them achievement is measured strictly in terms of thousands of feet, vertically, so that the mountain has no other dimension. Since the Himalayas offer the ultimate in altitude they also offer the ultimate in satisfaction. The very language of mountaineering is military in nature and reveals that climbers regard the mountain as an enemy for conquest. Thus the talk is of the expedition, assault, conquest and defeat. The camps on the mountainside are, like Army Corps, identified with Roman numerals from I to VIII (Nicolson, 1975). In many instances the enemy to be overcome is not only the mountain but competitors also trying to subjugate it. We may illustrate some of these motivations and characteristics of sea-level man at high altitude by briefly considering some famous climbs in history.

The ascent of the Matterhorn

By the middle of the nineteenth century the last great mountain for conquest in the Alps was the Matterhorn (4480 m) which stands on the Swiss-Italian frontier with Zermatt on the Swiss side (Fig. 33.1) and Breuil on the Italian. A local mountain-guide, Jean Antoine Carrel, was determined to conquer the peak from the Italian side to bring fame to himself, his family and his country. With some relatives he made unsuccessful attempts on the summit in 1858 and 1859 but this measured and patient approach was suddenly put at risk with the

Fig. 33.1   The Matterhorn (4480 m), which stands on the Swiss-Italian frontier, seen from the village of Zermatt. The east face of the mountain is seen to the left and the north face to the right. Between these faces is the north-east ridge up which the celebrated conquest of the summit was made on July 14th, 1865. The arrow points to the spot at which Hadow slipped on the descent causing the four leading climbers in the party to plunge to their deaths down the north face.

Fig. 33.2   A plaque dedicated to the mature Whymper on the wall of the Monte Rosa hotel in Zermatt. The artist has caught the determination that led to the conquest of the Matterhorn in his early manhood.

arrival in Breuil the following year of Edward Whymper (Fig. 33.2). This young Englishman had been born the son of a prosperous engraver in London in 1840. His skill as an illustrator led him to be commissioned to prepare the illustrations for a second series of *Peaks, Passes and Glaciers* and in this way he became familiar with the Alps and obsessed by the Matterhorn. Whymper invited Carrel to act as his guide in an attempt on the summit but the local man wanted the honour for himself alone and refused. He became alarmed when the young foreigner began the climb without him and hastily gathered his uncle to give chase (Clark, 1976). They had the satisfaction of overtaking Whymper on the mountain and reaching an altitude of 4020 m but the disappointment of failing once again to conquer the mountain. This persistent young Englishman came back in 1862 and the struggle continued into 1863, four further attempts being made. In two of these the adversaries came together very briefly for joint climbs. By June 1865 Whymper had failed in his sixth attempt. Yet again he sought out Carrel in Breuil to persuade him to try once more. This time, however, it emerged that the guide had offered his services for an Italian attempt under the leadership of the famous Alpinist, Felice Giordano (Sanuki and Yamada, 1974). To his alarm, Whymper found that this attempt from the Italian side was imminent.

At this very moment there was another young English climber in Breuil, Lord Francis Douglas, 18-years-old and the brother of the Marquess of Queensbury. With him was a 21-year-old guide, 'Young Peter' Taugwalder of Zermatt. Both joined Whymper and rushed back to the Swiss village. Here they found at the Monte Rosa hotel a highly experienced mountaineer, the Reverend Charles Hudson with a young, inexperienced climber, Douglas Hadow who was only 19-years-old. They had already engaged a highly experienced guide from Chamonix, Michel Croz, for an assault on the Matterhorn. The two English groups decided as a matter of some urgency to join forces. They also took 'Old Peter' Taugwalder, 45 years of age and the father of 'Young Peter'. They began their ascent urgently and chose the North East ridge (Fig. 33.1). To their delight they found this to be much less steep than appearances from Zermatt suggested. Indeed, when they bivouacked at 3350 m, Croz and 'Young Peter' were sent on to test the route and were encouraged to get a report that it showed no insuperable difficulties. So it proved and at 1.40 pm on July 14th, 1865 the English party stood in triumph on the roughly level ridge of 110 m which formed the summit. The Italians were still far below and appeared as mere dots.

After one hour the party descended, the order of descent being Croz, Hadow, Hudson, Douglas, 'Old Peter', Whymper and 'Young Peter'. The experienced Croz was anxious to aid the youthful Hadow in every way, holding his legs so that his feet gained exactly the right holds. Suddenly Hadow was gone (Fig. 33.1). He knocked over Croz and their combined weight pulled Hudson and Douglas off the mountainside. Whymper and the Taugwalders braced themselves for the strain but the rope between 'Old Peter' and Douglas snapped with a report that would echo around Europe. The leading four climbers fell to their death down the north face of the mountain onto the Matterhorn-gletscher some 1220 m below, sliding helplessly and quickly on their backs and dropping from precipice to precipice (Clark, 1976).

For half an hour the survivors were unable to move. They descended to Zermatt to face gossip, an enquiry and a controversy that has persisted to this day. Some said 'Old Peter' had cut the rope to save himself. It seems most unlikely that this was so but Whymper had already found to his horror that the rope linking Douglas and 'Old Peter' was the thinnest and weakest they had. The essence of the controversy has always been whether 'Old Peter' had deliberately linked himself to Lord Douglas by this rope so that, should the leading climbers fall, the rope would break and he at least would survive. Whymper set off on the following morning, a Sunday, with three English volunteers and guides to retrieve the bodies of Croz, Hadow and Hudson. The body of Douglas was never found. The rope binding the three men was thick and strong. Only the rope binding Douglas to 'Old Peter' had been selected to be thin and weak. However, the Court of Enquiry at Zermatt accepted the statement of 'Old Peter' Taugwalder that in his judgement the rope was stout enough (Sanuki and Yamada, 1974). He left Switzerland to live in the United States, returning to die in Zermatt in 1888. Whymper said he was 'done with the Alps now', but, like the guide, chose to end his days there, dying in Chamonix in 1911. In the Mountaineering Museum in the village of Zermatt today one can see the rope, photographs

of the main characters in the drama and a copy of the transcript of the Court of Enquiry in a room simply and powerfully labelled 'July 14, 1865'. This story of the ascent and descent of the Matterhorn has been told and retold but it regains a certain freshness when placed in the context of our consideration of man at high altitude in this book. This eventful climb underlined the fact that on the scene had appeared a new man at high altitude totally different in kind from that which had lived in his mountain home for centuries. Here was a type of sea-level man bringing with him to the mountainside his characteristics of competitive thrust and aggression so alien to the native highlander of the Andes and the Himalayas.

Conquest for a nation. The ascent of Nanga Parbat

The conquest of the Matterhorn introduced the new unhealthy factor of nationalism to climbing. The race up that mountain on the same day by parties representing different countries introduced a new theme to reappear time and time again in the history of climbing. The prolonged assault on Nanga Parbat (8125 m) illustrates the terrifying determination, regardless of cost, that can be brought to bear on problems when national pride is felt to be at stake. When the British became preoccupied with Mount Everest, German mountaineers adopted Nanga Parbat as their target. The first expedition of 1932 was led by Willy Merkl into all the attendant dangers of the Ice Barrier (Chapter 3) as they approached by a long glacier. Ice avalanches and blizzards made the ascent treacherous and the summit could not be reached. Undeterred, Merkl returned with a second expedition in 1934. On this climb Drexel died from 'pneumonia' which was in reality almost certainly high altitude pulmonary oedema (Chapter 15). Soon after, Merkl himself and Willi and Uli Welzenbach also died with no less than seven of the porters on this blizzard-ridden expedition. In spite of this appalling loss of life, yet another German expedition appeared in 1937 under the leadership of Dr Karl Wien. On July 18th of that year Uli Luft climbed to Camp IV to join the party. He found the camp buried under a vast avalanche. It was almost a month before a relief party reached the scene where seven climbers and nine porters had been overwhelmed. Sixteen years and one World War later, the Germans returned to finish the job under the leadership of Dr Karl Herligkaffer, the stepbrother of Willy Merkl. In the early morning of July 3rd, 1953 Hermann Buhl set out on his own from Camp V at 6900 m above sea level to reach the summit still four miles away and 1220 m higher up (Clark, 1976). Four and a half hours later he reached the summit. 'I simply slumped to the ground and lay there fighting a desperate battle for that essential commodity—air' (Clark, 1976). The old score had been settled. Nicolson (1975) reminds us that when the Italians reached the summit of K2 (8610 m), Ardito Desio, their leader, proclaimed 'Lift up your hearts, dear comrades. By your efforts you have won great glory for your native land'.

Death on the mountain

Climbers are exposed to the dangers associated with high altitude *per se* and elsewhere we have described these risks from high altitude pulmonary oedema, cerebral oedema, retinal haemorrhage and thrombosis. Cold is another major factor and accounts of severe frostbite affecting those who scale very high peaks abound in the literature on climbing. One of the celebrated ascents marred by this mishap was the conquest of Annapurna (8080 m) by Maurice Herzog and Louis Lachenal in 1950. On descending from the summit Herzog lost his gloves and Lachenal fell 300 feet and was rescued with difficulty. Both then had to spend the night in a shallow crevasse, were covered in huge masses of snow, and had to dig themselves out. As a result Lachenal lost all his toes from frostbite while Herzog lost toes and fingers. Following the successful ascent of Mount Everest in May 1976 Sergeant John Stokes had an operation for the removal of all his toes following frostbite. His fellow climber Corporal Lane was expected to lose his toes and the fingertips of his right hand (*Daily Telegraph*, 1976). Both men suffered frostbite to their fingers, feet and faces after three nights which they spent close to the summit of Mount Everest.

As we describe in Chapter 3, the higher reaches of the Nepal Himalayan range in contrast to the Western Himalayas are exposed to a high annual precipitation because of the Monsoon. As a result high altitude climbers operating in this region have to penetrate a formidable Ice Barrier with the risk of avalanches, traversing ice crevasses and the like. They are also exposed to devastating blizzards. Frostbite thus becomes an outstanding problem under such conditions.

The sudden coming down of appalling blizzards may lead to the disappearance of climbers on mountains. Some of these episodes have become legendary adding to the myths and folklore of climbing. Blizzards swept Mount Everest (8850 m) during the attempted ascent of 1924 by a British party led by General Bruce. Mallory and Irvine set off from Camp IV on June 6th accompanied by eight Sherpas. The following day they moved to Camp VI. On June 8th, Odell also made for Camp VI and, while climbing, saw what he thought were the figures of Mallory and Irvine near the summit, although it has been subsequently thought possible that they were choughs (Chapter 3). Snow was falling as he reached Camp VI. Then the mists rose but there was no sign of the two climbers. Odell descended to Camp IV but returned once again to Camp VI. There was still no sign of Mallory or Irvine and it has never been established whether they reached the summit of Mount Everest in 1924 or not. During the 1933 expedition under Hugh Ruttledge a Swiss ice axe believed to be that of Irvine was found 60 feet below the crest of the North East ridge.

A continuing tradition of death

Clark (1976) does not subscribe to the view that climbing offers the means for fulfilment of deep psychological goals. He finds worrying 'the overtones of high seriousness which today are sometimes tacked onto a simple sport'. He believes that 'the tendency to see in the sport dark depths that would have intrigued most of the pioneers and sent many of them into roars of laughter is probably a passing fad'. We find it difficult to accept this view and believe, on the contrary, that powerful psychological forces motivate the high altitude climber. A visit to the cemetery of Zermatt recalls the dramatic events of 1865 described earlier in this Chapter (Fig. 33.3) but also reveals the continuing tradition of death associated with attempts to scale the Alpine peaks. A visit to the Church of St. Peter's in this village demonstrates that English climbers

Fig. 33.3 A stone in the village graveyard of Zermatt commemorating the death of Croz during the descent following the conquest of the Matterhorn on July 14th, 1865.

Fig. 33.4 The English Church of St. Peter's at Zermatt where many young English climbers are buried.

have participated fully in it (Fig. 33.4). This ever-present risk of injury or death appears to be cheerfully accepted and may in fact be one of the fascinations of Alpine and high altitude climbing. 'It is true the great ridges sometimes demand their sacrifice, but the mountaineer would hardly forgo his worship though he knew himself to be the destined victim' (Mummery, 1894). The year after he wrote that Mummery died while trying to scale the summit of Nanga Parbat in the Himalayas.

REFERENCES

Clark, R. W. (1976) In: *Men, Myths and Mountains*. London: Weidenfeld and Nicholson.
Daily Telegraph (1976) Wednesday, August 11th.
Mummery, A. F. (1894) Quoted by Sanuki and Yamada (see below).
Nicolson, N. (1975) *The Himalayas. The World's Wild Places*. Amsterdam: Time-Life Books.

Ryn, Z. (1971) Psychopathology in alpinism. *Acta medica polona*, 12, 453.
Sanuki, M. & Yamada, K. (1974) *The Alps*. Tokyo: Kodansha International Ltd.
Ward, M. (1975) *Mountain Medicine. A Clinical Study of Cold and High Altitude* London: Crosby Lockwood Staples.

34

Mountain monsters

The gigantic monoliths of the great mountain ranges, shrouded in mist or covered in snow, have the potential for conveying a sense of awe and of the eternal to man. This may flower into the concept of a mountain deity if the background religious climate is sympathetic. In the Andes the overwhelming religious presence is the Roman Catholic Church which has no need or place for such deities. It is rather to the Himalayas that we must look for a society that includes the gods of the mountains in its beliefs. The Himalayas are central to the religious faith of the Hindus. Shiva is believed to sit on the celestial heights of the mountain peaks in a state of perpetual meditation, generating the spiritual force that sustains the cosmos (Nicholson, 1975). To the Tibetans the high places are the sacred ground of a multitude of gods and devils. Nicolson (1975) refers to their belief that on one peak dwells a red goddess who owns a nine-headed tortoise, while on another lives a brown goddess who rides a turquoise-maned horse. While these deities may appear fanciful and picturesque to Westerners, it is worth keeping in mind the practical import of such beliefs to projected mountaineering expeditions. In 1955 a British climbing team planned to ascend Kangchenjunga, 'The five treasures of the eternal snows', (these being gold, silver, copper, corn and sacred books). The expedition was immediately recognised as a threat to the divinity of the mountain God of Wealth and the governments of India, Nepal and Sikkim refused to allow the climb. The peoples of these countries feared that the god of Kangchenjunga might be provoked to exact retribution. A compromise was reached when the climbers agreed to halt a few metres beneath the summit of the mountain when the Sherpas could bury offerings. Some highlanders overcome their religious objections to act as porters for climbing expeditions because the work is so lucrative. However, as Nicolson (1975) says so strikingly, to the majority of highlanders 'the idea of climbing these peaks seems nearly as sacreligious as for a devout Catholic to scale St. Peter's'.

Mountain myths and monsters

The native highlander has a stable psyche and is at peace with his mountain environment. He accepts the daily hazards and difficulties of his mountain home in a matter-of-fact manner. However, he has another relationship with the mountains which is on an entirely different level. The isolated and desolate country in which he lives is a fertile breeding ground for strange myths which appear to meet deep psychological needs. It has led him to create mountain-monsters.

The Yeti

For countless generations the Sherpas have recounted stories of a Yeti, a wild, hairy humanoid creature of the Himalayas. The first mention of this mythical creature in Western literature appears to have been by Hodgson in 1832. He reported that his native hunters were frightened by a *rakshas* (Sanskrit: Demon) with long, dark hair and no tail. It is of interest to note that Hodgson considered this intruder to be an orang-utan. In 1921 Howard-Bury was shown tracks like those of a human foot and they were ascribed by his porters as those of 'The wild man of the snows', the *Metoh-Kangmi*, later translated as the Abominable Snowman. Four years later Tombazi reported that he had sighted a naked human being at a distance of some 300 yards. In 1939

Captain d'Auvergne related an extraordinary story that has become typical of the Yeti legend (Napier, 1976). He stated that he was in danger from exposure and snowblindness in the Himalayas when he was rescued by a Yeti which was fully nine feet in height. This creature was said to have carried him to its cave where it fed him and nursed him to complete recovery. Such was the public interest in this creature that in 1961 a special licence was issued by the Government of Nepal to authorise the hunting of Yetis at 400 pounds sterling per animal. Yeti-seeking expeditions in the Himalayas were undertaken in 1954 and 1957. These had been prompted by the finding of a giant footprint, $13'' \times 8''$, at 5490 m on the Menlung glacier by Shipton and Ward in 1951 during an Everest reconnaisance expedition. Photographs of this and a line of footprints at the edge of the glacier have never been explained and they stimulated great public interest in the Yeti myth.

Napier (1976) presents a fascinating account of the nature and behaviour of the Yeti based on 18 separate Sherpa reports gathered from the literature. It lives in caves high in the mountains between 4370 m and 6100 m although some live in the impenetrable thickets of the mountain forests at some 3050 m. It is nocturnal and carnivorous. It eats the small Himalayan mouse-hare, the Pika (see Chapter 3), but it has been reported as devouring yaks. On occasion it has been reported as raiding villages and carrying off human beings. It has a vile pungent smell and mews, yelps, whistles and roars. Yetis have tremendous physical strength and can uproot trees and lift and hurl boulders over vast distances. The breasts of the females are so large that they have to throw them over their shoulders when running or when bending down. They are impeded, when running down slopes, by their long head-hair, which falls over their eyes and thus blinds them. Yetis are inordinately fond of alcohol. So far as social behaviour is concerned, they are solitary.

The nature of the Yeti. Belief in the Yeti seems to have a deep psychological significance for the Sherpas and provides at the same time an opportunity for sensationalism and publicity for visitors to these isolated areas. It is likely that the creation of a mountain-monster creates a common danger and thus binds the group together. It may act as a traditional bugbear, used to frighten children by

Sherpa parents (Napier, 1976). This monster may be like the biblical scapegoat (Leviticus xvi, 7–10), one of a pair of goats driven into the wilderness bearing into oblivion the sins of the people. The Yeti may be a convenient respository for undesirable human qualities. A third explanation is that it represents a form of cultural grooming, providing a feeling of togetherness and comfort in the fact of a common external enemy, namely the adverse mountain environment. An interesting feature of the Yeti myth is the almost total absence of sexuality. Only once amongst the hundreds of eye-witness accounts of the Yeti (or its North American counterpart to be described subsequently) have the sexual organs of the male been described or referred to (Napier, 1976). This is of interest for sexuality often assumes considerable importance in folklore.

There appears to be no reasonable doubt that the Yeti is an idea rather than an animal. There is no hard evidence of its existence such as a skull, or a captive animal, or a photograph of it of unquestionable probity (Napier, 1976). The scalps and hair which have been presented as evidence have not been authenticated by the institution investigating them.

The two forms of soft evidence abound in quantity. There are numerous sightings of the Yeti in the Himalayas but individuals, especially newcomers, at high altitude are susceptible to the mental disturbances and even hallucinations referred to in Chapter 30. As with the Loch Ness monster, one cannot exclude always the desire for sensationalism and publicity in those who claim to have seen the phenomenon.

There have been many instances of footprints. As we have seen in Chapter 3, many mammals frequent high altitudes for shorter or longer periods and their footprints may be misinterpreted. Pilgrims and lamas undertake dedicated plodding through this region of Tibet, Bhutan and Sikkim. Certainly the excessive size of the feet is highly characteristic of the Yeti. A step of $1\frac{1}{2}$ to 2 feet would be acceptable for a man walking in snow, and a foot length of 10 to 12 inches is within the normal human range, but the crux comes in trying to explain away a foot width of eight inches. The broadness of the foot is what makes the photograph of Shipton and Ward, taken in 1951, so difficult to explain away. However, it has to be kept in mind that, when footprints melt and then resolidify, with the rising and setting of the sun, they enlarge. In Figures 34.1 to 34.4 we provide examples of how big footprints in snow at high altitude can be explained simply without invoking the existence of the Yeti. We are indebted for these photographs to a colleague Kevin Marsh who took them while he was climbing in the Karakorums. So characteristic of the mountain-monster is the large footprint that Napier (1976) refers to both the Yeti and its North American counterpart by the group-name 'Bigfoot'.

There have been attempts to relate the *idea* of the Yeti to an *animal*. Some still maintain that the Yeti is real and represents a relict form of *Gigantopithecus*. Others believe the animal exists as a form derived from this. It is believed that a species of orang-utan (Fig. 34.5) roamed Southern China and Tibet for the last quarter of a million years and may have acted as the physical expression of the Yeti legend of such psychological need and importance to the indigenous peoples of the area. Belief in the Yeti exists even to the present day. Loudon reports that in April 1978 the Government of Sikkim sent four expeditions to the heights of Kishong La following reports that the Yeti had killed several yaks in the area, hurling them for a distance of 200 yards. Mr. Ara Singh, a forest guard, stated that he saw a Yeti

Fig. 34.2 As already noted in Chapter 3 various species of bird frequent high altitudes and some of them may be responsible for the mysterious appearance of big footprints at the tops of mountains. To the left and right of the picture are typical bird-tracks in firm snow. In the centre area the snow was soft and the bird has sunk into it, replacing a footprint with a body-print. When freshly made, the foot marks can still be seen in the floor of the body-print but with melting and subsequent freezing a print of a 'Bigfoot' would be produced.

Fig. 34.1 Figures 34.1 to 34.4 provide examples of how big footprints in snow at high altitude can be explained simply without invoking the existence of the Yeti. In this figure a print has been made by a climbing boot and the surrounding snow has fallen in to produce a much larger footprint. Melting and subsequent freezing perpetuate the large print. These photographs of footprints were taken by Kevin Marsh at Camp II (4880 m) on Cherichor peak (6750 m) in the Karakorums, on September 28th, 1976.

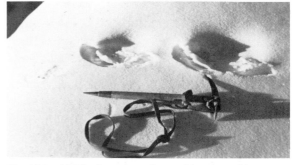

Fig. 34.3 (a) The footprints of a small quadruped in snow. (b) When this small animal runs across soft snow, the footprints (seen to the left) become converted into body-prints (seen in the centre and to the right). As we have already noted in Chapter 3, some species of small animals live at high altitude. The ice-axe is 21″ in length.

face-to-face in 1978. It was a 'man-like animal with brown hair on the body. It had a red face and red lips.'

The Sasquatch

It is now generally accepted that Mongoloid people emigrated into North America across the Bering Strait and the legend of 'Bigfoot' seems to have travelled with them to give rise to the legend of the Sasquatch, an Amerindian word meaning 'the wild man of the woods' (Fig. 34.6). On its introduction

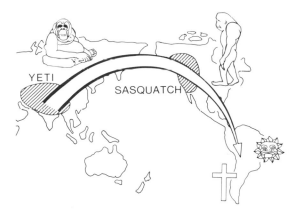

Fig. 34.4 Typical footprints produced by the bipedal gait of man (shown to the left) and the single line of large body-prints produced by an unknown animal in the snow at high altitude (shown to the right). The single nature of the 'Bigfoot' prints makes it clear that they have been produced as body-prints of a small animal leaping through the snow.

Fig. 34.6 With the emigration of Mongoloid peoples from Asia across the Bering Strait into the North American continent went the legend of 'Bigfoot'. The concept of the Yeti based on the Orang-utan appears to have been translated into the idea of the Sasquatch, itself a humanoid creature based on a North American animal, the bear, rather than the Asiatic Orang unknown to the population. In the Andes the Sun God of the Incas and the Cross of Roman Catholicism left no place for an Andean mountain deity.

into North America the *idea* had to be transferred to a *different kind of animal*. There are no orang-utans in the New World so the physical features of the Sasquatch had to be those of an animal native and common in the north-west of this continent, namely the bear. Thus the reddish shaggy fur changes to black, brown or grey, the partly quadrupedal gait of the Yeti has become wholly bipedal, and the cone-shaped head is lost.

Fantastic stories surround sightings of the Sasquatch and we may refer to one of these. In 1957 Albert Ostman recounted the story of an event that was said to have taken place some 33 years earlier. As a lumberman working in the Toba inlet in British Columbia in 1924, he was said to have been carried

Fig. 34.5 The face of an Orang-utan (*Pongo pygmaeus*). With its humanoid qualities it is easy to see why this species has been regarded by some authorities as forming the basis of the Yeti legend in the Himalayas.

for 25 miles inside his sleeping bag partially closed by an enormous hand. Finally he was deposited on the ground in the presence of four Sasquatches. He said he remained in this community for six days and was not molested. He described the animals as humanoid except for hairiness and an unusually big toe, with feet padded like those of a dog. The Sasquatch is a vegetarian but Ostman does not describe as part of the daily routine of this animal the foraging for large amounts of leafy food.

Napier (1976) presents an interesting profile of the Sasquatch based on the reports of 72 incidents from British Columbia, Alberta, Washington, Oregon and Northern California. Its estimated stature ranges between six and 11 feet. Its footprints range in length from 12 to 22 inches (the commonest quoted range being 14 to 18 inches), and the width being most frequently reported as 7 inches. The Sasquatch is seen most often during June and July (the holiday season) and October and November (the hunting season). Males are more commonly observed than females. A Sasquatch was captured on cinefilm by Roger Patterson in 1967 in the Bluff Creek Valley of Northern California. The validity of this controversial film record is examined at length by Napier (1976). The acquisition of frankly human attributes does not help the credibility of the Sasquatch as an animal species.

Los Muquis

In the Andean mines of Cerro de Pasco, Huancavelica and the district of Junin live 'Los Muquis', the spirits of the mines appearing as dwarfs with long white beards (Bravo, 1967). They wear ponchos trimmed with bright colours and have a liking for red or green hats. Los Muquis are nocturnal and noisy. They like to dance, sing, whistle and blow horns. Like the Yeti, however, they are not altogether pleasant for while they may save some mine-workers from death they allow others to perish (Bravo, 1967). The Quechuas who work in these high altitude mines treat Los Muquis with respect and depend for protection from their possibly malign influence on a combination of coca, pisco (Peruvian brandy) and prayer.

Alpine dragons

Lest one should adopt too superior an attitude towards Sherpa beliefs about the Yeti it is well to recall that until two or three centuries ago Europeans, including some members of the Royal Society, believed that the Alps were infested with dragons. Some had the bodies of serpents with the heads of cats; others had bats' wings and long hairy tails (Sanuki and Yamada, 1974). As the dragons flew through the mountains they dropped stones which had the power to effect miraculous cures of diseases such as dysentery and cholera. One of these supposed 'dragon-stones' is housed in the Museum of Natural History in Basle (Fig. 34.7). Jacob

Fig. 34.7 Dragon stone exhibited at Natural History Museum at Basle.

Scheuchzer of Zurich made nine journeys to the Alps from 1702 and came to the conclusion that the large numbers of caves in the region suitable for habitation by dragons constituted proof of their existence. He noted that, if dragon-stones are not found by good fortune, they have to be extracted from a live but unconscious dragon. Once the monster is awakened or killed, the stone loses its potency. Scheuchzer's study of dragons was dedicated to the Royal Society whose members subscribed so that his valuable book could be published and the data contained therein, collected from eye-witnesses of dragons, disseminated to the public (Sanuki and Yamada, 1974).

Monsters on the Matterhorn. When the young Whymper became obsessed by a desire to climb the Matterhorn he soon learned that many of the local villagers were convinced that the summit was inhabited by 'djinns and efreets'. They spoke of a ruined city there populated by spirits and demons. According to Sanuki and Yamada (1974), Sir Arnold Lunn in the last century reported that in the Italian valley of Breuil 'the Becca', that is the Matterhorn, had been regarded for centuries as the embodiment of supernatural terror. Mothers frightened their children by threats that 'the wild man of Becca' would carry them away. The parallel with the frightening of Sherpa children by their parents with stories of the Yeti is striking. Even today grotesque carvings of faces from the rough bark of trees abound in the Alpine valleys (Fig. 34.8). While produced nowadays for the tourist, they almost certainly have their origins in a belief in mountain monsters by the people of the alpine valleys of bygone days.

Fig. 34.8   One of the grotesque carvings of tree bark and roots which are to be found in the valleys of the Valais massif in Switzerland. The carving shown was seen in Zermatt at the foot of the Matterhorn, the summit of which was thought to be the site of a ruined city populated by demons.

REFERENCES

Bravo, C. A. E. (1967) In: *The man of Junin looking at his country and folklore*. Vol. II, p. 593. Lima, Peru: P. L. Villaneuva.

Hodgson, B. H. (1832) On the Mammalia of Nepal. *Journal of Asiatic Society of Bengal*, **1**

Loudon, B. (1978) Yak-hurling Yeti hunted in Sikkim. *Daily Telegraph* April 13, p. 4.

Napier, J. (1976) *Bigfoot. The Yeti and Sasquatch in myth and reality*. Abacus edition. London: Cox and Wyman.

Nicolson, N. (1975) *The Himalayas. The World's Wild Places*. Amsterdam: Time-Life Books.

Sanuki, M., and Yamada, K. (1974) *The Alps*. Tokyo: Kodansha International Ltd.

Index

High altitude pulmonary oedema (*contd.*)
prophylaxis of, 165
prognosis in, 167
radiological features, 154, 155, **155**, 156, **156**
recognition of as a clinical entity, 151
reversibility of, 166
skin in, 229
treatment of, 165–167
ultrastructure of, 159, 162
Highlanders
body composition in 33, **34**
features of, 24–27
HbF 93, **93**
leukocytes in, 94
longevity in, 34–36, **34**, 35
natural acclimatization in, 269
partial adaptation in, 269, 29–31
tolerance for maximal work in, 269
Hillary, Sir Edmund
on Mount Everest, 297
thrombosis on Makalu, 300
Himalayas
and diseases of Indian soldiers, 238
and migration of bar-headed goose, 274
and mountain gods, 324
and studies of blood coagulation, 99–101
effect of Monsoon on, 13, **13**, 14
high altitude in, 6, **6**
HAPO in, **153**
military operations in and acute mountain sickness, 2
mountain zones of, 15, **16**
Himalayan Rescue Association, 148
Hindu Kush
aridity of, 14
treatment of HAPO in, 166, 167
Histamine
and ultraviolet radiation, 230
mast cells and pulmonary hypertension, 105–108
Histidine carboxylase, 105
Hot radicals and ionizing radiation, 10
HPVD, hypoxic, 111
and Monge's disease, 174
causes of, 111
longitudinal muscle in pulmonary arteries in, 112
Hualiata, 61
'Huamanripu', 146
Huascaran, 15
height of, 14
Hübener-Thomsen-Friedenreich phenomenon, 25
Human geography
of high altitude, 16–18
Humboldt current, 15, **15**
and aridity of Andes, 15
Humidity, low
and frusemide treatment, 145, 166
and high altitude, 8
causing dehydration at high altitude, 299, 300
Hurtado, Alberto, **269**

Hyaline membranes in lung in HAPO, 157
Hydration of body
in native highlanders, 220
in newcomers to high altitude, 220
Hydrocolloid
use in dentistry at high altitude, 209
Hydroxycorticosteroids
at high altitude, 247
Hyperkeratosis and acclimatization, 231
Hyperons in ionizing radiation, 10
Hypertension
pulmonary, 103–117
and ADP, 107
and mast cells, 105–108
and pulmonary diffusing capacity, 46
and renin-angiotensin system, 249
and thrombosis, 100, 101
disorders of blood coagulation in, 99
genetic factors in cattle, 112, 113
immediate regression on O_2 inhalation, 284, **284**
individual variation in development of, 113, 114
in cattle at high altitude, 112, 113
in HAPO 157, 158
in highlanders (adult), 103, **103**, 104, 113
in highlanders (children), 103, 104
in Monge's disease, 173, **173**
in new communities at high altitude, 113
in the fetus, 104, 105
long term regression, 104
mechanism of production of, 105
reversibility of at high altitude, 104
reversal on descent from high altitude, 104, **104**, 283, 284, **284**
systemic
amelioration of in high altitude sojourners, 186, 187
and perivenous glomic tissue in the lung, 87
and the carotid bodies, 87
at high altitude and polycythaemia, 187
in Monge's disease, 187
rarity in Quechuas, 185, 186
Hyperventilation
and accommodation, 144
and acute mountain sickness, 136
and carbon dioxide, 42, **42**
and distinction from acute mountain sickness, 136
and lack of in the llama, 271
and peripheral chemoreceptors, 40
and respiratory alkalosis, 40
and respiratory rate, 40
and tidal volume, 40
as feature of acclimatization, 39, 40
in pregnancy at high altitude, 264, 265
sustained, 41, 42
Hypobaric chamber and acclimatization, 81, 82, 191

Hypocapnia
and cutaneous flow, 230
and retinal blood flow, 307, **307**
Hypokalaemia and ECG changes at high altitude, 193
Hypophagia, 215–217, **215**, **217**
Hypothalamus
in acute mountain sickness, **143**
thermostatic centre in, 224
Hypothermia
definition of, 299
ectopic beats in, 299
in high altitude stowaways, 302, **302**
myocardial ischaemia in, 299
ventricular fibrillation in, 299
Hypoventilation, alveolar
in Monge's disease, 173, 174
primary hypoventilation syndrome, 174
Hypoxaemia and pulmonary hypertension, 105
Hypoxia, acute
glycolytic enzymes in, 198
myocardial lipids in, 200
myocardial response to, 198
protein synthesis in, 198, 202
Hypoxia, chronic
and barometric pressure, 6, **7**
and intestinal absorption, 212
and lowering of systemic pressure, 185–187
and mast cells, 105, 106
and peptic ulceration, 210, 211
and pulmonary hypertension, 105
and relation to cold, 226
and spermatogenesis, 261, 262
and the developing lung, 32, **32**, 33, **33**
at high altitude, 6
direct action on pulmonary vascular smooth muscle, 107
individual hyper-reactivity to, 113, 114
influences on heart, indirect, 199
myocardial acclimatization to, 199
'myocardial anoxic resistance' in, 201
myocardial hypertrophy and protein synthesis in, 201
protective effect of, 201

Ibex, 22
Immersion hand (and foot), 298
Immune response at high altitude, 236–238
to bacteria, 237
to viruses, 237
Immunoglobulins at high altitude, 237–238
and platelet factor, 238
IgA, 238
IgG, 237
IgM, 237
Incas, 17
and fertility, 259
blood groups in, 26
extent of empire, 259
in concept of 'normal values', 280